Sustainable Agriculture and Advances of Remote Sensing
Volume 2: In Image Processing

Sustainable Agriculture and Advances of Remote Sensing
Volume 2: In Image Processing

Editors

Dimitrios S. Paraforos
Anselme Muzirafuti
Giovanni Randazzo
Stefania Lanza

MDPI • Basel • Beijing • Wuhan • Barcelona • Belgrade • Manchester • Tokyo • Cluj • Tianjin

Editors
Dimitrios S. Paraforos
Hochschule Geisenheim University
Germany

Anselme Muzirafuti
University of Messina
Italy

Giovanni Randazzo
University of Messina
Italy

Stefania Lanza
University of Messina
Italy

Editorial Office
MDPI
St. Alban-Anlage 66
4052 Basel, Switzerland

This is a reprint of articles from the Special Issue published online in the open access journal *Applied Sciences* (ISSN 2076-3417) (available at: https://www.mdpi.com/journal/applsci/special_issues/Agriculture).

For citation purposes, cite each article independently as indicated on the article page online and as indicated below:

LastName, A.A.; LastName, B.B.; LastName, C.C. Article Title. *Journal Name* **Year**, *Volume Number*, Page Range.

Volume 2
ISBN 978-3-0365-5339-9 (Hbk)
ISBN 978-3-0365-5340-5 (PDF)

Volume 1-2
ISBN 978-3-0365-5335-1 (Hbk)
ISBN 978-3-0365-5336-8 (PDF)

Cover image courtesy of Anselme Muzirafuti

© 2022 by the authors. Articles in this book are Open Access and distributed under the Creative Commons Attribution (CC BY) license, which allows users to download, copy and build upon published articles, as long as the author and publisher are properly credited, which ensures maximum dissemination and a wider impact of our publications.

The book as a whole is distributed by MDPI under the terms and conditions of the Creative Commons license CC BY-NC-ND.

Contents

About the Editors vii

Preface to "Sustainable Agriculture and Advances of Remote Sensing
Volume 2: In Image Processing" ix

Farzad Kiani, Giovanni Randazzo, Ilkay Yelmen, Amir Seyyedabbasi, Sajjad Nematzadeh,
Fateme Aysin Anka, Fahri Erenel, Metin Zontul, Stefania Lanza and Anselme Muzirafuti
A Smart and Mechanized Agricultural Application: From Cultivation to Harvest
Reprinted from: *Appl. Sci.* 2022, 12, 6021, doi:10.3390/app12126021 1

Loganathan Agilandeeswari, Manoharan Prabukumar, Vaddi Radhesyam,
Kumar L. N. Boggavarapu Phaneendra and Alenizi Farhan
Crop Classification for Agricultural Applications in Hyperspectral Remote Sensing Images
Reprinted from: *Appl. Sci.* 2022, 12, 1670, doi:10.3390/app12031670 23

Manar Ahmed Hamza, Fadwa Alrowais, Jaber S. Alzahrani, Hany Mahgoub,
Nermin M. Salem and Radwa Marzouk
Squirrel Search Optimization with Deep Transfer Learning-Enabled Crop Classification Model
on Hyperspectral Remote Sensing Imagery
Reprinted from: *Appl. Sci.* 2022, 12, 5650, doi:10.3390/app12115650 43

Yingxin Zhang, Mengqi Duan, Shimei Li, Xiaoguang Zhang, Xiangyun Song and Dejie Cui
Rational Sampling Numbers of Soil pH for Spatial Variation: A Case Study from Yellow River
Delta in China
Reprinted from: *Appl. Sci.* 2022, 12, 6376, doi:10.3390/app12136376 59

Wenjing Wang, Mengqi Duan, Xiaoguang Zhang, Xiangyun Song, Xinwei Liu and Dejie Cui
Determining Optimal Sampling Numbers to Investigate the Soil Organic Matter in a Typical
County of the Yellow River Delta, China
Reprinted from: *Appl. Sci.* 2022, 12, 6062, doi:10.3390/app12126062 71

Luis Josué Méndez-Vázquez, Rodrigo Lasa-Covarrubias, Sergio Cerdeira-Estrada
and Andrés Lira-Noriega
Using Simulated Pest Models and Biological Clustering Validation to Improve Zoning Methods
in Site-Specific Pest Management
Reprinted from: *Appl. Sci.* 2022, 12, 1900, doi:10.3390/app12041900 85

Muhammad Shahab Alam, Mansoor Alam, Muhammad Tufail, Muhammad Umer Khan,
Ahmet Güneş, Bashir Salah, Fazal E Nasir, Waqas Saleem and Muhammad Tahir Khan
TobSet: A New Tobacco Crop and Weeds Image Dataset and Its Utilization for Vision-Based
Spraying by Agricultural Robots
Reprinted from: *Appl. Sci.* 2022, 12, 1308, doi:10.3390/app12031308 115

Paolo Rommel Sanchez and Hong Zhang
Simulation-Aided Development of a CNN-Based Vision Module for Plant Detection: Effect of
Travel Velocity, Inferencing Speed, and Camera Configurations
Reprinted from: *Appl. Sci.* 2022, 12, 1260, doi:10.3390/app12031260 135

Gabriel Gaspar, Juraj Dudak, Maria Behulova, Maximilian Stremy, Roman Budjac,
Stefan Sedivy and Boris Tomas
IoT-Ready Temperature Probe for Smart Monitoring of Forest Roads
Reprinted from: *Appl. Sci.* 2022, 12, 743, doi:10.3390/app12020743 157

Muhammad Attique Khan, Abdullah Alqahtani, Aimal Khan, Shtwai Alsubai, Adel Binbusayyis, M Munawwar Iqbal Ch, Hwan-Seung Yong and Jaehyuk Cha
Cucumber Leaf Diseases Recognition Using Multi Level Deep Entropy-ELM Feature Selection
Reprinted from: *Appl. Sci.* **2022**, *12*, 593, doi:10.3390/app12020593 179

Almetwally M. Mostafa, Swarn Avinash Kumar, Talha Meraj, Hafiz Tayyab Rauf, Abeer Ali Alnuaim and Maram Abdullah Alkhayyal
Guava Disease Detection Using Deep Convolutional Neural Networks: A Case Study of Guava Plants
Reprinted from: *Appl. Sci.* **2022**, *12*, 239, doi:10.3390/app12010239 199

Rabia Saleem, Jamal Hussain Shah, Muhammad Sharif, Mussarat Yasmin, Hwan-Seung Yong and Jaehyuk Cha
Mango Leaf Disease Recognition and Classification Using Novel Segmentation and Vein Pattern Technique
Reprinted from: *Appl. Sci.* **2021**, *11*, 11901, doi:10.3390/app112411901 219

Razieh Pourdarbani, Sajad Sabzi, Mohammad H. Rohban, José Luis Hernández-Hernández, Iván Gallardo-Bernal, Israel Herrera-Miranda and Ginés García-Mateos
One-Dimensional Convolutional Neural Networks for Hyperspectral Analysis of Nitrogen in Plant Leaves
Reprinted from: *Appl. Sci.* **2021**, *11*, 11853, doi:10.3390/app112411853 231

Joon-Keat Lai and Wen-Shin Lin
Assessment of the Rice Panicle Initiation by Using NDVI-Based Vegetation Indexes
Reprinted from: *Appl. Sci.* **2021**, *11*, 10076, doi:10.3390/app112110076 247

Muaadh A. Alsoufi, Shukor Razak, Maheyzah Md Siraj, Ibtehal Nafea, Fuad A. Ghaleb, Faisal Saeed and Maged Nasser
Anomaly-Based Intrusion Detection Systems in IoT Using Deep Learning: A Systematic Literature Review
Reprinted from: *Appl. Sci.* **2021**, *11*, 8383, doi:10.3390/app11188383 259

Anna Stankiewicz
Optimal Versus Equal Dimensions of Round Bales of Agricultural Materials Wrapped with Plastic Film—Conflict or Compliance?
Reprinted from: *Appl. Sci.* **2021**, *11*, 10246, doi:10.3390/app112110246 283

About the Editors

Dimitrios S. Paraforos

Prof. Dr. Dimitrios S. Paraforos is serving as a Deputy Professor of Agricultural Engineering in Special Crops at the Hochschule Geisenheim University (Department of Technology, Von-Lade-Str. 1, D-65366 Geisenheim, Germany). He holds an MSc from the University of Thessaly (Greece) and, in 2016, obtained a PhD from the University of Hohenheim, both in Agricultural Engineering. His research focus is on precision farming, digital technologies in agriculture, and, more generally, on control systems, robotics, and automation applied to agriculture; in addition, he has obtained industry experience working as an automation engineer in the food industry. He is the author of more than 53 papers indexed by Scopus, with an important contribution in the field of sensors and ISOBUS technologies for the enhancement of agricultural practices. Currently, he is coordinating the ERA-NET ICT-Agri II European project iFAROS on developing methods for increasing the efficiency and precision of site-specific fertilizer application.

Anselme Muzirafuti

Dr. Anselme Muzirafuti, born in Rwanda, is an assistant professor at the University of Messina (Department of Mathematics, Computer Sciences, Physics and Earth Sciences, Via F. Stagno d'Alcontres, 31–98166 Messina, Italy). He holds a master's degree in Applied Geophysics and Geology Engineering obtained in 2015 from University of Moulay Ismail, Meknes (Morocco), and a PhD in Hydrogeophysics obtained in 2021 from the same University. Since 2009, he received multi-excellence scholarships for his higher education studies from different governments, including the Government of Rwanda, the Government of the Kingdom of Morocco, and the European Union. In 2016, 2018 and 2019, he participated in major conferences on climate change, sustainability and geoscience, namely, the 22nd Conference of Parties held in Marrakech (Morocco); the 24th International Sustainable Development Research Society Conference (action for a sustainable world: from theory to practice), held in Messina (Italy); and the 2019 European Geoscience Union General Assembly, held in Vienna (Austria). His research interest has been focused on the use of Structural Geology, Geomatics and Geophysics for sustainable management of territories. He worked on different projects, in Morocco and in Italy, related to geomorphological mapping and surveys using images acquired by satellites and drones. He recently worked as analysist of satellite images in the BESS project "(Pocket Beaches management and Remote Monitoring Systems)—Program Interreg VA Italia Malta 2014–2020". The results of his works have been presented at international conferences and published in international journals.

Giovanni Randazzo

Prof. Dr. Giovanni Randazzo is an associate professor of Coastal Geomorphology and Environmental Geology at the University of Messina (Department of Mathematics, Computer Sciences, Physics and Earth Sciences, Via F. Stagno d'Alcontres, 31–98166 Messina, Italy). He holds a PhD in Marine Environment and Resources obtained from University of Messina (Italy). Since 1987, his research interest has been focused on the study of the coastal area, on its management and protection. During these 30 years, he collaborated with the Smithsonian Institution of Washington D.C. in the study of the Nile Delta; with the Thai Geological Service in the study of the east coast of the local peninsula; and with ENEA (Italian National Agency for New Technologies, Energy and Sustainable Economic Development), he participated in the X Italian expedition in Antarctica. He has collaborated in the environmental assessment impact of various public works (especially in the coastal area), and he participated in the drafting of the Territorial Landscape Plan of the Province of Messina (Sicily, Italy). In recent years, he has actively participated in the debate on the emergence of waste, writing scientific articles, intervening in the local press, and participating in various debates, where he presented a scheme of management of the emergency alternative to those not acting for the Sicilian Region. In 2013, he founded Geologis s.r.l., a branch of the University of Messina, active in the field of territory surveys using aerial and marine drones equipped with RGB cameras, LiDAR sensors, and thermal imaging cameras. On behalf of the European Union, he has coordinated at national level and/or as a local unit several projects related to coastal management and territorial security. Since December 2017, he has been the lead partner of the Pocket Beaches management and Remote Monitoring Systems (BESS) project as part of the Interreg Italy—Malta Program. He is the author of more than 120 scientific publications.

Stefania Lanza

Dr. Stefania Lanza is an administrator of Geologis s.r.l, an Academic Spin-Off of the University of Messina (Via F. Stagno d'Alcontres, 31–98166 Messina, Italy). In 2007, she obtained a PhD in Geology from the University of Messina (Italy) with a thesis on "The risk assessment of coastal areas: from planning to monitoring". She worked on different projects related to sedimentology and geomorphology mapping. In 2008, she took the final exam of the master course with a thesis entitled: "Coastal monitoring of the coast of Badalona (Spain) contribution to the 2007–2008 survey campaign" supervised by Prof. Jordi Serra of the Autonomous University of Barcelona. In 2013, she co-founded Geologis s.r.l., active in the field of territory surveys using aerial and marine drones equipped with RGB cameras, LiDAR sensors, and thermal imaging cameras. She recently worked as Coordinator of the Geomorphological and Sedimentological Activities of the Project in the context of BESS project "Pocket Beach Management & Remote Surveillance System—Program Interreg VA Italia Malta 2014–2020". She is currently working as coordinator of field activities in the context of the "BIOBLU project—Robotic Bioremediation for Coastal Debris in Blue Flag Beach and in a Maritime Protected Area—Interreg V-A Italy—Malta 2014–2020 program".

Preface to "Sustainable Agriculture and Advances of Remote Sensing Volume 2: In Image Processing"

This Special Issue on "Sustainable Agriculture and Advances of Remote Sensing" falls within the scope of current efforts to mitigate and adapt to the changing climate. It has been launched with the aim of collecting and promoting recent scientific studies proposing and evaluating advances in remote sensing technology and agricultural engineering leading to sustainable agriculture. It is mainly addressed to the policy makers, entrepreneurs and academicians engaged in the fight against climate change, in zero hunger initiatives, in natural resource management and in environment protection research. A special thanks is addressed to the authors who submitted their manuscripts to contribute to these initiatives.

Dimitrios S. Paraforos, Anselme Muzirafuti, Giovanni Randazzo, and Stefania Lanza
Editors

Article

A Smart and Mechanized Agricultural Application: From Cultivation to Harvest

Farzad Kiani [1,†], Giovanni Randazzo [2], Ilkay Yelmen [3,*], Amir Seyyedabbasi [1], Sajjad Nematzadeh [4], Fateme Aysin Anka [5,‡], Fahri Erenel [5], Metin Zontul [6], Stefania Lanza [7] and Anselme Muzirafuti [2,*]

1. Software Engineering Department, Faculty of Engineering and Natural Science, Istinye University, Istanbul 34396, Turkey; farzad.kiyani@gmail.com (F.K.); amir.seyyedabbasi@gmail.com (A.S.)
2. Dipartimento di Scienze Matematiche e Informatiche, Scienze Fisiche e Scienze della Terra, Università degli Studi di Messina, Via F. Stagno d'Alcontres, 31, 98166 Messina, Italy; giovanni.randazzo@unime.it
3. R&D Center, Turkcell Technology, Istanbul 34854, Turkey
4. Computer Engineering Department, Faculty of Engineering and Architecture, Nisantasi University, Istanbul 34398, Turkey; nematzadeh.sajjad@gmail.com
5. Political Science and Public Administration Department, Faculty of Economics, Administrative and Social Sciences, Istinye University, Istanbul 34396, Turkey; fatemeh.dehghan@istinye.edu.tr (F.A.A.); fahri.erenel@istinye.edu.tr (F.E.)
6. Department of Computer Engineering, Faculty of Engineering, Istanbul Topkapı University, Istanbul 34093, Turkey; metinzontul@topkapi.edu.tr
7. GeoloGIS s.r.l., Dipartimento di Scienze Matematiche e Informatiche, Scienze Fisiche e Scienze della Terra, Università degli Studi di Messina, Via F. Stagno d'Alcontres, 98166 Messina, Italy; stefania.lanza@unime.it

* Correspondence: ilkay.yelmen@turkcell.com.tr (I.Y.); anselme.muzirafuti@unime.it (A.M.)
† Other publish name is Ferzat Anka.
‡ Other publish name is Fatemeh Dehghan Khangahi.

Citation: Kiani, F.; Randazzo, G.; Yelmen, I.; Seyyedabbasi, A.; Nematzadeh, S.; Anka, F.A.; Erenel, F.; Zontul, M.; Lanza, S.; Muzirafuti, A. A Smart and Mechanized Agricultural Application: From Cultivation to Harvest. *Appl. Sci.* 2022, *12*, 6021. https://doi.org/10.3390/app12126021

Academic Editor: José Miguel Molina Martínez

Received: 19 May 2022
Accepted: 11 June 2022
Published: 14 June 2022

Publisher's Note: MDPI stays neutral with regard to jurisdictional claims in published maps and institutional affiliations.

Copyright: © 2022 by the authors. Licensee MDPI, Basel, Switzerland. This article is an open access article distributed under the terms and conditions of the Creative Commons Attribution (CC BY) license (https://creativecommons.org/licenses/by/4.0/).

Abstract: Food needs are increasing day by day, and traditional agricultural methods are not responding efficiently. Moreover, considering other important global challenges such as energy sufficiency and migration crises, the need for sustainable agriculture has become essential. For this, an integrated smart and mechanism-application-based model is proposed in this study. This model consists of three stages. In the first phase (cultivation), the proposed model tried to plant crops in the most optimized way by using an automized algorithmic approach (Sand Cat Swarm Optimization algorithm). In the second stage (control and monitoring), the growing processes of the planted crops was tracked and monitored using Internet of Things (IoT) devices. In the third phase (harvesting), a new method (Reverse Ant Colony Optimization), inspired by the ACO algorithm, was proposed for harvesting by autonomous robots. In the proposed model, the most optimal path was analyzed. This model includes maximum profit, maximum quality, efficient use of resources such as human labor and water, the accurate location for planting each crop, the optimal path for autonomous robots, finding the best time to harvest, and consuming the least power. According to the results, the proposed model performs well compared to many well-known methods in the literature.

Keywords: autonomous robots; remote sensing; smart agriculture; climate change; environmental protection; drone; metaheuristic; Internet of Things

1. Introduction

Population growth in the world naturally causes an increase in food needs, and with the prediction that the world population will reach nine billion people by 2050, agricultural production should be increased by 70% [1]. Therefore, growing crops becomes very important. However, agricultural practices using traditional methods to meet people's food consumption needs can have quite inefficient results. Therefore, it becomes essential to manage agricultural activities using different advanced methods. Additionally, environmental protection and sustainability have also become basic needs [2]. In this regard, smart farming mechanisms using new technologies have become very popular [3]. It should not be forgotten that not only modern agriculture methods but also sustainable solutions should be proposed. With the creation of efficient solutions to meet identified needs, it may be possible to provide the next generation with a lifestyle at least equal to that of the current generation [4] and to use existing natural resources efficiently in this direction. Otherwise, we may face not only the problem of hunger but also challenges such as energy and migration crises. In summary, problems such as global climate change, increasing conflicts around the world, migration crises, improper use of existing agricultural lands, decrease in precipitation levels, and incorrect use of water all have made the use of modern agriculture systems mandatory.

Technologies such as remote sensing, the Internet of Things (IoT), intelligent agents, autonomous robots, unmanned aerial vehicles (UAVs) [5], Internet of Vehicles (IoVs), wireless ad-hoc networks, big data analytics, and deep learning (DL) have shed light on promising visions for a breakthrough in agricultural applications [6–9]. Smart farming can be applied to improve crop quality and profit and reduce costs by optimizing various processes such as environmental conditions, growth status, soil status, irrigation water, pest control, fertilizers, weed management, and greenhouse production environments [10]. For example, questions such as which product should be planted in a given region, how efficient the use of water resources should be, and accurate estimation of the harvest time all can be answered and monitored in smart agriculture. Thus, smart agriculture ensures green technology by eliminating the inefficient and faulty methods of traditional agriculture, and it can also further reduce problems such as leakage and emissions and their impact of climate change [11].

Since the role of smart systems in sustainable agriculture is increasing day by day, many technological methods have been used and recommended in recent years. Smart farming optimizes complex farming systems by applying new and modern technologies in agriculture. It aims to produce and collect more quality crops with less investment, irrigation, and human labor. In general, plant growth and harvest processes are critical issues in smart farming. Therefore, intelligent mechanism models will be very useful in solving identified problems and achieving specified goals. However, an approach that covers the whole process, not a single or specific goal(s), is important. In this regard, in this study an application model is proposed that covers this process from cultivation to harvesting. This model can be used to grow a variety of crops on any farmland. For this, software and hardware approaches are used together, making it possible to use them in the real world. Previous studies have been focused on certain perspectives and/or reaching a limited number of goals.

The main aim of this study is to be able to solve the problems with green and sustainable agricultural practices from the field to the table, and to ensure that people can access agricultural products, which are their basic needs, without interruption. In addition, it is aimed at reducing product losses, preventing waste, increasing productivity, and even reducing to zero waste due to faulty planting and similar causes at every stage of the food chain. Part of these efforts should be the aim of ensuring the full physical, mental, and social well-being of agricultural workers, by minimizing recently increasing work accidents and occupational diseases of those working in the agricultural sector with preventive and protective measures. All these concerns can be resolved in a sustainable way by providing a single mechanism system from planting to harvesting.

This study focuses on tracking and harvesting crops on large-scale farmlands in three stages. In the first phase (*cultivation*), planting the crops in the most optimized way is sought by using an algorithmic approach (Swarm Sand Cat Optimization (SCSO) algorithm) during the planting phase. In this way, the location of the crops that need to be planted (without the need for human or manual intervention) is determined in accordance with the specific requirements of any crops. An optimization approach can be a good solution since it is a nondeterministic polynomial-time hardness (NP-Hard) problem to meet this need with analytical and similar solutions. The SCSO algorithm could be a successful solution in determining the correct places of the products in accordance with its working mechanism. In this way, optimized results can be obtained without human intervention, with the lowest error rate and lower operation cost.

In the second stage (*control and monitoring*), the growing processes of the planted crops are tracked and monitored thanks to IoT devices. With the IoT devices, the status of the planted products is constantly checked, and the results are transferred to the next stage as an input for instant processing. The third phase (*harvesting*) is the product collection stage. At this stage, a metaheuristic method (Reverse Ant Colony Optimization) is proposed. The results from the third stage are presented as the optimum path for autonomous robots (IoVs) to harvest crops. In this third phase, a new fitness function is also defined. Since the second and third stages are carried out simultaneously, it is ensured that there will be a sustainable model. The Reverse ACO (RACO) is proposed as a new improvement to the ACO algorithm. Thanks to this algorithm, the results from the second stage can consistently and regularly provide the input for the third stage correctly. The performance of the proposed method was evaluated in different scenarios and situations and compared with similar studies in the literature. The following benefits were also obtained with the application developed in this study:

(1) The efficient use of resources such as human and natural resources by providing sustainable and smart agriculture. In other words, it allows farmers to constantly monitor crop variability and stress conditions. In addition, the occupational health and safety of the farmers can be increased, the best products can be harvested, and efficient resource consumption and profit increase will be possible.
(2) Assistance in gaining maximum profit with minimum consumed energy and time by considering several objectives in the fitness function. As an example, the proposed mechanism generates the optimized routes for each autonomous robot in harvesting crops at the annual harvest.
(3) Providing decision-makers with an infrastructure architecture for autonomous agricultural robots.
(4) Collecting crops in the required periods with a scheduling mechanism; thus, it prevents early or late harvesting. Meanwhile, it is adaptive to several crop types. In this study, tomatoes were considered as an example case.
(5) Maximum profit, maximum quality, efficient use of resources such as human labor and water, the accurate location for planting each crop, finding the best time to harvest, and consuming the least power.

In Section 2 of this paper, the literature review is discussed. The proposed model and application are explained in Section 3. In Section 4, the simulation results are evaluated. The last section of the study includes the conclusions and possible further studies.

2. Literature Review

This study focuses on the most efficient crop-harvesting methods in farmlands (tomato assumed as a case study). In this regard, studies and general perspectives about the new generation of agriculture in the literature are presented. As is known, there are many types of sowing, monitoring, and harvesting in agriculture. Although traditional methods are successful in growing crops, they have many disadvantages.

In traditional agriculture, farmers cannot prepare the soil properly because they only use hoes after burning the bush and clearing the field. In addition, they cannot get fertile results as they can only scratch the earth and mix the ashes into the soil [12]. The burning method damages the soil, and as a result, erosion is exacerbated as this burnt soil is left bare. Furthermore, roots cannot go deep enough to absorb water and mineral salts from the soil, and thus water (the most important and critical resource) is used inefficiently. The soil can become very poor in a short time due to the usual techniques applied incorrectly and inefficiently, and to solve this problem (i.e., the soil becomes fertile again), the field must be left fallow for two or three years [13]. Since the farmers cannot stay idle during this period, they move to another field, which is called alternating cultivation. As a result, large-scale lands will not be used, and bountiful crops will not be produced. Mechanism systems blended with new technologies can be the solution to these problems. On the other hand, it is also very important to harvest the crops efficiently, on time, with the maximum amount and quality and least use of energy and resources. However, traditional methods consume resources excessively, the rate of poorer-quality crops is higher, the profit rates are not fully maximized, and most importantly, they cannot offer sustainability [14]. These problems can be resolved with the transition to modern agriculture. In this regard, the IoT and mobile robots provide great contributions. In the solution to these complex situations, software techniques and methods are needed together with new technologies. According to this, artificial-intelligence-based smart methods are becoming more important day by day. In solving this type of complex problem, learning or heuristic-based solutions may be especially useful—we should not forget that some problems in this area may not have a deterministic solution. Finally, taking into account increasing human population and food demand, these problems will have dire consequences. For the reasons stated, we can conclude that intelligent agriculture is an essential requirement.

The IoT and similar technologies such as wireless sensor networks (WSNs) [15], which have become popular in recent years, are used to meet the needs in this field. The IoT in agriculture means using sensors and other devices to turn every element and action involved in farming into data [16]. Scientists believe that the IoT will lead the agriculture sector to agriculture 4.0 and even agriculture 5.0 in the future [17,18]. This new philosophy, data-driven agriculture, is also expressed in the literature with several different names: Agriculture 4.0, digital farming, or smart farming [19]. This smart farming can be combined with the precision agriculture concept in data management, leading to more accurate and efficient results [20]. In traditional methods, farmers had to go to the farmland and check the condition of their crops, and they would decide based on their experience whether it was time to harvest. New technologies such as the IoT are very useful in solving these problems, and more, to achieve greater efficiency, sustainability, and availability rates. Additionally, in traditional methods, the experienced farmer had a higher success rate, so younger farmers had a lower chance at succeeding. However, thanks to these new technologies, this problem can now be eliminated. On the other hand, savvy farmers can also adapt to new methods. IoT technologies are predicted to play a major role in the generation of large amounts of valuable information in all types of agriculture and the advancement of this sector [21]. In addition, the IoT is thought to be a potential solution to increasing agricultural productivity by 70% by 2050 [22]. In [23], the authors designed an IoT-based system to monitor air and soil parameters and develop mobile and web-based applications. They tried to monitor crop and yield forecasts in real time. In [24], the authors focused on farm management information systems to automate data acquisition and processing, monitoring, planning, decision-making, documenting, and managing farm operations. For this, they proposed architecture and implemented it in two different regions in Turkey. In another proposed example to monitor various components of the farmland with an IoT-based mechanism, an agro-meteorological system was developed using an Arduino [25]. Another study based on sensors, IoT, ZigBee, and Arduino focused on rural agriculture [26]. The authors tried to guide farmers in estimating crop suitability and other relevant factors by using various types of sensors. Another study explored how accurate analysis of agro-meteorological and weather parameters can help farmers improve crop production [27]. However, their proposed system is not portable, and may only be suitable for small-scale farms. Other IoT-based studies in modern agriculture in recent years have also been explored in [28–30].

Along with the IoT, the widespread use of autonomous robots such as UAVs increases productivity in agriculture. Recently, studies related to this subject have accelerated [31–35]. One of several studies in the literature is presented in [36]. The authors used UAVs to detect possible drainage pipes. Often, farmers need to repair or construct drain lines to efficiently remove water from the soil. Therefore, in this study, they wanted to decrease resource consumption and increase productivity in agriculture by focusing on this issue. In [37], the authors offered a combined application of UAVs and unmanned ground vehicles (UGVs) to monitor and manage crops. The authors proposed a system that can periodically monitor the condition of crops, capture multiple images of them, and determine the state of the crops. In addition to many UAV-based studies and products, recently the concepts of the IoT and autonomous robots have begun to be presented together. In this way, the data detected by the UAVs reaches the place where it needs to be sent instantly, the necessary actions can be taken on this data, and it can quickly provide a decision mechanism to the farmer or other technological devices. For example, in [33], the authors present a farm-monitoring system using UAV, IoT, and Long-Range Wide Area Network (LoRAWAN) technologies for efficient resource management and data delivery. In this study, they were monitoring water quality.

In general, most of the studies were aimed at increasing efficiency in the field with technological approaches. It is also very important to efficiently analyze and evaluate the data generated from new technologies such as IoT and UAV using recent technologies such as WebGIS [38]. At the same time, we may sometimes encounter complex situations for a service expected to be provided in smart agriculture. These problems become more difficult to solve as the dimension and number of uncertain parameters increase. Metaheuristic algorithms can be an appropriate mechanism for solving similar problems. Indeed, in recent years in the literature, metaheuristic-based approaches have been proposed for different purposes related to agriculture [39]. For example, in [40], three local search metaheuristic algorithms, which were simulated by annealing and Tabu search references, were used to calculate annual crop planning with a new irrigation mechanism. The objective function of this study was to maximize the gross benefits associated with the allocation of crops. The authors claim that the Tabu search method gave the best results in comparisons. In [41], an evolutionary algorithm was used for a complex strategic land-use problem based on the management of a farming system. This study pursued a multi-purpose strategy that fulfilled spatial constraints in the 50-year planning management of the farm. Although the study is comprehensive, the metaheuristic method used and proposed may not be a very high-performing and efficient solution. In [42], a bi-objective optimization model was proposed that minimizes cost and maximizes geographic diversity. In the test of the proposed method, a case was considered that showed new types of relationships in the food logistics chain. Although the proposed mechanism is interesting, the number of parameters it deals with is not complex enough. In [43], the authors introduced a smart-engine-based decision system focusing on the type of crop, time/month of harvest, type of plant required for the crop, type of harvest, and authorized rental budget. According to the results from this system, the best way to rent and share agricultural equipment was provided. The other metaheuristic-based method focused on economic crop planning at the tactical level or agricultural policy planning at the national level [44]. The supposed crop products could be for home consumption, export (cash crops), or to feed milk cows. The proposed method focused on optimal farm reconstructions that met four objectives (maximize profit and balance of soil structure, and minimize the soil nitrogen and human labor), and a set of stringent constraints. In [45], the authors tried to optimize the deployment of their sensor nodes to best monitor potato and wheat crops. In this regard, they proposed a Genetic Algorithm-based method. Other studies based on metaheuristic algorithms have been presented in the literature in the last few years [46–48].

In summary, many studies have been carried out in the field of smart and sustainable agriculture, which has become a trend in recent years. Some of these studies, along with their characteristics, are summarized in Table 1. This paper focuses on optimal solutions from cultivation to harvesting benefiting from the IoT, autonomous robots, and a metaheuristic approach in three stages.

Table 1. Characterization of current studies in the literature.

Study	Approach, Technique	Task and Goals
[31]	GIS system	Monitoring process based on map analysis and reduced data
[23]	IoT and autonomous robot	Monitoring process and yield forecasts by IoT devices and mobile/web app
[24]	IoT	Process management based on a farm management information system and architecture
[26]	IoT	Product management based on data analysis
[36]	Single autonomous robot	Monitoring process based on efficient water usage, controlling amounts of phosphate (PO4) and nitrate (N03), and detecting drainage pipes
[37,49]	Single autonomous robot	Monitoring process based on monitoring vegetation state
[32]	Multi autonomous robots	Monitoring process based on providing a multiple UAV system for aerial imaging
[33]	Single autonomous robot, metaheuristic	Spraying process based on spraying fruits and trees
[34]	Multi autonomous robots	Spraying process based on path-planning algorithm
[42]	Metaheuristic	Food logistics chain and crop planning based on minimizing cost and maximizing geographic diversity
[44]	Metaheuristic	Convenient and efficient planting and cultivation
[39]	Metaheuristic	Annual crop planning
[43]	Metaheuristic	Rent and share agricultural equipment based on a smart-engine-based decision system
[45]	Metaheuristic and IoT	Best planting model based on optimal deployment
[48]	Metaheuristic	Optimal plant

3. Materials and Methods

In this study, the technological and algorithmic aspects of sustainable agriculture are discussed. Towards this end, an intelligent and mechanized agricultural application is proposed which includes three phases: (1) cultivation, (2) control and monitoring, (3) harvesting. These stages affect each other like a lifecycle. The results from the first stage are used in the second stage, and the results from the second phase are constantly used as the input of the third stage. The working mechanism of the proposed method is described in Figure 1 with all its phases.

3.1. Cultivation Phase

As stated before, with the increase in the world population, the need for agricultural and food products is also increasing. Therefore, the importance of smart farming systems and methods has increased. In this context, the first step in the mechanism and a sustainable model is the planting (cultivation) phase. Here, it is very important to place each product in the correct location in order to minimize or remove manpower and manual interventions. The metaheuristic approach is useful because the solution to this problem is a type of NP-Hard problem to be solved systematically. Therefore, the SCSO algorithm was used in the first phase of this study. Due to its nature, this algorithm can provide good performance both locally and globally, as its transitions are balanced in the exploration and exploitation phases, and therefore this new algorithm was used to solve the first problem case in this study.

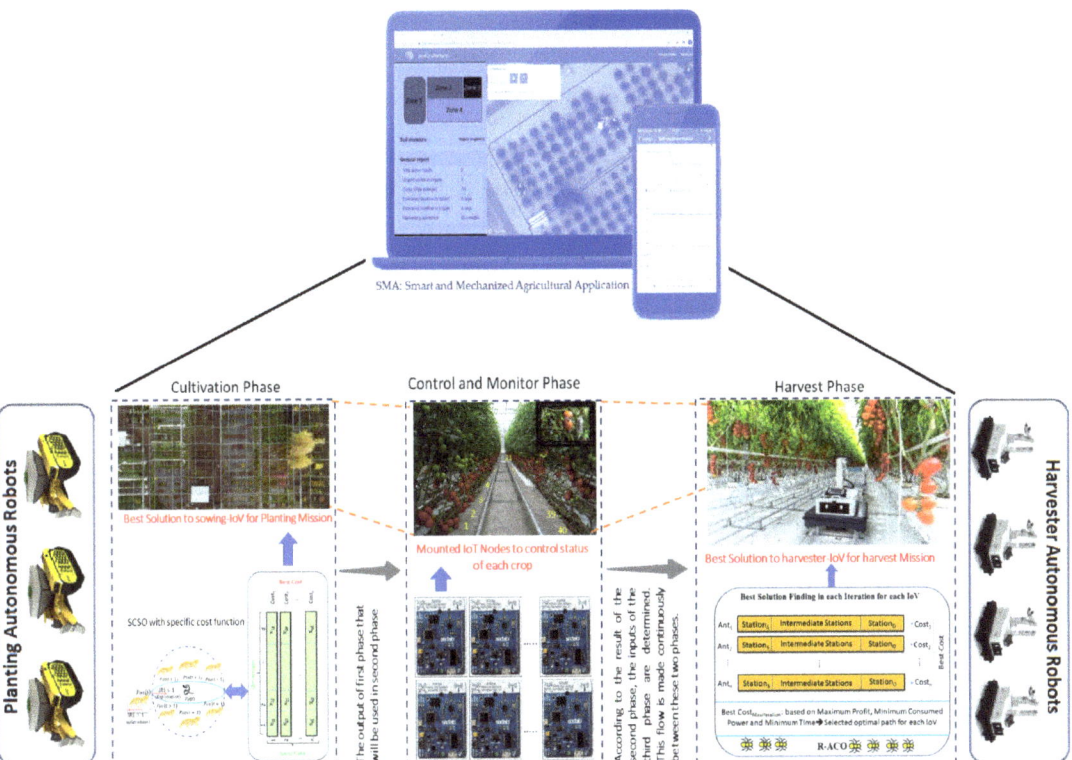

Figure 1. General scheme of all operations of the proposed model.

3.1.1. SCSO Algorithm

The Sand Cat Herding Optimization (SCSO) algorithm is inspired by the sand cat's foraging and hunting behavior [50]. In this algorithm, a low-frequency noise detection behavior mechanism is used to find the prey of sand cats. Since these cats can detect low frequencies below 2 kHz, they can find even the furthest prey in the shortest possible time and with very little movement. These cats also have an incredible ability to dig for prey. This algorithm has a successful performance in searching and hunting using these two great features. With an adaptive mechanism, SCSO can be good at solving many problems, as they are balanced in the exploration and exploitation phases. In addition to these features, this algorithm is preferred specifically for this problem, since it uses few parameters and is a simple implementation. The mathematical equations of this algorithm are given below (Equations (1)–(5)):

$$\vec{r_G} = s_M - \left(\frac{2 \times S_M \times iter_c}{iter_{Max} + iter_{max}} \right) \qquad (1)$$

$$\vec{R} = 2 \times \vec{r_G} \times rand\,(0,1) - \vec{r_G} \qquad (2)$$

$$\vec{r} = \vec{r_G} \times rand(0,1) \qquad (3)$$

where 'r_G' is a constant inspired by the listening ability of sand cats. It will linearly decrease from two to zero as the iterations progress to approach the prey (solution) it is looking for and not to lose or pass it (not to move away). The "S_M" value is inspired by the hearing characteristics of the sand cats, its value is assumed to be 2. The 'r' demonstrates sensitivity range of each cat. The 'r' is used for operations in exploration or exploitation phases while 'r_G' guides the 'R' parameter for transition control in these phases. 'R' is the main parameter in controlling the transition between exploration and exploitation phases. If '*R*' is lower than one, the sand cats are directed to attack their prey, otherwise, the cats are tasked with finding a new possible solution in the global area. The '$iter_c$' is the current iteration and '$iter_{max}$' indicates the maximum iterations. The '*t*' is the current time.

$$\vec{X}(t+1) = \begin{cases} \vec{Pos_b}(t) - \vec{Pos_{rnd}} \cdot \cos(\theta) \cdot \vec{r} & |R| \leq 1 \, ; exploitation \quad (a) \\ \vec{r} \cdot \left(\vec{Pos_{bc}}(t) - rand(0,1) \cdot \vec{Pos_c}(t) \right) & |R| > 1 \, ; exploration \quad (b) \end{cases} \quad (4)$$

Equation (4) is proposed to determine the next move of each cat. The 'Pos_b' indicates the best position and 'Pos_c' represents the current position. The 'Pos_{rnd}' indicates the random position and ensures that the cats involved can be close to prey. 'Pos_{rnd}' is obtained from Equation (5). In addition, SCSO also performs well in convergence behavior. The pseudocode of the relevant algorithm is shown in Algorithm 1.

$$\vec{Pos_{rnd}} = \left| rand(0,1) \cdot \vec{Pos_b}(t) - \vec{Pos_c}(t) \right| \quad (5)$$

Algorithm 1. Sand cat swarm optimization algorithm pseudocode.

Initialize the population
Calculate the fitness function based on the objective function
Initialize the *r*, r_G, *R*
While (*t* <= maximum iteration)
    For each search agent
        Get a random angle based on the Roulette Wheel Selection ($0° \leq \theta \leq 360°$)
        If (abs® <= 1)
            Update the search agent position; *Equation (4a)*
        Else
            Update the search agent position; *Equation (4b)*
        End
    End
    t = *t*++
End

3.1.2. Planting Crops Based on SCSO Algorithm

The SCSO is used for economical product planning and efficient resource consumption. Hereby, crops can be planted in the most optimized places in a mechanized way without human intervention during the cultivation (planting) phase. In this regard, the SCSO algorithm is used instead of an analytical solution to plant each crop with an optimized method. As mentioned earlier, since this action is an NP-Hard problem type, metaheuristic-based approaches can lead to fruitful results. In addition, restrictions and requirements were taken into account so that it can be applied in real farmlands. These criteria are, for example, biodiversity indices based on landscape ecology measures, diversity of land uses, and soil erosion. The criteria to be optimized here will mostly have a positive effect on the farmer's income from the harvest and the type and amount of employment. In addition, efficient land use will lead to optimum allocation of water resources. An optimized planting stage will also be useful for an autonomous robot to accurately position itself in crop fields to perform precision farming tasks effectively.

In this study, tomato (*Lycopersicon esculentum* Mill.) is taken as a case study. In this arrangement, the row spacing can be selected as 80 cm, and the seedling spacing as 40 cm ((3.1 m^2/unit)), 60 cm (2.1 m^2/unit), and 80 cm (1 m^2/unit) in tomato planting. It is recommended to choose 40 cm seedling spacing for pollination, temperature preservation, and to grow more product [51,52]. In addition, it is recommended that the soil's pH level be 6.0–6.5 when planting tomato seeds [53]. Single-row arrangement and the hanging method are preferred in tomato cultivation [51,52]. In line with the reasons and objectives mentioned above, the places of the products relative to each other are determined algorithmically, considering the distance and other requirements. Here, the process will be the planting of the products with the least error rate. The crops planted at the end of this phase were equipped with IoT devices in the next phase. The places where the products should be placed was given to the autonomous robots as a map with this algorithm. They would perform the mechanized planting work (saving manpower and other natural resources). This was the first mission of the autonomous robots. Their second and final task was in collecting (harvesting) items.

3.2. Control and Monitoring Phase

The second phase plans the timely determination of whether the planted crops are grown, their water and similar needs, and the realization of other purposes. At this stage, it is expected that the products will be monitored to increase productivity. This process can be a very costly and error-prone structure when performed by the farmer. However, being able to manage this whole process with technological solutions, and, in this context, offering farmers a user-friendly integrated application increases efficiency in every aspect.

In this phase, IoT devices were used to easily track each product and obtain query or in-demand-based data flow. The features of these devices may vary according to the purpose and expectations in the field. In this study, the tomato case study is focused on as an example. Tomatoes have commercial importance as one of the most-grown vegetables in the world [51]. Today, tomatoes are grown in open fields and greenhouses. The method of growing tomatoes in the open field is considered one of the traditional methods and we did not consider it. The growing period of tomatoes may be different depending on the parameters of light, water, minerals, and temperature [53]. In greenhouses, parameters such as temperature, humidity, amount of light, amount of CO_2, and amount of water can be controlled more easily, and therefore tomato cultivation in a greenhouse can be preferred [52,54].

For this purpose, KIANI-WSN kit nodes were used as IoT nodes [55] as shown in Figure 2. These nodes consist of sensing, communication, processing, and power units. Since the communication unit is equipped with a Wi-Fi card, it also fully meets the need for an IoT. These nodes are generally of two types. One of them is the CC1101 transmitter chip and the other is the nodes equipped with both CC1101 and CC1190 chips. The first node types were used in this study. The CC1101 is a low-cost sub-1GHz transceiver that can be used in very-low-power wireless-based applications [56]. It also supports packet processing, data buffering, burst transmissions, clear channel evaluation, link quality indication, and wake-on-radio. The CC1101's main operating parameters and 64-byte send/receive FIFOs can be controlled via the serial peripheral interface (SPI). These nodes are equipped with sensors such as humidity, temperature, and light. The ad-hoc feature of this device provided another advantage because it could transfer the required data to our system even when there was no internet. Thanks to this system, productivity in harvesting increased and resources such as water and manpower were used efficiently.

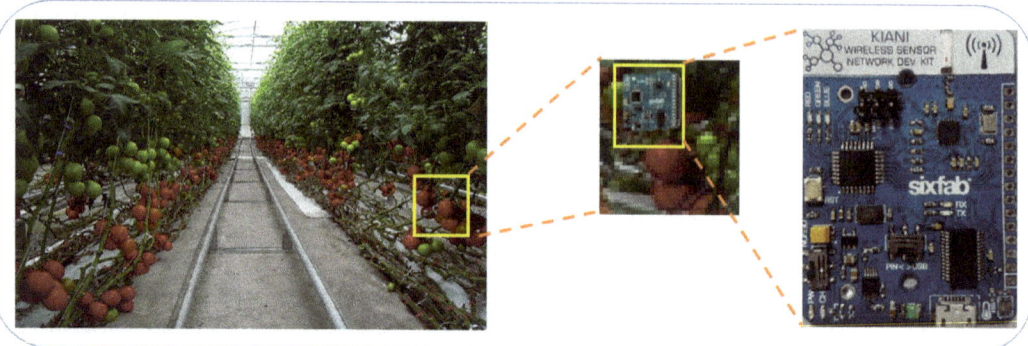

Figure 2. The device used as an IoT node in the field [5,55].

3.3. Harvesting Phase

The purpose of this phase is to come up with a road map for each autonomous robot. At this stage, the SCSO and similar algorithms may be not very successful in their rationality, as it is our goal to generate paths and find a good route. Therefore, a new method was proposed by utilizing the one of the best metaheuristic algorithms (ACO) for this concept. Based on this path, each robot will begin its task of picking (harvesting) crops in the most efficient manner. In other words, with this route, each robot collects the crops efficiently, making the most profit. Moreover, this goal is achieved in as short a time as possible and with the least power consumption. This new algorithm is proposed to achieve these three intermediate targets (*max-profit, min-energy, min-time*) together in a balanced way. In summary, the output (goal) of the harvest phase is finding the optimal path for each autonomous robot.

The ACO [57] algorithm is inspired by the real-life behavior of ants. Although ants are not insects that live in a swarm model, they help and guide each other by using chemical deposits called pheromones, and as a result, they act like a smart colony. In the classical ACO, each ant starts from a point, goes hunting, and returns to its home (start). That is, each ant starts eating from the nest and then returns to the nest. However, in RACO, each ant starts from a point and only tries to reach the prey, and the concept of returning to the nest is not in question in this algorithm. It is worth remembering that this algorithm is not a new ACO variant. This algorithm is inspired by the concept of ACO for solving just this problem and similar problems.

The second main difference between the proposed algorithm and ACO is based on the concept of pheromones. In the classic ACO, in each iteration each ant releases pheromones on each path it prefers, so the pheromone values of the chosen paths are constantly updated. Since the pheromone density is high in the most preferred route, the relevant route is chosen as the best route. Additionally, as the number of iterations increases, some pheromone evaporates from the amount present on each path. This prevents it from facing problems such as local optima. For this, a matrix is defined. However, the concept of pheromones is not fully similar in RACO, and instead, a similar matrix is defined, reflecting the increase in growth rates of the crops planted in each iteration. In addition, when each product is selected by the ant, it defines it as the product that has been collected, and the value of that product in the relevant matrix is decreased. In fact, the matrix defined in RACO shows the growth rates of crops. Items that are not collected increase in value as the iteration goes on (contrary to evaporation), and items that are collected are devalued (a very small number close to zero). In short:

(1) In classical ACO, pheromones evaporate as time progresses, in RACO the crops grow as time passes.

(2) In classical ACO, the pheromone ratio of the most preferred routes increases, in RACO, the product is collected and its value decreases in the current table of the product, while on the other hand, the value of the profit parameter increases.

Since this model is applied on a continuous field, the same process continues as the harvested crops begin to grow and mature again, and therefore the algorithm works smoothly. This presents a sustainable farming mechanism for all three phases described in our study. The main idea of this proposed algorithm is to model the problem as a search for the best path by constructing a path graph representing the states of the problem. Therefore, it can also be used to solve other problems similar to the product aggregation problem. In the classical ACO algorithm, the ants make the path selections according to Equation (6),

$$P_{i,j}^k(t) = \frac{[\tau_{i,j}(t)]^\alpha [\varphi_{i,j}(t)]^\beta}{\sum_{k \, allowed_k}[\tau_{i,k}(t)]^\alpha [\varphi_{i,k}(t)]^\beta} ; j \, \epsilon \, allowed_k \qquad (6)$$

where 'k' represents the number of ants used in the algorithm. The 'i' and 'j' denote the source and destination station, respectively, which indicate the points of each autonomous robot. $\tau_{i,j}$ is represented as a matrix and shows the instantaneous states of the crops. The $\varphi_{i,j}$ is the heuristic cost between the edges 'i' and 'j'. The 'α' and 'β' are related to the importance of trail and heuristic cost. In the proposed RACO, the related mathematical equations are defined as follows (Equations (7)–(12)), based on the problem in this study:

$$Profit^K (MaxProfit) = \sum_1^D SelectedCrop^K \qquad (7)$$

$$Power^K = (BatteryCapacity * DeliverCapacityPercentage) - \sum_i^j (distance * BatteryConsumed) \qquad (8)$$

$$Time^K (MinTime) = \sum_i^j (distance * TimeConsumed) \qquad (9)$$

where '$SelectedStatus$' shows the drop collected by each ant. The goal here is to get the most profit. '$BatteryCapacity$' indicates the maximum power capacity of the autonomous robots, generally related to Ah and V. '$DeliverCapacityPercentage$' represents the maximum amount of power each robot can use. The '$distance$' is the distance between two points in meters. '$BatteryConsumed$' is the amount of energy used between both points. In this study, battery capacity is assumed 8 Wh and the maximum battery usage rate is accepted as 90%. The amount of '$Power$' obtained here indicates the amount of energy that each robot will use, and this is desired to be minimum. It is worth mentioning that at the end of the algorithm, a route will be assigned for each of the robots. Since the concept of power is meaningful to them, autonomous robot is expressed here, but it actually means ant. The 'D' represents the total number of crops in the field. The '$Time$' and '$TimeConsumed$' parameters represent the total time used by each ant and the elapsed time between both points, respectively.

$$CostFunction^K = \frac{C_2 * Power^K + C_3 * Time^K}{C_1 * Profit^K}; \; C_1 + C_2 + C_3 = 1, \; C_1 > C_2 \geq C_3 \qquad (10)$$

$$BestCost = Min(CostFunction) \qquad (11)$$

where C_i are coefficients that can be adjusted according to the requirements of the problem. Optimal values for these constants can also be found by a metaheuristic method. In this study, the method in [58] is used to tune the values of these coefficients. Since these weights (coefficients) are not defined in the classical ACO, they are assumed to be of equal value. The relationship to C_i may differ depending on the purpose of the problem. For example, C_1 should be bigger than the others if we want the minimum answer, and vice versa if we are looking for the maximum. This proves the flexibility of the relevant cost function. At the end of each iteration, 'CostFunction' is calculated for each ant and the minimum among them is accepted as the best solution.

$$\Delta\tau_{i,j} = \frac{CropStatus - SelectedCrop}{\sum_1^D Crops} \qquad (12)$$

$$\tau_{i,j} = (1+\rho)*\tau_{i,j} + \Delta\tau_{i,j} \qquad (13)$$

where ρ is a coefficient that affects and controls the rate of the crops' status. 'CropStatus' shows the latest status of each crop. 'Crops' indicates the total number of crops in the field. The working mechanism of the proposed algorithm for finding the optimal path for each autonomous robot is shown in Figure 3. It is worth remembering that the starting and ending points of each ant do not have to be the same. At the end of each iteration the best solution is chosen. When the iterations are over, the best solution is given as a path to the relevant autonomous robot (IoV), and in this way each robot is assigned a task to collect crops in the real field. The pseudocode of the relevant algorithm is shown in Algorithm 2. When complex analysis of the algorithm is performed, it has proven to be $O(n^2)$, which means that it is successful compared to other classical and analytical solutions.

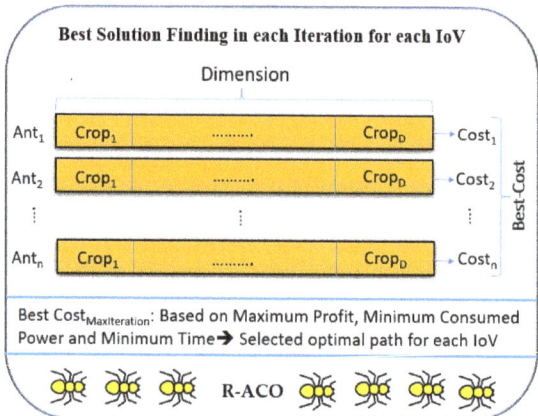

Figure 3. The working mechanism of the RACO.

Algorithm 2. Reverse ACO algorithm pseudocode.

Initialize the population
Calculate the fitness function based on the objective function
Initialize the input parameters
While (*t* <= maximum iteration)
    For each search agent
        Path election by ants; *Equation (6)*
        Pheromone (CropStatus) Updates; *Equations (12) and (13)*
    End
    t = *t*++
End
Find Best Solution; *Equations (10) and (11)*

4. Results and Discussion

This section presents the performance of the proposed model. The model results are analyzed and compared with the various methods. Comparison of the results is made in two scenarios. In the first, only the performances in the first phase (cultivation) are considered. Here, several algorithms that have been used have been successful in solving the refinement problem: Grey Wolf Optimization (GWO) [59,60], Moth-Flame Optimization

(MFO) [61], and Particle Swarm Optimization (PSO) [62]. In the other scenario, comparisons were made to cover all phases. Since the second and third phases are integrated, it would not be correct to analyze these phases separately. The methods used for comparison are a variant of Firefly Algorithm (FA) [63], Genetic Algorithm (GA) [64], Cuckoo Search (CS) [65], and Glowworm Swarm Optimization (GSO) [66] algorithms. In addition, the MAP-ACO [67] algorithm, which is based on ACO, was also used for comparison. In summary, a comparison and analysis covering all phases was performed to preserve the integrity of the results. The implantation and analysis presentation were performed in MATLAB. The algorithms proposed in this study were performed on a Core i7-5500 U 2.4 processor with 16 GB of RAM.

4.1. Simulation Setting

In the simulation, each of the algorithms were simulated under similar conditions, with 15 independent runs consisting of 30 search agents and 200 iteration numbers. These independent runs were performed to manage the effects generated from random parameters in the methods used. The size of the environment was 100 m ∗ 100 m. the simulation parameters are presented in Table 2. Time is assumed to be discrete ($t = 1, 2, \ldots$) and at each time step, every ant moves toward a neighbor node at a constant speed (m/s).

Table 2. The simulation parameters.

Method	Parameter	Value
SCSO	Sensitivity range (r_G)	[2, 0]
	Phase control (R)	[$-2r_G, 2r_G$]
GWO	a	[2, 0]
	A	[2, 0]
	C	2.rand (0, 1)
MFO	b	1
	t	[$-1, 1$]
PSO	C1 and C2 acceleration constants	1.7
	Maximum inertia weight (W_{max})	0.9
	Minimum inertia weight (W_{min})	0.2
	Maximum velocity (Vmax)	6
RACO	α	0.5
	β	0.5
	$\tau_{i,j}(0)$	0.008
	ρ	0.05
	C1, C2, C3	0.47, 0.34, 0.19
CS	p_a	0.25
FA	λ	2
GA	Crossover rate	0.1
GSO	p	0.4
	γ	0.6
	β	0.08
	n_t	5
	s	0.03
	L_0	5
MAP-ACO	α	0.5
	β	0.5
	$\tau_{i,j}(0)$	0.008
	ρ	0.05
	C1, C2, C3, C4	0.15, 0.2, 0.25, 0.4

4.2. Tomato Case Study

In this study, the performance of the proposed algorithms was analyzed, assuming tomato as a case study. It is worth emphasizing that the method proposed in this study can also be used in different agricultural terrain. Table 3 presents an example of the growth cycle of tomatoes through to the harvest process. After the fruit is tough, the fruit ripens over 45–70 days, depending upon the cultivar, climate, and growth conditions. The fruit continues growing until the stage of green ripeness. The stages of the tomatoes ripening are categorized into three steps, as shown in Table 3. These stages are valid from the beginning to the end of the first harvest.

Table 3. A typical example of a tomato growth cycle [52,53].

Growing Method	Germination Time of Tomato Seeds (Day)	First Flowering Time (Day)	Time from Planting to First Harvest (Day)	Starting Harvest (Day)	End of the Last Harvest (Day)	Ripening Stages	Average Root Medium Temperature (Centigrade)
Greenhouse	2–7	30	65	81	210	Breaker, pink, and red	20–35

In addition to the information above, also inspired by the information in [68,69], we try to propose a more realistic and accurate metaheuristic-based method by considering the tomato characteristics such as irrigation mechanism, growth pattern, and relations between seedling growth and the prevailing environment. In addition, we used all this information to impact the behavior of autonomous robots.

4.3. Analysis and Evaluation (Scenario 1, Crop Cultivation)

In this section, the performance of the proposed method at the planting stage of the crops is analyzed. Sowing, which is the first phase of smart and sustainable agriculture, has an important role. Here, the working mechanism and performance of each algorithm are presented in Table 4. The results of the proposed method at the planting stage are analyzed and also compared with other methods in the literature. Due to the nature of the problem type, the methods in the metaheuristic approach were used for comparison. The lower the error rate, the higher the performance of the relevant mechanism. In this study, the places where tomatoes should be planted were based on practical and analytical information (Table 3).

Table 4. The overall performance of each algorithm in the cultivation phase.

Algorithm	MSE	Algorithm	MSE
SCSO	**0.005022 m**	MFO	0.025599 m
GWO	0.018713 m	PSO	0.010692 m

Note: The best values of algorithms are written in bold.

The least amount of Mean Square Error (MSE) belongs to the SCSO based method. Here, data is presented in meters. According to the results, the method that could complete the placement task with the least error rate was our method based on the SCSO algorithm. It has 5.02×10^{-3} error, which is the minimum error rate between the algorithms used. The MSE values gained from various metaheuristics go down, while the iterations increase. The MSE convergence of each method is presented in Figure 4. The SCSO outperforms the other metaheuristic approaches, and its MSE value is 5.02×10^{-3}. In other words, SCSO shows very good results when the MSE value is around 4.17×10^{0} in the first iteration and converges at 5.02×10^{-3} in the final iteration. Among all the metaheuristic approaches, MFO has the worst performance, starting at 6.78×10^{0} in the first iteration and converging at 2.56×10^{-2} in the final iteration.

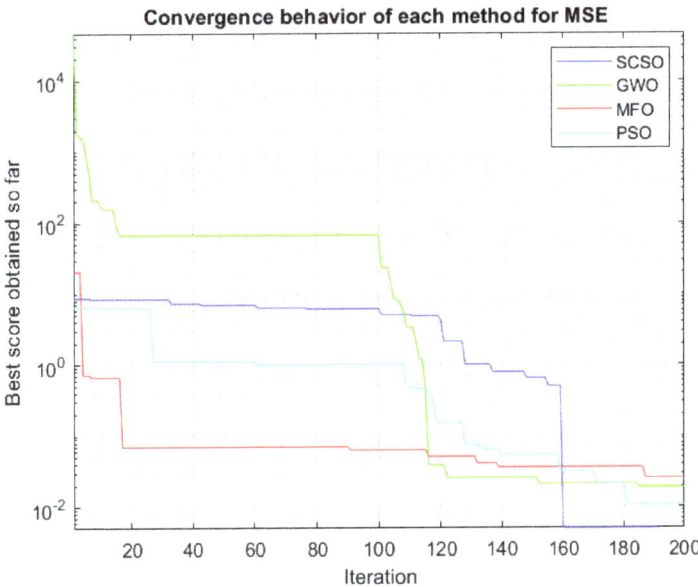

Figure 4. The MSE convergence of each method.

4.4. Analysis and Evaluation (Scenario 2)

At this stage, the performance analysis of the proposed method focused on the expected goals after the harvesting phase, based on all stages. In this study, some of the most important parameters in smart and sustainable agriculture are assumed. One of them is the profit rate. It is preferable that the collected crops offer more profit. In our real world, not only the growth of the crops but also the logistics are important in the collection of agricultural products. Therefore, it is possible to collect the crops before they are fully grown. In addition, the collected products start to grow again as time progresses. These two issues have been taken into account in our performance analyses. The second parameter is the amount of power consumed by each autonomous robot. The third parameter is that the robots perform the product-picking tasks in as short a time as possible. These three parameters will lead us to the main goal. This aim will generate an efficient and optimum roadmap for each robot. According to this map, each robot will be assigned the task of collecting items. The path that provides the maximum profit and consumes the least energy in the shortest time is attempted to be selected by the algorithm.

Here a route for the robot is revealed. At the same time, when the relevant route is selected by the algorithm, it is shown how much profit will be achieved, how long the robots will take to complete their tasks, and how much power they will use during this time. The results obtained from the simulation environment are presented. In this scenario, performances are considered from two dimensions. In Table 5, results including all phases are presented. Since the SCSO-based method gave the best results in the first phase, this approach was used for all the methods used. In the second, the results are obtained based on only the 2nd and 3rd phases, excluding the 1st phase, and are presented in Table 6. In this model, the planting was carried out with an analytical and classical approach. All results were evaluated on five different parameters. When an analysis including all phases is made, improvements are seen in all evaluation parameters (except profit). In other words, phase 1 has an effect on other parameters apart from the profit rate. The profit value does not change. This is because it has nothing to do with the realization of the first phase. According to the results in Tables 5 and 6, it can be seen that the proposed method (RACO) performs better than the others. Therefore, it was more realistic and efficient to continue the evaluation and analysis to include all phases.

Table 5. The simulation results of each algorithm (best) based on all phases.

Algorithm	Profit (%)	Time Consumed by Robot (m)	Traveled Distance (m)	Energy Consumed (%)	Simulation Time (s)	Path List
RACO	**83.77**	**16.57**	**332.55**	**12.46**	**34.82**	[15, 16, 17, 24, 30, 36, 34, 33, 32, 31, 25, 26, 27, 28, 29, 35, 23, 22, 21, 20, 14, 19, 13, 7, 1, 8, 2, 3, 9, 4, 5, 6, 12, 18, 11, 10]
MAP-ACO	67.82	22.78	371.7	17.09	35.51	[5, 4, 3, 26, 27, 33, 34, 29, 28, 35, 36, 30, 24, 22, 21, 32, 31, 25, 19, 20, 14, 8, 7, 13, 1, 2, 9, 15, 16, 10, 11, 17, 18, 23, 12, 6]
CS	72.59	22.5	394.33	16.87	37.6	[14, 21, 22, 28, 33, 34, 23, 29, 35, 36, 30, 24, 17, 16, 10, 9, 8, 7, 1, 13, 19, 20, 26, 27, 32, 31, 25, 2, 3, 5, 4, 6, 11, 18, 12, 15]
FA	70.13	23.94	398.71	17.95	36.65	[23, 24, 18, 17, 11, 6, 12, 16, 15, 14, 7, 8, 9, 4, 5, 10, 3, 2, 1, 20, 21, 27, 32, 26, 28, 34, 33, 31, 25, 13, 19, 22, 29, 36, 35, 30]
GA	68.99	27.68	443.11	20.76	38.07	[25, 26, 32, 33, 28, 29, 30, 36, 22, 27, 15, 21, 20, 8, 3, 31, 19, 14, 13, 7, 1, 2, 9, 10, 4, 5, 6, 11, 17, 12, 18, 24, 34, 35, 23, 16]
GSO	62.88	34.11	598.03	25.58	37.46	[29, 21, 22, 5, 11, 14, 19, 25, 26, 31, 20, 1, 16, 8, 2, 4, 9, 6, 12, 10, 3, 23, 24, 32, 33, 28, 18, 15, 17, 13, 7, 27, 34, 35, 30, 36]

The best values of algorithms are written in bold.

Table 6. Simulation results of each algorithm (best) based on phases 2 and 3.

Algorithm	Profit (%)	Time Consumed by Robot (m)	Travelled Distance (m)	Energy Consumed (%)	Simulation Time (s)	Path List
RACO	**83.77**	**29.89**	**429.11**	**19.26**	**89.13**	[5, 12, 13, 29, 14, 28, 26, 8, 21, 24, 25, 32, 20, 16, 15, 6, 22, 18, 31, 3, 36, 9, 19, 11, 10, 35, 33, 23, 30, 1, 2, 34, 4, 27, 7, 17]
MAP-ACO	67.82	34.17	441.41	23.05	92.48	[33, 23, 35, 6, 32, 5, 14, 12, 13, 30, 10, 8, 4, 24, 34, 9, 1, 11, 28, 36, 3, 19, 21, 7, 18, 25, 17, 29, 26, 27, 31, 22, 2, 16, 15, 20]
CS	72.59	34.11	464.31	22.56	97.22	[14, 15, 36, 18, 34, 7, 4, 12, 11, 30, 10, 9, 2, 25, 33, 8, 1, 20, 28, 35, 3, 19, 22, 23, 5, 21, 24, 26, 32, 16, 27, 31, 29, 6, 13, 17]
FA	70.13	34.91	457.13	22.59	93.19	[5, 4, 13, 27, 20, 30, 36, 28, 12, 35, 16, 31, 25, 17, 32, 19, 21, 6, 29, 34, 14, 8, 7, 23, 3, 2, 9, 15, 26, 10, 11, 18, 33, 22, 24, 1]
GA	68.99	37.66	521.12	26.51	96.03	[27, 15, 16, 11, 12, 13, 9, 8, 33, 30, 21, 14, 10, 1, 5, 32, 17, 26, 20, 31, 19, 22, 25, 29, 23, 18, 4, 28, 3, 36, 34, 24, 2, 6, 35, 7]
GSO	62.88	42.22	689.44	33.12	96.02	[16, 15, 9, 4, 32, 10, 8, 24, 1, 36, 5, 13, 2, 25, 30, 19, 34, 7, 6, 35, 14, 22, 11, 23, 12, 26, 33, 17, 28, 20, 27, 31, 3, 21, 18, 29]

The best values of algorithms are written in bold.

In Tables 5 and 6, the performances of all algorithms are presented in detail. At this stage, a route with the least cost was chosen for the robot as the optimal path. When electing this, we aimed for the robot to consume the least power in the shortest time and to collect products at a high rate. In this regard, the performance values of the best path chosen by each algorithm are given. Based on the results, all three goals were achieved, with the highest rate on the route chosen by RACO. Finally, the results obtained from this subsection (scenario 2) are ordered comparatively and presented in Table 7. According to Table 7, the RACO method took the first order in the "profit" parameter. The second place is the CS-based method. In the "Time Consumed by Robot" parameter, the results of the RACO method found the best solution according to the results from the first phase of our study. Other ranks have been determined by the numbers written in relevant columns of this table. In the "Traveled Distance" parameter, the RACO-based method found the best result. The minimum amount in this parameter is the best answer. The worst performance was observed to be the GSO method. The amount of energy to be used by the robots, whose task is defined thanks to the operation of each method, is the fourth evaluation parameter. In this analysis, the RACO method performed better. The last parameter is the "Simulation Time", and it is the running time of the simulation. This indicates how efficiently another resource was used. RACO performed the best and the GA method the worst.

Table 7. Ranking of each method in all evaluation parameters.

	Profit	Time Consumed by Robot	Traveled Distance	Energy Consumed	Simulation Time
RACO and SCSO (based on all phases)	1	1	1	1	1
R_ACO (based on phase 2 and 3)	1	6	5	5	7
MAP-ACO and SCSO (based on all phases)	5	2	2	3	2
MAP-ACO (based on phase 2 and 3)	5	9	7	9	8
CS and SCSO (based on all phases)	2	3	3	2	5
CS (based on phase 2 and 3)	2	7	8	7	11
FA and SCSO (based on all phases)	3	4	4	4	3
FA (based on phase 2 and 3)	3	10	9	8	9
GA and SCSO (based on all phases)	4	5	6	6	6
GA (based on phase 2 and 3)	4	11	10	11	12
GSO and SCSO (based on all phases)	6	8	11	10	4
GSO (based on phase 2 and 3)	6	12	12	12	10

In addition, these selected routes are presented visually in Figure 5. The convergence behavior of the methods used at this stage is also presented in Figure 6, and it is observed that the proposed method is also good in this regard. For the abovementioned reasons, the results including all phases have been taken into account. The convergence behavior and local optimum avoidance of the methods used at this stage are also presented in Figure 6, and it is observed that the proposed method is also good in this regard. The proposed algorithm starts from a large sensitivity range to discover more possible solutions and explore the whole search area. As the iterations progress, by decreasing the value of the sensitivity ranges search agents try to exploit and find the global optima.

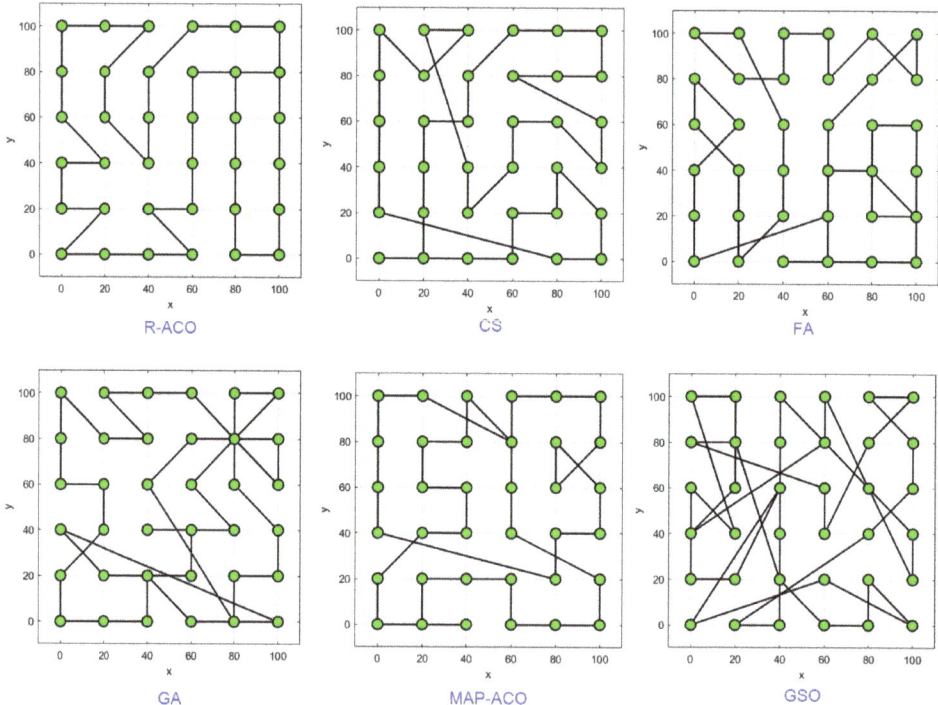

Figure 5. The elected path by each algorithm.

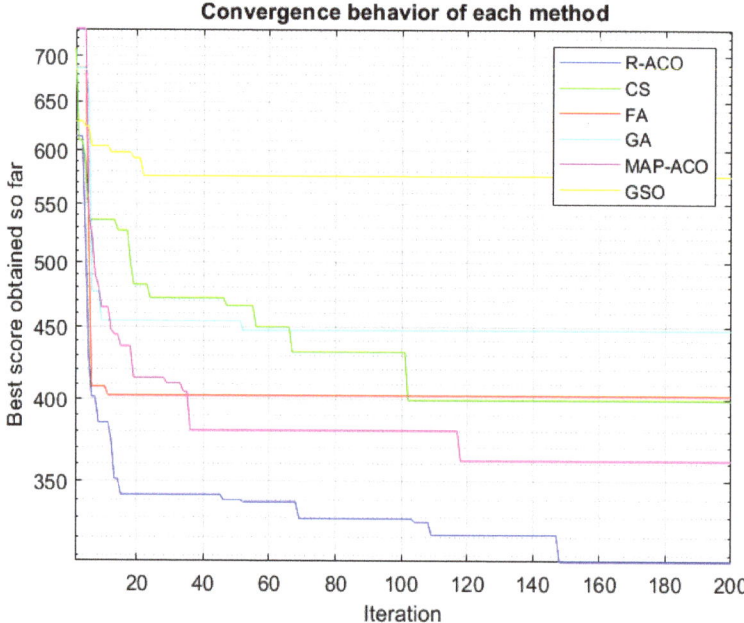

Figure 6. Convergence behavior of each method.

5. Conclusions

In this study, attention is drawn to the importance of smart and mechanical systems for the efficient use of natural resources such as water and nutrients, which are vital in human life. At the same time, it is not enough to just suggest smart systems or methods, it is also important to develop sustainable approaches. Therefore, we focused on an application that can meet many needs in agriculture, from planting to harvesting. Moreover, the health of the farmers can be protected and their heavy workload on the agricultural lands will be reduced. In addition, accident rates will automatically decrease. Therefore, the occupational health and safety rate can be increased. In addition, crop abundance can be experienced due to the mechanized and sustainable nature of the proposed model, and therefore possible forced migration events will also decrease.

The proposed mechanism consisted of three phases. In the first phase, an SCSO-based approach was presented to assign the cultivation task to the relevant autonomous robots, eliminating the human resource factor. It should be noted that in this study, we did not focus on the robotics field, but only on the development of an algorithmic-based model and the architecture of the simulation environment. In the second phase, we used IoT devices to instantly check the status of the crops planted by robots. The result from this phase was taken as the input for the third phase (harvesting). In the third stage, a new method was proposed, inspired by the working mechanism of the ACO algorithm. This method tries to get the maximum profit from the crops to be collected. In addition, it attempts to generate paths for the harvesting robots in a way that would consume the least power and time. According to the results, our proposed algorithms performed better than other methods. In this study, the tomato field was considered as an example, but the suggested mechanism could be used for different purposes in many agricultural lands.

Author Contributions: Conceptualization, F.K., A.S., S.N. and F.A.A.; methodology, F.K., A.S., S.N. and I.Y.; software, A.S. and I.Y.; validation, A.S., M.Z., G.R., S.L., A.M. and I.Y.; formal analysis, A.S. and I.Y.; investigation, A.S., F.A.A., F.E., I.Y., S.N. and F.K.; resources, F.K., F.E., I.Y., A.M. and M.Z.; data curation, M.Z. and I.Y.; writing—original draft preparation, F.K., F.E., S.N., F.A.A., G.R., S.L. and A.M.; writing—review and editing, F.K., F.A.A., I.Y., G.R., S.L. and A.M.; visualization, A.S., F.E., G.R., S.L. and A.M.; supervision, F.K., A.M. and I.Y.; project administration, F.K. and I.Y. All authors have read and agreed to the published version of the manuscript.

Funding: This research received no external funding.

Institutional Review Board Statement: Not applicable.

Informed Consent Statement: Not applicable.

Data Availability Statement: Available on request.

Conflicts of Interest: The authors declare no conflict of interest

References

1. Sylvester, G. *E-Agriculture in Action: Drones for Agriculture*; Food and Agriculture Organization of the United Nations and International Telecommunication Union: Bangkok, Thailand, 2018; pp. 11–22.
2. Kiani, F.; Seyyedabbasi, A.; Nematzadeh, S.; Candan, F.; Çevik, T.; Anka, F.A.; Randazzo, G.; Lanza, S.; Muzirafuti, A. Adaptive Metaheuristic-Based Methods for Autonomous Robot Path Planning: Sustainable Agricultural Applications. *Appl. Sci.* **2022**, *12*, 943. [CrossRef]
3. Quy, V.K.; Hau, N.V.; Anh, D.V.; Quy, N.M.; Ban, N.T.; Lanza, S.; Randazzo, G.; Muzirafuti, A. IoT-Enabled Smart Agriculture: Architecture, Applications, and Challenges. *Appl. Sci.* **2022**, *12*, 3396. [CrossRef]
4. United Nations Environment Programme. "Sustainability". Available online: https://www.unep.org/about-un-environment/sustainability (accessed on 29 September 2021).
5. Muzirafuti, A.; Cascio, M.; Lanza, S. UAV Photogrammetry-based Mapping the Pocket Beach of Isola Bella, Taormina (Northeastern Sicily). In Proceedings of the 2021 International Workshop on Metrology for the Sea; Learning to Measure Sea Health Parameters (MetroSea), Reggio Calabria, Italy, 4–6 October 2021; pp. 418–422.
6. Wolfert, S.; Ge, L.; Verdouw, C.; Bogaardt, M.-J. Big data in smart farming a review. *Agric. Syst.* **2017**, *153*, 69–80. [CrossRef]
7. Kiani, F.; Seyyedabbasi, A. Wireless Sensor Network and Internet of Things in Precision Agriculture. *Int. J. Adv. Comput. Sci. Appl.* **2018**, *9*, 99–103. [CrossRef]

8. Saiz-Rubio, V.; Rovira-Más, F. From Smart Farming towards Agriculture 5.0: A Review on Crop Data Management. *Agronomy* **2020**, *10*, 207. [CrossRef]
9. Kiani, F.; Seyyedabbasi, A.; Nematzadeh, S. Improving the performance of hierarchical wireless sensor networks using the metaheuristic algorithms: Efficient cluster head selection. *Sens. Rev.* **2021**, *41*, 368–381. [CrossRef]
10. Boursianis, A.D.; Papadopoulou, M.S.; Diamantoulakis, P.; Liopa-Tsakalidi, A.; Barouchas, P.; Salahas, G.; Karagiannidis, G.; Wan, S.; Goudos, S.K. Internet of Things (IoT) and Agricultural Unmanned Aerial Vehicles (UAVs) in smart farming: A comprehensive review. *Internet Things* **2020**, *18*, 100187. [CrossRef]
11. Wong, S. Decentralised, Off-Grid Solar Pump Irrigation Systems in Developing Countries-Are They Pro-poor, Pro-environment and Pro-women? In *Climate Change-Resilient Agriculture and Agroforestry*; Springer: Cham, Switzerland, 2019; pp. 367–382.
12. Singh, R.; Singh, G.S. Traditional agriculture: A climate-smart approach for sustainable food production. *Energy Ecol. Environ.* **2017**, *2*, 296–316. [CrossRef]
13. Goldstein, M.; Christopher, U. The Profits of Power: Land Rights and Agricultural Investment in Ghana. *J. Political Econ.* **2008**, *116*, 981–1022. [CrossRef]
14. Karikaya, A. Smart Farming-Precision Agriculture Technologies and Practices. *J. Sci. Perspect.* **2020**, *4*, 123–136. [CrossRef]
15. Kiani, F. A novel channel allocation method for time synchronization in wireless sensor networks. *Int. J. Numer. Model. Electron. Netw. Devices Fields* **2016**, *29*, 805–816. [CrossRef]
16. Radoglou-Grammatikis, P.; Sarigiannidis, P.; Lagkas, T.; Moscholios, I. A compilation of UAV applications for precision agriculture. *Comput. Netw.* **2020**, *172*, 107148. [CrossRef]
17. Evan, D.; Fraser, G.; Campbell, M. Agriculture 5.0: Reconciling Production with Planetary Health. *One Earth* **2019**, *1*, 278–280.
18. Araújo, S.O.; Peres, R.S.; Barata, J.; Lidon, F.; Ramalho, J.C. Characterising the Agriculture 4.0 Landscape—Emerging Trends, Challenges and Opportunities. *Agronomy* **2021**, *11*, 667. [CrossRef]
19. CEMA. Digital Farming: What Does It Really Mean? Available online: http://www.cema-agri.org/publication/digital-farming-what-does-it-really-mean (accessed on 21 September 2021).
20. Tzounis, A.; Katsoulas, N.; Bartzanas, T.; Kittas, C. Internet of Things in agriculture, recent advances and future challenges. *Biosyst. Eng.* **2017**, *164*, 31–48. [CrossRef]
21. Sarni, W.; Mariani, J.; Kaji, J. From Dirt to Data: The Second Green Revolution and IoT. Deloitte Insights. Available online: https://www2.deloitte.com/insights/us/en/deloitte-review/issue-18/second-greenrevolution-and-internet-of-things.html#endnote-sup-9 (accessed on 22 September 2021).
22. Sandeep, V.; Amol, D.; Vibhute, D.; Karbhari, V.; Suresh, K.; Mehrotra, C. An innovative IoT based system for precision farming. *Comput. Electron. Agric.* **2021**, *187*, 106291.
23. Köksal, Ö.; Tekinerdogan, B. Architecture design approach for IoT-based farm management information systems. *Precis. Agric.* **2019**, *20*, 926–958. [CrossRef]
24. Jangam, A.R.; Kale, K.V.; Gaikwad, S.; Vibhute, A.D. Design and development of IoT based system for retrieval of agrometeorological parameters. In Proceedings of the International Conference on Recent Innovations in Electrical, Electronics & Communication Engineering (ICRIEECE), Bhubaneswar, India, 27–28 July 2018; pp. 804–809.
25. Robles, J.R.; Martin, Á.; Martin, S.; Ruipérez-Valiente, J.; Castro, M. Autonomous sensor network for rural agriculture environments, low cost, and energy self-charge. *Sustainability* **2020**, *12*, 5913. [CrossRef]
26. Sawant, S.; Durbha, S.S.; Jagarlapudi, A. Interoperable agro-meteorological observation and analysis platform for precision agriculture: A case study in citrus crop water requirement estimation. *Comput. Electron. Agric.* **2017**, *138*, 175–187. [CrossRef]
27. Ayaz, M.; Ammad-Uddin, M.; Sharif, Z.; Mansour, A.; Aggoune, E.M. Internet-of-Things (IoT)-Based Smart Agriculture: Toward Making the Fields Talk. *IEEE Access* **2019**, *7*, 129551–129583. [CrossRef]
28. Nabi, F.; Jamwal, S.; Padmanbh, K. Wireless sensor network in precision farming for forecasting and monitoring of apple disease: A survey. *Int. J. Inform. Technol.* **2020**, *14*, 769–780. [CrossRef]
29. Vitali, G.; Francia, M.; Golfarelli, M.; Canavari, M. Crop Management with the IoT: An Interdisciplinary Survey. *Agronomy* **2021**, *11*, 181. [CrossRef]
30. Tayari, E.; Jamshid, A.R.; Goodarzi, H.R. Role of GPS and GIS in precision agriculture. *J. Scient. Res. Dev.* **2015**, *2*, 157–162.
31. Del Cerro, J.; Cruz Ulloa, C.; Barrientos, A.; de León Rivas, J. Unmanned Aerial Vehicles in Agriculture: A Survey. *Agronomy* **2021**, *11*, 203. [CrossRef]
32. Dai, B.; He, Y.; Gu, F.; Yang, L.; Han, J.; Xu, W. A vision-based autonomous aerial spray system for precision agriculture. In Proceedings of the 2017 IEEE International Conference on Robotics and Biomimetics (ROBIO), Macau, China, 5–8 December 2017; pp. 507–513.
33. Behjati, M.; Mohd Noh, A.B.; Alobaidy, H.A.H.; Zulkifley, M.A.; Nordin, R.; Abdullah, N.F. LoRa Communications as an Enabler for Internet of Drones towards Large-Scale Livestock Monitoring in Rural Farms. *Sensors* **2021**, *21*, 5044. [CrossRef]
34. Li, X.; Zhao, Y.; Zhang, J.; Dong, Y. A hybrid PSO algorithm based flight path optimization for multiple agricultural uavs. In Proceedings of the 2016 IEEE 28th International Conference on Tools with Artificial Intelligence (ICTAI), San Jose, CA, USA, 6–8 November 2016; pp. 691–697.
35. Allred, B.; Eash, N.; Freeland, R.; Martinez, L.; Wishart, D. Effective and efficient agricultural drainage pipe mapping with uas thermal infrared imagery: A case study. *Agric. Water Manag.* **2018**, *197*, 132–137. [CrossRef]

36. Vasudevan, A.; Kumar, D.A.; Bhuvaneswari, N.S. Precision farming using unmanned aerial and ground vehicles. In Proceedings of the 2016 IEEE Technological Innovations in ICT for Agriculture and Rural Development (TIAR), Chennai, India, 15–16 July 2016; pp. 146–150.
37. Memmah, M.M.; Lescourret, F.; Yao, X.; Lavigne, C. Metaheuristics for agricultural land use optimization. A review. *Agron. Sustain. Dev.* **2015**, *35*, 975–998. [CrossRef]
38. Randazzo, G.; Italiano, F.; Micallef, A.; Tomasello, A.; Cassetti, F.P.; Zammit, A.; D'Amico, S.; Saliba, O.; Cascio, M.; Cavallaro, F.; et al. WebGIS Implementation for Dynamic Mapping and Visualization of Coastal Geospatial Data: A Case Study of BESS Project. *Appl. Sci.* **2021**, *11*, 8233. [CrossRef]
39. Chetty, S.; Adewumi, A.O. Three new stochastic local search metaheuristics for the annual crop planning problem based on a new irrigation scheme. *J. Appl. Math.* **2013**, *2013*, 158538. [CrossRef]
40. Chikumbo, O.; Goodman, E.; Deb, K. Approximating a multidimensional Pareto front for a land use management problem: A modified MOEA with an epigenetic silencing metaphor. In Proceedings of the Evolutionary Computation (CEC), Brisbane, Australia, 10–15 June 2012; pp. 1–9.
41. Adenso-Díaz, B.; Villa, G. Crop Planning in Synchronized Crop-Demand Scenarios: A Biobjective Optimization Formulation. *Horticulturae* **2021**, *7*, 347. [CrossRef]
42. Rakhra, M.; Singh, R.; Lohani, T.; Shabaz, M. Metaheuristic and Machine Learning-Based Smart Engine for Renting and Sharing of Agriculture Equipment. *Math. Probl. Eng.* **2021**, *2021*, 5561065. [CrossRef]
43. Pal, P.; Sharma, R.P.; Tripathi, S.; Kumar, C.; Ramesh, D. Genetic algorithm optimized node deployment in IEEE 802.15.4 potato and wheat crop monitoring infrastructure. *Sci. Rep.* **2021**, *11*, 8231. [CrossRef]
44. Chetty, S.; Adewumi, A.O. Comparison Study of Swarm Intelligence Techniques for the Annual Crop Planning Problem. *IEEE Trans. Evol. Comput.* **2014**, *18*, 258–268. [CrossRef]
45. Dwivedi, A.; Jha, A.; Prajapati, D.; Sreenu, N.; Pratap, S. Meta-heuristic algorithms for solving the sustainable agro-food grain supply chain network design problem. *Mod. Supply Chain. Res. Appl.* **2020**, *2*, 161–177. [CrossRef]
46. Sabzi, S.; Abbaspour-Gilandeh, Y.; García-Mateos, G. A fast and accurate expert system for weed identification in potato crops using metaheuristic algorithms. *Comput. Ind.* **2018**, *98*, 80–89. [CrossRef]
47. Khan, M.S.; Anwar ul Hassan, C.H.; Sadiq, H.A.; Ali, I.; Rauf, A.; Javaid, N. A New Meta-heuristic Optimization Algorithm Inspired from Strawberry Plant for Demand Side Management in Smart Grid. In *International Conference on Intelligent Networking and Collaborative Systems*; Springer: Cham, Switzerland, 2018; pp. 143–154.
48. Jiang, S.; Zhang, H.; Cong, W.; Liang, Z.; Ren, Q.; Wang, C.; Zhang, F.; Jiao, X. Multi-Objective Optimization of Smallholder Apple Production: Lessons from the Bohai Bay Region. *Sustainability* **2020**, *12*, 6496. [CrossRef]
49. Sharma, V.; Tripathi, A.K. A systematic review of meta-heuristic algorithms in IoT based application. *Array* **2022**, *14*, 1–8. [CrossRef]
50. Seyyedabbasi, A.; Kiani, F. Sand Cat swarm optimization: A nature-inspired algorithm to solve global optimization problems. *Eng. Comput.* **2022**, 1–25. [CrossRef]
51. Heuvelink, E. *Tomato Growth and Yield: Quantitative Analysis and Synthesis*; Oxford University Press: Oxford, UK, 1996.
52. Gupta, M.K.; Chandra, P.; Samuel, D.V.K.; Singh, B.; Singh, A.; Garg, M.K. Modeling of Tomato Seedling Growth in Greenhouse. *Agric. Res.* **2012**, *1*, 362–369. [CrossRef]
53. Haifa Group. Available online: https://www.haifa-group.com/tomato-fertilizer/crop-guide-tomato (accessed on 4 October 2021).
54. Giniger, M.S.; McAvoy, R.J.; Giacomelli, G.A.; Janes, H.W. Computer Simulation of a Single Truss Tomato Cropping System. *Trans. Am. Soc. Agric. Eng.* **1998**, *31*, 1176–1179. [CrossRef]
55. Kiani, F.; Nematzadehmiandoab, S.; Seyyedabbasi, A. Designing a dynamic protocol for real-time Industrial Internet of Things-based applications by efficient management of system resources. *Adv. Mech. Eng.* **2019**, *11*, 1687814019866062. [CrossRef]
56. Kiani, F. AR-RBFS: Aware-routing protocol based on recursive best-first search algorithm for wireless sensor networks. *J. Sens.* **2016**, *2016*, 8743927. [CrossRef]
57. Dorigo, M.; Birattari, M.; Stutzle, T. Ant colony optimization. *IEEE Comput. Intell. Mag.* **2006**, *1*, 28–39. [CrossRef]
58. Nematzadeh, S.; Kiani, F.; Torkamanian-Afshar, M.; Aydin, N. Tuning hyperparameters of machine learning algorithms and deep neural networks using metaheuristics: A bioinformatics study on biomedical and biological cases. *Comput. Biol. Chem.* **2022**, *97*, 107619. [CrossRef]
59. Mirjalili, S.; Mirjalili, S.M.; Lewis, A. Grey wolf optimizer. *Adv. Eng. Softw.* **2014**, *69*, 46–61. [CrossRef]
60. Seyyedabbasi, A.; Kiani, F. I-GWO and Ex-GWO: Improved algorithms of the Grey Wolf Optimizer to solve global optimization problems. *Eng. Comput.* **2021**, *37*, 509–532. [CrossRef]
61. Mirjalili, S. Moth-flame optimization algorithm: A novel nature-inspired heuristic paradigm. *Knowl. Based Syst.* **2015**, *89*, 228–249. [CrossRef]
62. Zhou, Z.; Li, F.; Abawajy, J.H.; Gao, C. Improved PSO Algorithm Integrated with Opposition-Based Learning and Tentative Perception in Networked Data Centres. *IEEE Access* **2020**, *8*, 55872–55880. [CrossRef]
63. Bazi, S.; Benzid, R.; Bazi, Y.; Rahhal, M.M.A. A Fast Firefly Algorithm for Function Optimization: Application to the Control of BLDC Motor. *Sensors* **2021**, *21*, 5267. [CrossRef]
64. Yang, X.S. *Nature-Inspired Metaheuristic Algorithms*, 2nd ed.; Luniver: Bristol, UK, 2010.

65. Joshi, A.S.; Kulkarni, O.; Kakandikar, G.M.; Nandedkar, V.M. Cuckoo search optimization-a review. *Mater. Today Proc.* **2017**, *4*, 7262–7269. [CrossRef]
66. Krishnanand, K.N.; Ghose, D. Glowworm swarm optimization for simultaneous capture of multiple local optima of multimodal functions. *Swarm Intell.* **2009**, *3*, 87–124. [CrossRef]
67. Seyyedabbasi, A.; Kiani, F. MAP-ACO: An efficient protocol for multi-agent pathfinding in real-time WSN and decentralized IoT systems. *Microprocess. Microsyst.* **2020**, *79*, 103325. [CrossRef]
68. Dianfan, Z.; Meinke, H.; Wilson, M.; Leo, F.; Marcelis, M. Towards delivering on the sustainable development goals in greenhouse production systems. *Resour. Conserv. Recycl.* **2021**, *169*, 105379.
69. Chahal, I.; Van Eerd, L.L. Cover crops increase tomato productivity and reduce nitrogen losses in a temperate humid climate. *Nutr. Cycl. Agroecosyst.* **2021**, *119*, 195–211. [CrossRef]

Article

Crop Classification for Agricultural Applications in Hyperspectral Remote Sensing Images

Loganathan Agilandeeswari [1], Manoharan Prabukumar [1,*], Vaddi Radhesyam [2], Kumar L. N. Boggavarapu Phaneendra [2] and Alenizi Farhan [3]

[1] School of Information Technology Engineering (SITE), Vellore Institute of Technology, Vellore 632014, India; agila.l@vit.ac.in
[2] Department of Information Technology, Velagapudi Ramakrishna Siddhartha Engineering College, Vijayawada 520007, India; syam.radhe@vrsiddhartha.ac.in (V.R.); phaneendra.b@vrsiddhartha.ac.in (K.L.N.B.P.)
[3] Electrical Engineering Department, Prince Sattam Bin Abdulaziz University, Al-Kharj 16278, Saudi Arabia; fa.alenizi@psau.edu.sa
* Correspondence: mprabukumar@vit.ac.in; Tel.: +91-9894699058

Citation: Agilandeeswari, L.; Prabukumar, M.; Radhesyam, V.; Phaneendra, K.L.N.B.; Farhan, A. Crop Classification for Agricultural Applications in Hyperspectral Remote Sensing Images. *Appl. Sci.* 2022, 12, 1670. https://doi.org/10.3390/app12031670

Academic Editors: Dimitrios S. Paraforos and Anselme Muzirafuti

Received: 19 December 2021
Accepted: 28 January 2022
Published: 5 February 2022

Publisher's Note: MDPI stays neutral with regard to jurisdictional claims in published maps and institutional affiliations.

Copyright: © 2022 by the authors. Licensee MDPI, Basel, Switzerland. This article is an open access article distributed under the terms and conditions of the Creative Commons Attribution (CC BY) license (https://creativecommons.org/licenses/by/4.0/).

Abstract: Hyperspectral imaging (HSI), measuring the reflectance over visible (VIS), near-infrared (NIR), and shortwave infrared wavelengths (SWIR), has empowered the task of classification and can be useful in a variety of application areas like agriculture, even at a minor level. Band selection (BS) refers to the process of selecting the most relevant bands from a hyperspectral image, which is a necessary and important step for classification in HSI. Though numerous successful methods are available for selecting informative bands, reflectance properties are not taken into account, which is crucial for application-specific BS. The present paper aims at crop mapping for agriculture, where physical properties of light and biological conditions of plants are considered for BS. Initially, bands were partitioned according to their wavelength boundaries in visible, near-infrared, and shortwave infrared regions. Then, bands were quantized and selected via metrics like entropy, Normalized Difference Vegetation Index (NDVI), and Modified Normalized Difference Water Index (MNDWI) from each region, respectively. A Convolutional Neural Network was designed with the finer generated sub-cube to map the selective crops. Experiments were conducted on two standard HSI datasets, Indian Pines and Salinas, to classify different types of crops from Corn, Soya, Fallow, and Romaine Lettuce classes. Quantitatively, overall accuracy between 95.97% and 99.35% was achieved for Corn and Soya classes from Indian Pines; between 94.53% and 100% was achieved for Fallow and Romaine Lettuce classes from Salinas. The effectiveness of the proposed band selection with Convolutional Neural Network (CNN) can be seen from the resulted classification maps and ablation study.

Keywords: band selection; CNN; NDVI; hyperspectral imaging; crops; agriculture

1. Introduction

Due to advancements in remote sensing image acquisition mechanisms and the growing availability of rich spectral and spatial information by using a variety of sensors, hyperspectral imaging has gained importance. In particular, Hyperspectral Image (HSI) classification has become a prominent source for practical applications in fields like agriculture, environment, forestry, mineral mapping, etc. [1–5].

The present paper focuses on analyzing and using HSI in the agriculture field. Accurate information about growing crops with different climate conditions and agricultural resources and with different timestamps (before, during, and after cultivation) is extremely important and useful for agricultural development. Traditional methods, like field surveys and other statistical-based analyses, are very time-consuming. Advanced remote sensing technology, including HSI, provides a suitable solution and can fill the gap [6–10] with solutions like crop classification.

The problem of crop classification using hyperspectral images has been addressed by researchers with various methods [11,12]. A method based on regression analysis was used to classify the variety of sugarcane crops in Brazil. This HSI data was captured using the EO-1 satellite [13]. The method, proposed in [14], is a combination of Support Vector Machine (SVM) and linear spectral models and was used successfully on the data captured from the Hyperion satellite. This method was also used to classify litchi crops in Guangzhou. Crops in the Karnataka area were classified using the Spectral Angular Mapper (SAM) classifier method for the Hyperion data [15].

The HSI sensor called Airborne Visible Infra-Red Imaging Spectrometer (AVIRIS) has recently become important in the remote sensing community. AVIRIS has high spectral bands (224 bands) and spatial resolution (20 m for Indian Pines and 3.7 m for Salinas datasets) with a wavelength range of 380–2500 nm covering VIS, NIR, and SWIR regions and hence is known to be crucial for agricultural applications [16].

Some of the crop classification methods in this connection are as follows. Combined linear and nonlinear SVM algorithms were used to classify corn crops on AVIRIS data. This method obtained moderate accuracy. Soybeans and wheat crops were classified using SVM and Markov Random Field with good accuracy for the AVIRIS HSI data [17]. Unmanned Aerial Vehicle (UAV) datasets also experimented with classifying crops like cabbage, cotton, and strawberry. High accuracy was noticed using Conditional Random fields [18,19]. Further, the Salinas data set was tested to classify different crops using a support vector machine. This method achieved a moderate level of accuracy [20]. Other methods, including spatial context support vector machines, had reasonable accuracy [21].

The above methods are insufficient to extract the required information, and it is difficult to obtain commendable results [22–24]. Several successful band selection methods have been introduced (to perform before classification) in the literature over the last few decades, including ranking-based approaches, clustering-based approaches, searching-based approaches, relative entropy, and information entropy-based approaches [25–28]. The optimal clustering framework was introduced in [29] for band selection and successfully applied the novel objective function with several constraints. In these methods, bands are clustered initially and then ranked according to different measures to select the representative bands from the image. All these methods work well for selecting the informative bands and hence produce high classification accuracy. However, these methods may not be suitable for application-specific classification problems. For example, in the present context, we consider crop classification as our target. It is important to adopt the band section strategy in the view of agricultural phenology, where biophysical properties of plants are also taken into consideration. This is shown in Figure 1 (Source for Figure 1a: Vegetation analysis: using vegetation indices in ENVI [16]).

The details of the measures used for the band selection, mathematical equations, range of the measures, and BS procedure are discussed in the Section 2.

The contributions of the paper are shown below:

- There are different successful methods in the literature for HSI classification. However, not all methods are suitable for all the available application areas to perform classification. In the present paper, we designed a band selection model for crop classification based on the physical and biological properties of plants.
- A new framework for informative band selection is proposed by partitioning the original hyperspectral cube based on the reflectance nature in the visible, near-infrared, and short wave infrared regions of the electromagnetic spectrum. This further uses measures such as entropy, NDVI, and NDWI, respectively, for band quantization.
- A two-dimensional convolution network for hyperspectral image classification is designed and implemented for the accurate classification of agriculture crops with the selected bands. Detailed analysis of the results in crop classification is showcased.

The following is a breakdown of the paper's structure: materials and methods are presented in Section 2, which consists of a technical description of the proposed method. Dataset description is described in Section 3, followed by Section 4, which includes HSI classification, experimental results, and analysis. Finally, the conclusions are presented in Section 5.

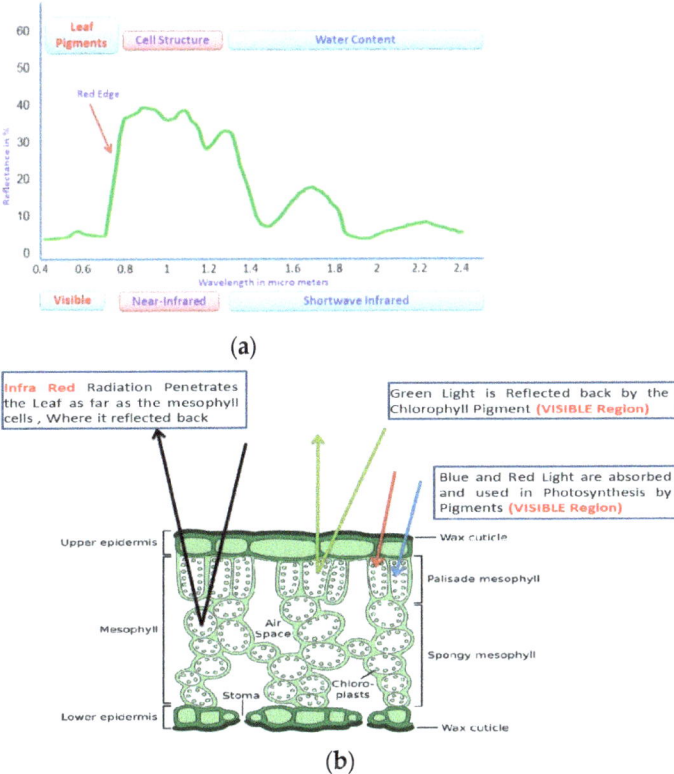

Figure 1. (a): Spectral Reflectance Properties of Vegetation Spectrum, (b): Leaf cell structure showing the interaction of light with VIS, NIR, and SWIR regions of the electromagnetic spectrum.

2. Materials and Methods

Crop classification using HSI consists of two steps. The first is band selection, and the second is classification. Partition-based band selection is proposed in this paper. Initially, the partition is performed based on properties of the vegetation spectrum, and three partitions are created. These are termed VIS, NIR, and SWIR partitions (Figure 1). Further bands are selected from these three partitions based on three relevant metrics, entropy, NDVI, and MNDWI, respectively. In the second step to perform classification, the concatenated bands are given as input to the designed CNN model. The CNN model consists of a series of convolution and fully connected layers. Finally, crop classification can be achieved for the selected input data. The proposed architecture for the classification of crops using HSI data is shown in Figure 2.

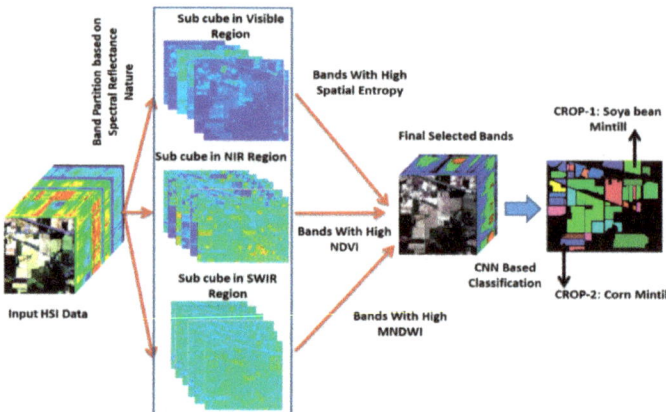

Figure 2. Architecture of the proposed partition-based band selection and CNN-based classification for crop classification.

Let H be hyperspectral data with h rows, w columns and d represents number of spectral bands of H. Each denotes one band in the dataset. According to spectral reflectance properties in various regions of the electromagnetic spectrum, HSI data can be separated into n partitions Partition#1, Partition#2 ... Partition#n with bands $d_1, d_2, \ldots, d_n$ respectively. Here:

$$d_1 + d_2 + \cdots d_n = d \qquad (1)$$

In the present context, n value is considered as 3 with Partition#1 denoting bands in the VIS region, Partition#2 denotes bands in the NIR region, and Partition#3 denotes bands in the SWIR region. Three metrics are then chosen to select bands from each partition.

Information Entropy (IE) is a criterion to measure spatial information in the HSI bands. For a particular band X_j, Information Entropy is defined as:

$$IE(X_j) = -\sum_{x \in X_j} P(x) * \log P(x) \qquad (2)$$

where $P(x)$ is the probability of number of grey level of the histogram of the band x in the image.

Based on IE, each band in the VIS region is quantified, and top $m_1 (< d_1)$ bands are selected from Partition#1 based on the threshold limit value of $\delta 1$.

Normalized Difference Vegetation Index (NDVI) measures plant health in terms of greenness density, as shown in Equation (3). This is a widely used vegetation index in the remote sensing community. The NDVI ranges from +1 to −1. Dead plants have −1 as NDVI value, and healthy plants have values between 0.65 and 1.

$$NDVI = \frac{(NIR - RED)}{(NIR + RED)} \qquad (3)$$

Based on NDVI, each band in the NIR region is quantified, and top $m_2 (< d_2)$ bands are selected from Partition#2 based on the threshold limit value of $\delta 2$.

Modified Normalized Difference Water Index (MNDWI) measures the open water enhanced identification and is computed using Equation (4). This will suppress noise generated by vegetation and soil and, at the same time, improve the open water features. MNDWI ranges from +1 to 1.

$$MNDWI = \frac{(GREEN - SWIR)}{(GREEN + SWIR)} \qquad (4)$$

Based on MNDWI, each band in the SWIR region is quantified, and top $m_3 (< d_3)$ bands are selected from Partition#3 based on the threshold limit value of $\delta 3$.
Here:
$$m_1 + m_2 + m_3 = m \text{ and strictly } m < d. \quad (5)$$

The proposed band selection algorithm is presented as Algorithm 1 below.

Algorithm 1: Proposed Band Selection Approach for Crop Classification

---

Input: $H \in R^{h \times w \times d}$ be the Hyperspectral image Data, R: Red band, G: Green band, thresholds: $\delta 1, \delta 2$, and $\delta 3$
Output: H_{BS}, finer sub cube with informative and selected bands
Step 1: Partition the image H with d bands into three sub cubes based on light properties and biological conditions
Step 2: Let the number of bands in each of the sub cubes be d_1, d_2 and d_3 bands respectively from visible, near Infrared, and shortwave infrared regions
Step 3: for i:1 to d_1, Compute the entropy, E_i, of each band using Equation (2)
Step 4: Generate finer sub cube H_{vis} with m_1 bands for those bands from d_1 whose $E_i > \delta 1$
Step 5: for i = 1 to d_2, Compute the Normalized Difference Vegetation Index, $NDVI_i$, using Equation (3)
Step 6: Generate finer sub cube H_{NIR} with m_2 bands for those bands from d_2 whose $NDVI_i > \delta 2$
Step 7: for i = 1 to d_3, Compute the Modified Normalized Difference Water Index, $MNDWI_i$, using Equation (4)
Step 8: Generate finer sub cube H_{SWIR} with m_3 bands for those bands from d_2 whose $MNDwI_i > \delta 3$
Step 9: Combine the sub cubes H_{vis}, H_{NIR} and H_{SWIR} as $H_{BS} = H_{vis} \cup H_{NIR} \cup H_{SWIR}$ which satisfies Equation (5)

There are two modules in the proposed methodology for the band selection task known as Partition and Ranking. The partition module is focused on the Biophysical properties of plants in Visible, NIR, and SWIR regions. These regions are named Partition#1, Partition#2, and Partition#3. In the ranking module, agricultural phenology metrics such as Entropy, NDVI, and MNDWI are computed using Equations (2)–(4) for the three partitions, respectively. Then, the computed values and selected representative bands are ranked using an adaptive threshold for each of the measures. The parameter tunning is shown in the experimental section.

Let $H_{BS} \in R^{h \times w \times m}$ denote hyperspectral data cube after the selection of m spectral bands [30–32]. This data can be split into two parts. One is for training and the other for testing. Let imagine χ as a training vector that will be input to the CNN model. The first layer is the convolution layer which follows according to Equation (6). Here $\otimes$ denotes convolution operator, filter is denoted F and (i, j) denotes the corresponding spatial location.

$$\text{CONVOLUTION}_{i,j} = \sigma((F \otimes \chi)_{i,j} + b) \quad (6)$$

Here $\sigma(.)$ denotes activation function. For better convergence, the ReLU activation function is used in the present model as in Equation (7). This function gives output as same as input or zero.

$$\sigma(x) = \max(0, x) \quad (7)$$

After a series of convolution layers, the feature vectors are converted into a single Flatten Vector (FV), which will be given as input to Fully Connected (FC) layers. In FC layers, two operations, pre-activation and activation, will be performed at every node. All the FC layers use the ReLU activation function. However, the last layer uses the softmax activation function, as shown in Equation (8).

$$\text{Softmax}(x_i) = \frac{e^{x_i}}{\sum_j e^{x_j}} \quad (8)$$

This function gives output probabilities of each crop, and hence classification is possible.

The proposed method performance was tested with state-of-the-art methods (a brief description of the methods is given in Section 4). The experimental works were carried out using MATLAB R 2018b and Python with Google co-laboratory. The hardware utilized for the work was a personal computer with Intel(R) Core(TM) i5-6500 CPU with 3.20 GHz and 8 GB RAM.

3. Dataset Description

In the present work, experiments were conducted using two popular AVIRIS sensor-based datasets. These datasets are freely available and can be downloaded from [33]. The first hyperspectral data used is Indian Pines. These data were captured in the agricultural area of northwestern Indiana, USA, on 12 June 1992. The number of pixels is 145*145, collected in the form of 224 bands by covering the electromagnetic spectrum in the range of 400–2500 nm. Agriculture crops like corn and soya are covered in this data as 64% area. The vegetation types of grass and pastures are covered with 25% area [34–36].

The second hyperspectral data used is Salinas. These data were captured in the agricultural area of the Salinas Valley region, CA, USA, on 9 October 1998. The number of pixels is 512*217, collected in the form of 224 bands by covering the electromagnetic spectrum in the range of 400–2500 nm. Agriculture crops like vineyard fields, broccoli weeds, celery, fallow, and lettuce crops are covered with 100% area [37–39]. The class description and pixel samples information for the two datasets is shown in Table 1 [40–42].

The major type of classes that exist in both Indian Pines and Salinas are shown in Figure 3. As per our interest for the present paper, different types of crops that exist in both Indian Pines and Salinas are also shown in Figure 4 [43–46].

The Indian Pines data belongs to the agricultural area of northwestern Indiana, USA. The area includes major portions of the Indian Creek and Pine Creek watersheds. Indiana is the tenth-largest farming state in the USA. More than 80% of the land in Indiana is dedicated to farms, forests, and woodland. The Salinas data belongs to one of the efficacious agricultural areas located in the central coast region of California, called the Salinas Valley, USA. This site is famous for producing most of the agricultural activities in the county due to its rich soil and plentiful underground water supplies [47].

The nomenclature for the individual classes is set according to the type of land, growing type, and spectral properties. For example, the classes "Brocoli_green_weeds_1" and "Brocoli_green_weeds_2" of Salinas's data belong to the same type of land cover representing broccoli weeds, but they have different spectral properties due to their different conditions. The crop "Lettuce_romaine_4wk" refers to the lettuce crop that grows in the fourth week. The same terminology is used for the classes "Lettuce_romaine_5wk", "Lettuce_romaine_6wk", and "Lettuce_romaine_7wk". From the Indian Pines data, "Corn-notill" represents cultivation without tillage and "Corn-mintill" represents cultivation with minimum tillage. The same terminology is used for the classes "Soybean-notill" and "Soybean-mintill". These two classes, along with "Soybean clean", represent different growing periods of the same soybean crop.

More information about the data and subclass regions in both datasets is presented in Figure 5.

On the whole, it can be concluded that the selected data is more suitable for crop classification applications. Indian Pines data consists of 64% pixel regions as different types of crops, and for Salinas data, crop regions are found to be 56%.

Table 1. Number of samples per class and description of each class for two standard data sets of AVIRIS Sensor.

Indian Pines			Salinas		
Name of the Class	Number of Pixel Samples per Band	Class Description	Name of the Class	Number of Pixel Samples per Band	Class Description
Alfalfa	46	Type of green grass	Brocoli_green_weeds_1	2009	Green colored vegetable
Corn-notill	1428	Corn crop cultivation without tillage	Brocoli_green_weeds_2	3726	Green colored vegetable
Corn-mintill	830	Corn crop cultivation with minimum tillage	Fallow	1976	Land Region
Corn	237	Corn crop	Fallow_rough_plow	1394	Land Region
Grass-pasture	483	Type of green grass	Fallow_smooth	2678	Land Region
Grass-trees	730	Type of green grass	Stubble	3959	Land Region
Grass-pasture-mowed	28	Type of green grass	Celery	3579	Green colored Plant region
Hay-windrowed	478	Row of cut small grain grass	Grapes_untrained	11,271	Type of Vineyard
Oats	20	Plant in Brown color	Soil_vinyard_develop	6203	Soil Region
Soybean-notill	972	Soya crop cultivation without tillage	Corn_senesced_green_weeds	3278	Green colored Plant
Soybean-mintill	2455	Soya crop cultivation with minimum tillage	Lettuce_romaine_4wk	1068	lettuce crop that grows in the fourth week
Soybean-clean	593	Soya Plant	Lettuce_romaine_5wk	1927	lettuce crop that grows in the fifth week
Wheat	205	Brown colored Wheat Plant	Lettuce_romaine_6wk	916	lettuce crop that grows in the sixth week
Woods	1265	Type of Tree	Lettuce_romaine_7wk	1070	lettuce crop that grows in the seventh week
Buildings-Grass-Trees-Drives	386	Building area	Vinyard_untrained	7268	Type of Vineyard
Stone-Steel-Towers	93	Tower area	Vinyard_vertical_trellis	1807	Type of Vineyard
Background	10,776			56,975	

Figure 3. Different classes of AVIRIS Sensor data (**a**) Indian Pines (**b**) Salinas.

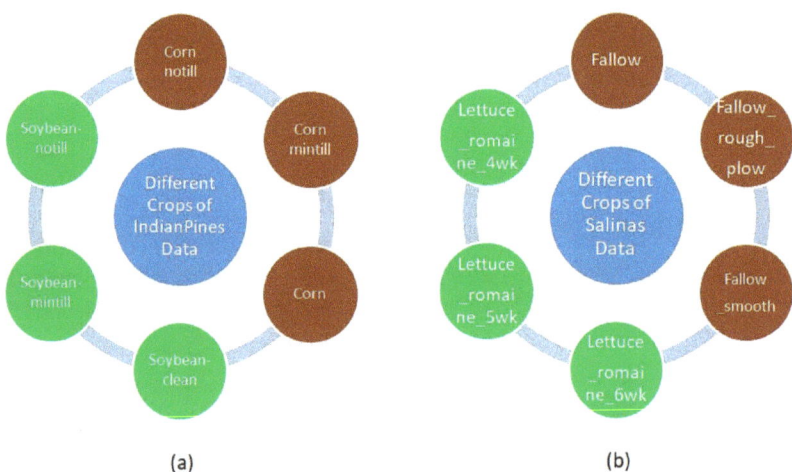

Figure 4. Different crops of AVIRIS Sensor data (**a**) Indian Pines (**b**) Salinas.

Figure 5. Different class regions (**a**) Indian Pines (**b**) Salinas.

4. Experimental Results and Analysis

This section consists of parameter tuning, complete experimental results, and analysis. Crop classification results are evaluated against standard metrics.

4.1. Subsection Parameter Tuning in Band Selection

AVIRIS Sensor data is observed to be spread over three partitions. The first partition is the visible region (VIS) in the wavelength region of 400–700 nm and is equipped with

32 bands (d_1). The second partition is the near-infrared region (NIR) in the wavelength region of 700–1000 nm and is equipped with 40 bands (d_2). The third partition is the short wave infrared region (SWIR) in the wavelength region of 1000–2500 nm and is equipped with 148 bands (d_3).

The threshold value to select bands from the VIS region is set as 4.6 (δ_1), and the number of bands selected is 15 (m_1) for Indian Pines data, and for Salinas data, these values are set as 4 and 12, respectively. The threshold value to select bands from the NIR region is set as 0.2 (δ_2), and the number of bands selected is 15 (m_2) for Indian Pines data, and for Salinas data, these values are set as 0.11 and 16, respectively. The threshold value to select bands from the SWIR region is set as 0.62 (δ_3), and the number of bands selected is 15 (m_3) for Indian Pines data, and for Salinas data, these values are set as 0.99 and 15, respectively.

The red band and green band information is essential to calculate NDVI and MNDWI. These are shown in Figure 6 for both datasets [48,49].

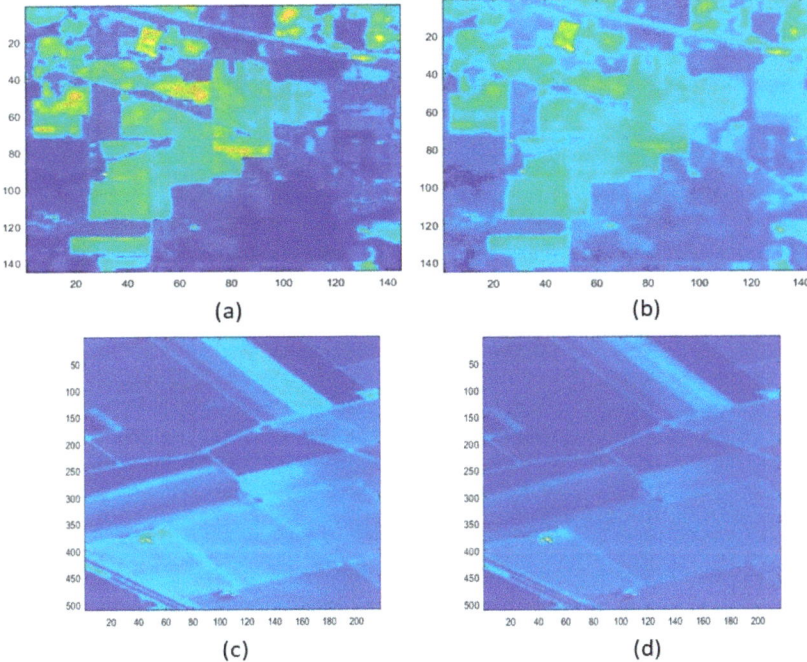

Figure 6. Different bands of Indian Pines data. (**a**) Red band with band number 29, (**b**) Green band with band number 15. For Salinas data (**c**) Red band with band number 29, (**d**) Green band with band number 15.

The δ_1, δ_2, and δ_3 values were set as shown in Figure 7. Using the number of available bands (224) from the AVIRIS Sensor, we considered 1/5th of the number of the bands. Accordingly, we set m_1, m_2, and m_3 values.

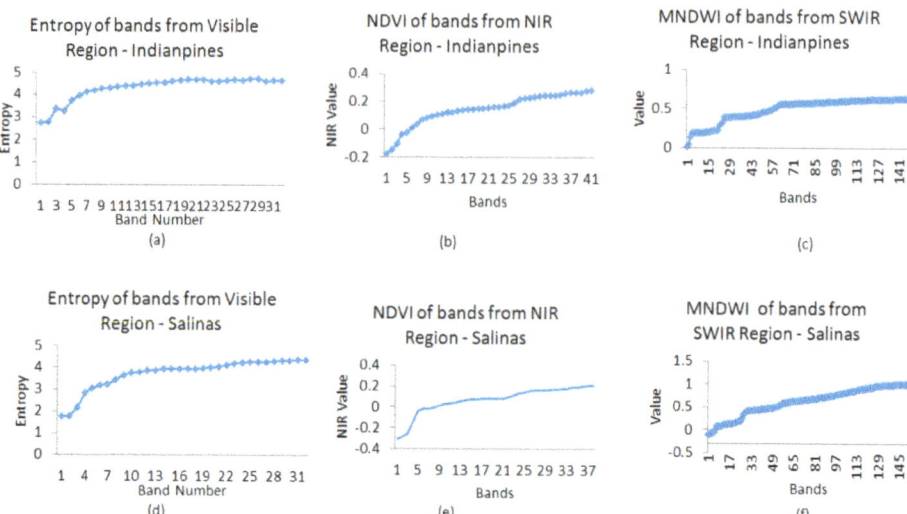

Figure 7. Row-1: Indian Pines data (**a**) Entropy of visible region bands (**b**) NDVI of NIR region bands (**c**) MNDWI of SWIR region bands. Row-2: Salinas's data (**d**) Entropy of visible region bands (**e**) NDVI of NIR region bands (**f**) MNDWI of SWIR region bands (Bands on x-axis for (**b**,**c**,**e**,**f**) are the chosen bands from the respective region).

From Indian Pines data, a total of 45 (m) bands were chosen with following band numbers: 18 19 20 21 22 23 24 25 26 27 28 29 30 31 32 37 38 39 41 42 43 44 45 47 48 49 50 51 52 53 151 152 153 154 155 156 157 158 159 160 161 162 163 219 220.

From Salinas data, a total of 42 (m) bands were chosen with following band numbers: 21 22 23 24 25 26 27 28 29 30 31 32 40 41 42 44 45 46 47 48 50 51 52 53 54 55 56 108 109 110 154 155 156 157 158 159 160 161 162 163 164 165.

The graphical representation, which shows the entropy of visible region bands, NDVI of NIR region bands, and MNDWI of SWIR region bands, is in Figure 7 for both datasets.

4.2. Parameter Tuning in Classification Using CNN

After the selection of representative bands, it is necessary to perform classification. The detail of the split between training and testing samples of both datasets is presented in Table 2.

Table 2. Details of training and testing samples per band for the data sets.

Dataset Name	Total Number of Pixel Samples	Number of Training Pixels	Number of Test Pixels	Number of Background Pixels
Indian Pines	10,249	7679	2569	10,776
Salinas	54,129	40,593	13,536	56,975

Parameter tuning for common parameters of the CNN model is presented in Table 3.

Table 3. Selected Parameters for CNN Model for two data sets.

S.No	Parameter Type	Value
1	Loss Function	Categorical cross entropy
2	Model	Sequential
3	Epochs	100
4	Optimizer	Adam
5	Batch size	$2^{**}3$
6	Convolution Layer: filters_1	$2^{**}6$
7	Convolution Layer: filters_2	$2^{**}7$
8	Dropout rate	0.35
9	Fully Connected Layer:Units_1	$2^{**}8$
10	Fully Connected Layer:Units_2	$2^{**}7$
11	Fully Connected Layer:Units_3	$2^{**}6$
12	Fully Connected Layer:Units_4	$2^{**}5$

CNN network model for Indian Pines data is presented in Table 4, and the CNN network model for Salinas data is presented in Table 5.

Table 4. CNN Model used for Indian Pines data.

Layer (Type)	Output Shape	Param #
conv2d_10 (Conv2D)	(None, 5, 5, 64)	10,432
conv2d_11 (Conv2D)	(None, 5, 5, 128)	73,856
dropout_5 (Dropout)	(None, 5, 5, 128)	0
flatten_5 (Flatten)	(None, 3200)	0
dense_23 (Dense)	(None, 256)	819,456
dense_24 (Dense)	(None, 128)	32,896
dense_25 (Dense)	(None, 64)	8256
dense_26 (Dense)	(None, 32)	2080
dense_27 (Dense)	(None, 16)	528

Table 5. CNN Model used for Salinas data.

Layer (Type)	Output Shape	Param #
conv2d_6 (Conv2D)	(None, 5, 5, 64)	11,584
conv2d_7 (Conv2D)	(None, 5, 5, 128)	73,856
dropout_3 (Dropout)	(None, 5, 5, 128)	0
flatten_3 (Flatten)	(None, 3200)	0
dense_16 (Dense)	(None, 128)	32,896
dense_17 (Dense)	(None, 64)	8256
dense_18 (Dense)	(None, 32)	2080
dense_19 (Dense)	(None, 16)	528
conv2d_6 (Conv2D)	(None, 5, 5, 64)	11,584

4.3. Results with Accuracy Measures

In this work, we used both quantitative and qualitative result analysis to show the performance of the proposed approach for crop classification. Three standard evaluation quantitative metrics were used [42] and are defined as:

- Overall Accuracy (OA): The percentage of correctly labeled pixels in the crop classification;
- Average Accuracy (AA): Average percentage of correctly labeled pixels for each crop;
- Class-wise Accuracies (CA): Percentage of correctly labeled pixels in each crop.

The OA, AA, and CA values of Indian Pines and Salinas, along with execution time, are shown in Table 6.

Table 6. Details of class-wise accuracy of Indian Pines and Salinas data.

Indian Pines		Salinas	
Class ID	Accuracy	Class ID	Accuracy
1	0.6667	1	1.0000
2	0.9664	2	0.9388
3	0.9712	3	0.9453
4	0.9667	4	1.0000
5	1.0000	5	0.9925
6	0.9891	6	0.9970
7	0.9000	7	0.9966
8	1.0000	8	0.9411
9	1.0000	9	0.9974
10	0.9794	10	0.9817
11	0.9935	11	0.9652
12	0.9597	12	0.9751
13	1.0000	13	1.0000
14	0.9937	14	0.9366
15	0.9485	15	0.9004
16	1.0000	16	0.9912
AA	95.84%	AA	97.24%
OA	97.62%	OA	96.08%
Execution Time (in seconds)	460.45	Execution Time (in seconds)	1024.35

The qualitative metric used in this work is the classification map. Figure 7 shows the classification map of the Indian Pines data, and Figure 8 shows the classification map of the Salinas data [49–51].

For Indian Pines data, the highest classification result was obtained for two crop classes: (Corn-mintill: class number "3") with an accuracy of 97.12%, and (Soybean-mintill: class number "11") with an accuracy of 99.35%. These two crop class regions are circled in Figure 8d of the classification map result.

For Salinas data, the highest classification result was obtained for two crop classes: (Fallow_rough_plow: class number "4") with an accuracy of 100%, and (Lettuce_romaine_6wk: class number "13") with an accuracy of 100%. These two crop class regions are pointed out in Figure 9d of the classification map result.

4.4. Discussion

This subsection discusses the classification accuracies of classes from the two datasets and also elaborates on the comparison of the proposed framework with the state-of-the-art methods.

From the Indian Pines data, it was observed that there are 6 types of crops, Corn-notill, Corn-mintill, Corn, Soybean-notill, Soybean-mintill, and Soybean-clean, with class numbers 2, 3, 4, 10, 11, and 12. The robust implementation of the classification method resulted in accuracies of 96.64%, 97.12%, 96.67%, 97.94%, 99.67%, and 95.97%, respectively. The accuracy range was found to be 95.97–99.35%.

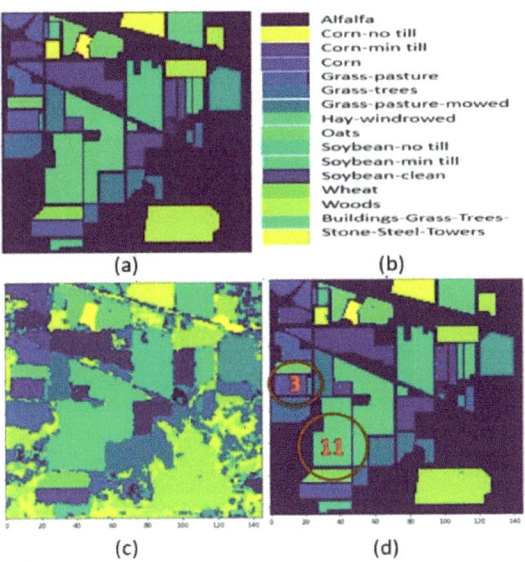

Figure 8. Indian Pines data (**a**) Ground truth (**b**) Category label (**c**) Crop classification result on entire data (**d**) Crop classification result.

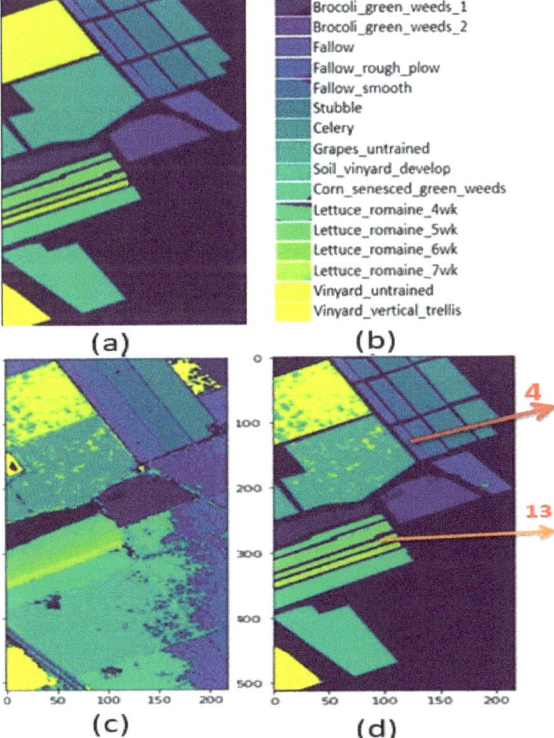

Figure 9. Salinas data (**a**) Ground truth (**b**) Category label (**c**) Crop classification result on entire data (**d**) Crop classification result.

From the Salinas data, it was observed that there are 6 types of crops, Fallow, Fallow_rough_plow, Fallow_smooth, Lettuce_romaine_4wk, Lettuce_romaine_5wk, and Lettuce_romaine_6wk, with class numbers 3, 4, 5, 11, 12, and 13. The robust implementation of the classification method resulted in accuracies of 94.53%, 100%, 99.25%, 96.52%, 97.51%, and 100%, respectively. The accuracy range was found to be 94.53–100%.

The proposed method for HSI classifications was compared with four state-of-the-art methods, including 3DGSVM [52], CNN-MFL [53], SS3FC [54], and WEDCT-MI [30], to prove the effectiveness in classifying the different crop regions.

The first method used for the comparison is the integration of 3-dimensional discrete wavelet transform and Markov random field for hyperspectral image classification called 3DGSVM [52]. In this work, more importance was given to spatial information. 3DDWT is used to extract spatial features. Probabilistic SVM coupled with MRF-based post-processing was used for HSI classification.

The second method used for the comparison is Hyperspectral Image Classification Using Convolutional Neural Networks and Multiple Feature Learning called CNN-MFL [53]. In this work, multiple features were extracted first, followed by several CNN blocks for each set of features. Here, geometric features were incorporated using attribute profiles. This is a novel technique that takes advantage of multiple feature learning and CNN to perform accurate HSI classification.

The third method used for the comparison is Spectral–Spatial Exploration for Hyperspectral Image Classification via the Fusion of Fully Convolutional Networks called SS3FC [54]. This method used spectral, spatial, and semantic information along with Fusion of Fully Convolutional Networks for HSI classification. A novel technique for the balanced splitting of the training/test dataset was introduced to solve the insufficient training samples problem.

The fourth method used for the comparison is unsupervised band selection based on weighted information entropy and 3D discrete cosine transform for hyperspectral image classification called WEDCT-MI [30]. In this work, original HSI data was first converted in discrete cosine transform-based coefficient matrices. The weighted entropy was calculated to quantify each band. Then, top-ranked bands were selected. Finally, SVM was used for classification.

Figure 10 shows the crop classification from Indian Pines and Salinas datasets. It can be seen from Figure 10 that the proposed band selection approach is effective in extracting the bands which contain much information about the crops. This is due to the inclusion of the physical properties of light in partitioning the bands and the biological properties of plants in band quantization.

4.5. Crop-Wise Analysis

The crop-wise analysis [55] on the two datasets is shown in this subsection.

4.5.1. Corn Crops

The first crop used for the comparison is "Corn-notill," class number 2 from the Indian Pines data. For this crop, the proposed method outperformed the state-of-the-art methods 3DGSVM, SS3FC, and WEDCT-MI with higher accuracy of 96.64%. The CNN-MFL method had 94.82% accuracy, which is on par with the proposed method. The second crop used for the comparison is "Corn-mintill", class number 3 from the Indian Pines data. For this crop, the proposed method outperformed the state-of-the-art methods 3DGSVM, SS3FC, and WEDCT-MI with higher accuracy of 97.12%. The CNN-MFL method had 96% accuracy, which is on par with the proposed method. The third crop used for the comparison is "Corn", class number 4 from the Indian Pines data. For this crop, the proposed method outperformed the state-of-the-art methods SS3FC and WEDCT-MI with higher accuracy of 96.67%. The methods 3DGSVM and CNN-MFL had 96.64% and 96% accuracy, respectively, which are on par with the proposed method.

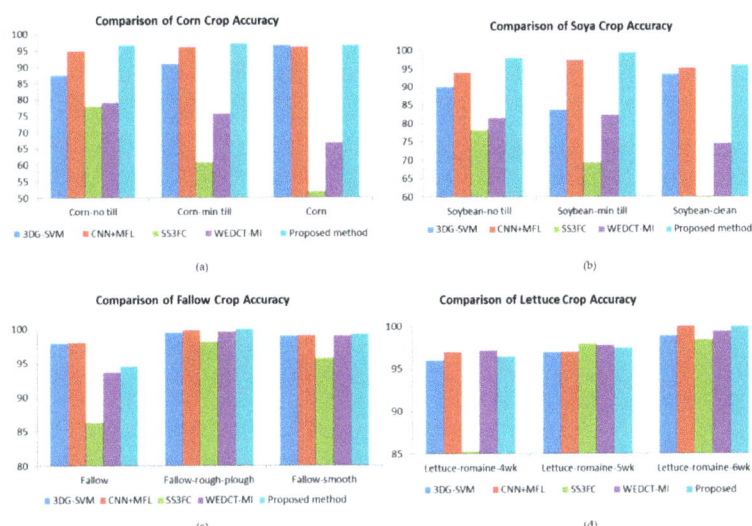

Figure 10. Comparison of crop classification accuracy Indian Pines: (**a**) Corn (**b**) Soya Salinas (**c**) Fallow (**d**) Lettuce.

4.5.2. Soya Crops

The first crop used for the comparison is "Soybean-no till", class number 10 from the Indian Pines data. For this crop, the proposed method outperformed the state-of-the-art methods 3DGSVM, CNN-MFL, SS3FC, and WEDCT-MI with higher accuracy of 97.94%. The second crop used for the comparison is "Soybean-min till", class number 11 from Indian Pines data. For this crop, the proposed method outperformed the state-of-the-art methods 3DGSVM, SS3FC, and WEDCT-MI with higher accuracy of 99.35%. The third crop used for the comparison is "Soybean-clean", class number 12 from the Indian Pines data. For this crop, the proposed method outperformed the state-of-the-art methods 3DGSVM, SS3FC, and WEDCT-MI with higher accuracy of 95.97%. The method CNN-MFL had 95% accuracy, which is on par with the proposed method.

4.5.3. Fallow Crops

The first crop used for the comparison is "Fallow", class number 3 from the Salinas data. For this crop, the proposed method outperformed the state-of-the-art methods SS3FC and WEDCT-MI with higher accuracy of 94.53%. The methods 3DGSVM and CNN-MFL were slightly more accurate than the proposed method. The second crop used for the comparison "Fallow-rough-plough", class number 4 from the Salinas data. For this crop, the proposed method outperformed the state-of-the-art methods 3DGSVM, CNN-MFL, SS3FC, and WEDCT-MI with higher accuracy of 100%. The third crop used for the comparison was "Fallow-smooth", class number 5 from the Salinas data. For this crop, the proposed method outperformed the state-of-the-art methods 3DGSVM, CNN-MFL, and SS3FC with higher accuracy of 99.25%. The method WEDCT-MI had 99.03% accuracy, which is on par with the proposed method.

4.5.4. Lettuce Romaine Crops

The first crop used for the comparison is "Lettuce-romaine-4wk", class number 11 from the Salinas data. For this crop, the proposed method outperformed the state-of-the-art methods 3DGSVM and SS3FC with higher accuracy of 96.52%. Methods CNN-MFL and WEDCT-MI were slightly more accurate than the proposed method. The second crop used for the comparison is "Lettuce-romaine-5wk", class number 12 from the Salinas data. For this crop, the proposed method outperformed the state-of-the-art methods

3DGSVM, CNN-MFL, and WEDCT-MI with higher accuracy of 97.51%. Method SS3FC was slightly more accurate than the proposed method. The third crop used for the comparison is "Lettuce-romaine-6wk", class number 13 from the Salinas data. For this crop, the proposed method and CNN-MFL had 100% accuracy and outperformed the remaining state-of-the-art methods.

4.6. Ablation Study

We conducted an ablation study to verify the effectiveness of the proposed band selection. There are two modules in the proposed methodology known as Partition and Ranking. We conducted experiments by varying the modules (replace Proposed partition based on VIS, NIR, and SWIR regions into Coarser partition into 3 regions and replace Proposed ranking based on Entropy, NDVI, and NDWI into random selection). The ablation analysis for Indian Pines and Salinas data is shown in Table 7. The accuracy values show that the proposed band selection method has a positive impact on the hyperspectral classification.

Table 7. Ablation experiments of proposed network.

	Partition		Ranking				
Coarser Partition in to 3 Regions	Proposed Partition Based on VIS, NIR, and SWIR Regions	Entropy Based Ranking	NDVI Based Ranking	NDWI Based Ranking	OA	AA
			Indian Pines data				
	✓	✓	✓	✓	97.62%	95.84%
✓		✓	✓	✓	77.76%	79.31%
	✓				82.51%	84.45%
✓					69.72%	62.81%
			Salinas data				
	✓	✓	✓	✓	96.08%	97.24%
✓		✓	✓	✓	80.5%	74.6%
	✓				83.9%	86.76%
✓					48.92%	51.33%

4.7. Application of Proposed Methodology with Other Satellite Data

In order to test the adaptability and the effectiveness as in [56], we applied the proposed framework on the WHU-Hi-HongHu dataset [57–59]. This data consists of a complex agricultural area with a variety of crops. The data were acquired in Honghu City, China. The number of pixels is 940 × 475, with 270 bands acquired from 400 to 1000 nm wavelength. The data were acquired using an unmanned aerial vehicle (UAV)-borne hyperspectral system with high spatial resolution of 0.043 m. Out of 22 class regions, 15 classes were found with class numbers 4, 6, 7, 8, 9, 10, 11, 12, 13, 14, 16, 17, 18, 19, 20. Table 8 shows class-wise accuracies after the application of the proposed framework on the WHU-Hi-Hong Hu dataset. We used all the parameters similar to the Indian Pines and Salinas datasets. It can be concluded that the proposed framework was successful in classifying, with an average accuracy of 98.56% for 15 crop classes.

Table 8. Ablation experiments of the proposed network.

Class No	Class Name	Accuracy in %
1	Red roof	99.63
2	Road	94.76
3	Bare soil	99.43
4	**Cotton**	**99.82**
5	Cotton firewood	96.85
6	**Rape**	**99.62**
7	Chinese cabbage	97.81
8	Packchoi	97.93
9	Cabbage	99.59
10	Tuber mustard	98.45
11	Brassica parachinesis	97.75
12	Brassica chinesis	96.38
13	Small brassica chinesis	97.81
14	Latuca sativa	99.02
15	Celuce	97.21
16	Film covered lettuce	99.72
17	Romaine lettuce	98.41
18	Carrot	99.63
19	White radish	97.70
20	Garlic sprout	98.82
21	Broad bean	94.88
22	Tree	99.01

5. Conclusions

In this paper, an approach for crop classification for HSI images is proposed. Firstly, the bands are selected based on agricultural phenology, where biophysical properties of plants are also taken into consideration. Then, a two-dimensional CNN is trained with the extracted bands. The proposed method is tested and validated on two benchmark datasets. The average accuracy of the crops corn and soya from Indian Pines is 96.81% and 97.75%. For the Salinas dataset, the average accuracy of the crops fallow and lettuce romaine is 97.93% and 98.01%, respectively. The results clearly show the effectiveness of the proposed band selection in selecting the required features necessary for crop classification. In the future, we can incorporate spatial information along with spectral information and extend it to more crops from the datasets.

Author Contributions: Conceptualization, L.A.; Data curation, L.A.; Funding acquisition, L.A. and A.F.; Methodology, V.R. and K.L.N.B.P.; Project administration, M.P.; Resources, A.F.; Software, A.F.; Supervision, M.P.; Validation, V.R. and K.L.N.B.P.; Writing—original draft, V.R. and K.L.N.B.P.; Writing—review & editing, M.P. All authors have read and agreed to the published version of the manuscript.

Funding: This research received no external funding.

Institutional Review Board Statement: Not applicable.

Informed Consent Statement: Not applicable.

Data Availability Statement: Not applicable.

Acknowledgments: The authors thank the Vellore Institute of Technology, Vellore for providing a VIT seed grant for carrying out this research work.

Conflicts of Interest: The authors declare no conflict of interest.

References

1. Hoffer, R.M. Biological and physical considerations in applying computer-aided analysis techniques to remote sensor data. In *Remote Sensing: The Quantitative Approach*; Swain, P.H., Davis, S.M., Eds.; McGraw-Hill Book Company: New York, NY, USA, 1978; pp. 227–289.
2. Karlovska, A.; Grinfelde, I.; Alsina, I.; Prieditis, G.; Roze, D. Plant Reflected Spectra Depending on Biological Characteristics and Growth Conditions. In Proceedings of the 7th International Scientific Conference Rural Development 2015, Kaunas, Lithuania, 19–20 November 2015. [CrossRef]
3. Prabukumar, M.; Sawant, S.; Samiappan, S.; Agilandeeswari, L. Three-dimensional discrete cosine transform-based feature extraction for hyperspectral image classification. *J. Appl. Remote Sens.* 2018, *12*, 046010. [CrossRef]
4. Sawant, S.S.; Prabukumar, M. Band Fusion Based Hyper Spectral Image Classification. *Int. J. Pure Appl. Math.* 2017, *117*, 71–76.
5. Sawant, S.; Manoharan, P. A hybrid optimization approach for hyperspectral band selection based on wind driven optimization and modified cuckoo search optimization. *Multimedia Tools Appl.* 2020, *80*, 1725–1748. [CrossRef]
6. Prabukumar, M.; Shrutika, S. Band clustering using expectation–maximization algorithm and weighted average fusion-based feature extraction for hyperspectral image classification. *J. Appl. Remote Sens.* 2018, *12*, 046015. [CrossRef]
7. Sawant, S.S.; Prabukumar, M. A review on graph-based semi-supervised learning methods for hyperspectral image classification. *Egypt. J. Remote Sens. Space Sci.* 2020, *23*, 243–248. [CrossRef]
8. Sawant, S.; Manoharan, P. Hyperspectral band selection based on metaheuristic optimization approach. *Infrared Phys. Technol.* 2020, *107*, 103295. [CrossRef]
9. Sawant, S.S.; Manoharan, P. New framework for hyperspectral band selection using modified wind-driven optimization algorithm. *Int. J. Remote Sens.* 2019, *40*, 7852–7873. [CrossRef]
10. Sawant, S.S.; Manoharan, P.; Loganathan, A. Band selection strategies for hyperspectral image classification based on machine learning and artificial intelligent techniques—Survey. *Arab. J. Geosci.* 2021, *14*, 646. [CrossRef]
11. Makantasis, K.; Karantzalos, K.; Doulamis, A.; Doulamis, N. Deep supervised learning for hyperspectral data classification through convolutional neural networks. In Proceedings of the 2015 IEEE International Geoscience and Remote Sensing Symposium (IGARSS), Milan, Italy, 26–31 July 2015; pp. 4959–4962.
12. He, K.; Zhang, X.; Ren, S.; Sun, J. Spatial Pyramid Pooling in Deep Convolutional Networks for Visual Recognition. *IEEE Trans. Pattern Anal. Mach. Intell.* 2015, *37*, 1904–1916. [CrossRef] [PubMed]
13. Galvao, L.S.; Formaggio, A.R.; Tisot, D.A. Discrimination of sugarcane varieties in Southeastern Brazil with EO-1 Hyperion data. *Remote Sens. Environ.* 2005, *94*, 523–534. [CrossRef]
14. Li, D.; Chen, S.; Chen, X. Research on method for extracting vegetation information based on hyperspectral remote sensing data. *Trans. Chin. Soc. Agric. Eng.* 2010, *26*, 181–185.
15. Bhojaraja, B.E.; Hegde, G. Mapping agewise discrimination of are canut crop water requirement using hyperspectral remote-sensing. In Proceedings of the International Conference on Water Resources, Coastal and Ocean Engineering, Mangalore, India, 12–14 March 2015; pp. 1437–1444.
16. Vegetation Analysis: Using Vegetation Indices in ENVI. Available online: https://www.l3harrisgeospatial.com/Learn/Whitepapers/Whitepaper-Detail/ArtMID/17811/ArticleID/16162/Vegetation-Analysis-Using-Vegetation-Indices-in-ENVI (accessed on 17 December 2021).
17. Tarabalka, Y.; Fauvel, M.; Chanussot, J.; Benediktsson, J.A. SVM- and MRF-Based Method for Accurate Classification of Hyperspectral Images. *IEEE Geosci. Remote Sens. Lett.* 2010, *7*, 736–740. [CrossRef]
18. Wei, L.; Yu, M.; Zhong, Y.; Zhao, J.; Liang, Y.; Hu, X. Spatial–Spectral Fusion Based on Conditional Random Fields for the Fine Classification of Crops in UAV-Borne Hyperspectral Remote Sensing Imagery. *Remote Sens.* 2019, *11*, 780. [CrossRef]
19. Liang, L.I.U.; Xiao-Guang, J.I.A.N.G.; Xian-Bin, L.I.; Ling-Li, T.A.N.G. Study on Classification of Agricultural Crop by Hyperspectral Remote Sensing Data. *J. Grad. Sch. Chin. Acad. Sci.* 2006, *23*, 484–488.
20. Wei, L.; Yu, M.; Liang, Y.; Yuan, Z.; Huang, C.; Li, R.; Yu, Y. Precise Crop Classification Using Spectral-Spatial-Location Fusion Based on Conditional Random Fields for UAV-Borne Hyperspectral Remote Sensing Imagery. *Remote Sens.* 2019, *11*, 2011. [CrossRef]
21. Li, C.H.; Kuo, B.C.; Lin, C.T.; Huang, C.S. A Spatial–Contextual Support Vector Machine for Remotely Sensed Image Classification. *IEEE Trans. Geosci. Remote Sens.* 2012, *50*, 784–799. [CrossRef]
22. Zhao, C.-I.; Luo, G.; Wang, Y.; Chen, C.; Wu, Z. UAV Recognition Based on Micro-Doppler Dynamic Attribute-Guided Augmentation Algorithm. *Remote Sens.* 2021, *13*, 1205. [CrossRef]
23. Singh, J.; Mahapatra, A.; Basu, S.; Banerjee, B. Assessment of Sentinel-1 and Sentinel-2 Satellite Imagery for Crop Classification in Indian Region During Kharif and Rabi Crop Cycles. In Proceedings of the IGARSS 2019—2019 IEEE International Geoscience and Remote Sensing Symposium, Yokohama, Japan, 28 July–2 August 2019; pp. 3720–3723.
24. Wei, L.; Wang, K.; Lu, Q.; Liang, Y.; Li, H.; Wang, Z.; Wang, R.; Cao, L. Crops Fine Classification in Airborne Hyperspectral Imagery Based on Multi-Feature Fusion and Deep Learning. *Remote Sens.* 2021, *13*, 2917. [CrossRef]
25. Chang, C.-I.; Du, Q.; Sun, T.-L.; Althouse, M.L.G. A joint bandprioritization and band-decorrelation approach to band selection forhyperspectral image classification. *IEEE Trans. Geosci. Remote Sens.* 1999, *37*, 2631–2641. [CrossRef]
26. MartÌnez-UsÓ, A.; Pla, F.; Sotoca, J.M.; García-Sevilla, P. Clustering-based hyperspectral band selection using information measures IEEE Trans. *Geosci. Remote Sens.* 2007, *45*, 4158–4171. [CrossRef]

27. Xie, L.; Li, G.; Peng, L.; Chen, Q.; Tan, Y.; Xiao, M. Bandselection algorithm based on information entropy for hyperspectralimage classification. *J. Appl. Remote Sens.* **2017**, *11*, 026018. [CrossRef]
28. Sheffield, C. Selecting band combinations from multi spectral data. *Photogramm. Eng. Remote Sens.* **1985**, *58*, 681–687.
29. Wang, Q.; Zhang, F.; Li, X. Optimal Clustering Framework for Hyperspectral Band Selection. *IEEE Trans. Geosci. Remote Sens.* **2018**, *56*, 5910–5922. [CrossRef]
30. Sawant, S.S.; Manoharan, P. Unsupervised band selection based on weighted information entropy and 3D discrete cosine transform for hyperspectral image classification. *Int. J. Remote Sens.* **2020**, *41*, 3948–3969. [CrossRef]
31. Thenkabail, P.S.; Smith, R.B.; De Pauw, E. Hyperspectral Vegetation Indices and Their Relationships with Agricultural Crop Characteristics. *Remote Sens. Environ.* **2000**, *71*, 158–182. [CrossRef]
32. Du, Y.; Zhang, Y.; Ling, F.; Wang, Q.; Li, W.; Li, X. Water Bodies' Mapping from Sentinel-2 Imagery with Modified Normalized Difference Water Index at 10-m Spatial Resolution Produced by Sharpening the SWIR Band. *Remote Sens.* **2016**, *8*, 354. [CrossRef]
33. Hyperspectral Remote Sensing Scenes. Available online: http://www.ehu.eus/ccwintco/index.php/Hyperspectral_Remote_Sensing_Scenes (accessed on 14 December 2021).
34. Manoharan, P.; Boggavarapu, L.N.P.K. Improved whale optimization based band selection for hyperspectral remote sensing image classification. *Infrared Phys. Technol.* **2021**, *119*, 103948. [CrossRef]
35. Boggavarapu, L.N.P.K.; Manoharan, P. Hyperspectral image classification using fuzzy-embedded hyperbolic sigmoid nonlinear principal component and weighted least squares approach. *J. Appl. Remote Sens.* **2020**, *14*, 024501. [CrossRef]
36. Boggavarapu, L.N.P.K.; Manoharan, P. A new framework for hyperspectral image classification using Gabor embedded patch based convolution neural network. *Infrared Phys. Technol.* **2020**, *110*, 103455. [CrossRef]
37. Boggavarapu, L.N.P.K.; Manoharan, P. Whale optimization-based band selection technique for hyperspectral image classification. *Int. J. Remote Sens.* **2021**, *42*, 5105–5143. [CrossRef]
38. Boggavarapu, L.N.P.K.; Manoharan, P. Survey on classification methods for hyper spectral remote sensing imagery. In Proceedings of the 2017 International Conference on Intelligent Computing and Control Systems (ICICCS), Madurai, India, 15–16 June 2017; pp. 538–542. [CrossRef]
39. Boggavarapu, L.N.P.K.; Manoharan, P. Classification of Hyper Spectral Remote Sensing Imagery Using Intrinsic Parameter Estimation. In Proceedings of the ISDA 2018: International Conference on Intelligent Systems Design and Applications, Vellore, India, 6–8 December 2018. [CrossRef]
40. Vaddi, R.; Manoharan, P. Hyperspectral image classification using CNN with spectral and spatial features integration. *Infrared Phys. Technol.* **2020**, *107*, 103296. [CrossRef]
41. Vaddi, R.; Manoharan, P. CNN based hyperspectral image classification using unsupervised band selection and structure-preserving spatial features. *Infrared Phys. Technol.* **2020**, *110*, 103457. [CrossRef]
42. Vaddi, R.; Manoharan, P. Hyperspectral remote sensing image classification using combinatorial optimisation based un-supervised band selection and CNN. *IET Image Process.* **2020**, *14*, 3909–3919. [CrossRef]
43. Baisantry, M.; Sao, A.K.; Shukla, D.P. Two-Level Band Selection Framework for Hyperspectral Image Classification. *J. Indian Soc. Remote Sens.* **2021**, *49*, 843–856. [CrossRef]
44. Vaddi, R.; Manoharan, P. Comparative study of feature extraction techniques for hyper spectral remote sensing image classification: A survey. In Proceedings of the 2017 International Conference on Intelligent Computing and Control Systems (ICICCS-17), Madurai, India, 15–17 May 2019.
45. Vaddi, R.; Manoharan, P. Probabilistic PCA Based Hyper Spectral Image Classification for Remote Sensing Applications. In *ISDA 2018 2018: Intelligent Systems Design and Applications, Proceedings of the ISDA 2018: International Conference on Intelligent Systems Design and Applications, Vellore, India, 6–8 December 2018*; Advances in Intelligent Systems and Computing; Springer Nature Switzerland AG: Cham, Switzerland, 2020; Volume 941, pp. 1–7. [CrossRef]
46. Manoharan, P.; Vaddi, R. Wavelet enabled ranking and clustering-based band selection and three-dimensional spatial feature extraction for hyperspectral remote sensing image classification. *J. Appl. Remote Sens.* **2021**, *15*, 044506. [CrossRef]
47. Baumgardner, M.F.; Biehl, L.L.; Landgrebe, D.A. 220 Band AVIRIS Hyperspectral Image Data Set: 12 June 1992 Indian Pine Test Site 3. *Purdue Univ. Res. Repos.* **2015**, *10*, R7RX991C. [CrossRef]
48. Roy, S.K.; Kar, P.; Hong, D.; Wu, X.; Plaza, A.; Chanussot, J. Revisiting Deep Hyperspectral Feature Extraction Networks via Gradient Centralized Convolution. *IEEE Trans. Geosci. Remote Sens.* **2021**, *60*. [CrossRef]
49. Roy, S.K.; Hong, D.; Kar, P.; Wu, X.; Liu, X.; Zhao, D. Lightweight Heterogeneous Kernel Convolution for Hyperspectral Image Classification with Noisy Labels. *IEEE Geosci. Remote Sens. Lett.* **2021**, *19*, 1–5. [CrossRef]
50. Wang, Y.; Song, T.; Xie, Y.; Roy, S.K. A probabilistic neighbourhood pooling-based attention network for hyperspectral image classification. *Remote Sens. Lett.* **2022**, *13*, 65–75. [CrossRef]
51. Melgani, F.; Bruzzone, L. Classification of hyperspectral remote sensing images with support vector machines. *IEEE Trans. Geosci. Remote Sens.* **2004**, *42*, 1778–1790. [CrossRef]
52. Cao, X.; Xu, L.; Meng, D.; Zhao, Q.; Xu, Z. Integration of 3-dimensional discrete wavelet transform and Markov random field for hyperspectral image classification. *Neurocomputing* **2017**, *226*, 90–100. [CrossRef]
53. Gao, Q.; Lim, S.; Jia, X. Hyperspectral Image Classification Using Convolutional Neural Networks and Multiple Feature Learning. *Remote Sens.* **2018**, *10*, 299. [CrossRef]

54. Zou, L.; Zhu, X.; Wu, C.; Liu, Y.; Qu, L. Spectral–Spatial Exploration for Hyperspectral Image Classification via the Fusion of Fully Convolutional Networks. *IEEE J. Sel. Top. Appl. Earth Obs. Remote Sens.* **2020**, *13*, 659–674. [CrossRef]
55. Sawant, S.S.; Manoharan, P. A survey of band selection techniques for hyperspectral image classification. *J. Spectr. Imaging* **2020**, *9*, a5. [CrossRef]
56. Pan, E.; Ma, Y.; Fan, F.; Mei, X.; Huang, J. Hyperspectral Image Classification across Different Datasets: A Generalization to Unseen Categories. *Remote Sens.* **2021**, *13*, 1672. [CrossRef]
57. Felegari, S.; Sharifi, A.; Moravej, K.; Amin, M.; Golchin, A.; Muzirafuti, A.; Tariq, A.; Zhao, N. Integration of Sentinel 1 and Sentinel 2 Satellite Images for Crop Mapping. *Appl. Sci.* **2021**, *11*, 10104. [CrossRef]
58. Zhong, Y.; Hu, X.; Luo, C.; Wang, X.; Zhao, J.; Zhang, L. WHU-Hi: UAV-borne hyperspectral with high spatial resolution (H2) benchmark datasets and classifier for precise crop identification based on deep convolutional neural network with CRF. *Remote Sens. Environ.* **2020**, *250*, 112012. [CrossRef]
59. Zhong, Y.; Wang, X.; Xu, Y.; Wang, S.; Jia, T.; Hu, X.; Zhao, J.; Wei, L.; Zhang, L. Mini-UAV-Borne Hyperspectral Remote Sensing: From Observation and Processing to Applications. *IEEE Geosci. Remote Sens. Mag.* **2018**, *6*, 46–62. [CrossRef]

Article

Squirrel Search Optimization with Deep Transfer Learning-Enabled Crop Classification Model on Hyperspectral Remote Sensing Imagery

Manar Ahmed Hamza [1,*], Fadwa Alrowais [2], Jaber S. Alzahrani [3], Hany Mahgoub [4], Nermin M. Salem [5] and Radwa Marzouk [6]

1. Department of Computer and Self Development, Preparatory Year Deanship, Prince Sattam bin Abdulaziz University, Al-Kharj 16278, Saudi Arabia
2. Department of Computer Sciences, College of Computer and Information Sciences, Princess Nourah bint Abdulrahman University, P.O. Box 84428, Riyadh 11671, Saudi Arabia; falrowais@pnu.edu.sa
3. Department of Industrial Engineering, College of Engineering at Alqunfudah, Umm Al-Qura University, Mecca 24382, Saudi Arabia; jzahrani@uqu.edu.sa
4. Department of Computer Science, College of Science & Arts at Mahayil, King Khalid University, Muhayel Aseer 62529, Saudi Arabia; hmahgoub@kku.edu.sa
5. Department of Electrical Engineering, Faculty of Engineering and Technology, Future University in Egypt, New Cairo 11835, Egypt; nafawzy@fue.edu.eg
6. Department of Information Systems, College of Computer and Information Sciences, Princess Nourah bint Abdulrahman University, P.O. Box 84428, Riyadh 11671, Saudi Arabia; rmarzouk@pnu.edu.sa
* Correspondence: ma.hamza@psau.edu.sa

Abstract: With recent advances in remote sensing image acquisition and the increasing availability of fine spectral and spatial information, hyperspectral remote sensing images (HSI) have received considerable attention in several application areas such as agriculture, environment, forestry, and mineral mapping, etc. HSIs have become an essential method for distinguishing crop classes and accomplishing growth information monitoring for precision agriculture, depending upon the fine spectral response to the crop attributes. The recent advances in computer vision (CV) and deep learning (DL) models allow for the effective identification and classification of different crop types on HSIs. This article introduces a novel squirrel search optimization with a deep transfer learning-enabled crop classification (SSODTL-CC) model on HSIs. The proposed SSODTL-CC model intends to identify the crop type in HSIs properly. To accomplish this, the proposed SSODTL-CC model initially derives a MobileNet with an Adam optimizer for the feature extraction process. In addition, an SSO algorithm with a bidirectional long-short term memory (BiLSTM) model is employed for crop type classification. To demonstrate the better performance of the SSODTL-CC model, a wide-ranging experimental analysis is performed on two benchmark datasets, namely dataset-1 (WHU-Hi-LongKou) and dataset-2 (WHU-Hi-HanChuan). The comparative analysis pointed out the better outcomes of the SSODTL-CC model over other models with a maximum of 99.23% and 97.15% on test datasets 1 and 2, respectively.

Keywords: hyperspectral remoting sensing; crop mapping; image classification; deep transfer learning; hyperparameter optimization

1. Introduction

Due to advancements in remote sensing image acquisition mechanisms and the increasing availability of rich spatial and spectral data by means of various sensors, hyperspectral imaging has become more prominent [1]. Especially, hyperspectral remote sensing image (HSI) classification has become a major source for real-time application in fields such as mineral mapping, agriculture, environment, and forestry, etc. [2,3]. Usually, the HIS is taken at a large number of contiguous narrow spectral wavelengths for the improved analysis of the earth object. Since the spectral resolution could be in nm, the hyperspectral sensor offers significant facility in data analysis [4] for many humanitarian tasks, including precision agriculture for improved farming practices, discrimination amongst vegetation classes for better treatment, etc. [5]. The current study emphasizes using and analyzing HSI in the agriculture area. Conventional techniques, such as statistical-based analyses and field surveys, are time-consuming [6]. Cutting-edge remote sensing technologies involving HSI provide an appropriate solution and might fill the gap with solutions such as crop classification. In the HSI framework, the classification has the common objective of automatically labeling the pixel (spectral pattern or signature) into a predetermined class [7]. The classification is implemented either by utilizing the transformed feature or the original feature. An HSI has numerous features and is hard to adapt to a single convolutional kernel size. When the number of model layers is increased, many useful features are lost [8–10].

The authors of [11] proposed a rotation-invariant local binary pattern-based weighted generalized closest neighbor (RILBP-WGCN) approach for an HSI classifier. The presented RILBP is an improved texture-based classifier paradigm, which employs LBP filters to any designated bands to generate a wide sketch of spatial texture data. Similarly, the presented WGCN approach effectually maintained the spatial uniformity amongst the adjacent pixel employing a local weight method and point-to-set distances. Meng et al. [12] concentrated on a DL-based crop mapping, utilizing one-shot hyperspectral satellite imagery, whereas three CNN techniques, such as 1D-CNN, 2D-CNN, and 3D-CNN, were executed for end-to-end crop mapping. Furthermore, a manifold learning-based visualized method, i.e., t-distributed stochastic neighbor embedding (t-SNE), was established for demonstrating the discriminative capability of deep semantic feature extracting by the distinct CNN approaches.

In [13], a hybrid model was established for estimating the chlorophyll content from the crops utilizing HIS segmentation with active learning, which contains two important stages. First, it can utilize a sparse multinomial logistic regression (SMLR) method for learning the class posterior probability distribution with quadratic programming or joint probability distributions. Second, it can utilize the data developed from the preceding step for segmenting the HSI utilizing a Markov random field segment. Farooq et al. [14] examine patch-based weed identification utilizing HSI. A CNN was estimated and correlated to a histogram of oriented gradients (HoG) for this solution. Appropriate patch sizes were examined. The restriction of RGB imagery was established. In [15], a deep one-class crop (DOCC) structure that contains a DOCC extracting element and an OCC extraction loss element was presented for large-scale OCC mapping. The DOCC structure takes only the instances of one target class as input for extracting the crop of interest by positive and unlabeled learning and automatically extracts the feature for OCC mapping.

In [16], a low altitude UAV hyperspectral remote sensing platform was created for collecting higher spatial resolution remote sensing images of degraded grassland. The GDIF-3D-CNN classifier method was utilized for classifying the pure pixel and every pixel data set, whose accuracy and performance were enhanced by optimizing the eight parameters of the method. Wei et al. [17] present a fine classifier approach dependent upon multi-feature fusion and DL. During this case, the morphological profiles, GLCM texture, and endmember abundance features were leveraged to exploit the spatial data of HIS. Next, the spatial data were fused with original spectral data to generate a classifier outcome by utilizing a DNN with a conditional random field (DNN + CRF) method. In detail, the DNN is a deep detection method that extracts depth features and mines the potential data.

For smaller samples and higher-dimension HSIs, it becomes very complex to learn wide-ranging image features; subsequently, it becomes hard to precisely recognize complex HSI. The UAV-borne HSIs have rich spatial data, and the spatial resolution reaches centimeter level; however, the higher spatial resolution causes serious spatial heterogeneity and spectral variability. Nowadays, the deep learning (DL) method is extensively employed in image processing because of its effective feature learning abilities [9]. Currently, the most common DL-based network framework is the convolution neural network (CNN). CNN has the features of parameter sharing, equivariant mapping, and sparse interaction, which reduce the training parameter size and complexity of the network. Such features permit the algorithm to generate a certain degree of invariance in scaling, shifting, and distortion and also create fault tolerance and stronger robustness [10]. Consequently, CNN has been extensively employed in HSI classification.

This article introduces a novel squirrel search optimization with a deep transfer learning-enabled crop classification (SSODTL-CC) model on HSIs. The proposed SSODTL-CC model initially derives a MobileNet with an Adam optimizer for the feature extraction process. The utilization of the Adam optimizer allows for effectual adjustment of the hyperparameters of the MobileNet model. In addition, a bidirectional long-short term memory (BiLSTM) method is employed for crop type classification. To enhance the classifier efficiency of the BiLSTM model, the SSO algorithm is employed for hyperparameter optimization, which shows the novelty of the work. To demonstrate the better performance of the SSODTL-CC model, a wide-ranging experimental analysis is performed on a benchmark dataset.

2. Materials and Methods

In this article, a new SSODTL-CC model has been developed to identify the crop type in HSIs properly. To do so, the proposed SSODTL-CC model performed feature extraction using MobileNet with an Adam optimizer. In addition, the BiLSTM model received feature vectors and performed crop type classification. To enhance the classifier efficiency of the BiLSTM model, the SSO algorithm was employed for hyperparameter optimization. Figure 1 illustrates the block diagram of the SSODTL-CC technique.

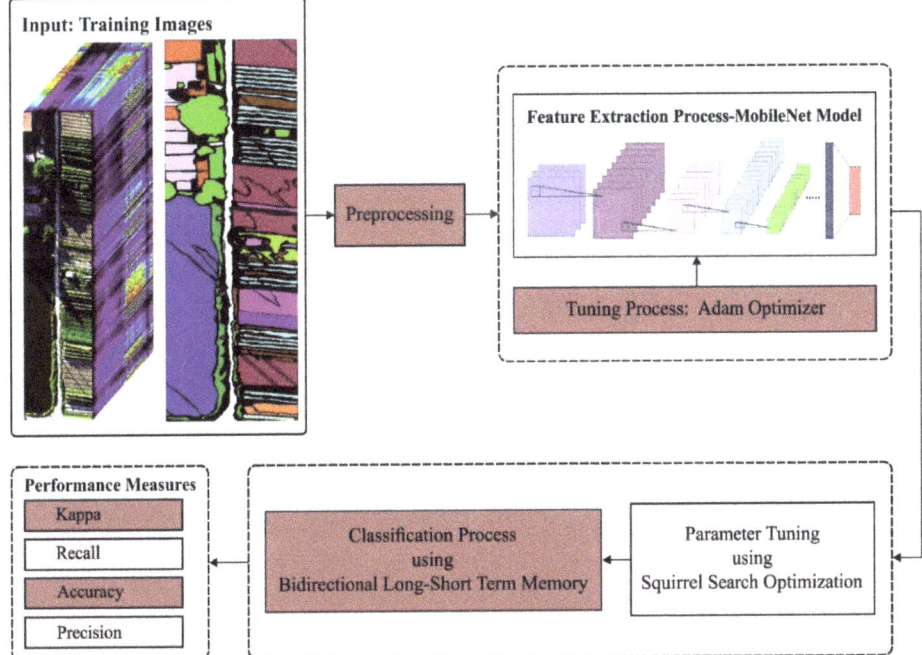

Figure 1. Block diagram of the SSODTL-CC technique.

2.1. Data Collection

In this section, the experimental validation of the proposed model is performed against two datasets [18], namely dataset-1 (WHU-Hi-LongKou) and dataset-2 (WHU-Hi-HanChuan). The dataset-1 comprises a total of 9000 samples with nine class labels, holding 1000 samples under each class. In addition, dataset-2 comprises a total of 16,000 samples with 16 class labels, holding 1000 samples under each class. Figure 2 shows the sample HSIs from various classes, such as water spinach, soybean, strawberry, corn, sesame, and broad-leaf soybean.

Figure 2. Sample images: (**a**) water spinach, (**b**) soybean, (**c**) strawberry, (**d**) corn, (**e**) sesame, and (**f**) broad-leaf soybean.

2.2. Feature Extraction: MobileNet Model

During the feature extraction process, the HSIs were passed into the MobileNet model to generate feature vectors. MobileNet is a CNN-based technique that is extensively applied in classifier procedures. The most important benefit of utilizing the presented method is that the model needs moderately low computation work in comparison with the CNN, which makes it appropriate to operate with a mobile device and a computer that operates with lower computational capabilities. The presented method is a fundamental architecture that combines convolution layers that are applied to efficiently distinguish details according to two controllable attributes that change between parameter precision and potential. The presented method is valuable in diminishing the size of the system.

The MobileNet structure is very effective with the least amount of attributes, namely Palmprint detection. This concerns a depth-wise convolution. The fundamental architecture is dependent on discrete abstracted layers, i.e., a module of dissimilar convolution layers that seem to be the quantal structure that measures a typical in-depth complication [19]. The resolution multiplier variable ω is added to minimize the measurement of the input dataset and inner layer representation with the analogous variable.

The feature vector map of size $F_m \times F_m$, and the filter is of size $F_s \times F_s$. The input variable is embodied by p, and the output variable is denoted by q. For the basic abstract layer of the structure, the whole computation work is considered as variable c_e, and it could be evaluated as follows:

$$c_e = F_s \cdot F_s \cdot \omega \cdot \alpha F_m \cdot \alpha F_m + \omega \cdot \rho \cdot \alpha F_m \cdot \alpha F_m \tag{1}$$

The ω multiplier value can be considered within one to n. The variable resolution multiplier is known as α. The computational effort is recognized as the variable $cost_e$ and is evaluated by the following equation:

$$cost_e = F_s \cdot F_s \cdot \omega \cdot \rho \cdot F_m \cdot F_m \tag{2}$$

The proposed approach incorporates the pointwise and depth-wise convolutions that are circumscribed by the reduction variable known as the variable d, which is evaluated in the following:

$$d = \frac{F_s \cdot F_s \cdot \omega \cdot \alpha F_m \cdot \alpha F_m + \omega \cdot \rho \cdot \alpha F_m \cdot \alpha F_m}{F_s \cdot F_s \cdot \omega \cdot \rho \cdot F_m F_m} \tag{3}$$

The two hyper characteristics, resolution and width multipliers, enable changing the optimal window size for accurate prediction based on the context. The third values suggest that it contains three input channels. The principle under the MobileNet structure replaced the complicated convolutional layer, which comprises a convolutional layer with 3×3 buffers for the input dataset, along with a pointwise convolutional layer of size 1×1 that combines the filtered variable to construct an element.

To optimally tune the hyperparameters related to the MobileNet model, the Adam optimizer is exploited. Furthermore, the hyperparameter optimized by the MobileNetv2 approach utilizes the Adam optimizer. It can be utilized for estimating an adoptive learning value, whereas the parameter was implemented for training the parameter of the DNN approach [20]. It can be a well-designed and effective approach for the 1st-order gradient with constraints stored for stochastic optimization. At this point, the newly presented approach was utilized to resolve the ML problem with the maximum dimensional parameter space, and the massive data set measures the rate of learning for different features with approximations of 1st and 2nd order moments. Additionally, the Adam optimizer was heavily utilized depending upon the gradient descent (GD) and momentum technique and a variety of intervals. Therefore, the 1st momentum is attained utilizing Equation (4):

$$m_i = \beta_1 m_{i-1} + (1 - \beta_1) \frac{\partial C}{\partial w}. \tag{4}$$

The 2nd momentum is expressed as:

$$v_i = \beta_2 v_{i-1} + (1 - \beta_2) \left(\frac{\partial C}{\partial w}\right)^2. \tag{5}$$

$$w_{i+1} = w_i - \eta \frac{\hat{m}_i}{\sqrt{\hat{v}_i + \epsilon}}, \tag{6}$$

in which $\hat{m}_i = m_i / (1 - \beta_1)$ and $\hat{v}_i = v_i / (1 - \beta_2)$.

2.3. Crop Type Classification: BiLSTM Model

At the time of image classification, the extracted feature vectors are fed into the BiLSTM model. The BiLSTM approach receives the feature vector as input and executes the detection method. The LSTM signifies a different RNN method, which solves the problem of gradient vanishing of RNN by offering a threshold method and memory unit [21]. However, x denotes the network input at different times, y refers to the network outcome, h stands for the hidden layer (HL), u refers to the weighted input to HLs, w demonstrates the weighted input of the previous node HL to the existing node HL, and v signifies the weighted input in HL to the output layer.

During the actual implementation of the LSTM technique, the LSTM unit was upgraded at time t as:

$$i_t = \sigma(W_i h_{t-1} + U_i x_t + b_t) \quad (7)$$

$$f_t = \sigma\left(W_j h_{t-1} + U_f x_t + b_f\right) \quad (8)$$

$$\tilde{c} = tanh(W_c h_{t-1} + U_c x_t + b_c) \quad (9)$$

$$c_t = f_t \odot c_{t-1} + i_t \odot \tilde{c}_t \quad (10)$$

$$o_t = \sigma(W_o h_{t-1} + U_o x_t + b_o) \quad (11)$$

$$h_t = o_{t-1} \odot tanh(c_t) \quad (12)$$

At this point, $\odot$ stands for the equal product of elements, and σ denotes the sigmoid function. x_t signifies the input vector at time t. h_t refers to the HL vector named as the output vector and the storage of all the data at time t and the preceding time. b_t, b_f, b_c, b_o demonstrates the offset vector. W_i, W_f, W_c, W_o implies the weight of various gates to the HL vector h_t. U_i, U_f, U_c, U_o stands for the weighted input vector. x_t stands for the input, forgotten, unit, and output gates, correspondingly. Utilizing the 3-gates infrastructure, the LSTM permits the recurrent network to maintain the useful data of the task from the memory units at the time of the trained method, therefore evading the problem of the RNN disappearing but reaching an extensive range of data.

In addition to processing the series data, the BLSTM presents more backward estimate procedures, for instance, different normal LSTM cases. This process employs the subsequent data of sequences. At last, the forward and reverse estimations are executed. The values were resultant of the output layer simultaneously; thus, as the outcome, all of the sequence data are reached in 2×2 directions, which is utilized to complete a variety of natural language processing tasks.

2.4. Hyperparameter Tuning: SSO Algorithm

For enhancing the classifier efficiency of the BiLSTM model, the SSO algorithm is employed for hyperparameter optimization. The SSO technique is proposed by the foraging behavior of a flying squirrel; subsequently, an effectual method employed small animals for migration. According to the food foraging hierarchy of squirrels [22], the optimum SSO algorithm is iteratively developed in an arithmetical model. There are important characteristics in SSA, that is, population sizes NP, maximal value of iteration $Iter_{max}$, the predator existence possibility P_{dp}, decision variables value n, gliding constants G_c, scaling factors sf, upper and lower limits to decision variable FS_U and FS_L. They are given in the following. The position of the squirrel is randomly loaded from the searching space:

$$FS_{i,j} = FS_L + rand(\ ) * (FS_U - FS_L), i = 1, 2, \ldots, NP, j = 1, 2, \ldots, n \quad (13)$$

However, $rand(\ )$ denotes an arbitrary value in $[0, 1]$. The fitness measure $f = (f_1 f_2, f_{NP})$ of a squirrel position was processed by replacing the decision variable with FF:

$$f_i = f_i(FS_{i,1}, FS_{i,2}, \ldots, FS_{i,n}), i = 1, 2, \ldots, NP \quad (14)$$

Next, the quality of food sources is evaluated by the fitness measure of a squirrel position as follows:

$$[sorted\_f, sorte\_index] = sort(f) \quad (15)$$

In addition, the organization of food sources was processed, which comprised hickory trees, normal trees, and oak trees (acorn nuts). The optimal food source (lower fitness) was assumed to be the hickory nut tree (FS_{hr}), the successive food sources that exist are denoted as acorn nut trees (FS_{ar}), and the rest are called normal trees (FS_{nt}):

$$FS_{ht} = FS(sorte - index(1)) \qquad (16)$$

$$FS_{at}(1:3) = FS(sorte - index(2:4)) \qquad (17)$$

$$FS_{nt}(1:NP-4) = FS(sorte - index(5:NP)) \qquad (18)$$

The three states that denote the dynamic gliding approach of squirrels are described in the following.

Scenario 1. The squirrel resides in an acorn nut tree and jumps to a hickory nut tree. A novel location can be given as follows:

$$FS_{at}^{new} = \begin{cases} FS_{at}^{old} + d_g G_c \left(FS_{ht}^{old} - FS_{at}^{old} \right) & if R \geq P_{dp} \\ random\ location & otherwise \end{cases} \qquad (19)$$

Now d_g indicates the gliding distance, R_1 denotes a function that proceeds the measured value of a uniform distribution value within 0 and 1, and G_c denotes a gliding constant.

Scenario 2. The squirrel resides in a normal tree and moves to acorn nut trees for gathering needed food. A novel location can be determined by:

$$FS_{nt}^{new} = \begin{cases} FS_{nt}^{old} + d_g G_c (FS_{at}^{old} - FS_{nt}^{old}) & if R \geq P_{dp} \\ random\ location, & otherwise \end{cases} \qquad (20)$$

Here, R_2 indicates a function that provides a measure of uniform distribution value in $[0, 1]$.

Scenario 3. Squirrels on normal trees go to hickory nut trees once they meet the routine objectives. Now, a novel position of squirrel can be determined by:

$$S_{nt}^{new} = \begin{cases} FS_{nt}^{old} + d_g G_c \left(FS_{ht}^{old} - FS_{nt}^{old} \right) & if R \geq P_{dp} \\ random\ location & otherwise \end{cases} \qquad (21)$$

where R_3 shows a function that suggests the measure of uniform distribution amongst $[0, 1]$. Hence, this measure is a maximum that invokes high perturbation. For achieving an appropriate method, a scaling factor (sf) is employed as a divisor of d_g.

The foraging nature of flying squirrels depends on the season, which varies frequently. Therefore, the seasonal observation must be implemented; thus, the trapping is removed in the local optimal result. The seasonal constant Sc and minimal value can be given as:

$$S_c^t = \sqrt{\sum_{k=1}^{n} \left(FS_{at,k}^t - FS_{ht,k} \right)^2}, t = 1, 2, 3 \qquad (22)$$

$$S_{c\min} = \frac{10E - 6}{365^{Iter/(Iter_{max})/2.5}} \qquad (23)$$

For $S_c^t < S_{c\min}$, the winter becomes the highest, the squirrel loses its exploring ability, and the method of searching for food sources and locations changes:

$$FS_{nt}^{new} = FS_L + \text{Lévy}(n) \times (FS_U - FS_L) \qquad (24)$$

Now the Lévy distribution is employed to improve the global search to an enhanced method:

$$\text{Lévy}(x) = 0.01 \times \frac{r_a \times \sigma}{|r_b|^{1/\beta}} \qquad (25)$$

$$\sigma = \left(\frac{\Gamma(1+\beta) \times \sin(\pi\beta/2)}{\Gamma((1+\beta)/2) \times \beta \times 2^{((\beta-1)/2)}} \right)^{1/\beta} \qquad (26)$$

This approach stops when the maximal constraint is fulfilled. If not, the nature of creating a novel location and approving the seasonal observation need to be repeatedly followed.

3. Experimental Validation

3.1. Result Analysis of SSODTL-CC Model

This section investigates the performance of the proposed model on test images.

Figure 3 showcases the sample classification results obtained by the SSODTL-CC model. The figure implies that the proposed model has obtained effective classification results. In addition, some of the misclassified regions by the SSODTL-CC model are marked in blue circles.

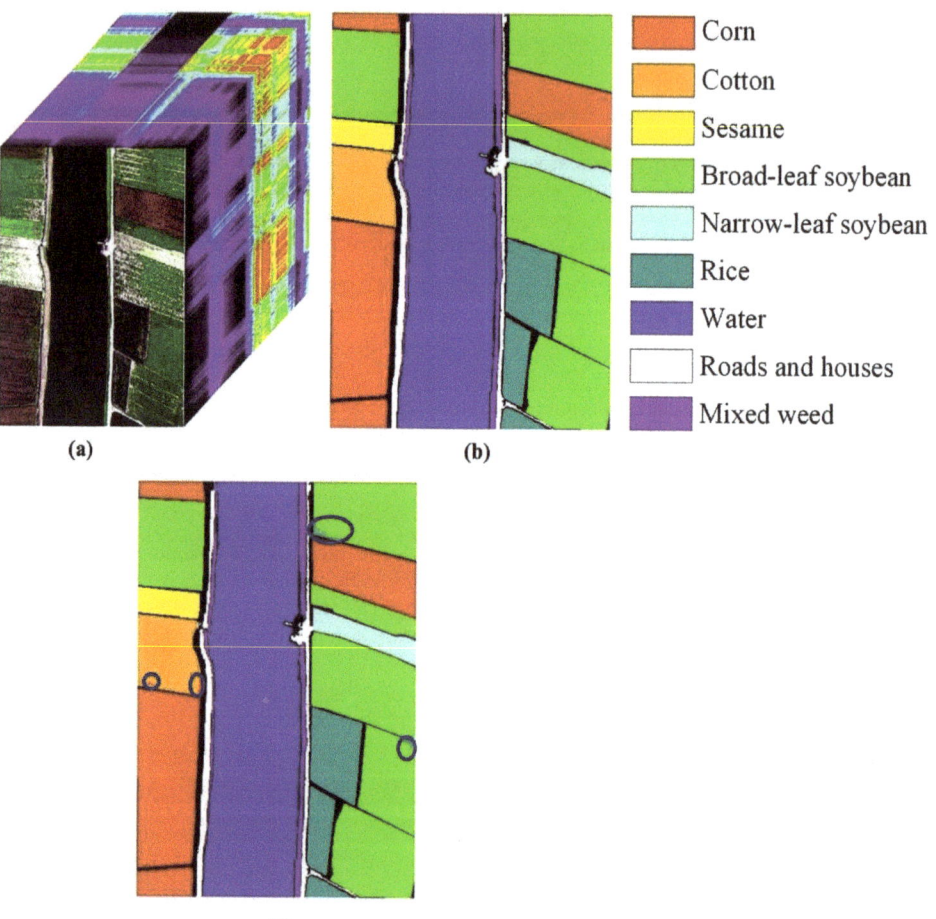

Figure 3. Sample classification result of the SSODTL-CC technique under dataset-1: (**a**) input image, (**b**) class labels, and (**c**) classification output.

Figure 4 inspects the confusion matrices created by the SSODTL-CC model on the classification of nine classes under dataset-1. The figure reports that the SSODTL-CC model has categorized all the classes under different sets of datasets. For the entire dataset, the SSODTL-CC model recognized 956 samples under corn, 975 samples under cotton, 971 samples under sesame, 971 samples under broad-leaf soybean, 964 samples under narrow-leaf soybean, 949 samples under rice, 965 samples under water, 958 samples under roads and houses, and 967 samples under mixed weed. Similarly, the SSODTL-CC model has categorized the class labels proficiently on 70% of the training samples and 30% of the testing samples on dataset-1.

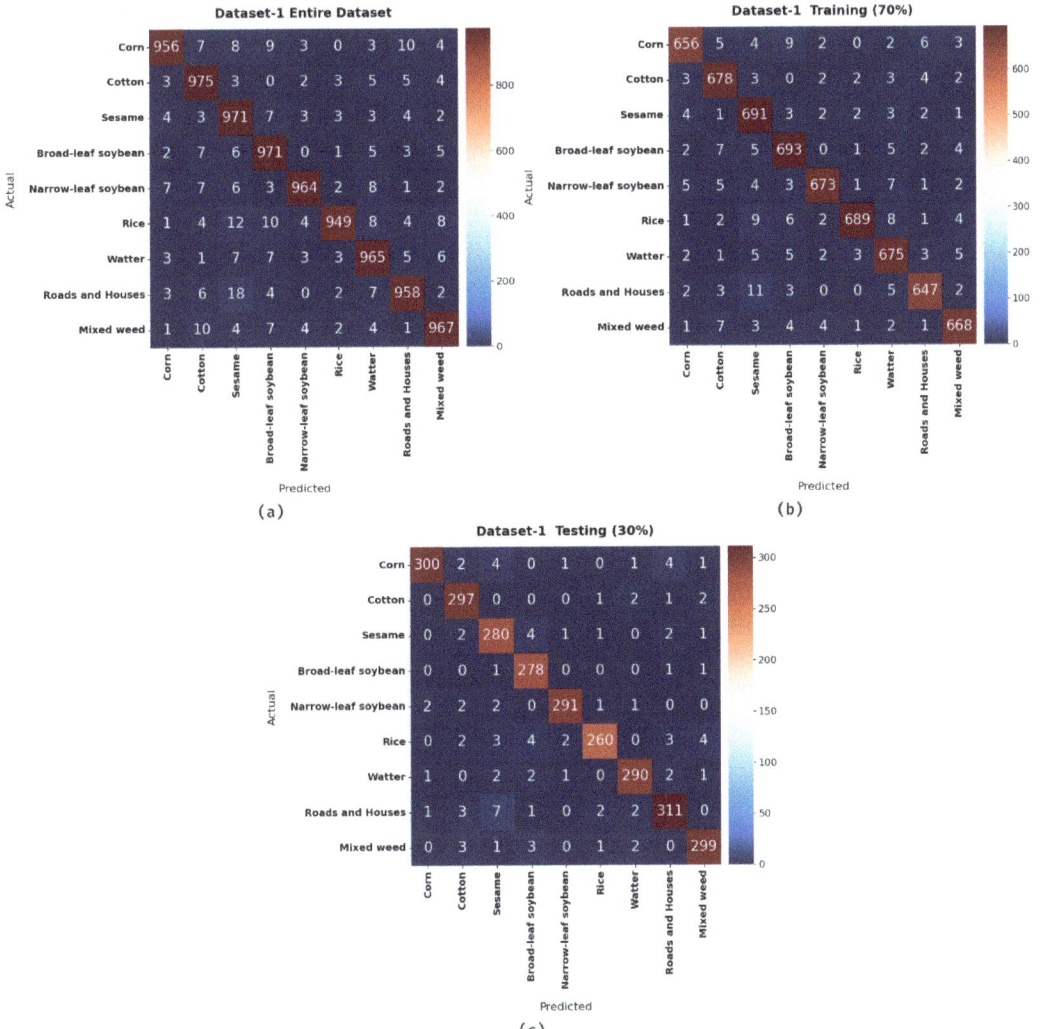

Figure 4. Confusion matrix of the SSODTL-CC technique under dataset-1. (**a**) Entire dataset-1. (**b**) 70% of Training dataset-1 and (**c**) 30% of Testing dataset-1.

Table 1 reports detailed crop classification outcomes of the SSODTL-CC model on all of dataset-1. The experimental values indicated that the SSODTL-CC model gained effectual outcomes under every individual class. For instance, in the corn class, the SSODTL-CC model offered $accu_y$, $prec_n$, and $reca_l$ of 99.24%, 97.55%, and 95.60%, respectively. Similarly, on the mixed weed class, the SSODTL-CC model reached $accu_y$, $prec_n$, and $reca_l$ of 99.27%, 96.70%, and 96.70%, respectively. Overall, the SSODTL-CC model showed a maximum average $accu_y$, $prec_n$, and $reca_l$ of 99.20%, 96.43%, and 96.40%, respectively.

Table 2 depicts a brief crop classification outcome of the SSODTL-CC approach on 70% of training dataset-1. The experimental values stated that the SSODTL-CC method gained effectual outcomes under every individual class. For instance, in the corn class, the SSODTL-CC model offered $accu_y$, $prec_n$, and $reca_l$ of 99.19%, 97.04%, and 95.49%, respectively. In addition, in the mixed weed class, the SSODTL-CC system obtained $accu_y$,

$prec_n$, and $reca_l$ of 99.27%, 96.67%, and 96.67%, respectively. Overall, the SSODTL-CC model demonstrated maximum average $accu_y$, $prec_n$, and $reca_l$ of 99.19%, 96.38%, and 96.35%, correspondingly.

Table 1. Result analysis of the SSODTL-CC technique with distinct classes under all of dataset-1.

	Entire Dataset Samples			
Class Labels	Accuracy	Precision	Recall	Kappa Score
Corn	99.24	97.55	95.60	-
Cotton	99.22	95.59	97.50	-
Sesame	98.97	93.82	97.10	-
Broad-leaf soybean	99.16	95.38	97.10	-
Narrow-leaf soybean	99.39	98.07	96.40	-
Rice	99.26	98.34	94.90	-
Water	99.13	95.73	96.50	-
Roads and Houses	99.17	96.67	95.80	-
Mixed weed	99.27	96.70	96.70	-
Average	**99.20**	**96.43**	**96.40**	**95.95**

Table 2. Result analysis of the SSODTL-CC technique with distinct classes under 70% of training dataset-1.

	Training Samples (70%)			
Class Labels	Accuracy	Precision	Recall	Kappa Score
Corn	99.19	97.04	95.49	-
Cotton	99.21	95.63	97.27	-
Sesame	99.02	94.01	97.46	-
Broad-leaf soybean	99.06	95.45	96.38	-
Narrow-leaf soybean	99.33	97.96	96.01	-
Rice	99.32	98.57	95.43	-
Water	99.03	95.07	96.29	-
Roads and Houses	99.27	97.00	96.14	-
Mixed weed	99.27	96.67	96.67	-
Average	**99.19**	**96.38**	**96.35**	**95.89**

Table 3 defines the detailed crop classification outcomes of the SSODTL-CC model on 30% of testing dataset-1. The experimental values indicated that the SSODTL-CC model gained effectual outcomes under every individual class. For instance, in the corn class, the SSODTL-CC approach presented $accu_y$, $prec_n$, and $reca_l$ of 99.37%, 98.68%, and 95.85%, correspondingly. Furthermore, in the mixed weed class, the SSODTL-CC methodology reached $accu_y$, $prec_n$, and $reca_l$ of 99.26%, 96.76%, and 96.76%, respectively. Overall, the SSODTL-CC model portrayed enhanced average $accu_y$, $prec_n$, and $reca_l$ of 99.23%, 96.54%, and 96.53%, correspondingly.

Figure 5 illustrates the confusion matrices created by the SSODTL-CC approach on the classification of sixteen classes under dataset-2. The figure reveals that the SSODTL-CC model categorized all the classes under different sets of datasets. On the entire dataset, the SSODTL-CC model recognized 783 samples under class 1, 757 samples under class 2, 766 samples under class 3, 728 samples under class 4, 721 samples under class 5, 774 samples under class 6, 764 samples under class 7, 788 samples under class 8, 779 samples under class 9, 779 samples under class 10, 733 samples under class 11, 806 samples under class 12, 771 samples under class 13, 829 samples under class 14, 733 samples under class 15, and 821 samples under class 16. Similarly, the SSODTL-CC approach categorized the class labels proficiently on 70% of the training samples and 30% of the testing samples on dataset-2.

Table 3. Result analysis of the SSODTL-CC technique with distinct classes under 30% of testing dataset-1.

Class Labels	Testing Samples (30%)			
	Accuracy	Precision	Recall	Kappa Score
Corn	99.37	98.68	95.85	-
Cotton	99.26	95.5	98.02	-
Sesame	98.85	93.33	96.22	-
Broad-leaf soybean	99.37	95.21	98.93	-
Narrow-leaf soybean	99.52	98.31	97.32	-
Rice	99.11	97.74	93.53	-
Water	99.37	97.32	96.99	-
Roads and Houses	98.93	95.99	95.11	-
Mixed weed	99.26	96.76	96.76	-
Average	**99.23**	**96.54**	**96.53**	**96.08**

Table 4 demonstrates the detailed crop classification outcomes of the SSODTL-CC model on all of dataset-2. The experimental values exposed that the SSODTL-CC model gained effectual outcomes under every individual class. For instance, in class 1, the SSODTL-CC algorithm obtained $accu_y$, $prec_n$, and $reca_l$ of 97.39%, 79.57%, and 78.30% correspondingly. In addition, in class 16, the SSODTL-CC model gained $accu_y$, $prec_n$, and $reca_l$ of 97.17%, 74.98%, and 82.10%, correspondingly. Overall, the SSODTL-CC model outperformed higher average $accu_y$, $prec_n$, and $reca_l$ of 97.13%, 74.98%, and 82.10%, respectively.

Table 5 reports a brief crop classification outcome of the SSODTL-CC model on 70% of training dataset-2. The experimental values exposed that the SSODTL-CC model gained effectual outcomes under every individual class. For instance, in class 1, the SSODTL-CC model offered $accu_y$, $prec_n$, and $reca_l$ of 97.43%, 79.74%, and 79.29%, respectively. In addition, in class 16, the SSODTL-CC model reached $accu_y$, $prec_n$, and $reca_l$ of 97.21%, 74.08%, and 81.65%, respectively. Overall, the SSODTL-CC methodology exhibited maximal average $accu_y$, $prec_n$, and $reca_l$ of 97.13%, 77.08%, and 77.05%, correspondingly.

Table 6 defines the detailed crop classification outcome of the SSODTL-CC technique on 30% of testing dataset-2. The experimental values indicated that the SSODTL-CC algorithm gained effectual outcomes under every individual class. For sample, in class 1, the SSODTL-CC model offered $accu_y$, $prec_n$, and $reca_l$ of 97.29%, 79.15%, and 75.93%, correspondingly. In the same way, in class 16, the SSODTL-CC system reached $accu_y$, $prec_n$, and $reca_l$ of 97.06%, 76.80%, and 82.99%, respectively. Overall, the SSODTL-CC approach showed maximal average $accu_y$, $prec_n$, and $reca_l$ of 97.15%, 77.25%, and 77.09%, correspondingly.

3.2. Discussion

To ensure the improved crop classification results of the SSODTL-CC model, a comparison study with recent models on two datasets is given in Table 7 [22,23].

Figure 6 investigates a comparative classification outcome of the SSODTL-CC model with existing models on dataset-1. The results indicated that the SVM model gained an ineffectual outcome with the least $accu_y$ of 95.98%. In line with this, the FNEA-OO model certainly accomplished increased performance with an $accu_y$ of 97.07%. In addition, the SVRFMC, CNN, and CNN-CRF models depicted closer $accu_y$ values of 98.20%, 98.08%, and

98.80%, respectively. However, the SSODTL-CC model demonstrated superior performance with an $accu_y$ of 99.23%.

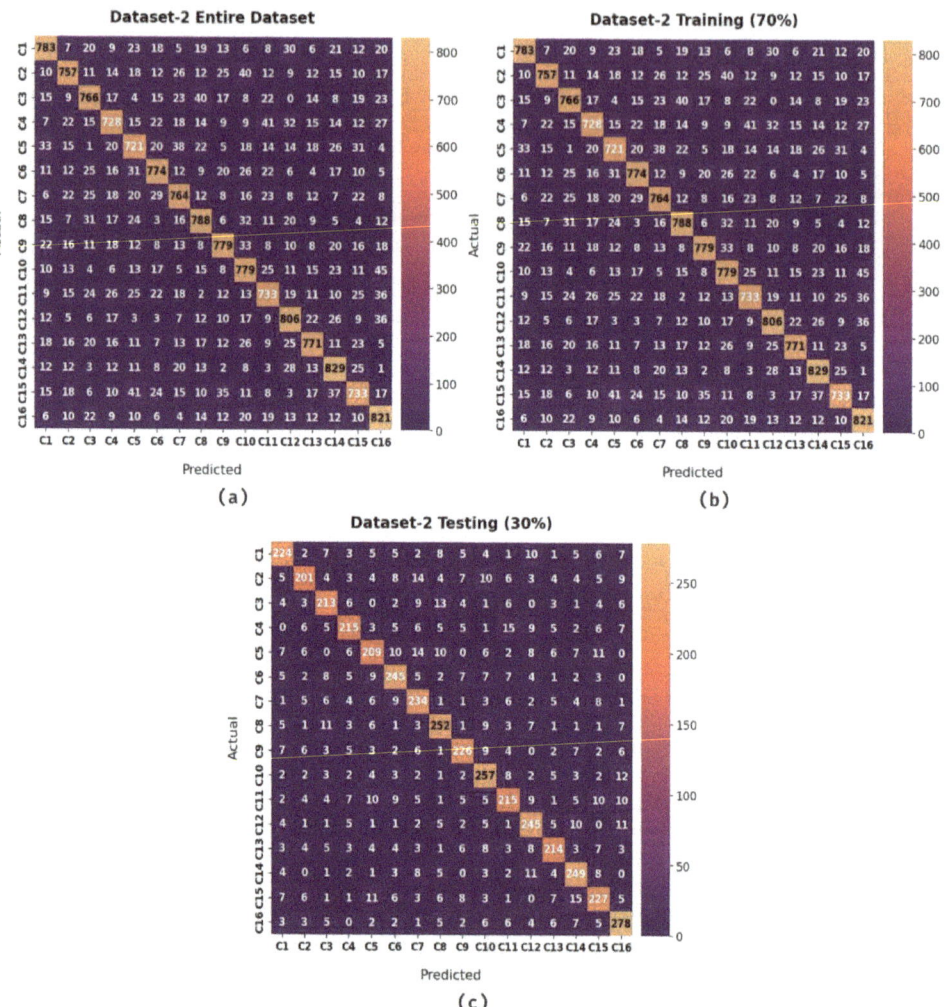

Figure 5. Confusion matrix of the SSODTL-CC technique under dataset-2. (**a**) Entire dataset-2. (**b**) 70% of Training dataset-2, and (**c**) 30% of Testing dataset-2.

Figure 7 examines a comparative classification outcome of the SSODTL-CC model with existing approaches on dataset-2. The outcomes indicated that the SVM model gained an ineffectual outcome with the least $accu_y$ of 77.34%. Likewise, the FNEA-OO model certainly accomplished an increased performance with an $accu_y$ of 86.49%. Then, the SVRFMC, CNN, and CNN-CRF models depicted closer $accu_y$ values of 86.95%, 87.72%, and 94.67%, correspondingly. At last, the SSODTL-CC methodology demonstrated superior performance with an $accu_y$ of 97.15%.

From these results and discussions, it is evident that the SSODTL-CC model has the capability of attaining improved crop classification outcomes on HSIs.

Table 4. Result analysis of the SSODTL-CC technique with distinct classes under all of dataset-2.

Class Labels	Entire Dataset Samples			
	Accuracy	Precision	Recall	Kappa Score
Class-1	97.39	79.57	78.30	-
Class-2	97.24	79.18	75.70	-
Class-3	97.14	77.37	76.60	-
Class-4	96.89	76.39	72.80	-
Class-5	96.63	73.42	72.10	-
Class-6	97.25	78.34	77.40	-
Class-7	97.07	76.63	76.40	-
Class-8	97.31	78.25	78.80	-
Class-9	97.41	80.06	77.90	-
Class-10	96.85	73.35	77.90	-
Class-11	96.87	75.80	73.30	-
Class-12	97.36	77.95	80.60	-
Class-13	97.39	80.40	77.10	-
Class-14	97.36	76.69	82.90	-
Class-15	96.84	75.41	73.30	-
Class-16	97.17	74.98	82.10	-
Average	**97.13**	**77.11**	**77.08**	**75.55**

Table 5. Result analysis of the SSODTL-CC technique with distinct classes under 70% of training dataset-2.

Class Labels	Training Samples (70%)			
	Accuracy	Precision	Recall	Kappa Score
Class-1	97.43	79.74	79.29	-
Class-2	97.31	78.98	78.42	-
Class-3	97.04	77.56	76.28	-
Class-4	96.77	75.11	72.77	-
Class-5	96.63	72.73	73.35	-
Class-6	97.29	78.6	76.89	-
Class-7	97.11	77.94	75.28	-
Class-8	97.29	78.02	77.91	-
Class-9	97.35	79.91	77.78	-
Class-10	96.69	72	75.65	-
Class-11	96.94	76.06	74.21	-
Class-12	97.4	78.79	80.03	-
Class-13	97.36	80.84	77.25	-
Class-14	97.37	76.72	82.98	-
Class-15	96.89	75.86	73.02	-
Class-16	97.21	74.08	81.65	-
Average	**97.13**	**77.06**	**77.05**	**75.5**

Table 6. Result analysis of the SSODTL-CC model with distinct classes under 30% of testing dataset-2.

	Testing Samples (30%)			
Class Labels	**Accuracy**	**Precision**	**Recall**	**Kappa Score**
Class-1	97.29	79.15	75.93	-
Class-2	97.06	79.76	69.07	-
Class-3	97.38	76.90	77.45	-
Class-4	97.19	79.63	72.88	-
Class-5	96.63	75.18	69.21	-
Class-6	97.15	77.78	78.53	-
Class-7	96.98	73.82	79.05	-
Class-8	97.33	78.75	80.77	-
Class-9	97.54	80.43	78.20	-
Class-10	97.23	76.26	82.90	-
Class-11	96.71	75.17	71.19	-
Class-12	97.27	76.09	81.94	-
Class-13	97.48	79.26	76.70	-
Class-14	97.33	76.62	82.72	-
Class-15	96.71	74.43	73.94	-
Class-16	97.06	76.80	82.99	-
Average	**97.15**	**77.25**	**77.09**	**75.64**

Table 7. Comparative analysis of the SSODTL-CC technique with recent algorithms in terms of $accu_y$.

Methods	Dataset-1	Dataset-2
SVM	95.98	77.34
FNEA-OO	97.07	86.49
SVRFMC	98.20	86.95
CNN	98.08	87.72
CNN-CRF	98.80	94.67
SSODTL-CC	99.23	97.15

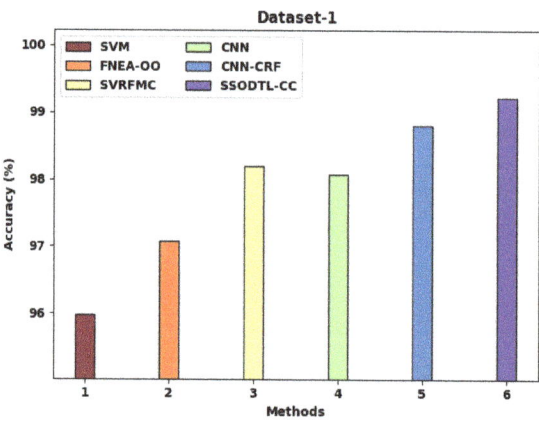

Figure 6. Comparative analysis of the SSODTL-CC technique under dataset-1.

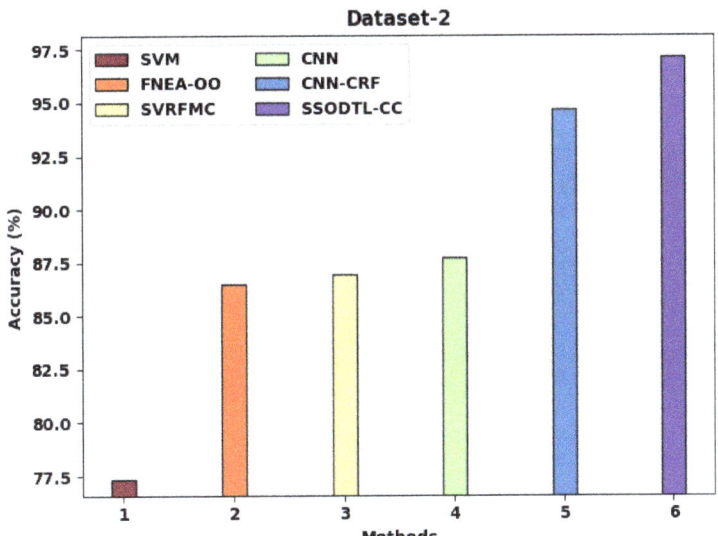

Figure 7. Comparative analysis of the SSODTL-CC technique under dataset-2.

4. Conclusions

In this article, a new SSODTL-CC model was developed to properly identify the crop type in HSIs. To do so, the proposed SSODTL-CC model performed feature extraction using MobileNet with an Adam optimizer. In addition, the BiLSTM model received feature vectors and performed crop type classification. To enhance the classifier efficiency of the BiLSTM model, the SSO algorithm was employed for hyperparameter optimization. To demonstrate the better performance of the SSODTL-CC model, a wide-ranging experimental analysis was performed on two benchmark datasets, namely dataset-1 (WHU-Hi-LongKou) and dataset-2 (WHU-Hi-HanChuan). The comparative analysis pointed out the better outcomes of the SSODTL-CC model over the recent approaches, with a maximum of 99.23% and 97.15% on test datasets 1 and 2, respectively. Therefore, the SSODTL-CC model can be utilized for effective crop type classification on HSIs. In the future, the classification performance of the SSODTL-CC model can be enhanced by the design of hybrid DL models.

Author Contributions: Conceptualization, M.A.H.; Data curation, F.A.; Formal analysis, F.A. and J.S.A.; Investigation, J.S.A.; Methodology, M.A.H.; Project administration, H.M.; Resources, H.M.; Software, N.M.S.; Supervision, N.M.S.; Validation, R.M.; Visualization, R.M.; Writing—original draft, M.A.H. All authors have read and agreed to the published version of the manuscript.

Funding: This research was funded by King Khalid University, grant number RGP 2/46/43, Princess Nourah bint Abdulrahman University, grant number PNURSP2022R77 and Umm al-Qura University, grant number 22UQU4340237DSR24.

Institutional Review Board Statement: Not applicable.

Informed Consent Statement: Not applicable.

Data Availability Statement: Data sharing not applicable to this article as no datasets were generated during the current study.

Acknowledgments: The authors extend their appreciation to the Deanship of Scientific Research at King Khalid University for funding this work through Large Groups Project under grant number (46/43). Princess Nourah bint Abdulrahman University Researchers Supporting Project number (PNURSP2022R77), Princess Nourah bint Abdulrahman University, Riyadh, Saudi Arabia. The authors would like to thank the Deanship of Scientific Research at Umm Al-Qura University for supporting this work by Grant Code: 22UQU4340237DSR19.

Conflicts of Interest: The authors declare no conflict of interest.

References

1. Bhosle, K.; Musande, V. Evaluation of deep learning CNN model for land use land cover classification and crop identification using hyperspectral remote sensing images. *J. Indian Soc. Remote Sens.* **2019**, *47*, 1949–1958. [CrossRef]
2. Wang, C.; Chen, Q.; Fan, H.; Yao, C.; Sun, X.; Chan, J.; Deng, J. Evaluating satellite hyperspectral (Orbita) and multispectral (Landsat 8 and Sentinel-2) imagery for identifying cotton acreage. *Int. J. Remote Sens.* **2021**, *42*, 4042–4063. [CrossRef]
3. Wei, L.; Yu, M.; Liang, Y.; Yuan, Z.; Huang, C.; Li, R.; Yu, Y. Precise crop classification using spectral-spatial-location fusion based on conditional random fields for UAV-borne hyperspectral remote sensing imagery. *Remote Sens.* **2019**, *11*, 2011. [CrossRef]
4. Wei, L.; Yu, M.; Zhong, Y.; Zhao, J.; Liang, Y.; Hu, X. Spatial–spectral fusion based on conditional random fields for the fine classification of crops in UAV-borne hyperspectral remote sensing imagery. *Remote Sens.* **2019**, *11*, 780. [CrossRef]
5. Uddin, M.P.; Mamun, M.A.; Hossain, M.A. PCA-based feature reduction for hyperspectral remote sensing image classification. *IETE Tech. Rev.* **2021**, *38*, 377–396. [CrossRef]
6. Papp, L.; Van Leeuwen, B.; Szilassi, P.; Tobak, Z.; Szatmári, J.; Árvai, M.; Mészáros, J.; Pásztor, L. Monitoring invasive plant species using hyperspectral remote sensing data. *Land* **2021**, *10*, 29. [CrossRef]
7. Singh, P.; Pandey, P.C.; Petropoulos, G.P.; Pavlides, A.; Srivastava, P.K.; Koutsias, N.; Deng, K.A.K.; Bao, Y. Hyperspectral remote sensing in precision agriculture: Present status, challenges, and future trends. In *Hyperspectral Remote Sensing*; Elsevier: Amsterdam, The Netherlands, 2020; pp. 121–146.
8. Zhong, Y.; Wang, X.; Wang, S.; Zhang, L. Advances in spaceborne hyperspectral remote sensing in China. *Geo-Spat. Inf. Sci.* **2021**, *24*, 95–120. [CrossRef]
9. Vangi, E.; D'Amico, G.; Francini, S.; Giannetti, F.; Lasserre, B.; Marchetti, M.; Chirici, G. The new hyperspectral satellite PRISMA: Imagery for forest types discrimination. *Sensors* **2021**, *21*, 1182. [CrossRef] [PubMed]
10. Lassalle, G. Monitoring natural and anthropogenic plant stressors by hyperspectral remote sensing: Recommendations and guidelines based on a meta-review. *Sci. Total Environ.* **2021**, *788*, 147758. [CrossRef] [PubMed]
11. Sharma, M.; Biswas, M. Classification of hyperspectral remote sensing image via rotation-invariant local binary pattern-based weighted generalized closest neighbor. *J. Supercomput.* **2021**, *77*, 5528–5561. [CrossRef]
12. Meng, S.; Wang, X.; Hu, X.; Luo, C.; Zhong, Y. Deep learning-based crop mapping in the cloudy season using one-shot hyperspectral satellite imagery. *Comput. Electron. Agric.* **2021**, *186*, 106188. [CrossRef]
13. Nandibewoor, A.; Hegadi, R. A novel SMLR-PSO model to estimate the chlorophyll content in the crops using hyperspectral satellite images. *Clust. Comput.* **2019**, *22*, 443–450. [CrossRef]
14. Farooq, A.; Hu, J.; Jia, X. Weed classification in hyperspectral remote sensing images via deep convolutional neural network. In Proceedings of the 2018 IEEE International Geoscience and Remote Sensing Symposium, Valencia, Spain, 22–27 July 2018; pp. 3816–3819.
15. Lei, L.; Wang, X.; Zhong, Y.; Zhao, H.; Hu, X.; Luo, C. DOCC: Deep one-class crop classification via positive and unlabeled learning for multi-modal satellite imagery. *Int. J. Appl. Earth Obs. Geoinf.* **2021**, *105*, 102598. [CrossRef]
16. Pi, W.; Du, J.; Bi, Y.; Gao, X.; Zhu, X. 3D-CNN based UAV hyperspectral imagery for grassland degradation indicator ground object classification research. *Ecol. Inform.* **2021**, *62*, 101278. [CrossRef]
17. Wei, L.; Wang, K.; Lu, Q.; Liang, Y.; Li, H.; Wang, Z.; Wang, R.; Cao, L. Crops fine classification in airborne hyperspectral imagery based on multi-feature fusion and deep learning. *Remote Sens.* **2021**, *13*, 2917. [CrossRef]
18. Zhong, Y.; Hu, X.; Luo, C.; Wang, X.; Zhao, J.; Zhang, L. WHU-Hi: UAV-borne hyperspectral with high spatial resolution (H2) benchmark datasets and classifier for precise crop identification based on deep convolutional neural network with CRF. *Remote Sens. Environ.* **2020**, *250*, 112012. [CrossRef]
19. Wang, W.; Li, Y.; Zou, T.; Wang, X.; You, J.; Luo, Y. A novel image classification approach via dense-MobileNet models. *Mob. Inf. Syst.* **2020**, *2020*, 7602384. [CrossRef]
20. Bock, S.; Weiß, M. A proof of local convergence for the Adam optimizer. In Proceedings of the 2019 International Joint Conference on Neural Networks (IJCNN), Budapest, Hungary, 30 September 2019; pp. 1–8.
21. Hameed, Z.; Garcia-Zapirain, B. Sentiment classification using a single-layered BiLSTM model. *IEEE Access* **2020**, *8*, 73992–74001. [CrossRef]
22. Jain, M.; Singh, V.; Rani, A. A novel nature-inspired algorithm for optimization: Squirrel search algorithm. *Swarm Evol. Comput.* **2019**, *44*, 148–175. [CrossRef]
23. Zhong, Y.; Wang, X.; Xu, Y.; Wang, S.; Jia, T.; Hu, X.; Zhao, J.; Wei, L.; Zhang, L. Mini-UAV-borne hyperspectral remote sensing: From observation and processing to applications. *IEEE Geosci. Remote Sens. Mag.* **2018**, *6*, 46–62. [CrossRef]

Article

Rational Sampling Numbers of Soil pH for Spatial Variation: A Case Study from Yellow River Delta in China

Yingxin Zhang [1,†], Mengqi Duan [1,†], Shimei Li [2,*], Xiaoguang Zhang [1,3,*], Xiangyun Song [1] and Dejie Cui [1]

1. Department of Resources and Environment, Qingdao Agricultural University, Qingdao 266109, China; zyx1031644968@163.com (Y.Z.); duanmengqi28@163.com (M.D.); songxiangyun2000@126.com (X.S.); cuidejie@163.com (D.C.)
2. Department of Landscape Architecture and Forestry, Qingdao Agricultural University, Qingdao 266109, China
3. Qingdao Engineering Research Center for Remote Sensing Application in Agriculture, Qingdao 266109, China
* Correspondence: li_shimei@163.com (S.L.); zhangxg_66@sina.com (X.Z.); Tel.: +86-158-6309-1676 (S.L.); +86-151-9200-3056 (X.Z.)
† These authors contributed equally to this work.

Abstract: Spatial variation of soil pH is important for the evaluation of environmental quality. A reasonable number of sampling points has an important meaning for accurate quantitative expression on spatial distribution of soil pH and resource savings. Based on the grid distribution point method, 908, 797, 700, 594, 499, 398, 299, 200, 149, 100, 75 and 50 sampling points, which were randomly selected from 908 sampling points, constituted 12 sample sets. Semi-variance structure analysis was carried out for different point sets, and ordinary Kriging was used for spatial prediction and accuracy verification, and the influence of different sampling points on spatial variation of soil pH was discussed. The results show that the pH value in Kenli County (China) was generally between 7.8 and 8.1, and the soil was alkaline. Semi-variance models fitted by different point sets could reflect the spatial structure characteristics of soil pH with accuracy. With a decrease in the number of sampling points, the Sill value of sample set increased, and the spatial autocorrelation gradually weakened. Considering the prediction accuracy, spatial distribution and investigation cost, a number of sampling points greater than or equal to 150 could satisfy the spatial variation expression of soil pH at the county level in the Yellow River Delta. This is equivalent to taking at least 107 sampling points per 1000 km^2. The results in this study are applicable to areas with similar environmental and soil conditions as the Yellow River Delta, and have reference significance for these areas.

Keywords: sampling; soil pH; spatial variation; ordinary kriging

1. Introduction

Soil pH is an important index for evaluation of land quality [1]. Soils with acid and alkali over the national standard are not conducive to the utilization of land resources [2]. Therefore, it is very important to understand the spatial distribution of soil pH. At the same time, it is of great guiding significance to accurately grasp the spatial variation characteristics of soil pH for evaluating the salinization and acidification of the soil environment, rational fertilization and efficient utilization of nutrients [3]. However, sampling number affects the accuracy of soil properties and their spatial variation information and the degree of quantitative expression. The layout pattern and sampling number must be fully considered to ensure the accuracy of spatial interpolation in any study on spatial variation of soil pH [4–6]. Generally speaking, the larger the sampling density, the smaller the sample error and the higher the accuracy of the research results; however, this means the work cycle will be prolonged, and huge manpower, material resources and financial resources will be consumed [7]. Sampling costs also limit the sampling density to a large extent. If the number of sampling points is reduced, the interpolation accuracy of soil pH spatial variation will be difficult to guarantee, and local features may not be displayed [8,9]. Therefore, it is of great significance to study the reasonable number of sampling points for accurate quantitative expression on spatial distribution of soil pH and resource savings.

Using spatial geostatistics is one of the most accurate spatial prediction methods, and is often used to model the spatial variability for soil properties and evaluate their spatial uncertainty [10]. At present, a method combining geostatistics and GIS is used to study the spatial distribution of regional soil properties from the perspective of spatial prediction, and is also used to analyze the influence of sampling density on the spatial variation of soil properties [11]. In recent years, many scholars have carried out much of geostatistical research on the influences of sampling number on spatial variability for different soil properties in different areas, such as soil organic matter [12–14], nitrogen [15], exchangeable potassium, calcium, magnesium [16], heavy metals [17,18] and salt. [19]. The reasonable sampling quantity of different soil properties is basically different for their different characteristics. Even for the same specific soil index, the results of different research are different. For example, for soil organic matter, the reasonable sampling points of spatial variation in typical areas of Yangtze River Delta were 91 per 1000 km^2 [20], and 547 per 1000 km^2 in Fei County, which is a typical county of North China Plain [21]. For soil organic carbon, 908 sampling points were reasonable in typical gully areas of the Loess Plateau, which means that 178 sampling points were needed per 1000 km^2. These studies show that there were differences in the number of reasonable sampling points even with the same soil index because of the differences in the natural geographical environment, such as topography and geomorphology in different areas.

In addition, survey scale has a certain influence on reasonable sampling number. For example, in relatively consistent geomorphic units, such as Fujian Province and its counties, the reasonable sampling points of spatial distribution for soil organic matter were 10,000 and 11,000 for every 1000 km^2, respectively [12]. The existing research methods and conclusions still need to be tested because of the different evaluation indexes, the different natural geographical conditions and the different influence of human activities in different study areas.

To sum up, although many scholars have carried out relevant research on reasonable sampling numbers, at present there are few reports on related research for soil pH. Especially, research in county-level areas is lacking on the influence of different sampling points on the spatial variation of soil pH, and because of the fragile and salinized soil environment in the Yellow River Delta region and great attention from the government, it is of great significance to monitor and master the spatial distribution status of soil pH for green development in agriculture on the premise of clarifying the reasonable sampling numbers. Therefore, Kenli County in the Yellow River Delta was selected as the study area, and the spatial distribution of soil pH with 12 different sampling sets was predicted using geostatistical Kriging interpolation. The overall objectives of this research were (1) to assess the influence of different sampling numbers on the prediction accuracy of spatial distribution for soil pH, and (2) to determine reasonable sampling densities to determine the spatial variation of soil pH at the county scale.

2. Materials and Methods

2.1. Study Area

Kenli County in the Yellow River Delta region was selected as the study area (Figure 1). This county is located at the mouth of the Yellow River in the Yellow River Delta of northern Shandong Province in China, between 37°24′–38°10′ N and 118°15′–119°19′ E [22], with a total area of 2331 km^2. It has a temperate monsoon climate, with high annual temperature and uneven spatial-temporal distribution of precipitation. The terrain is fan-shaped and slightly inclined from southwest to northeast [23]. The altitude ranges from 2 m to 11.61 m. The parent material is loess. The mechanical composition of the soil is mainly sandy loam. The main soil types in Kenli County are fluvo-aquic soil and coastal saline soil in the soil genetic classification of China. The corresponding soil group names from WRB are Cambisols and Solonchaks, respectively. The typical crops are cotton, rice and wheat-corn rotation. Kenli County has rich soil resources, and is one of the most abundant land reserve resources in the coastal areas of eastern China, with great potential for agricultural development.

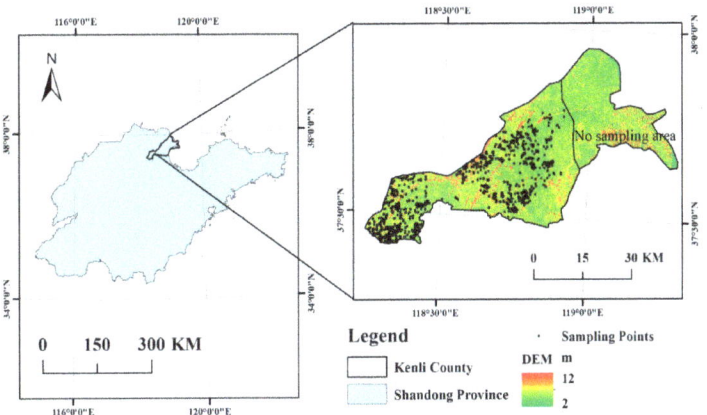

Figure 1. Overview of the study area.

2.2. Collection and Processing of Sample Data

Based on the grid distribution point method, 1140 sampling points were used in this research. Some points on non-agricultural land were deleted and 1000 points were left. These sampling points were mainly distributed in cultivated land, and the interval between sample points was about 1000 m. In the actual sampling process, we adjusted the specific positions of these sampling points for road accessibility and crop planting. The topsoil from 0–20 cm was taken with a shovel, and then was put into a plastic bag and brought back to the laboratory for analysis. Although 926 samples were sampled, there were many uncertainties and complexity in field sampling. Consequently, we eliminated some points that were not standardized in the collection process, and removed some outliers. Finally, 908 samples were obtained, and 797, 700, 594, 499, 398, 299, 200, 149, 100, 75 and 50 sampling points were randomly selected from these 908 sampling points (Figure 2). The above real numbers represent the approximate number sets of 900, 800, 700, 600, 500, 400, 300, 200, 150, 100, 75, 50, respectively. Each sampling point was extracted from the last sampling point set to compare the characteristics of different sample sets. For example, the 700 sample set was extracted from the 797 sample set, and the 299 sample set was extracted from the 398 sample set. Combining all the sample data, a total of 12 sample sets were formed. The sampling process was conducted by the "Geostatistical Analyst" module of arcgis 10.0. The pH of each sample was measured in 1:2.5 mixtures of soil and deionized water with a pH meter by a potentiometric method [24].

2.3. Spatial Prediction and Verification Method

Geostatistical methods are widely used to predict the spatial distribution of soil properties [25–29]. In this paper, we chose the Ordinary Kriging (OK) method to predict the spatial distribution of soil pH [30]. The OK method satisfies the intrinsic hypothesis, and the average value of the regionalized variables is an unknown constant [31]. OK is a linear estimation of regionalized variables, which is similar to weighted moving average in the process of interpolation research. However, the weights of weighted moving average are determined from different sources. The weighted sliding average weight values are derived from known spatial functions, while the weights of ordinary kriging are derived from spatial data analysis [31]. It is necessary to verify the prediction results after spatial distribution prediction. In this paper, an independent verification method was adopted.

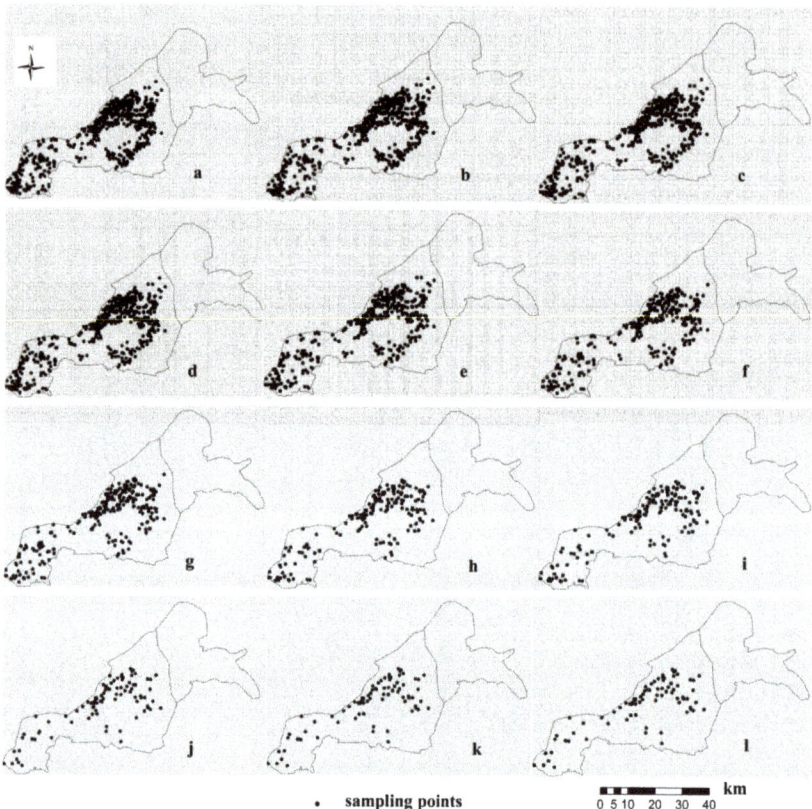

Figure 2. Distribution map of soil pH at different point sets. The letters (**a**–**l**) represent the spatial variation expression of soil pH at 908, 797, 700, 594, 499, 398, 299, 200, 149, 100, 75 and 50 sampling points.

The OK interpolation results were verified. The verification method of independent data set extracts some samples from all samples as independent data sets, takes the remaining samples as simulation data sets without repetition, and takes each sample in independent data sets as an inspection point [32]. In this paper, after 797 samples were extracted from 908 samples, the remaining 111 samples were taken as independent verification sets. The spatial prediction results of 12 sample sets were verified. In the verification method of the independent data set, the most representative evaluation indexes are root mean square error (RMSE), mean error (ME) and average standard error (ASE), which were chosen to evaluate the accuracy of prediction.

$$\text{RMSE} = \sqrt{\frac{1}{N}\sum_{i=1}^{n}\left[Z(X_i) - Z'(X_i)\right]^2} \tag{1}$$

$$\text{ME} = \frac{1}{N}\sum_{i=1}^{n}\left[Z(X_i) - Z'(X_i)\right] \tag{2}$$

$$\text{ASE} = \sqrt{\frac{1}{N}\sum_{i=1}^{n}\left[Z'(X_i) - \sum_{i=1}^{n}(Z'(X_i))/N\right]^2} \tag{3}$$

where N is the number of known samples, the actual value is $Z(X_i)$, and the estimated value is $Z'(X_i)$.

The smaller the RMSE, the closer the ME to zero, indicating that the accuracy of spatial prediction is higher. The average standard error was used to measure the uncertainty of the Kriging prediction value.

3. Results
3.1. Descriptive Statistics Characteristics of Soil pH for Different Sets of Sample Points

Descriptive statistics analysis was made on 12 sample sets, and the results are shown in Table 1. The soil pH values of 908 sampling points in the study area ranged from 7.00 to 8.80, with an average value of 7.85. The soil was weakly alkaline. The coefficient of variation was 5.35%, and the variability was weak, indicating that the alkalization degree was very concentrated. The skewness coefficient was −0.31, and the kurtosis coefficient was 2.12. The results of normal test on skewness coefficient and kurtosis coefficient showed that the soil pH values of different sampling points were in accordance with normal distribution.

Table 1. Descriptive statistics characteristics of soil pH for different sets of sample points.

Sampling Point Number	Min (g/kg)	Max (g/kg)	Average (g/kg)	Standard Deviation (g/kg)	Skewness	Kurtosis	Median (g/kg)	Variation (%)
908	7.00	8.80	7.85	0.42	−0.31	2.12	7.90	5.35
797	7.00	8.80	7.84	0.42	−0.32	2.15	7.90	5.36
700	7.00	8.80	7.85	0.41	−0.33	2.18	7.90	5.22
594	7.00	8.80	7.85	0.42	−0.35	2.18	7.90	5.35
499	7.00	8.80	7.86	0.41	−0.37	2.19	7.90	5.22
398	7.00	8.80	7.88	0.41	−0.37	2.18	7.90	5.20
299	7.00	8.80	7.88	0.40	−0.38	2.30	7.90	5.08
200	7.00	8.80	7.89	0.42	−0.43	2.33	7.90	5.32
149	7.00	8.60	7.90	0.42	−0.59	2.46	8.00	5.32
100	7.10	8.80	7.90	0.40	−0.35	2.40	7.95	5.06
75	7.00	8.60	7.88	0.40	−0.55	2.43	8.00	5.08
50	7.10	8.60	7.85	0.41	−0.15	2.07	7.90	5.22

The minimum value of soil pH for 100 and 50 sampling points was 7.10 g/kg, and the minimum value of other sampling points was 7.00 g/kg. The maximum values of 908, 797, 700, 594, 499, 398, 299, 200 and 100 sample points were all 8.80 g/kg, and the maximum values of only 149, 75 and 50 sample points were 8.60 g/kg. However, they were still very similar. Among the 11 sub-samples, the average value and standard deviation of soil pH also fluctuated around the average value and standard deviation of the complete set, which indicated that although the number of sampling points decreased, the 11 samples could still represent the complete set. The coefficient of variation of soil pH ranged from 5.06% to 5.36% among the 11 sub-samples, and the variability was weak. To sum up, the analysis of each index of each subset showed that the selected subsets were all representative.

3.2. The Influence of Different Sampling Points on the Semi-Variance Structure of Soil pH

Semi-variance analysis of soil pH was carried out by a geostatistical method. Table 2 shows the semi-variance function values of soil pH under different sampling points. The spatial variation structure of soil pH at other sampling densities conformed to the exponential model, except for 75 sampling points. The level of decision coefficient represented the effect of fitting the variogram by the model. The higher the decision coefficient, the better the effect of fitting the variogram by the model [33]. The determination coefficients of different sampling points were between 0.39 and 0.69, indicating that the model could reflect the spatial structure characteristics of soil pH with accuracy.

The ratio of Nugget to base Sill (C0 + C) reflects the degree of spatial autocorrelation of variables. This is considered a strong spatial autocorrelation when the ratio is less than 25%, has moderate spatial autocorrelation when the ratio is between 25% and 75%, and has weak spatial autocorrelation when the ratio is greater than 75%. The Nugget/Sill of the total sample set (908 sample points) in this study was less than 25%, showing a strong spatial autocorrelation. The Nugget/Sill of 200–800 samples in the other 11 sample subsets was less than 25%, which indicates that these sample sets had strong spatial autocorrelation. However, when the number of samples was less than 200, the Nugget/Sill ranged from 25% to 75% (with moderate spatial autocorrelation). When the number of samples was less than 150, the Nugget/Sill reached 40%, which indicates that the spatial autocorrelation of these samples was weakened.

Table 2. Parameters of semi-variance function of soil pH under different sampling points.

Sampling Point Number	Model	Nugget (C0)	Sill (C0 + C)	C0/Sill (%)	Range (km)	Determination	Residual
926	exponential model	0.0350	0.1786	19.60	4.80	0.68	0.00
801	exponential model	0.0360	0.1778	20.25	5.13	0.72	0.00
704	exponential model	0.0360	0.1752	20.55	4.77	0.68	0.00
598	exponential model	0.0360	0.1780	20.22	4.75	0.69	0.00
502	exponential model	0.0330	0.1742	18.94	4.82	0.69	0.00
401	exponential model	0.0350	0.1750	20.00	4.70	0.60	0.00
300	exponential model	0.0350	0.1658	21.11	4.81	0.60	0.00
201	exponential model	0.0295	0.1810	16.30	4.87	0.44	0.00
150	exponential model	0.0442	0.1736	25.44	6.54	0.39	0.00
100	exponential model	0.0952	0.1914	49.74	26.52	0.64	0.00
75	spherical model	0.0869	0.1748	49.71	13.77	0.49	0.00
50	exponential model	0.1247	0.3174	39.29	160.59	0.61	0.00

The variable range represents the autocorrelation range of the variogram [34], and can reflect the size of the autocorrelation range in the variable space. In this paper, the fitting range of soil pH under a different number of points was more than 4 km, indicating that the spatial autocorrelation distance was relatively large. Among them, when the number of sampling points was reduced to 100, the range increased, reaching 26.52, which was about five times that of the 200 sampling points.

3.3. The Influence of Different Sampling Points on the Spatial Prediction Accuracy of Soil pH

For each sample set, Kriging interpolation was used to carry out the spatial prediction of soil pH. Root mean square error (RMSE), mean error (ME) and average standard error (ASE) were used to measure the prediction accuracy of soil pH under different sampling points.

It can be seen from Figure 3 that ME of 75, 100, 150, 200 and 300 was greater than 0, and that of other sampling points was less than 0. The ME value varies with the number of sampling points, but the variation does not follow the law that the ME decreases with the increase of sample number. When the sampling points were 50 and 300, the ME values were close to 0. In theory, the closer the ME to 0, the higher the accuracy of spatial prediction. In this case, the value of ME is generally calculated from statistical methods. It is possible that the independent verification results of each sample point were poor, but the residual errors after addition and averaging were smaller. When ME is close to 0, the prediction has low accuracy. Therefore, the ME index cannot indicate the accuracy very well at this point. On the other hand, a single index cannot indicate the interpolation accuracy well, and multiple indexes may be more accurate to judge the interpolation accuracy. With the decreased of sample size, the distance of ME deviating from X axis first increased, then decreased and then slightly increased. It basically surrounded the X axis except for 75 and 100 sample points. Therefore, when the number of sample points was too small (75, 100), the ME of predicted values would become larger.

The RMSE was slightly larger with 50 samples. There was no obvious change trend with the decreased sample size, and it could remained at a certain value. This shows that the RMSE had no obvious difference in the spatial prediction accuracy of different sample subsets, i.e., the accuracy of Kriging interpolation had no significant difference with decreased sample size.

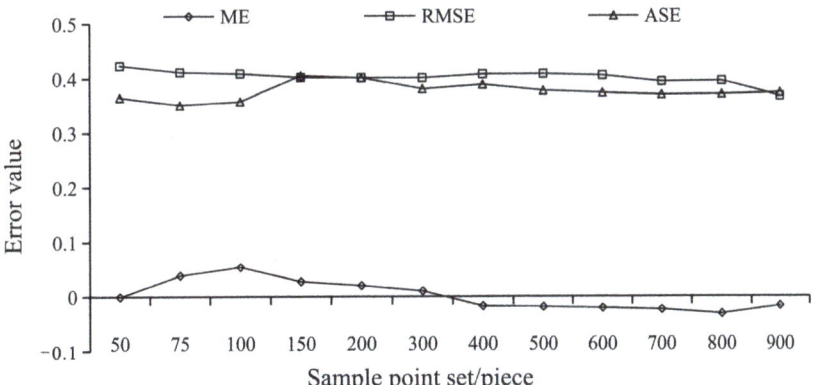

Figure 3. Prediction error of soil pH at different point sets.

The uncertainty of Kriging prediction was measured by the ASE. The closer the ASE the RMSE, the more accurate the prediction of attribute value. In this study, when there were 150, 200 and 900 samples, ASE and RMSE were basically equal, indicating that the spatial variability of the earth was properly estimated. When the number of sampling points was 300–800, the ASE was smaller than the RMSE, and the difference between them was very small, indicating that these sampling points overestimated the spatial variability. However, when the number of sampling points was less than 150, the ASE was obviously less than the RMSE, so these sample sets could not reasonably predict the variability.

Considering the ME, RMSE and ASE, 107 sampling points per 1000 km^2 could meet the needs of spatial variation expression of soil pH in the Yellow River Delta. According to the above analysis, we also determined that it was not enough to evaluate Kriging prediction accuracy only by a single evaluation index. Therefore, it was necessary to use a variety of evaluation indexes and combine them to accurately evaluate the prediction results.

3.4. Effects of Different Sampling Points on Spatial Distribution of Soil pH

To more intuitively show the influence of different sampling point sets on the spatial distribution of soil pH, the Kriging method was used to carry out a spatial interpolation operation (Figure 4). The eastern of Kenli County was not sampled because the area is in the Yellow River Delta National Nature Reserve. However, for the continuity and completeness of pictures, Kriging interpolation was extended to the entire Kenli County. When analyzing the spatial variation characteristics of soil pH, the Yellow River Delta National Nature Reserve in the east of Kenli County was not a focus of analysis.

In the study area, the yellow part shown on the map (Figure 4) had a large area: that is, the soil pH in Kenli County of the Yellow River Delta was generally between 7.8 and 8.1, which proved that the soil in this area was alkaline. The areas with high soil pH value (orange-red) were distributed in the middle and southwest of study area. With the decreased of the number of sample points, the ability to describe details was gradually weakened. When the number of sampling points was reduced to 100, the details of soil pH in the middle and north of the study area were no longer detailed, and only a general distribution trend could be seen. Therefore, considering the precision and research funds, it is suggested that the reasonable sampling number in the Yellow River Delta should not be less than 150.

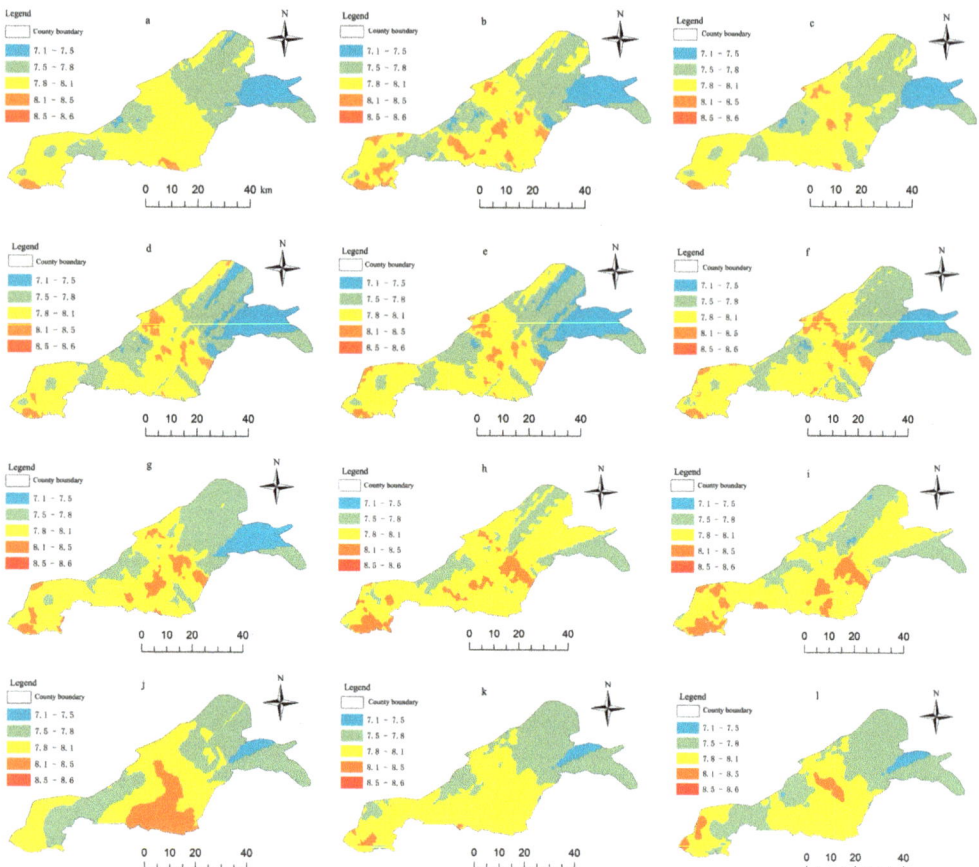

Figure 4. Expression diagram of soil pH spatial variation of different point sets. The letters (**a–l**) represent the spatial variation expression of soil pH at 908, 797, 700, 594, 499, 398, 299, 200, 149, 100, 75 and 50 sampling points respectively. Note: The eastern of Kenli County was not sampled because the area is in the Yellow River Delta National Nature Reserve. However, for the continuity and completeness of maps, the result of the unsampled area was deduced and extended by the Kriging interpolation based on the near sampled soils.

4. Discussion

Spatial variation of soil properties is controlled by various structural and random factors. The larger the ratio of Nugget to Sill, the more obvious the influence of human activities, such as irrigation, fertilization and cultivation. On the contrary, structural factors such as soil parent material, climate, biology, topography and other natural factors play a major role [35,36]. The ratio of Nugget to Sill of almost all sample sets in this study was less than 25% and showed strong spatial autocorrelation, indicating that the spatial variation of soil pH in the study area was mainly affected by structural components such as topography, climate and soil parent material. When the sampling point was less than 150, the ratio of Nugget to Sill reached 40%, indicating that the spatial autocorrelation of these sample sets was weakened, and was influenced by structural components and random factors. With the decreased of the number of sampling points, the small-scale structural factors and random factors gradually increased, while the influence of large-scale structural factors such as parent material, topography and soil type on soil pH gradually weakened, which caused the soil pH to change strongly with the small number of sampling points.

According to the analysis of spatial prediction accuracy of soil pH with different sampling sets, the ME of predicted values gradually increased when the number of sampling points was too small. The RMSE could basically be maintained at a certain value with a decrease of sampling numbers, except for being slightly larger at 50 sample points. Interpolation prediction error and interpolation of the distribution map had a certain synergistic relationship with the change of sample numbers. When there were less than 150 samples, the difference between ASE and RMSE was large. Correspondingly, the spatial distribution maps of soil pH with less than 150 sampling points were too smooth to show the spatial distribution of soil pH accurately. However, when the number of samples was increased to 150 or more, details of the spatial variation expression diagram of soil pH were more obvious, and could show the spatial variation expression of soil pH more accurately. In addition, at 150 or more sampling points, the prediction error maps of RMSE, ME and ASE of soil pH and the expression maps of spatial variation of soil pH in different point sets could better predict the spatial variation of soil pH. Therefore, it was shown that the prediction error of soil pH was consistent with the spatial distribution of soil pH.

In this paper, the influence of the number of different sampling points on the spatial distribution of soil pH in the Yellow River Delta region was investigated by using the ordinary Kriging method. It was concluded that the number of sampling points most suitable for Kenli County in the Yellow River Delta region should be no less than 150; that is, at least 107 sampling points should be taken every 1000 km^2. This paper also compared the existing literature investigating reasonable sampling numbers of soil pH. Study [37] showed that the rational sampling number at the county scale was about 4900 samples per 1000 km^2, which was inconsistent with our findings. This may be because in that study the area was located in a more undulating mountainous and hilly region resulting in a reduced spatial autocorrelation of soil pH. While this result was similar to the most reasonable number of sampling points needed for soil salinity in Kenli County of the Yellow River Delta studied by Zhang et al. [19], it revealed that the number of sampling points for spatial variation expression of different soil properties in areas with similar environmental conditions may get closer.

5. Conclusions

In this paper, Kenli County in the Yellow River Delta was selected as the research area. Twelve sample sets consisting of 908, 797, 700, 594, 499, 398, 299, 200, 149, 100, 75 and 50 sampling points were selected to study the influence of sampling number on the spatial variation of soil pH. A reasonable sampling number was determined, which provides for the collection of soil samples with minimum human, material and financial resources for research on soil pH.

With the decreasing the sampling points number, the C0/Sill value of the sample set increased, and the spatial autocorrelation decreased gradually. The variation range of soil pH fitted by different numbers of points was greater than 4 km, and the spatial autocorrelation distance was relatively large. When the number of sampling points was reduced to 100, the range increased significantly, reaching 26.52, which was about 5 times that of 200 sampling points. Comprehensive analysis of ME, RMSE and ASE showed that when the number of sampling points was 150, prediction accuracy was the highest, which can satisfy the spatial variation expression of soil pH in the Yellow River Delta region.

The pH value in Kenli County was generally between 7.8 and 8.1, and the soil was alkaline. Areas with high soil pH value were distributed in the middle and southwest of the study area. With a decreased number of sampling points, the detailed characteristics of spatial variation of soil pH gradually disappear. When the number of sampling points was 150, it could not only describe the spatial distribution of soil pH in detail, but also accurately describe the spatial distribution pattern of soil pH.

Therefore, considering prediction accuracy, spatial distribution and research funding, it is suggested that the reasonable sampling number should be no less than 150 in the Yellow River Delta region, which is equivalent to at least 107 sampling points for 1000 km^2.

The results of this study are applicable to areas with environmental and soil conditions similar to those in the Yellow River Delta and have reference significance for these areas. However, the number of sampling points will be different in other areas to reasonably express the spatial distribution of soil pH, and needs to be analyzed in combination with local environmental conditions.

Author Contributions: Conceptualization, Y.Z. and X.Z.; data curation, Y.Z. and M.D.; investigation, S.L. and X.Z.; methodology, M.D.; project administration, X.Z.; software, M.D.; supervision, S.L., X.Z. and X.S.; validation, M.D.; visualization, Y.Z.; writing–original draft, Y.Z.; writing–review & editing, X.Z., X.S. and D.C.; funding acquisition, X.S. and D.C. All authors have read and agreed to the published version of the manuscript.

Funding: This research was funded by the National Key Research and Development Program of China (grant number 2021YFD1900900), the Key Research and Development Program of Shandong Province, China (grant number 2021CXGC010801, 2021CXGC010804), Shandong Province Modern Agricultural Industry Technology System Cotton Post Innovation Team (grant number SDAIT-03-06), and the Talent Fund of Qingdao Agricultural University, China (grant number 1114344).

Conflicts of Interest: The authors declare no conflict of interest.

References

1. Xu, H.; Zhang, C. Investigating spatially varying relationships between total organic carbon contents and pH values in European agricultural soil using geographically weighted regression. *Sci. Total Environ.* **2020**, *752*, 141977. [CrossRef] [PubMed]
2. Kang, E.; Li, Y.; Zhang, X.; Yan, Z.; Wu, H.; Li, M.; Yan, L.; Zhang, K.; Wang, J.; Kang, X. Soil pH and nutrients shape the vertical distribution of microbial communities in an alpine wetland. *Sci. Total Environ.* **2021**, *774*, 145780. [CrossRef]
3. Vanderlinden, K.; Polo, M.J.; Ordóez, R.; Giráldez, J. Spatiotemporal evolution of soil pH and zinc after the Aznalcóllar mine spill. *J. Environ. Qual.* **2006**, *35*, 37–49. [CrossRef] [PubMed]
4. Zhang, Z.; Yu, D.; Shi, X.; Weindorf, D.; Wang, X.; Tan, M. Effect of sampling classification patterns on soc variability in the red soil region, China. *Soil Tillage Res.* **2010**, *110*, 2–7. [CrossRef]
5. Sun, W.; Zhao, Y.; Huang, B.; Shi, X.; Darilek, J.L.; Yang, J.; Wang, Z.; Zhang, B. Effect of sampling density on regional soil organic carbon estimation for cultivated soils. *J. Plant Nutr. Soil Sci.* **2012**, *175*, 671–680. [CrossRef]
6. Duan, M.; Zhang, X. Using remote sensing to identify soil types based on multiscale image texture features. *Comput. Electron. Agric.* **2021**, *187*, 106272. [CrossRef]
7. Molla, A.; Zuo, S.; Zhang, W.; Qiu, Y.; Ren, Y.; Han, J. Optimal spatial sampling design for monitoring potentially toxic elements pollution on urban green space soil: A spatial simulated annealing and k-means integrated approach. *Sci. Total Environ.* **2022**, *802*, 149728. [CrossRef]
8. Bhunia, G.; Shit, P.; Maiti, R. Comparison of GIS-based interpolation methods for spatial distribution of soil organic carbon (SOC). *J. Saudi Soc. Agric. Sci.* **2016**, *17*, 114–126. [CrossRef]
9. Bogunovic, I.; Mesic, M.; Zgorelec, Z.; Jurisic, A.; Bilandzija, D. Spatial variation of soil nutrients on sandy-loam soil. *Soil Tillage Res.* **2014**, *144*, 174–183. [CrossRef]
10. Fisonga, M.; Wang, F.; Mutambo, V. The estimation of sampling density in improving geostatistical prediction for geotechnical characterization. *Int. J. Geotech. Eng.* **2018**, *15*, 724–731. [CrossRef]
11. Wang, J.; Yang, R.; Feng, Y. Spatial variability of reconstructed soil properties and the optimization of sampling number for reclaimed land monitoring in an opencast coal mine. *Arab. J. Geosci.* **2017**, *10*, 46. [CrossRef]
12. Long, J.; Liu, Y.; Xing, S.; Qiu, L.; Huang, Q.; Zhou, B.; Shen, J.; Zhang, L. Effects of sampling density on interpolation accuracy for farmland soil organic matter concentration in a large region of complex topography. *Ecol. Indic.* **2018**, *93*, 562–571. [CrossRef]
13. Zhao, Z.; Yang, Q.; Sun, D.; Ding, X.; Meng, F. Extended model prediction of high-resolution soil organic matter over a large area using limited number of field samples. *Comput. Electron. Agric.* **2020**, *169*, 105172. [CrossRef]
14. Zhang, Z.; Sun, Y.; Yu, D.; Mao, P.; Xu, L. Influence of Sampling Point Discretization on the Regional Variability of Soil Organic Carbon in the Red Soil Region, China. *Sustainability* **2018**, *10*, 3603. [CrossRef]
15. Sckh, A.; Djb, B. How many sampling points are needed to estimate the mean nitrate-n content of agricultural fields? A geostatistical simulation approach with uncertain variograms—Sciencedirect. *Geoderma* **2021**, *385*, 114816.
16. Behera, S.; Shukla, A. Spatial distribution of surface soil acidity, electrical conductivity, soil organic carbon content and exchangeable potassium, calcium and magnesium in some cropped acid soils of india. *Land Degrad. Dev.* **2015**, *26*, 71–79. [CrossRef]
17. Song, X.; Zhang, G.; Liu, F.; Li, D.; Zhao, Y. Characterization of the spatial variability of soil available zinc at various sampling densities using grouped soil type information. *Environ. Monit. Assess.* **2016**, *188*, 600. [CrossRef]
18. Liu, D.; Wang, X.; Nie, L.; Liu, H.; Wang, W. Comparison of geochemical patterns from different sampling density geochemical mapping in Altay, Xinjiang Province, China. *J. Geochem. Explor.* **2021**, *6*, 106761. [CrossRef]

19. Zhang, X.; Wang, Z.; Song, X.; Liu, P.; Li, S.; Yang, X. Effect of sampling on spatial variability in soil salinity in the Yellow River Delta Area. *Resour. Sci.* **2016**, *12*, 2375–2382.
20. Hai, N.; Zhao, Y.; Tian, K.; Huang, B.; Sun, W.; Shi, X. Effect of number sampling sites on characterization of spatial variability of soil organic matter. *Acta Pedol. Sin.* **2015**, *52*, 783–791.
21. Zhao, Q.; Zhao, Y.; Jiang, H.; Li, M.; Tang, J. Study on Spatial Variability of Soil Nutrients and Reasonable Sampling Number at County Scale. *J. Nat. Resour.* **2012**, *27*, 1382–1391.
22. Li, Y.; Zhang, H.; Chen, X.; Chen, T.; Luo, Y.; Christie, P. Distribution of heavy metals in soils of the yellow river delta: Concentrations in different soil horizons and source identification. *J. Soil Sediments* **2014**, *14*, 1158–1168. [CrossRef]
23. Jiang, M.; Xu, L.; Chen, X.; Zhu, H.; Fan, H. Soil quality assessment based on a minimum data set: A case study of a county in the typical river delta wetlands. *Sustainability* **2020**, *12*, 9033. [CrossRef]
24. Hong, S.; Gan, P.; Chen, A. Environmental controls on soil pH in planted forest and its response to nitrogen deposition. *Environ. Res.* **2019**, *172*, 159–165. [CrossRef]
25. Rossel, R.; Mcbratney, A.B. Soil chemical analytical accuracy and costs: Implications from precision agriculture. *Aust. J. Exp. Agric.* **1998**, *38*, 765–775. [CrossRef]
26. Emadi, M.; Shahriari, A.R.; Sadegh-Zadeh, F.; Seh-Bardan, B.J.; Dindarlou, A. Geostatistics-based spatial distribution of soil moisture and temperature regime classes in mazandaran province, northern Iran. *Arch. Agron. Soil Sci.* **2016**, *62*, 502–522.
27. Hamidi, M.E.; Larabi, A.; Faouzi, M.; Souissi, M. Spatial distribution of regionalized variables on reservoirs and groundwater resources based on geostatistical analysis using GIS: Case of Rmel-Oulad Ogbane aquifers (Larache, Nw Morocco). *Arab. J. Geosci.* **2018**, *11*, 104. [CrossRef]
28. Randazzo, G.; Italiano, F.; Micallef, A.; Tomasello, A.; Cassetti, F.P.; Zammit, A.; D'Amico, S.; Saliba, O.; Cascio, M.; Cavallaro, F.; et al. WebGIS Implementation for Dynamic Mapping and Visualization of Coastal Geospatial Data: A Case Study of BESS Project. *Appl. Sci.* **2021**, *11*, 8233. [CrossRef]
29. Ozsahin, E.; Ozdes, M.; Smith, A.C.; Yang, D. Remote Sensing and GIS-Based Suitability Mapping of Termite Habitat in the African Savanna: A Case Study of the Lowveld in Kruger National Park. *Land* **2022**, *11*, 803. [CrossRef]
30. Bogunovic, I.; Kisic, I.; Mesic, M.; Percin, A.; Zgorelec, Z.; Bilandžija, D.; Jonjic, A.; Pereira, P. Reducing sampling intensity in order to investigate spatial variability of soil pH, organic matter and available phosphorus using co-kriging techniques. A case study of acid soils in Eastern Croatia. *Arch. Agron. Soil Sci.* **2017**, *63*, 1852–1863. [CrossRef]
31. Oliver, M.A.; Webster, R. A tutorial guide to geostatistics: Computing and modelling variograms and kriging. *Catena* **2014**, *113*, 56–69. [CrossRef]
32. Burgess, T.M.; Webster, R. Optimal interpolation and isarithmic mapping of soil properties. II. Block kriging. *Eur. J. Soil Sci.* **2010**, *31*, 333–341. [CrossRef]
33. Yan, P.; Peng, H.; Yan, L.; Zhang, S.; Chen, A.; Lin, K. Spatial variability in soil pH and land use as the main influential factor in the red beds of the Nanxiong Basin, China. *PeerJ* **2019**, *7*, e27156v1. [CrossRef] [PubMed]
34. Kerry, R.; Oliver, M.A. Comparing sampling needs for variograms of soil properties computed by the method of moments and residual maximum likelihood. *Geoderma* **2007**, *140*, 383–396. [CrossRef]
35. Cambardella, C.A.; Moorman, T.B.; Novak, J.M.; Parkin, T.B.; Konopka, A.E. Field-scale variability of soil properties in central iowa soils. *Soil Sci. Soc. Am. J.* **1994**, *58*, 1501–1511. [CrossRef]
36. Sokal, R.R.; Thomson, J.D. *Applications of Spatial Autocorrelation in Ecology*; Springer: Berlin/Heidelberg, Germany, 1987; pp. 431–466.
37. Wang, J.G.; Zhou, W.J.; Wang, B.W.; Chen, C. Soil Sample Density and Interpolation Accuracy on County Scale—A Case Study on Soil pH. *Hunan Agric. Sci.* **2011**, *21*, 27–30.

Article

Determining Optimal Sampling Numbers to Investigate the Soil Organic Matter in a Typical County of the Yellow River Delta, China

Wenjing Wang [1,2], Mengqi Duan [1,2], Xiaoguang Zhang [1,2,*], Xiangyun Song [1,2], Xinwei Liu [1,2,*] and Dejie Cui [1,2]

1 College of Resources and Environment, Qingdao Agricultural University, Qingdao 266109, China; sxxzwwj1997@163.com (W.W.); duanmengqi28@163.com (M.D.); xsong@qau.edu.cn (X.S.); cuidejie@163.com (D.C.)
2 Qingdao Agricultural Remote Sensing Application Engineering Research Center, Qingdao 266109, China
* Correspondence: zhangxg_66@163.com (X.Z.); sdxw@163.com (X.L.); Tel.: +86-151-9200-3056 (X.Z.); +86-158-6305-6712 (X.L.)

Citation: Wang, W.; Duan, M.; Zhang, X.; Song, X.; Liu, X.; Cui, D. Determining Optimal Sampling Numbers to Investigate the Soil Organic Matter in a Typical County of the Yellow River Delta, China. *Appl. Sci.* **2022**, *12*, 6062. https://doi.org/10.3390/app12126062

Academic Editors: Dimitrios S. Paraforos, Giovanni Randazzo, Anselme Muzirafuti and Stefania Lanza

Received: 4 May 2022
Accepted: 9 June 2022
Published: 15 June 2022

Publisher's Note: MDPI stays neutral with regard to jurisdictional claims in published maps and institutional affiliations.

Copyright: © 2022 by the authors. Licensee MDPI, Basel, Switzerland. This article is an open access article distributed under the terms and conditions of the Creative Commons Attribution (CC BY) license (https://creativecommons.org/licenses/by/4.0/).

Abstract: Soil organic matter (SOM) plays a crucial role in promoting soil tillage, improving soil fertility and providing crop nutrients. Investigation and sampling are the premise and basis for understanding the spatial distribution of SOM. The number of sampling points will affect the accuracy of spatial variation of SOM. Therefore, it is important scientific work to determine a reasonable number of sampling points under the premise of ensuring accuracy. In this study, Kenli County, a typical area of the Yellow River Delta in China, was taken as an example to investigate the effect of different sampling points on spatial-variation expression of SOM. A total of 12 sample subsets (including 900 samples) were randomly sampled at equal intervals from the 900 sample points, using geographic information system (GIS) technology and geostatistical analyses to explore the optimal number of samples. The results showed that the SOM content in the study area had a lower-middle degree of variation. As the number of sample points decreased, the spatial distribution of SOM showed the gradual weakening of detail-characterization ability; and when the number of sample points was too small (<100), there was a wrong expression that was not consistent with the actual situation. The value of RMSE has no obvious regularity with the change of sample number. The values of both ME and ASE showed a significant inflection point when the number of samples was 150 and remained around 0 and 4 as the number of samples increased, respectively. Combined with the three indicators of ME, RMSE and ASE, collecting at least 150 samples can satisfy the spatial-variation expression of SOM, equivalent to 107 sample points within the area of 1000 km^2. The research results could provide important references for investigation of SOM content in areas with similar natural geographical conditions.

Keywords: soil attribute; GIS; ordinary Kriging; rational sampling numbers; spatial heterogeneity

1. Introduction

Soil organic matter (SOM) is one of the important components of soil [1]. It has the functions of providing crop nutrients, improving soil cultivability and promoting microbial activities, and it plays an important role in the quality of soil fertility. Due to the long-term influence of natural factors such as parent material, topography, climate and biology, as well as the intervention of human factors such as irrigation, fertilization and farming, the distribution of SOM within a certain area will show corresponding spatial differences [2,3].

For spatial-variation analysis of regional soil properties, spatial prediction and evaluation are often carried out through soil survey sampling and indoor laboratory assays. The accuracy of spatial prediction is related to the layout and density of sampling points. Theoretically, under the same spatial-distribution mode, the more sampling numbers and the greater the density, the better the spatial-prediction results and the higher the accuracies. Many studies have also confirmed this claim that the spatial-prediction accuracy of SOM exhibits a tendency to increase with the number of sample points [4–6]. Nevertheless, in practice, too high a sampling density can cause a waste of manpower, material resources and financial resources [7,8]. Moreover, some studies believe that blindly increasing the number of samples may not always improve the accuracy of spatial prediction [9]. Therefore, setting a reasonable number of sampling points within the region is of great significance for saving sampling costs, improving sampling efficiency and spatial-prediction accuracy and achieving efficient and sustainable utilization of soil resources.

In view of the above problems, some scientists have conducted a series of studies on the impact of soil-sampling number on the spatial variability of soil properties in different regions. When performing the spatial prediction of SOM in Guangdong Province, China, Lai et al. found that regression Kriging combined with 800 soil-sampling points could economically and accurately provide spatial-distribution information of SOM, equivalent to collecting 5 samples per 1000 km^2 [10]. Marques Jr et al. showed that during a spatial-variation study against micronutrients and aluminum in the state of Sao Paulo, Brazil, based on standardized variograms, the minimum number of sampling points available for this region is 400 (equivalent to 2 sample points collected per 1000 km^2) [11]. In addition, they believed that the minimum sampling spacing varied across topographic units. Long et al. took Fujian Province with an area of 124,000 km^2 in Southeast China as the study area [6]. They found that a reasonable number of samples per 1000 square kilometer were 11,000, 10,000 and 9000, respectively, for SOM sampling studies in different terrain areas such as valley–basin, hill–mountain, and plain–platform. Moreover, this study also showed that the spatial interpolation accuracy of the SOM was more sensitive to the sampling density under relatively simple topographic conditions. Wang et al. in typical coal-mining areas at the Loess Plateau showed that 40 soil-sampling points could guarantee the accuracy of SOM and total nitrogen spatial expression in the region (equivalent to 90,900 sample points collected per 1000 km^2 [5]. Furthermore, in the other two studies, Pang et al. and Zhang et al. analyzed the effect of sampling density changes on spatial-interpolation accuracy, using soil copper and SOM, respectively [12,13]. The results of both suggested that increasing sampling density and selection of reasonable interpolation methods facilitated accurate estimation of spatial variation in soil properties. These studies indicate that the number of reasonable sampling points required in different regional geographical environment conditions were different. Therefore, seeking an accurate and economical optimal sampling number of SOM in specific natural geographical conditions is still an issue that we currently need to pay attention to.

The Yellow River Delta (about 5400 km^2) is a region of high soil salinity along the eastern coastal areas of China, whose unique natural geographical conditions determine the uniqueness of SOM changes [14,15]. Therefore, based on soil sampling data, this paper conducts a study in Kenli County located in the Yellow River Delta, to explore effects of different sampling numbers on the spatial variability expression of soil organic matter and to further clarify the optimal number of sampling points for predicting the spatial distribution of SOM in the estuarine plain.

2. Materials and Methods

2.1. Study Area

Kenli County is located in the Yellow River Delta region in the northeast of Shandong Province, China (118°15′–119°19′ E, 37°24′–38°10′ N), affiliated with Dongying City. It has jurisdiction over 7 townships with a land area of about 2331 km^2 (Figure 1). Due to the frequent swing of the tail section of the Yellow River in history, a typical delta landform has been formed, and the terrain is slightly inclined from southwest to northeast. The county is located in the lower reaches of the Yellow River and has low terrain (most areas between 6 m~8 m), coupled with high groundwater level, and is affected by seawater infiltration, the soil salinization within the county being more serious [16]. The main soil types in

Kenli County are fluvo-aquic soil and coastal saline soil, and the corresponding soil group names of WRB are Cambisols and Solonchaks, respectively. The soil texture is mainly sandy loam, and the texture of the plough layer is sandy loam, light loam, medium loam, heavy loam and clay, among which sandy loam and light loam are the most widely distributed, accounting for more than 70% of the area. The county has a temperate monsoon climate but is significantly affected by the continental monsoon. The northwest wind prevails in winter and the southeast wind prevails in summer. The local crops mainly include winter wheat, corn, rice and cotton [17–19].

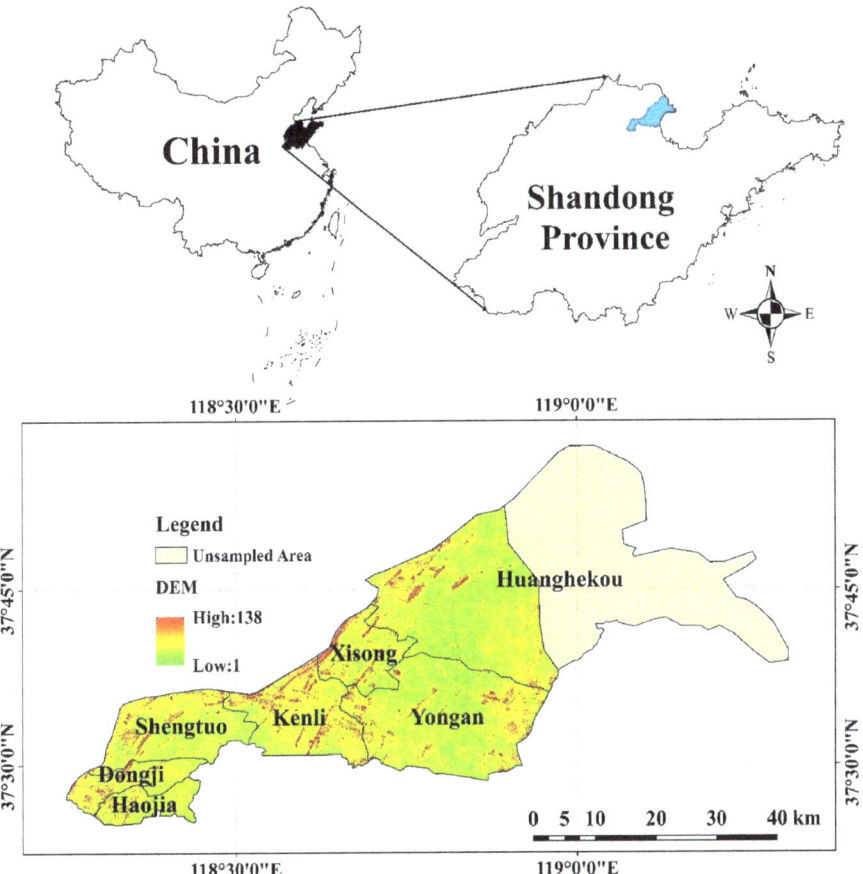

Figure 1. Location and elevation of Kenli County in Yellow River Delta.

2.2. Soil Sampling and Laboratory Analysis

Firstly, a total of 1000 sample points were uniformly deployed according to the pattern of the grid distribution, and then samples of 0~20 cm tillage layer were collected. In the actual sampling process, the number of samples in each grid was determined according to the main crop types in the grid. When a crop type was in the grid, we sampled at the center of the plot. When there were multiple crop types in the grid, we selected the plot with the larger type for sampling. At the same time, some sample points were slightly adjusted by referencing the land use and road traffic situation. The central plot of the sampling site was sampled by "S shape", and 5~10 points were collected from each site and fully mixed. 1 kg soil samples were retained by quartering method and loaded into the sample bag for soil analysis. Global Positioning System (GPS) was used to precisely locate each sampling site. The eastern area was not sampled because of conservation policy from the Yellow River Delta Nature Reserve. Finally, a total of 926 soil samples were collected in fall 2009 before and after crop harvest, and the sampling area was approximately 1400 km^2.

The collected soil samples were taken back to the laboratory for air drying, grinding and passing through a 0.25 mm nylon sieve for analysis. The soil organic carbon (SOC) was determined by the $K_2Cr_2O_7$ oxidation-titration method [20]. The content of SOM was obtained by multiplying the content of SOC by the van Benmmelen coefficient (1.724).

2.3. Data Processing and Analysis

Some outliers were removed by adding and subtracting 3 times standard deviation and yielding 900 SOM samples after removing the outliers. For analyzing the effect of different sampling-point numbers on spatial prediction, 800, 700, 600, 500, 400, 300, 200, 150, 100, 75 and 50 samples were extracted from 900 samples according to the random-sampling method at equal intervals. Adding the original set of 900 sample points, 12 different set series of sample points were formed. These sets were used to study the spatial-distribution characteristics of SOM under different sampling numbers and to study the suitable number of sampling points within the study area.

The SPSS13.0 was used to conduct descriptive statistical analysis of SOM content in 12 sample sets, which obtained multiple statistical features including mean value, median value, standard deviation, skewness coefficient, kurtosis coefficient and coefficient of variation. K-S test was performed to verify whether the SOM data conformed to normal distribution.

In predicting the spatial distribution of SOM, the semivariance model is needed to infer its spatial variation structure. In this paper, the sampling point data were calculated to obtain the theoretical model of the semivariance function by GS+ 7.0 software, and further to reflect the proportion of the random component to the structural component in the spatial variation through the Nugget ratio value.

2.4. Spatial Prediction and Validation

Expansion from point to surface was performed using the ordinary Kriging interpolation method in ArcGIS10.0, and the spatial distribution of SOM was obtained in different numbers of point.

To measure the accuracy of the SOM spatial-interpolation results at different numbers of point, validation of the spatial-interpolation results was needed. The common verification methods include cross-validation and independent validation. Because cross-validation is simple and fast, some scholars chose this method to validate the result, namely excluding a sample point and then using remaining points to predict the value on this position [21–23]. However, cross-validation could not accurately describe the prediction error of spatial interpolation in many cases, so the independent validation method was selected in this study. After extracting 800 samples from the entire 900 sample-point sets, the remaining 100 samples were taken as the validation dataset for verifying the spatial-interpolation results of the 12 sample-point sets, respectively. The mean error (ME), root-mean-square error (RMSE) and average standard error (ASE) were selected to measure the accuracy of spatial interpolation. The closer the absolute value of the ME is to 0, the smaller RMSE; and the closer the RMSE to the ASE, the higher the accuracy of spatial prediction. The calculation formulas for each verification index are as follows:

$$\text{ME} = \frac{\sum_{i=1}^{n}[z(x_i) - z^*(x_i)]}{n} \quad (1)$$

$$\text{RMSE} = \sqrt{\frac{\sum_{i=1}^{n}[z(x_i) - z^*(x_i)]^2}{n}} \quad (2)$$

$$\text{ASE} = \sqrt{\frac{\sum_{i=1}^{n} \sigma(x_i)}{n}} \quad (3)$$

where $z(x_i)$ = SOM observations, $z^*(x_i)$ = SOM predictions, $\sigma(x_i)$ = prediction standard error at the point x_i and n = the number of sampling points in the validation dataset (in this paper, n = 100).

3. Results

3.1. Descriptive Statistics of SOM

From the descriptive statistics of the SOM content at all sample points (Table 1), the minimum and maximum SOM content in the study area were 5.00 g/kg and 27.30 g/kg, respectively, and the average content was 10.96 g/kg, which belonged to the lower-middle level of soil fertility. The coefficient of variation (CV) was applied to clarify the total heterogeneity of the variables. According to the criteria proposed by Nielsen and Bouma, the coefficient of variation can be divided into low variability (CV < 0.1), moderate variability (CV 0.1–1) and high variability (CV > 1) [24]. The coefficient of variation was 35.80%, so it had a lower-middle degree of variation, indicating the small fluctuation and dispersion degree of the SOM content at the sampling site. This phenomenon further revealed that there is no particularly high variation of SOM in this study area, which may be related to the overall location of Kenli County in the Yellow River Delta and little difference in topography. Moreover, the parent material of the Yellow River Delta was transported from the Loess Plateau by the Yellow River. Under the both influence of the Yellow River water and sea water, similar fluvo-aquic soil and saline soil were formed. Thus, the same parent material type and similar soil type may also be responsible for the low degree of variability.

Table 1. Descriptive statistics of SOM content in different sampling number.

Number of Samples	Minimum /(g/kg)	Maximum /(g/kg)	Mean /(g/kg)	Standard Deviation /(g/kg)	Skewness	Kurtosis	Median /(g/kg)	Coefficient of Variation/(%)
900	5.00	27.30	10.96	3.93	0.98	4.13	10.30	35.80
800	5.00	27.30	10.89	3.99	1.05	4.30	10.20	36.60
700	5.00	27.30	10.93	4.05	1.06	4.33	10.25	37.05
600	5.00	27.30	10.97	3.99	0.96	4.09	10.35	36.40
500	5.00	25.50	10.99	3.87	0.86	3.76	10.40	35.18
400	5.00	24.60	11.10	3.80	0.67	3.25	10.60	34.27
300	5.00	24.60	11.22	3.81	0.70	3.41	10.80	33.99
200	5.00	24.60	11.23	3.89	0.82	3.65	10.80	34.63
150	5.00	23.20	11.03	3.79	0.70	3.20	10.60	34.38
100	5.20	24.60	11.49	4.18	0.88	3.47	10.60	36.42
75	5.20	24.60	11.61	4.33	0.90	3.45	10.80	37.29
50	5.80	24.60	11.56	3.99	0.99	3.93	10.75	34.49

From the extracted 11 subsets of sample points (800, 700, 600, 500, 400, 300, 200, 150, 100, 75 and 50), the maximum value of SOM at 150 samples was 23.20 g/kg, slightly smaller than the maximum at other samples, and the minimum value of SOM at 50 sample points was 5.80 g/kg, slightly greater than the minimum value of other sample points. Furthermore, for each subset, the minimum, maximum, mean, standard deviation, coefficient of variation and other statistical indicators of SOM content were closely similar to the results at 900 sample points. This indicated that the sample subsets selected in this study can still be highly representative of the whole and could essentially satisfy the requirements for the overall descriptive estimation of SOM content in Kenli County.

The mean values of all sample sets were distributed between 10.89 g/kg–11.61 g/kg, and the median values were between 10.20 g/kg–10.80 g/kg. The difference between the mean and the median value of each sample set was not large. The skewness coefficients of each sample-point set were generally distributed between 0.67–1.06, and the K-S test showed that the SOM content data under different numbers of sample points were in a moderate-skew distribution state. After logarithmic transformation, it obeyed the normal distribution (Table 2). The histogram of the normal distribution for 900 sample points is listed (Figure 2).

Table 2. Descriptive statistics of SOM content after logarithmic transformation.

Number of Samples	Minimum /(g/kg)	Maximum /(g/kg)	Mean /(g/kg)	Standard Deviation /(g/kg)	Skewness	Kurtosis	Median /(g/kg)	Coefficient of Variation/(%)
900	1.61	3.31	2.33	0.35	0.10	2.53	2.33	14.85
800	1.61	3.31	2.33	0.35	0.15	2.55	2.32	15.09
700	1.61	3.31	2.33	0.35	0.16	2.54	2.33	15.23
600	1.61	3.31	2.33	0.35	0.07	2.48	2.34	15.19
500	1.61	3.24	2.34	0.35	0.01	2.49	2.34	14.81
400	1.61	3.20	2.35	0.34	−0.10	2.42	2.36	14.68
300	1.61	3.20	2.36	0.34	−0.13	2.53	2.38	14.51
200	1.61	3.20	2.36	0.34	−0.03	2.56	2.38	14.52
150	1.61	3.14	2.34	0.35	0.10	2.43	2.36	15.12
100	1.65	3.20	2.38	0.35	0.10	2.43	2.36	14.89
75	1.65	3.20	2.39	0.36	0.11	2.45	2.38	15.17
50	1.76	3.20	2.39	0.33	0.23	2.47	2.37	13.76

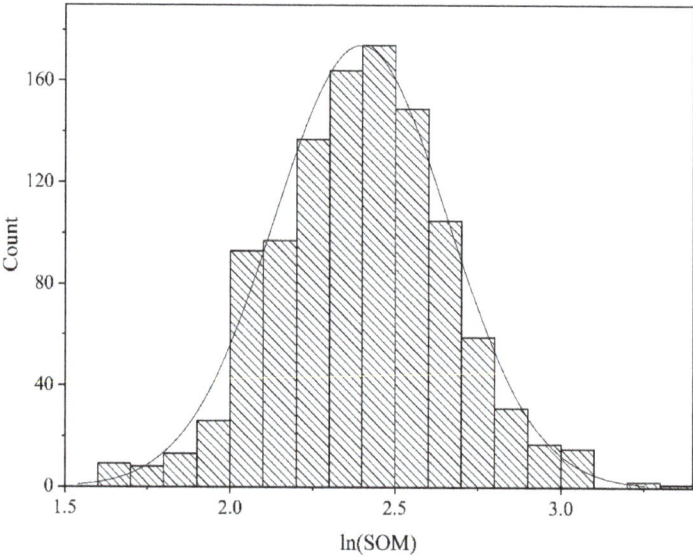

Figure 2. Histogram of normal distribution of SOM at 900 sample points after logarithmic transformation.

3.2. Characteristics of the Spatial-Variation Structure of SOM

The spatial-variability structure of the SOM can be fitted using the semivariance function. The ratio of C_0 to $(C_0 + C)$ is nugget coefficients, which indicate the proportion of the random components in the spatial-variation structure. The higher the nugget coefficient, the weaker the structural components and the stronger the random components [25]. By fitting the semivariance function of SOM under different sampling points, it was found that when the number of samples was large, the system could also greatly satisfy the spatial autocorrelation. Taking a subset of 800 sample points as an example (Table 3), the determining coefficient (R^2) of the semivariance function was 0.22. The nugget value was 9.09 and the sill value was 13.02. The nugget coefficient was reached at 69.81%, indicating that the system also had a moderate degree of autocorrelation. However, when the number of sample points decreased, the overall fitting accuracy of the semivariance function of each subset was low; the R^2 was mostly below 0.22. The value of nugget coefficient was overall higher, mostly close to 80%, and the value of nugget coefficient did not change much with the decrease in sample numbers. Therefore, we are not going to discuss the impact of the number of points on the spatial variation structure in SOM.

Table 3. The semivariance model of SOM content at 800 sample points and its fitting parameters.

Number of Samples	Model Type	Nugget (C_0)	Sill ($C_0 + C$)	Nugget Ratio (C_0/sill)	Range/(m)	R^2	RS
800	Spherical	9.09	13.02	69.81%	2780.00	0.22	14.60

3.3. Effect of Different Sampling-Point Numbers on the Prediction Accuracy of SOM Content

The closer the ME approaches 0, the higher the spatial prediction accuracy of SOM. As can be seen from Figure 3, when the number of sampling points was greater than 150, the value of ME were almost all close to 0 and varied little. The ME of the predicted value reached a minimum of −0.12 when the number of samples was 150. The value of the ME increased significantly when the number of samples was less than 150, and the ME of the predicted values at both 100 and 75 was approximately close to 1. This showed that when the number of sample points was greater than or equal to 150, Kriging interpolation method could better preserve the accuracy on the spatial prediction.

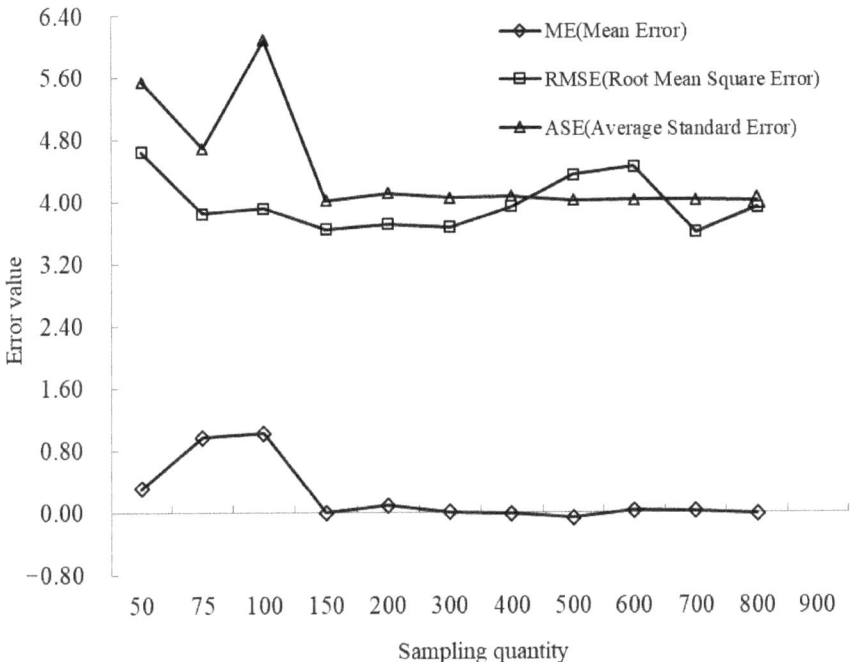

Figure 3. The spatial prediction error of SOM in different number of sampling points.

The smaller the value of RMSE, the higher the spatial prediction accuracy of the SOM. There was no obvious change trend in RMSE with the increase in sampling numbers, and all values fluctuated between 3.60 and 4.65. The value of RMSE reached the maximum when the number of sample points was 50; when the number of sample points changed from 75 to 400, the value of RMSE was relatively stable and essentially maintained at around 3.7; but when the number of sample points was greater than 400, the value of RMSE showed a certain degree of volatility. However, it did not directly explain the relationship between the increase or decrease in sample points and the spatial-prediction error of SOM. This showed that the single verification index of RMSE did not indicate the result of spatial-prediction accuracy.

ASE is an indicator used to evaluate the variability of a predicted value. If the value of ASE is equal to the value of RMSE, it means that the Kriging interpolation correctly estimates the spatial variation of SOM. The variability of spatial prediction is overestimated if ASE is greater than RMSE, and underestimated if ASE is less than RMSE. As can be seen from Figure 3, when the number of sampling points was less than 150, the values of ASE were mostly greater than 4, which were much higher than the value of RMSE. The large difference between the ASE and RMSE indicated that the variability of SOM spatial prediction was overestimated at this time, and the spatial prediction results had great uncertainty. When the number of sampling points was between 150 and 800, the value of ASE dropped to around 4.0 and remained relatively stable, and the difference between ASE and RMSE was relatively small in general. It reflected that the spatial prediction of the Kriging interpolation method was relatively stable and the prediction results were relatively accurate when the number of samples was more than or equal to 150.

Based on the above three verification indexes of spatial-prediction accuracy, it can be inferred that collecting at least 150 soil samples could meet the demand of spatial-variation expression for SOM within the Kenli County of the Yellow River Delta. It was equivalent to at least 107 sample points to be set on every 1000 km^2 area.

3.4. Effect of Sampling-Point Numbers on the Expression of Spatial Distribution of SOM Content

In order to further understand the effect of sampling-point numbers on the spatial-distribution characteristics of SOM content in the study area, an ordinary Kriging interpolation method was conducted for the 12 sample-point sets to obtain the spatial-distribution maps of SOM at different sampling-point densities (Figure 4). Since the sampling site did not cover the Yellow River Delta National Nature Reserve in the east of the study area, we did not analyze the eastern region when studying the spatial variability of the SOM.

According to Figure 4, when the numbers of samples were between 400 and 900, the SOM in the study area showed a similar spatial distribution pattern: soils with a medium content of SOM (9 g/kg~12 g/kg) occupied most of the study area; soils with relatively low content of SOM (<9 g/kg) were distributed in the southern part of the study area with the form of spots; while soils with relatively high SOM content (12 g/kg~14 g/kg) were mainly distributed in the central and northern parts with the form of discontinuous patches. Because of the smooth effect of the Kriging interpolation method, soils with SOM content greater than 14 g/kg formed narrow band regions with high value in the northeast when the number of sample points were 500 and 600. When the number of sample points dropped to 300~150, the ability to depict details was slightly weakened. Furthermore, when the blocky low-value area in the southern part of the study area was no longer obvious, the overall trend of SOM spatial variation could still be kept largely unchanged. When the sampling point number was 100, the distribution proportion of soils with relatively high SOM content (12 g/kg to 14 g/kg) was increased significantly within the study area. A large area of soils with relatively high SOM content appeared in the southern part where the SOM content was actually low, and the distribution continuity of the patches also increased. When the number of sample points dropped to 75, the SOM spatial-distribution information with the highest or lowest value could not displayed in the map. The expression information of SOM spatial distribution was relatively simple, and many details were difficult to be described. When only 50 sample points were left, there was a marked local overestimation in the southern area. Combined with the case of 100 and 75 sampling-point sets, it can be shown that when the number of sample points was less than 150, the Kriging interpolation method overestimated the variability of SOM spatial prediction, especially in the truly low-value region. These characteristics of the spatial expressions were consistent with the performance of spatial-prediction errors.

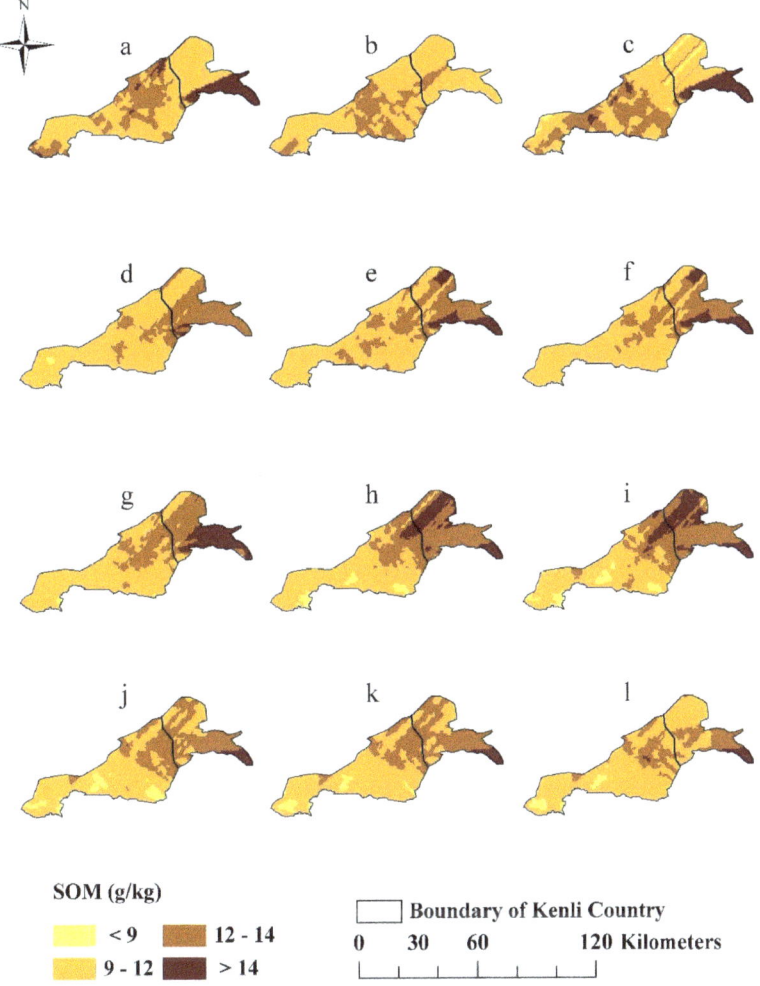

Figure 4. Spatial distribution of the SOM content under 12 sampling-point sets in Kenli County. Note: The number of sampling point sets represented by (**a–l**) is 50, 75, 100, 150, 200, 300, 400, 500, 600, 700, 800 and 900, respectively.

Overall, with the decrease in sample numbers, the spatial distribution of SOM showed the gradually weakened ability of detail characterization. Moreover, with the further reduction of the number of sample points, the predicted spatial distribution of SOM by Kriging interpolation might also have appeared to be a false characterization that was inconsistent with the actual situation. Combined with the independent verification index of SOM spatial prediction, when predicting the spatial distribution of SOM in Kenli County, it essentially required at least 150 sample points to ensure the accuracy of spatial prediction and the correctness of SOM spatial-distribution trend.

4. Discussion

4.1. Comparison of the Rational Sampling Numbers in Different Regions

In the spatial-prediction process of soil attribute, the number of sampling points or sampling-point density were important factors related to prediction accuracy and prediction cost [13]. In the area where the terrain was relatively flat and the soil salinization was more severe, such as Kenli County, it was believed that the minimum number of sample points was 150. It was equivalent to sampling more than 107 points on the area of 1000 km^2 for satisfying the prediction accuracy of SOM. This differs from the results found by other scholars at different regions and scales. For example, in a small karst-basin scale with an area of 75 km^2 in Guizhou Province, southern China, 357 sample points could better express the spatial variation of 0–20 cm soil organic carbon, equivalent to 4760 sample points collected on an area of 1000 km^2 [26]. In a small undulating hilly area with an area of 184 hm^2 in São Paulo state of southwestern Brazil, collecting one sample point per 3.75 hm^2 and one sample point per 7.2 hm^2 could satisfy the expression of the spatial variability of SOM and clay, which was equivalent to 26,667 samples and 13,889 samples collected on 1000 km^2, respectively [27]. Gascuel-Odoux et al. found that sampling 200 points could better describe spatial distribution of soil salinity within a 288 hm^2 scale in the Senegal River valley, correspond to sampling 69,444 soil samples per 1000 km^2 [28]. These studies all showed that a high number of samples were needed to express spatial variability of soil properties. It may be due to the fact that these study areas were located with undulating terrain and complex topography (e.g., hills, valleys, mountains, etc.), which resulted in high spatial variability of soil properties. Therefore, more sampling points were needed for spatial-distribution prediction.

The results in the Yangtze River Delta region showed that the reasonable sampling numbers of SOM were about 170 sample points per 1000 km^2 [29,30]. Although the topography of the Yangtze River Delta was similar to the Yellow River Delta, it had developed agriculture, high farming intensity and high geographical complexity, so the minimum sampling number was slightly higher than that of the Yellow River Delta region. In a typical agricultural planting area of Huang-Huai-Hai Plain in North China (Yucheng County), Sun et al. concluded that 82 samples points per 1000 km^2 area could satisfy the expression of the spatial variability of soil organic carbon [31]. Yucheng County belongs to a temperate monsoon climate zone and is located in the alluvial plains of the middle and lower reaches of the Yellow River. The county has similar climate and topographic characteristics as Kenli County, the study area of this paper, so the rational number of sampling points from Yucheng County was similar to that of this paper. At the same time, Yucheng County is located in the hinterland of the Huang-Huai-Hai plain, and the salinity degree was light at present, while Kenli County in this paper was affected by seawater infiltration and hydrogeology, and the soil salinity degree was relatively high. Thus, the environmental conditions of Kenli County were more complex than Yucheng, and furthermore, the minimum sampling number of SOM spatial variation was slightly more than that of Yucheng.

Therefore, the minimum sampling number of 107 sample points per 1000 km^2 determined in this paper was only suitable for areas with similar geographical environment to the Kenli County. Under other different natural geographical conditions and different agricultural production models, how many sampling points required need to be adjusted according to local actual conditions when carrying out the spatial prediction of soil properties.

4.2. Number of Sampling Points and Spatial-Prediction Error

The error of the spatial prediction was related to the number of sampling points. Some studies had shown that the spatial prediction errors of soil properties decreased as the samples number increases [13,29,32]. In this paper, the prediction accuracy of SOM was evaluated by three indicators: mean error (ME), root-mean-square error (RMSE) and mean standard error (MSE). Although the spatial-prediction error value of the SOM was missing at 900 sampling points, the trend of the predicted error changing from 50 to 800 sampling points still showed that the sample point numbers had some effect on the prediction accuracy of SOM.

The values of ME achieved the optimal prediction accuracy when the number of sample points was greater than or equal to 150, essentially keeping about 0, and there was no obvious trend with the changing of the sample point numbers. The values of RMSE did not have a particularly obvious trend with the increase in the number of samples. It had a certain degree of volatility when the sample numbers were greater than 400; in particular, the values of RMSE at 500 and 600 samples were significantly higher than those of in other sample sets. This showed that the single RMSE index did not well-indicate the prediction accuracy of SOM in this paper. The value of ASE remained around 4 when the sample-point numbers were greater than or equal to 150, and there was less impact on its trend with the increase in sampling-point numbers. At this time, the differences between ASE and RMSE were small, and the accuracies of spatial prediction were improved. It indicated that the accuracies of spatial prediction were stable at an acceptable level when the number of samples increased to a certain value, and the accuracies of spatial prediction were not significantly improved by the further increase in the number of samples. According to the study of Wu et al., this situation may be due to the fact that when the number of sample points was small, the sample points could not be ensured to be distributed in "key areas" that could effectively reflect the spatial-distribution characteristics of soil properties [33]. When the sample points reached a certain number, the integrity and rationality of sample points in spatial distribution would be improved, and the spatial prediction accuracy of soil properties may have improved to a certain extent and maintained at a reasonable value.

4.3. Influencing Factors of Spatial Prediction of SOM

SOM is a very sensitive to time and space. This paper mainly studies the relationship between the number of soil samples and the spatial-prediction accuracy of SOM. However, factors such as topography [6,21], land-use patterns [34,35], vegetation types [36] and human management [2] can all affect the spatial variability of SOC or SOM. Therefore, in the future research, we should further explore the factors that affect the spatial variation of SOM in Kenli County from the perspective of different influencing factors.

5. Conclusions

Spatial distribution of SOM showed a moderate variability in Kenli County, which is located in the Yellow River Delta, and both the spatial-prediction accuracy and spatial-distribution characteristics of SOM were influenced by the number of sampling points. The mean value of the SOM content in the study area was 10.96 g/kg, with a lower-middle fertility level. In terms of spatial variability, soils with SOM content of 9 g/kg~12 g/kg occupied most of the study area; soils with SOM content <9 g/kg were distributed in the southern part of Kenli County with the shape of spots, and soils with SOM content of 12 g/kg~14 g/kg were mainly distributed in the middle and northern parts of the study area. With the decrease in sampling-point numbers, the spatial variability characteristics of SOM showed the phenomenon of gradually weakened detail characterization, information loss and wrong description. From the independent verification results of the predicted values of SOM under different sampling numbers, the RMSE value did not have obvious regularity with the changes in the sampling numbers. Both the ME and ASE showed an obvious inflection point when the number of samples was 150, and remained at about 0 and 4, respectively, as the sampling numbers increased.

Combined with the independent verification index and spatial-distribution characteristics, when predicting the spatial distribution of SOM in Kenli County, the rational sample points to ensure the spatial prediction accuracy of SOM numbered 150. It was equivalent to 107 sample points per 1000 km^2.

Author Contributions: Conceptualization, X.Z.; methodology, X.Z. and W.W.; software, M.D. and W.W.; validation, M.D.; formal analysis, X.Z., W.W. and M.D.; investigation, X.Z., X.S., X.L. and D.C.; resources, X.Z.; data curation, X.Z., W.W. and M.D.; writing—original draft preparation, W.W.; writing—review and editing, X.Z., M.D., W.W. and X.S.; visualization, M.D.; supervision, X.Z. and

X.L.; project administration, X.Z.; funding acquisition, X.Z., X.S., X.L. and D.C. All authors have read and agreed to the published version of the manuscript.

Funding: This research was funded by the National Key Research and Development Program of China [grant number 2021YFD1900900]; the Key Research and Development Program of Shandong Province, China [grant number 2021CXGC010801, 2021CXGC010804]; Shandong Province Modern Agricultural Industry Technology System Cotton Post Innovation Team [grant number SDAIT-03-06]; the Talent Fund of Qingdao Agricultural University, China [grant number 1114344].

Institutional Review Board Statement: Not applicable.

Informed Consent Statement: Not applicable.

Acknowledgments: The authors acknowledge all the funds that provided funding during the writing process of the paper.

Conflicts of Interest: The authors declare no conflict of interest.

References

1. Medina-Méndez, J.; Volke-Haller, V.; Cortés-Flores, J.; Galvis-Spínola, A.; Santiago-Cruz, M.D.J. Soil Organic Matter and Grain Yield of Rainfed Maize in Luvisols of Campeche, México. *Agric. Sci.* **2019**, *10*, 1602–1613. [CrossRef]
2. Huang, B.; Sun, W.X.; Zhao, Y.C.; Zhu, J.; Yang, R.Q.; Zou, Z.; Ding, F.; Su, J.P. Temporal and spatial variability of soil organic matter and total nitrogen in an agricultural ecosystem as affected by farming practices. *Geoderma* **2007**, *139*, 336–345. [CrossRef]
3. Baskan, O.; Dengiz, O.; Gunturk, A. Effects of toposequence and land use-land cover on the spatial distribution of soil properties. *Environ. Earth Sci.* **2016**, *75*, 448. [CrossRef]
4. Yu, D.S.; Zhang, Z.Q.; Yang, H.; Shi, S.Z.; Tan, M.Z.; Sun, W.S.; Wang, H.J. Effect of Soil Sampling Density on Detected Spatial Variability of Soil Organic Carbon in a Red Soil Region of China. *Pedosphere* **2011**, *21*, 207–213. [CrossRef]
5. Wang, J.M.; Yang, R.X.; Bai, Z.K. Spatial variability and sampling optimization of soil organic carbon and total nitrogen for Minesoils of the Loess Plateau using geostatistics. *Ecol. Eng.* **2015**, *82*, 159–164. [CrossRef]
6. Long, J.; Liu, Y.L.; Xing, S.H.; Qiu, L.X.; Huang, Q.; Zhou, B.Q.; Shen, J.Q.; Zhang, L.M. Effects of sampling density on interpolation accuracy for farmland soil organic matter concentration in a large region of complex topography. *Ecol. Indic.* **2018**, *93*, 562–571. [CrossRef]
7. Lütticken, R.E. Automation and standardisation of site specific soil sampling. *Precis. Agric.* **2000**, *2*, 179–188. [CrossRef]
8. Kerry, R.; Oliver, M.A. Comparing sampling needs for variograms of soil properties computed by the method of moments and residual maximum likelihood. *Geoderma* **2007**, *140*, 383–396. [CrossRef]
9. Li, Y. Can the spatial prediction of soil organic matter contents at various sampling scales be improved by using regression kriging with auxiliary information? *Geoderma* **2010**, *159*, 63–75. [CrossRef]
10. Lai, Y.Q.; Wang, H.L.; Sun, X.L. A comparison of importance of modelling method and sample size for mapping soil organic matter in guangdong, china. *Ecol. Indic.* **2021**, *126*, 107618. [CrossRef]
11. Marques, J.J.; Alleoni, L.R.F.; Teixeira, D.D.B.; Siqueira, D.S.; Pereira, G.T. Sampling planning of micronutrients and aluminium of the soils of São Paulo, Brazil. *Geodermal Reg.* **2015**, *4*, 91–99. [CrossRef]
12. Pang, S.; Li, T.X.; Wang, Y.D.; Yu, Y.H.; Li, X. Spatial interpolation and sample size optimization for soil copper(cu) investigation in cropland soil at county scale using cokriging. *Agric. Sci. China* **2009**, *8*, 1369–1377. [CrossRef]
13. Zhang, Z.Q.; Yu, D.S.; Shi, X.Z.; Wang, N.; Zhang, G.X. Priority selection rating of sampling density and interpolation method for detecting the spatial variability of soil organic carbon in China. *Environ. Earth Sci.* **2015**, *73*, 2287–2297. [CrossRef]
14. Lv, Z.Z.; Liu, G.M.; Yang, J.S.; Zhang, M.M.; He, L.D.; Shao, H.B.; Yu, S.P. Spatial Variability of Soil Salinity in Bohai Sea Coastal Wetlands, China: Partition into Four Management Zones. *Plant Biosyst. Int. J. Deal. Asp. Plant Biosyst.* **2013**, *147*, 1201–1210. [CrossRef]
15. Zhang, X.G.; Wang, Z.G.; Song, X.Y.; Liu, P.R.; Li, S.M.; Yang, X. Effect of sampling on spatial variability in soil salinity in the Yellow River Delta Area. *Resour. Sci.* **2016**, *38*, 2375–2382. (In Chinese) [CrossRef]
16. Zhao, G.X.; Yan, M. Remote sensing image based information extraction for land salinized degradation and its evolution—A case study in Kenli County of the Yellow River Delta. In Proceedings of the Sixth International Conference on Natural Computation, ICNC 2010, Yantai, China, 10–12 August 2010.
17. Jiang, M.; Xu, L.; Chen, X.; Zhu, H.; Fan, H. Soil Quality Assessment Based on a Minimum Data Set: A Case Study of a County in the Typical River Delta Wetlands. *Sustainability* **2020**, *12*, 9033. [CrossRef]
18. Liu, J.; Zhang, L.; Dong, T.; Wang, J.; Fan, Y.; Wu, H.; Geng, Q.; Yang, Q.; Zhang, Z. The Applicability of Remote Sensing Models of Soil Salinization Based on Feature Space. *Sustainability* **2021**, *13*, 13711. [CrossRef]
19. Chang, C.; Lin, F.; Zhou, X.; Zhao, G. Hyper-spectral response and estimation model of soil degradation in Kenli County, the Yellow River Delta. *PLoS ONE* **2020**, *15*, e0227594. [CrossRef]
20. Nelson, D.W.; Sommers, L.E. Total carbon, organic carbon and organic matter. In *Methods of Soil Analysis, Part 2*, 2nd ed.; Page, A.L., Miller, R.H., Keeney, D.R., Eds.; ASA and SSSA: Madison, WI, USA, 1982; Agronomy Monograph; Volume 9, pp. 534–580.

21. Hu, K.L.; Wang, S.Y.; Li, H.; Huang, F.; Li, B.G. Spatial scaling effects on variability of soil organic matter and total nitrogen in suburban Beijing. *Geoderma* **2014**, *226–227*, 54–63. [CrossRef]
22. Goovaerts, P. Geostatistical modelling of uncertainty in soil science. *Geoderma* **2001**, *103*, 3–26. [CrossRef]
23. Cerri, C.E.P.; Bernoux, M.; Chaplot, V.; Volkoff, B.; Victoria, R.L.; Melillo, J.M.; Paustian, K.; Cerri, C.C. Assessment of soil property spatial variation in an Amazon pasture: Basis for selecting an agronomic experimental area. *Geoderma* **2004**, *123*, 51–68. [CrossRef]
24. Nielsen, D.R.; Bouma, J. Soil spatial variability. In Proceedings of the Workshop of the International Society of Soil Science and Soil Science Society of America, Las Vegas, NV, USA, 30 November–1 December 1984; Pudoc: Wageningen, The Netherlands, 1985; p. 243.
25. Goovaerts, P. Geostatistics in soil science: State-of-the-art and perspectives. *Geoderma* **1999**, *89*, 1–45. [CrossRef]
26. Zhang, Z.M.; Zhou, Y.C.; Huang, X.F. Exploring the optimal sampling density to characterize spatial heterogeneity of soil carbon stocks in a Karst Region. *Agron. J.* **2020**, *113*, 99–110. [CrossRef]
27. Nanni, M.R.; Povh, F.P.; Demattê, J.A.M.; Oliveira, R.B.D.; Chicati, M.L.; Cezar, E. Optimum size in grid soil sampling for variable rate application in site-specific management. *Sci. Agric.* **2011**, *68*, 386. [CrossRef]
28. Gascuel-Odoux, C.; Boivin, P. Variability of variograms and spatial estimates due to soil sampling: A case study. *Geoderma* **1994**, *62*, 165–182. [CrossRef]
29. Hai, N.; Zhao, Y.C.; Tian, K.; Huang, B.; Sun, W.X.; Shi, X.Z. Effect of number of sampling sites on characterization of spatial variability of soil organic matter. *Acta Pedol. Sin.* **2015**, *52*, 783–791. (In Chinese) [CrossRef]
30. Wang, Z.G.; Zhao, Y.C.; Huang, B.; Darilek, J.L.; Sun, W.X. Effects of sample size on spatial characterization of soil fertility properties in an agricultural area of the Yangtze River delta region, China. *Soils* **2010**, *42*, 421–428. (In Chinese) [CrossRef]
31. Sun, W.X.; Zhao, Y.C.; Huang, B.; Shi, X.Z.; Darilek, J.L.; Yang, J.S.; Wang, Z.G.; Zhang, B.E. Effect of sampling density on regional soil organic carbon estimation for cultivated soils. *J. Plant Nutr. Soil Sci.* **2012**, *175*, 671–680. [CrossRef]
32. Heim, A.; Wehrli, L.; Eugster, W.; Schmidt, M. Effects of sampling design on the probability to detect soil carbon stock changes at the swiss carboeurope site lägeren. *Geoderma* **2009**, *149*, 347–354. [CrossRef]
33. Wu, Z.F.; Zhao, Y.F.; Cheng, D.Q.; Chen, J. Influences of sample size and spatial distribution on accuracy of predictive soil mapping on a county scale. *Acta Pedol. Sin.* **2019**, *56*, 1321–1335. (In Chinese) [CrossRef]
34. Chuai, X.; Huang, X.; Wang, W.; Zhang, M.; Lai, L.; Liao, Q. Spatial Variability of Soil Organic Carbon and Related Factors in Jiangsu Province, China. *Pedosphere* **2012**, *22*, 404–414. [CrossRef]
35. Zhang, Z.Q.; Yu, D.S.; Shi, X.Z.; Weindorf, D.C.; Wang, X.X.; Tan, M.Z. Effect of sampling classification patterns on soc variability in the red soil region, china. *Soil Tillage Res.* **2010**, *110*, 2–7. [CrossRef]
36. Wang, Y.Q.; Zhang, X.C.; Zhang, J.L.; Li, S.J. Spatial variability of soil organic carbon in a watershed on the Loess Plateau. *Pedosphere* **2009**, *19*, 486–495. [CrossRef]

Article

Using Simulated Pest Models and Biological Clustering Validation to Improve Zoning Methods in Site-Specific Pest Management

Luis Josué Méndez-Vázquez [1], Rodrigo Lasa-Covarrubias [2], Sergio Cerdeira-Estrada [3] and Andrés Lira-Noriega [4,*]

[1] Red de Estudios Moleculares Avanzados, Instituto de Ecología A. C., Xalapa 91073, Mexico; luis.mendez@posgrado.ecologia.edu.mx
[2] Red de Manejo Biorracional de Plagas y Vectores, Instituto de Ecología A. C., Xalapa 91073, Mexico; rodrigo.lasa@inecol.mx
[3] Comisión Nacional para el Conocimiento y Uso de la Biodiversidad, Mexico City 14010, Mexico; scerdeira@conabio.gob.mx
[4] CONACyT Research Fellow, Red de Estudios Moleculares Avanzados, Instituto de Ecología A. C., Xalapa 91073, Mexico
* Correspondence: andres.lira@inecol.mx

Abstract: Site-specific pest management (SSPM) is a component of precision agriculture that relies on spatially enabled agronomic data to facilitate pest control practices within management zones rather than whole fields. Recent integration of high-resolution environmental data, multivariate clustering algorithms, and species distribution modeling has facilitated the development of a novel approach to SSPM that bases zone delineation on environmentally independent subfield units with individual potential to host pest populations (eSSPM). Although the potential benefits of eSSPM are clear, methods currently described for its implementation still demand further evaluation. To offer clear insight into this matter, we used field-level environmental data from a Tahiti lime orchard and realistic simulations of six citrus pests to: (1) generate a series of virtual (i.e., controlled) infestation scenarios suitable for methodological testing purposes, (2) evaluate the utility of nested (i.e., within-cluster) partitioning essays to improve the accuracy of current eSSPM methods, and (3) implement two biological clustering validators to evaluate the performance of 10 clustering algorithms and choose appropriate numbers of management zones during field partitioning essays. Our results demonstrate that: (1) nested partitioning essays outperform zoning methods previously described in eSSPM, (2) more than one clustering algorithm tend to be necessary to generate field partition models that optimize site-specific pest control practices within crop fields, and (3) biological clustering validation is an essential addition to eSSPM zoning methods. Finally, the generated evidence was integrated into an improved workflow for within-field zone delineation with pest control purposes.

Keywords: algorithms; clustering; modeling; pest control; precision agriculture; site-specific; virtual pests

1. Introduction

Site-specific pest management (SSPM) is a component of precision agriculture (PA) that relies on spatially explicit agronomic data to facilitate pest control practices within homogeneous sub-field units (i.e., management zones or MZ) rather than whole fields. Although the integration of precision inputs such as satellite imagery and climatic records into modern-day agriculture is relatively new, the first explorations of SSPM date back to the middle 1990s when data generated in the field of integrated pest management (IPM; e.g., pest samples, estimations of pest-induced crop damage) was geographically enabled by global positioning systems (GPS) and cutting edge variable rate technologies (VRT; e.g., automated tractors, planters). Such a fusion of tools and concepts facilitated

the consolidation of site-specific insect pest management (SSIPM) [3], a multidisciplinary approach to IPM which aims to partition infested crop fields into "treatment" and "no treatment" zones based on interpolated maps of within-field pest densities and economic thresholds of tolerance to pest-induced crop damages [4–7].

Recently, the integration of high-resolution environmental data, multivariate clustering techniques, and species distribution modeling (SDM) has led to the development of "ecological site-specific pest management" (eSSPM), an ecologically oriented approach to IPM which bases the delineation of MZ on environmentally independent sub-field units with individual potential to host pest populations [8]. SDM implements a variety of statistical mechanisms (i.e., modeling algorithms) to infer the spatial distribution of species based on correlations between their known geographic occurrences and the environmental conditions associated with them [9]. eSSPM field partitioning essays (i.e., delineation of MZ) are based on the following sequential steps: (1) description of cause-effect relationships between mapped environmental variables and within-field pest distributional patterns, (2) partitioning of a target crop field into a maximum number of MZ, (3) redefinition of MZ via SDM algorithms, (4) validation of environmental independence between MZ, and (5) classification of MZ based on their potential to host pest populations [8]. Although the prospective benefits of eSSPM are relevant and straightforward (e.g., controlled pesticide use, increase of crop value), field partition models generated by this approach are prone to show different sub-optimal results such as presence zones nested within absence clusters, presence zones insensitive to differentiated levels of pest infestation, more than one pest absence zones, and inaccurately delimited MZ [8].

Different factors explain eSSPM current limitations. First, the development of field partitioning essays based on single-time implementations of multivariate clustering (MC) algorithms, since pest absence zones within a crop field can consist of more than one environment equally unsuitable for pest establishment but still recognizable as independent MZ [8]. Second, the lack of clear-cut criteria to select appropriate MC algorithms during field partitioning essays, which is essential because the final topology of field partition models used in PA is highly influenced by the clustering approach used to compute them [10]. Third, the redefinition of sub-field units using SDM algorithms, due to SDM's tendency to generate zonal models with different degrees of spatial overlap between some MZ and incomplete representation of others [8]. Finally, the selection of optimal numbers of MZ based on measurements of environmental overlap between sub-field units (i.e., Schoener's D) rather than true clustering validation indexes (CVI). CVI are equations designed to evaluate the results of clustering analyses based on the degree of congruence between natural groups and the data used to create them (i.e., internal validation) or between natural groups and some other external reference (i.e., external validation) [10,11].

The development of this paper was based on two assumptions. First, to overcome the methodological limitations currently reported for eSSPM, field partitioning essays should consist of a two-steps process (i.e., nested field partitioning) where a rough distinction between pest presence and pest absence zones (i.e., binary field partitioning) precedes the subdivision of resulting sub-field units (i.e., complementary field partitioning). Second, external validation of eSSPM field partitioning essays based on biologically interpretable CVI should facilitate the selection of optimal MC algorithms to be used and appropriate numbers of MZ to be delineated. To prove these statements: (1) we used high-resolution environmental data from a Tahiti lime orchard and realistic simulations of six common citrus pests to generate a series of virtual infestation scenarios suitable for methodological testing purposes; (2) we implemented a series of nested field partitioning essays to test their capability to minimize sub-optimal zoning results reported for current eSSPM methods, and (3) we used two biologically meaningful CVI to compare the performance of 10 MC algorithms and to determine appropriate numbers of MZ to be considered during field partitioning essays. The use of simulated pest data allowed the development of testing essays under controlled virtual scenarios, an advised condition to assess modeling methodologies in ecology since it grants researchers unrestricted access to statistical processes and evidence necessary for drawing robust conclusions about natural mechanisms [12–14].

Five main contributions are presented in this paper. First, a clear distinction between SSIPM and eSSPM as conceptually and methodologically independent implementations of SSPM. Second, robust empirical evidence regarding the utility of nested field partitioning essays on overcoming the limitations reported for current eSSPM zoning approaches. Third, solid empirical evidence regarding the utility of biologically meaningful CVI to test the performance of MC algorithms and select adequate numbers of MZ during eSSPM zoning essays. Fourth, the first precedent on the simultaneous use of multiple MC algorithms to delineate MZ within the context of PA. Finally, an up-to-date workflow that considerably improves the accuracy of zoning methods presently implemented in eSSPM (Figure 1).

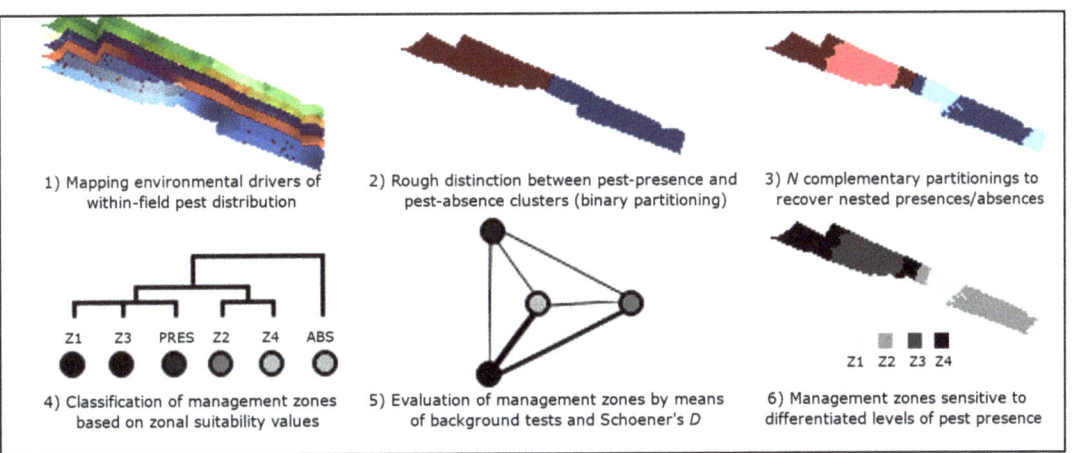

Figure 1. New workflow proposed to delineate management zones with pest control purposes.

2. Materials and Methods

2.1. Summary

The development of this work was based on the following methodological steps: (1) environmental representation of the experimental orchard by means of precision sampling tools (i.e., unmanned aerial vehicle, multispectral camera, data loggers, georeferenced soil samples, georeferenced pest samples), (2) probabilistic modelling of six virtual pests within the boundaries of the experimental orchard (i.e., phytopathogenic nematode, bacterial canker, fungal foot rot, insect-transmitted disease, invasive weed, phytophagous mite), (3) "binary field partitioning essays" to distinguish between pest presence and pest absence zones within the experimental orchard, (4) "complementary field partitioning essays" to distinguish differentiated levels of pest presence and to identify presence zones nested within absence clusters, (5) evaluation of field partition models by means of biologically meaningful CVI (i.e., biological homogeneity index or BHI and biological stability index or BSI), and (6) refinement of field partition models by means of hierarchical dendrograms (of environmental relationships between zones), bubble charts (of zonal suitability values) and visual networks (of zone environmental independence). The resulting observations were used to update zoning methods currently described in eSSPM.

2.2. Study Site

Environmental data were collected from a nine years old, artificially irrigated orchard (6.5 ha) dedicated to the commercial production of Tahiti lime (variety "Cucho," *Citrus aurantium* × *Citrus latifolia*, 6 m × 4 m between individuals) in the central region of Veracruz, Mexico. Climatic conditions associated with this region are warm humid, with abundant summer rains and annual temperatures between 24 °C and 26 °C [15]. Predominant soil types range from sandy-clay to sandy-loam, with pH values ranging from 6.62 to 6.99 [16]. Surrounding landscapes are represented by agricultural plantations (e.g., sugar cane, corn, beans) and small remnants of dry forest, which was the primary ecosystem of the region before recent agricultural expansion [17]. The selection of this study area (i.e., Carrillo Puerto municipality) was based on its contribution to regional citrus activities and abundance of small commercial plantations. In contrast, the experimental orchard was chosen on account of its available historical data (i.e., environmental, production-related).

2.3. Data Sets

Four data sets were used to represent environmental conditions within the experimental orchard (1 m^2/pixel): multispectral aerial imagery, microclimatic data logs, presence-absence data of citrus pests, and georeferenced soil samples.

Multispectral imagery was captured using an unmanned aerial system (UAS) integrated by a low-cost quadcopter (3DR Solo, discontinued in 2019) and a cheap sports camera (Ekken 4K, 60 FPS) modified with a planar lens designed for vegetation analysis (i.e., NDVI-7; red-edge 750 nm, green 500–565 nm, blue 450–485 nm). Recent studies have used similar platforms to approach vegetation biophysical features [8,18,19]. The UAS was deployed once over the experimental orchard on October 17th of 2018 approximately at zenith (between 14:00 and 14:30 h. local time) to guarantee maximum radiation conditions and minimum shadow effects. This UAS followed a photogrammetric route designed to take images with 60% side overlap and 80% vertical overlap at a flight altitude of 50 meters above the takeoff site. A set of 12 fixed ground control points (e.g., georeferenced vinyl squares on the ground) was used to facilitate aerial image spatial referencing.

Microclimatic data was sampled using nine Arduino-based data loggers assembled and programmed in the Biogeography laboratory of the Instituto de Ecología, A.C. (INECOL). Arduino is an open-source electronics prototyping platform based on simple, customizable hardware and software [20]. Data loggers were installed evenly across the orchard below fully grown trees, while ambient temperature and humidity sensors were placed 15 cm above the ground as in Méndez-Vázquez et al. [8]. Loggers were set to record information with a frequency of 60 minutes for 20 days, from October 28th to November 17th of the year 2018.

Georeferenced soil samples and presence-absence data of *Phytophthora* sp. "foot rot" and "brown rot" were collected from 73 randomly selected Tahiti lime trees. Two stages of foot rot were considered. Resinous wounds and callus tissue near the grafting area were associated with "active" and "inactive" foot rot infections, respectively [21]. After careful inspection of each tree, a soil sample of approximately 400 gr was collected from the uppermost 30 cm of the topsoil, where roots of citrus trees mainly develop [22]. Once in the laboratory, 100 gr of each available sample were used to prepare 1:5 dilutions in demineralized water as in Méndez-Vázquez et al. [8]. Such dilutions were used to measure soil pH and electrical conductivity (EC) values using a multipurpose sensor for monitoring water quality (YERYI TDS/EC/PH/TEMP meter).

2.4. Environmental Predictors of Virtual Pests

Multispectral images captured via UAS were used to generate three outputs: one multispectral orthomosaic (blue, green, and red edge bands), one digital surface model (DSM), and one digital terrain model (DTM). These products were created using Agisoft Photoscan (V1.2 for Debian Linux distributions), an image analysis software widely used in PA that facilitates the creation of maps from UAS imagery [23,24]. The native resolution of such outputs was re-scaled from 20 cm^2/pixel to 1 m^2/pixel to avoid computationally heavy processes and still achieve very high spatial resolution in our results. Four sub-products were obtained from the manipulation of spectral data. Red edge, green and blue bands of the orthomosaic were used to calculate a vegetation index highly correlated to plant metabolism and stress (i.e., single-band normalized differences vegetation index or SI-NDVI [21]). SI-NDVI (NDVI from now on) values facilitated the estimation of fractional vegetation cover (FVC) based on methods described by Thorp, Hunsaker, and French [25]. This measurement of plant area per surface unit is closely related to leaf area index and evapotranspiration [26,27]. The available DTM was used to compute maps of flow accumulation and topographic roughness index (TRI) using algorithms implemented in functions "r.terraflow" [28] and "r.tri" [29] of GRASS GIS 7 [30]. Within-field patterns of crop cover and terrain features are known environmental drivers of different agricultural pest species [27,31].

Microclimatic data logs were summarized into averages, maxima, and minima of every sampled variable (i.e., ambient temperature, relative humidity). Using logger coordinates and the inverse distance weights algorithm (IDW) implemented in the function "v.surf" of GRASS GIS 7, three raster maps were created to represent the experimental orchard in terms of average temperature, mean relative humidity, and vapor pressure deficit (VPD). This last variable is closely related to evapotranspiration and ecosystem function [32] and was calculated by implementing Allen's equation based on temperature and humidity data [33]. IDW is a statistical interpolation technique historically used to generate surface models of climatic variables and within-field pest distributions [34,35]. Although methods based on semivariograms (e.g., kriging) are a more standard approach to interpolate point-based data in PA [11,36,37], IDW is much simpler to implement, is less demanding in computing power, and is readily available in practically every GIS software today. The resulting interpolated maps were not accurate representations of the environment associated with the experimental orchard but reductionist models that simplified the evaluation of complex environment-pest interactions in geographic space. The influence exerted by climatic conditions over the distributional patterns of pests is well known [38–40].

Soil EC, soil pH and presence-absence data collected from citrus trees were also spatially interpolated using the IDW algorithm as in Corwin and Lesch [41] and Méndez-Vázquez et al. [8]. Soil pH and EC are relevant variables for agriculture due to their close relationship to crop productivity and soil physical properties, respectively [42,43]. Citrus foot rot and brown rot are different manifestations of *Phytophthora* sp. infections that affect crop productivity and facilitate the establishment of secondary diseases [44].

Twelve digital maps of environmental features were generated (in TIFF format). Variable names, codes, and methods used to compute them are presented in Table 1.

Table 1. Environmental predictors, their corresponding codes, and the estimation methods used to compute them.

Code	Variable	Estimation Method
aFRot	active citrus foot rot	IDW interpolation of presence-absence data
flowAccum	flow accumulation	"r.terraflow" function of GRASS GIS 7
cropFVC	fractional vegetation cover	FVC = (1 + NDVI)/(1 − NDVI) × NDVI^0.5
iFRot	inactive citrus foot rot	IDW interpolation of presence-absence data
relHum	mean relative humidity	IDW interpolation of data logs
sunRad	mean sub-canopy radiation	IDW interpolation of data logs
cropNDVI	single image NDVI	SI-NDVI = (NIR − BLUE)/(NIR + BLUE)
soilEC	soil electrical conductivity	IDW interpolation of soil samples
soiPH	soil pH	IDW interpolation of soil samples
TRI	topographic roughness index	"r.tri function" of GRASS GIS 7
VPD	vapor-pressure deficit	VPD = esm − ea

IDM: All IDW interpolation essays were executed using the "v.surf" function of GRASS GIS 7. esm: esm = (esmn + esmx)/2; esmn = $0.6108 \times \exp((17.27 \times \min \text{Temp})/(\min \text{Temp} + 273.3))$; esmx = $0.6108 \times \exp((17.27 \times \max \text{Temp})/(\max \text{Temp} + 273.3))$. ea = (mean RH/100) × esm.

2.5. Within-Field Distribution of Virtual Pests

Six pairs of uncorrelated predictor variables were used to simulate known distributional patterns of six virtual pests within the experimental orchard. Distributional maps of each pest are presented in Figure 2, whereas specific environmental ranges considered during pest design are shown in Table 2. Pest virtualization was performed using R statistical software's "virtualspecies" package [45].

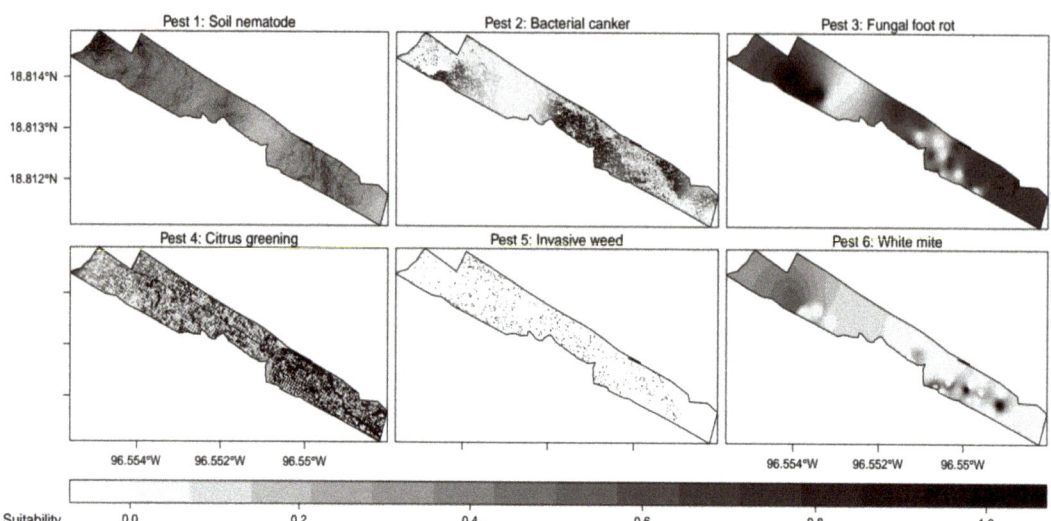

Figure 2. Distributional patterns of virtual pests (1–6) simulated within the experimental orchard. Values close to 1 (dark colors) represent regions of higher pest suitability.

Table 2. Environmental predictors, response functions, and parameterization values used to design virtual pests.

Pest	Variable 1	Fun. Var 1	Range Var 1	Variable 2	Fun. Var 2	Range Var 2
1	VPD	normal	m = 0.55, sd = 0.25	TRI	normal	m = 0.2, sd = 0.15
2	flowDir	quadratic	a = 3, b = 1, c = 0.25	sunRad	custom	m = 195, diff = 55, prob = 0.95
3	relHum	quadratic	a = 3, b = 1, c = 0.25	aFRot	logistic	beta = 0.3, alpha = 0.25
4	soilPH	logistic	beta = 10, alpha = 1	cropNDVI	normal	m = 0.05, sd = 0.1
5	cropHeight	normal	m = 1.5, sd = 0.1	ambTemp	quadratic	a = 3, b = 1, c = 0.25
6	iFRot	logistic	beta = 0.75, alpha = 0.05	soilEC	normal	m = 155, sd = 35

Distributional patterns of pest 1 were driven by temperature–humidity interactions (i.e., vapor pressure) and terrain features (i.e., topographic roughness) known to facilitate the proliferation of phytopathogenic nematodes specialized in citrus crops (i.e., *Tylenchulus semipenetrans*) [46].

Pest 2 responded to the existence of places prone to flooding (i.e., direction of flow accumulations) and low exposition to sunlight (i.e., sun radiation), where canker-producing bacteria (i.e., *Xanthomonas axonopodis*) can survive for days [47,48]. Pest 3 was inspired by fungal diseases (i.e., *Phytophthora* sp. foot rot/brown rot) that become active during the most humid months of the year (i.e., relative humidity) [21]. The distribution of pest 4 was based on an insect-transmitted disease (i.e., citrus greening) that proliferates better on citrus trees (i.e., NDVI) already exposed to physiological stress (i.e., pH) [49]. Pest 5 mimicked a generic undesired weed that invades bare soil areas of citrus orchards (i.e., FVC) when warm microclimates occur (i.e., ambient temperature). Finally, distributional patterns of pest 6 were based on those of a phytophagous mite (i.e., white/broad mite, *Polyphagotarsonemus latus*) whose populations thrive on trees damaged by previous diseases (i.e., inactive *Phytophthora* sp. foot rot) and highly stressing environmental conditions (i.e., electrical conductivity) [21].

2.6. Nested Field Partitioning Essays

As mentioned before, the field partitioning approach implemented by current eSSPM methods shows relevant shortcomings during the delineation of MZ, such as presence zones nested within absence clusters, presence zones insensitive to differentiated levels of pest infestation, more than one pest absence zones, and inaccurately delimited MZ [8]. To avoid these scenarios, MZ delineation essays implemented in this work were based on a two-step approach that we denominated nested field partitions. The first step consisted of binary field partitions where 500 random and spatially independent points representative of the experimental orchard (in terms of environmental factors relevant to the distribution of each simulated pest) were used to develop clustering essays that facilitated a rough distinction between pest presence and pest absence zones.

The second step consisted of complementary field partitions where 500 representations of presence-only and absence-only zones were used to implement within-cluster field partitions useful to identify differentiated levels of pest infestation and nested pest presence/absence zones. The main difference between binary and complementary field partitioning essays is that the former aims to partition the target crop field into two sub-field units (i.e., pest presence and pest absence clusters). In contrast, the latter seeks to partition binary sub-field units into several MZ that facilitate whether the "rescue" of nested presence/absence zones or the recognition of differentiated levels of pest infestation.

Ten MC algorithms were used to partition the experimental orchard binarily or complementarily (Table 3). These algorithms were implemented using R statistical software's "clValid" package [50] and were selected on account of their known capability to group biological data sets [10,50]. A more profound explanation of such clustering approaches is presented in Appendix A.

Table 3. Multivariate clustering (MC) algorithms compared during field partitioning essays developed in this work.

Method	Acronym	Class	Reference	Package
Average linkage	AL	hierarchical	[51]	fastcluster
Clustering large applications	CLA	partitioning	[52]	cluster
Complete linkage	CL	hierarchical	[51]	fastcluster
Divisive analysis	DIA	hierarchical	[52]	cluster
Fuzzy analysis	FNY	partitioning	[52]	cluster
Model-based clustering	MCL	model-based	[53]	mclust
Partitioning around medioids	PAM	partitioning	[52]	cluster
Self-organizing maps	SOM	machine learning	[54]	kohonen
Single linkage	SL	hierarchical	[51]	fastcluster
Ward's linkage	WL	hierarchical	[55]	fastcluster

Since it was not always possible to generate "perfect" field partition models (i.e., containing completely homogeneous clusters), binary partitions of the experimental crop field yielded one of four possible scenarios: (1) the partition model included clusters that clearly distinguished between pest presence and pest absence zones, (2) the model included one accurate absence cluster and a presence cluster that incorrectly hosted absence zones, (3) the model included one accurate presence cluster and an absence cluster that incorrectly hosted presence zones, and (4) both clusters in the model included a mixture of presence and absence zones. Complementary partitioning essays were implemented on presence-only clusters for scenarios (1) and (2). This facilitated the differentiation of pest levels (scenario 1) and the isolation of nested pest absence zones (scenario 2), depending on the case. Scenarios (3) and (4) demanded the partitioning of both pest presence and pest absence clusters, which facilitated the isolation of nested pest presence zones (scenario 3) and the distinction between pest presence and pest absence zones (scenario 4).

2.7. Validation of Field Partition Models

Selection of best field partition models (i.e., testing the performance of MC algorithms) and appropriate numbers of MZ were based on BHI and BSI indexes (0-1). BHI is an external measure for genetic clustering validation proposed by Datta and Datta [56] that determines how homogeneous clusters in a partition model are (higher values meaning a higher homogeneity) in terms of biologically meaningful categories called "functional classes" (i.e., genetic functions). In our case, surrogate environmental classes (e.g., high presence, low presence, pest absence) were generated by applying different suitability thresholds to distribution maps representative of the simulated pests. The specific number of presence levels and thresholds used to define them varied according to the case. The package "clValid" calculates BHI based on the following equation:

$$BHI(C,B) = \frac{1}{K}\sum_{k=1}^{K}\frac{1}{n_k(n_k-1)}\sum_{i\neq j \in C_k} I(B(i) = B(j)), \quad (1)$$

where n_k equals $n(C_k \cap B)$, which is the number of annotated categories (i.e., pest levels) in statistical cluster C_k, $B(i)$ is the functional class containing category i, and $B(j)$ is the functional class containing category j.

A second evaluation based on BSI measures was performed in cases where more than one field partition model shared the highest BHI values. BSI tests clustering consistency for observations with similar biological functionality (higher values meaning higher stability) [50]. To do so, new clustering essays are developed by removing one sample at a time (from the clustered data set) and cluster membership of observations with similar functional annotation is compared with cluster memberships observed during essays based on all available samples. BSI can also be calculated by the "clValid" package through the following equation:

$$BSI(C,B) = \frac{1}{F}\sum_{k=1}^{F}\frac{1}{n(B_k)(n(B_k)-1)M}\sum_{l=1}^{M}\sum_{i\neq j \in B_k}\frac{n\left(C^{i,0} \cap C^{j,l}\right)}{n\left(C^{i,0}\right)}, \quad (2)$$

where F is the total number of functional classes, $C^{i,0}$ is the statistical cluster containing observation i, and $C^{j,l}$ is the statistical cluster containing observation j when column l is removed.

After the best partition models were selected (i.e., MC algorithms and number of MZ that maximized BHI/BSI values), cluster membership numbers were interpolated within the experimental orchard using the IDW algorithm included in GRASS GIS 7 (function "v.surf"). Maps resulting from these interpolations were reclassified to eliminate decimal values and produce management zones containing unique cluster membership numbers.

2.8. Classification of Management Zones

After binary and complementary field partition models were fused to generate preliminary field partition models, management zones were categorically classified (e.g., absence, low presence, high presence) based on a decision support system consisting of hierarchical dendrograms, bubble charts, and cartographic projections.

Individual dendrograms were generated by hierarchically clustering (i.e., AL) environmental values representative of management zones included in a preliminary field partition model and true presence/absence zones known to operate within the experimental orchard. True presence and absence zones were delineated by selecting a presence-absence suitability threshold (PAST) for each virtual pest and reclassifying all values in their distribution models. All suitability values below the PAST established for a given pest were reclassified to 0, whereas those equal or above such a PAST were reclassified to 1. The resulting dendrograms were used to represent the existing relationships between MZ included in preliminary field partition models and the environmental closeness of such MZ to true presence/absence zones delimited for their corresponding pests.

Bubble charts were computed based on the mean suitability values observed within MZ included in preliminary field partition models. Zonal suitability averages were calculated by overlaying the distribution model generated for a target pest (see "Within field distribution of virtual pests") and MZ included in its corresponding partition model. Bubble size and color corresponded with their represented values (bigger/darker bubbles meant higher suitability values).

Cartographic representations of preliminary field partition models and the known distribution of their corresponding pests facilitated the interpretation of hierarchical dendrograms and bubble charts previously described.

2.9. Validation of Management Zones

According to PA theory, after a target crop field has been partitioned into n sub-field units, the appropriateness of MZ needs to be evaluated to determine whether there are real differences between them (or not) in terms of the agricultural phenomenon to be managed (e.g., soil properties, yield). Historically, this task has been accomplished by implementing strategies as simple as ANOVA models or as complex as mixed linear models (MLM). However, no standard method has been described to this date [10,11].

Since in our case MZ are expected to show individual potential to host pest populations, their represented environments are also likely to be differentiated from one another. This condition can be evaluated with SDM background tests, which are tools initially designed to measure the level of environmental overlap between SDM models generated for two species (pairwise comparisons) using Schoener's D. This index (0–1) is sensitive to ecological similarities (between geographic entities) given by diet and microhabitat variables [57,58].

In this work, the generation of zonal SDM models and the implementation of background tests was based on the "ENMTools" package for R [59], which estimates the spatial distribution of compared species based on the Maxent (i.e., maximum entropy) algorithm [60] and estimates Schoener's D with the following equation:

$$D = 1 - \frac{1}{2}\left(\sum_{ij}|Z_{1ij} - Z_{2ij}|\right), \quad (3)$$

where Z_{1ij} and Z_{2ij} represent the occupancy of entities 1 and 2, respectively.

In practice, environmental and geographic samples (i.e., environmental values, geographic coordinates; n = 500) of management zones included in preliminary field partition models were used to implement pairwise background tests that generated individual matrices of between-zone overlaps (one for each pest). Such matrices were used to feed a set of visual networks that represented management zones as labeled nodes, the environmental similarity between zones as numbers next to each link (i.e., D), and the statistical significance of a particular nexus value as the link's width. In cases where preliminary field partition models considered more than one absence zones, these were fused into a single absence cluster before MZ networks were computed.

The threshold used to determine similarities between MZ was an environmental overlap equal to or greater than 10% ($D \geq 0.1$). Regardless of the observed similarity between zones, statistically significant environmental relationships (α = 0.05) were represented as "thick" links between nodes. In contrast, statistically insignificant ones were drawn as "slim" (low similarity values below the alpha level) and "normal" (high similarity values below the alpha level) links between nodes.

The results of this exercise (i.e., environmental relationships between MZ) were used to generate a series of final field partition models that were regarded as the best possible options to facilitate pest management practices within the experimental orchard from an eSSPM perspective.

3. Results

3.1. Nested Field Partitioning Essays (Binary)

All BHI and BSI values used to evaluate binary partition models (roughly distinguish between pest presence and pest absence clusters) are presented in Figure 3. These results show that the best performing MC algorithms for pests 1 to 6 were, respectively: CL, MCL, SL, CL, SL, and WL (Figure 4). Binary field partition models that displayed BHI values approaching 1 showed excellent capability to distinguish between pest presence and pest absence zones (i.e., pest 5, pest 6). The exception to this pattern was observed in pest 2 which generated a field partition model with a BHI value of 0.93 but was unable to distinguish accurately between pest presence and pest absence zones on one half of the experimental orchard (i.e., south). It is worth noticing that the BSI value displayed by the binary partition model generated for pest 2 (i.e., 0.69) was significantly lower than BSI values observed in partition models developed for pests 5 and 6 (0.97 and 0.89 respectively).

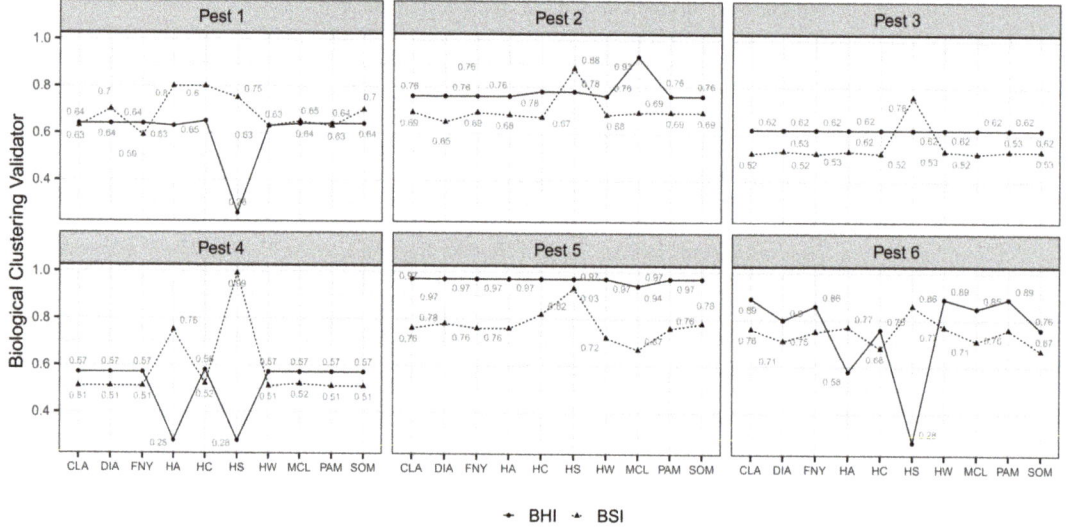

Figure 3. Biological homogeneity index (BHI) and biological suitability index (BSI) values calculated for binary partitions modeled by the compared algorithms. These indexes base partition selection on the highest observed values.

3.2. Nested Field Partitioning Essays (Complementary, Presence-Only)

Best performing MC algorithms during field partitioning essays developed within presence-only clusters are shown in Figure 5 (BHI) and Figure 6 (BSI). For pests 1 to 6 best performing algorithms were: SOM, SL, CL, MCL, DIA, and MCL (Figure 7). In this case, partitioning essays developed over homogeneous geographic entities (i.e., BHI approaching 1) were prone to recognize highly homogeneous zones (pest 2, pest 4). An exception to this pattern was observed in pest 6, where partitioning of a homogeneous presence-only cluster resulted in poorly homogeneous zones (BHI: 0.67).

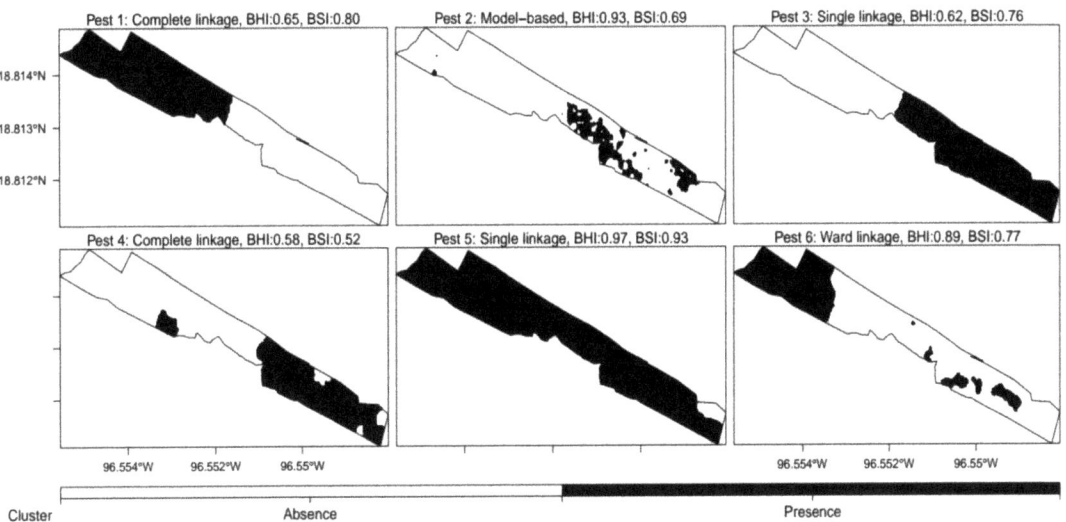

Figure 4. Best binary partition models generated for the six virtual pests simulated within the experimental orchard.

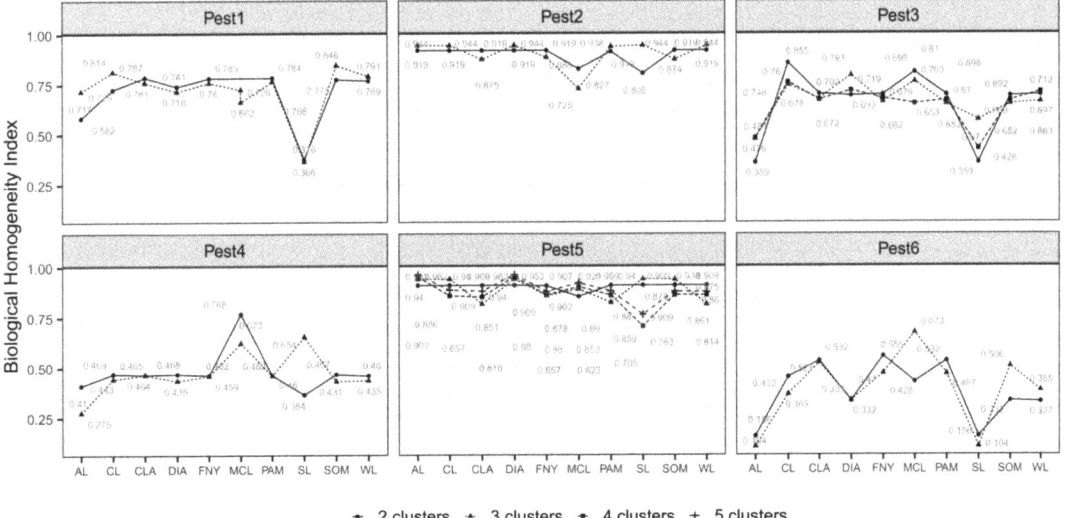

Figure 5. Biological homogeneity index (BHI) values calculated for complementary partition essays implemented over presence-only clusters of binary models. This index bases partition selection on the highest observed values.

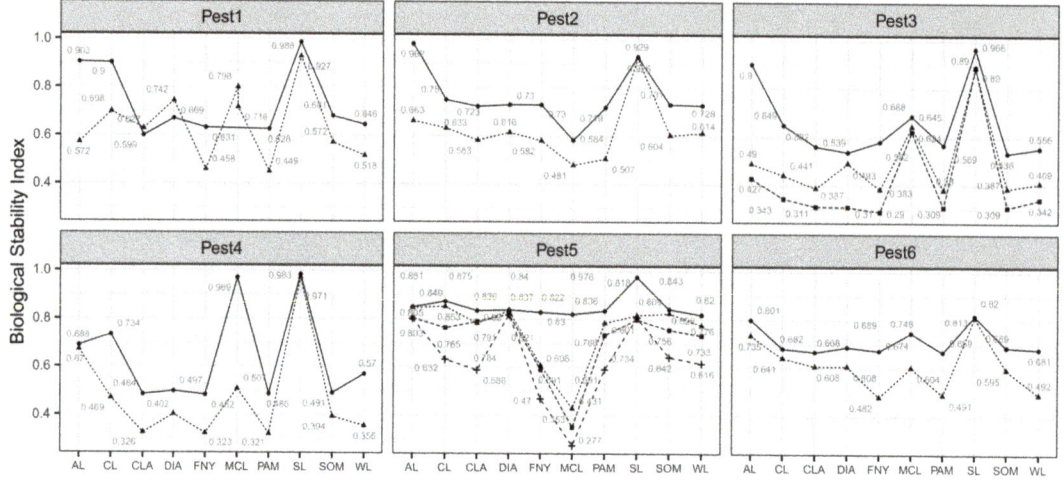

Figure 6. Biological suitability index (BSI) values calculated for complementary partition essays implemented over presence-only clusters of binary models. This index bases partition selection on the highest observed values.

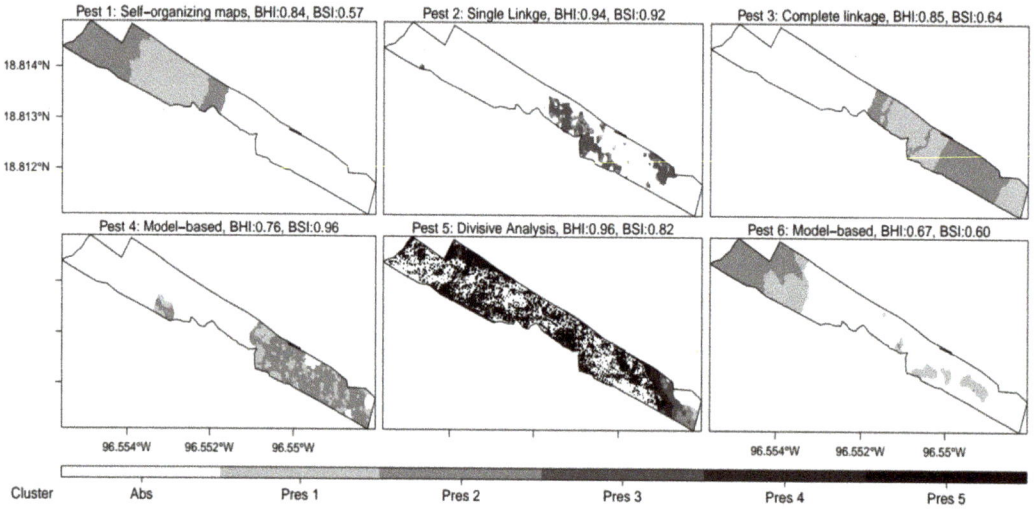

Figure 7. Best partition models of presence-only clusters used to identify differentiated levels of pest presence and isolated nested absences within the experimental orchard.

3.3. Nested Field Partitioning Essays (Complementary, Absence-Only)

Implementation of field partitioning essays within absence-only clusters was pertinent only for binary partitions of pests 1 to 4. Figures 8 and 9 show that best performing MC algorithms for these pests were: WL, SL, MCL and MCL (Figure 10). In this case, all generated field partition models displayed relatively high BHI values, even those that were developed within poorly homogeneous absence-only zones (pests 1, 3, and 4).

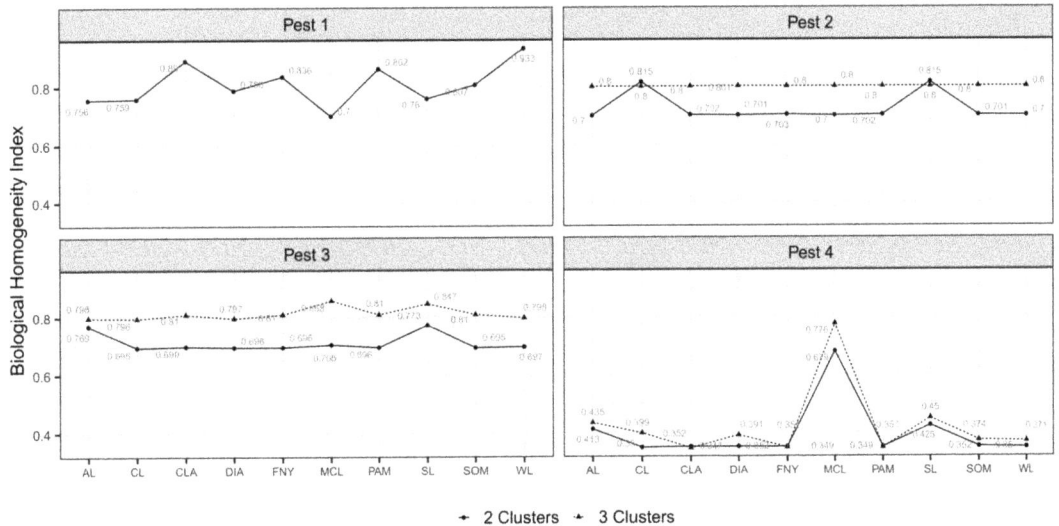

Figure 8. Biological homogeneity index (BHI) values calculated for complementary partition essays implemented over absence-only clusters of binary models. This index bases partition selection on the highest observed values.

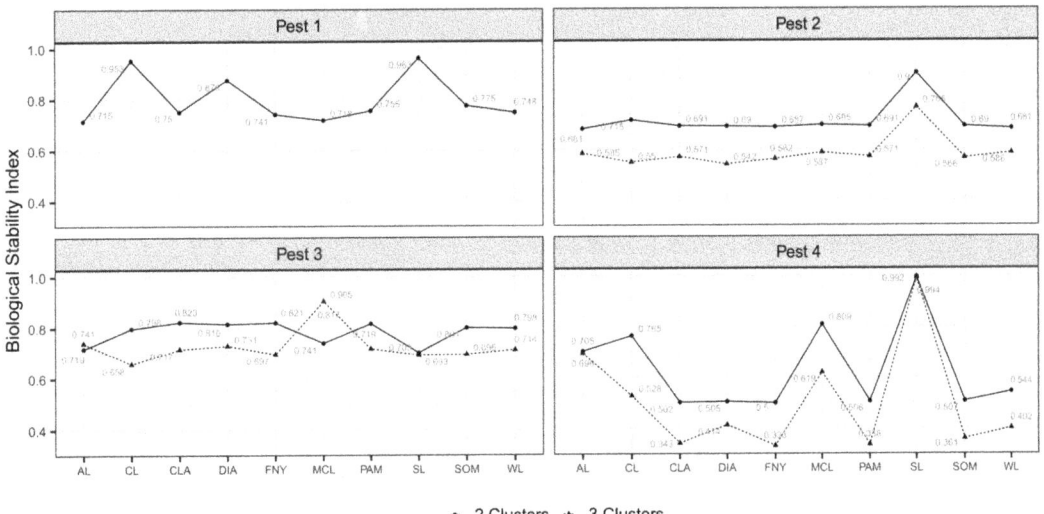

Figure 9. Biological stability index (BSI) values calculated for complementary partition essays implemented over absence-only clusters of binary models. This index bases partition selection on the highest observed values.

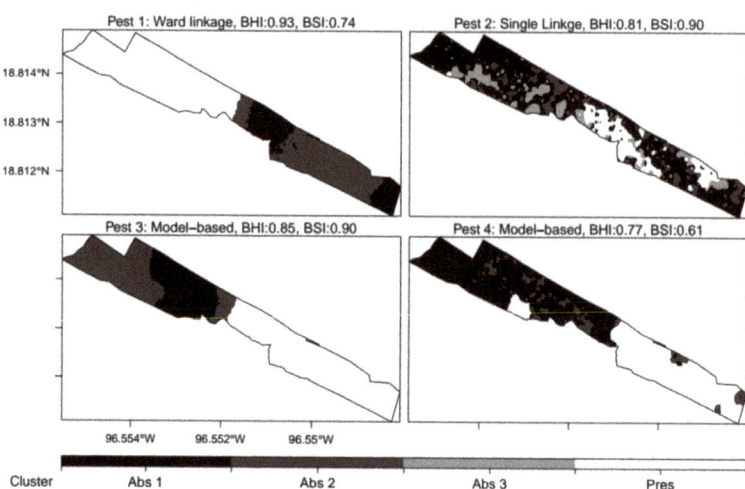

Figure 10. Best partition models of absence-only clusters used to isolate nested presences and absences from inadequate parent clusters.

3.4. Classified Management Zones

The preliminary field partition model generated for pest 1 showed good visual agreement with its corresponding distribution map (Figure 11, left). Moreover, only one absence zone was recognized (i.e., zone 1, suitability: 0.23), as well as three zones with differentiated levels of pest presence (i.e., zone 2, suitability: 0.41; zone 3, suitability: 0.43; zone 4, suitability: 0.59). This partition model also showed statistically supported environmental agreement with presence and absence zones included in the reclassified distribution model generated for pest 1. In the case of pest 2 (Figure 11, right), visual agreement between its preliminary field partition model and its corresponding reclassified distribution model was also good but partial, since a significant area of the orchard where the target pest was known to be present (roughly one-quarter of the total field) was associated with suitability values below the established threshold for presence-absence discrimination (i.e., 0.25). Three absence zones were identified (i.e., zone 1, suitability: 0.15; zone 2, suitability: 0.15; zone 4, suitability: 0.16) as well as three more zones of differentiated pest presence (i.e., zone 3, suitability: 0.41; zone 5: suitability: 0.46; zone 6, suitability: 0.57). Both sets of zones displayed high environmental agreement with true presence and absence zones included in the reclassified distribution model generated for pest 2.

The preliminary partition model generated for pest 3 (Figure 12, left) showed good visual and environmental agreement with its corresponding reclassified distribution model. Only one pest absence zone was recognized (suitability: 0.25), although a portion of it was incorrectly included in a pest presence management zone (i.e., zone 3). The highest mean suitability was observed in zone 2 (0.71), whereas those of zones 3 and 4 ranged from 0.55 to 0.64. In the case of pest 4 (Figure 12, right), the generated field partition model showed good agreement (both environmental and visual) with its corresponding reclassified pest distribution map; nevertheless, none of the delineated MZ could be classified as absence-only. Instead, four pest presence levels were recognized (zone 1, suitability: 0.32; zone 2, suitability: 0.45; zone 3, suitability: 0.45, zone 4, suitability: 0.67), two of which displayed identical mean suitability values (MZ 2 and 3) but significantly differentiated environmental conditions.

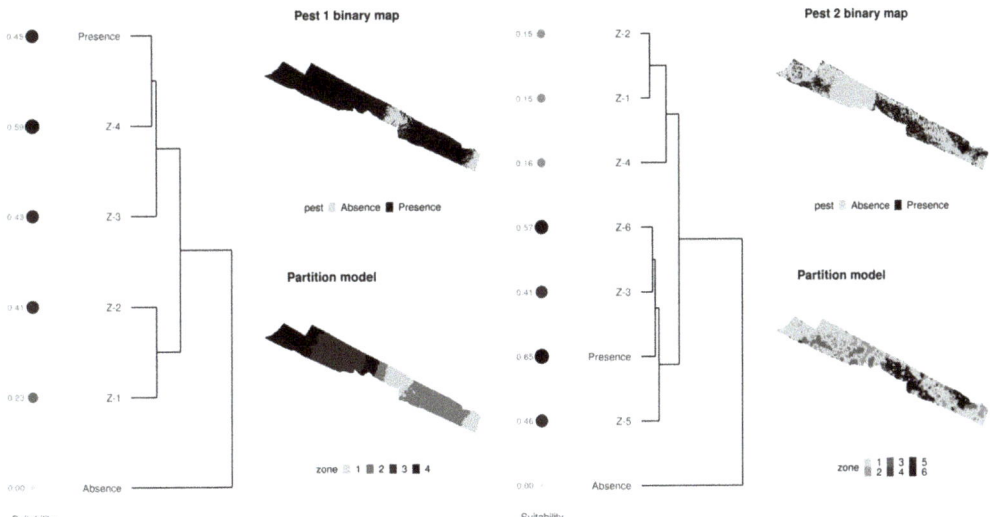

Figure 11. Multivariate chart used to classify MZ included in partition models that facilitate within-field management of pests 1 and 2. Environmental dendrogram based on Euclidean distances and average linkage.

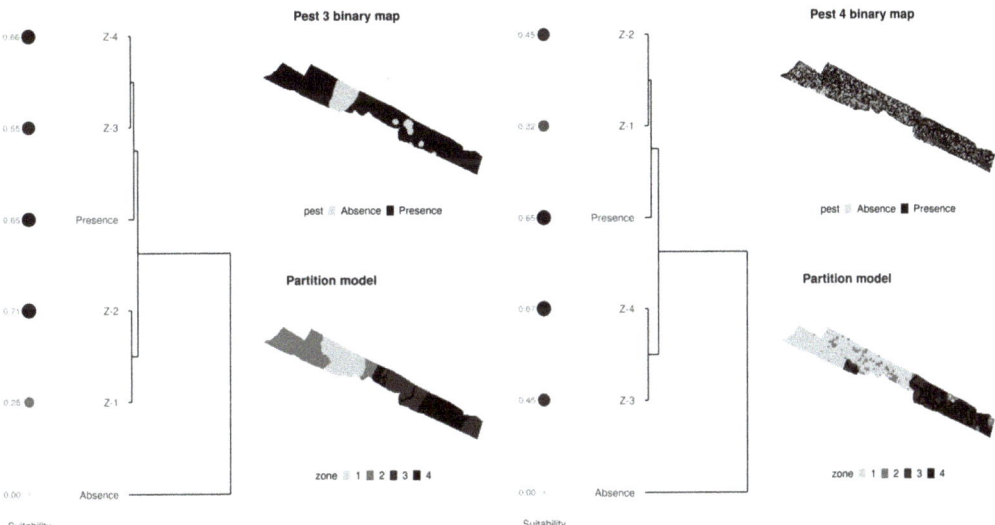

Figure 12. Multivariate chart used to classify MZ included in partition models that facilitate within-field management of pests 3 and 4. Environmental dendrogram based on Euclidean distances and average linkage.

Preliminary field partition models generated for pests 5 and 6 displayed good agreement with their corresponding reclassified pest maps. In the case of pest 5 (Figure 13, left), six MZ were delineated within the experimental orchard, with zones 1 through 4 showing mean suitability values below the presence-absence threshold previously established (suitability: 0.0). In an exceptional scenario, MZ that corresponded with known presences of pest 5 (MZ 5 and 6) also showed mean suitability values considerably below the presence-absence threshold (0.03 and 0.07 respectively), apparently as a result of natural distributional features of pest weeds. The field partition model developed for pest 6 (Figure 13, right) included three MZ, one pest absence zone (zone 1, suitability: 0.08) and two differentiated levels of presence (zone 2, suitability: 0.30; zone 3, suitability: 0.38).

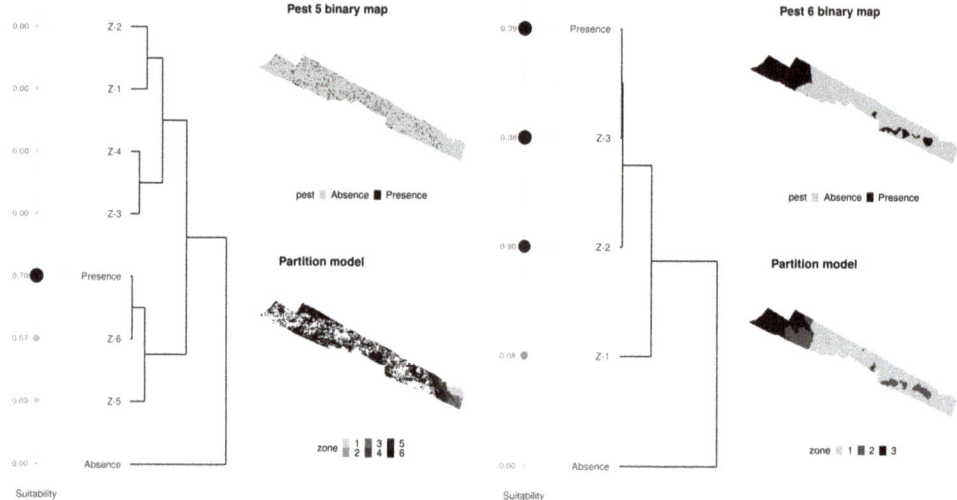

Figure 13. Multivariate chart used to classify MZ included in partition models that facilitate within-field management of pests 5 and 6. Environmental dendrogram based on Euclidean distances and average linkage.

Corrected versions of preliminary partition models (i.e., zone number according to mean suitability values, fused redundant zones) generated for all evaluated pests is presented in Figure 14.

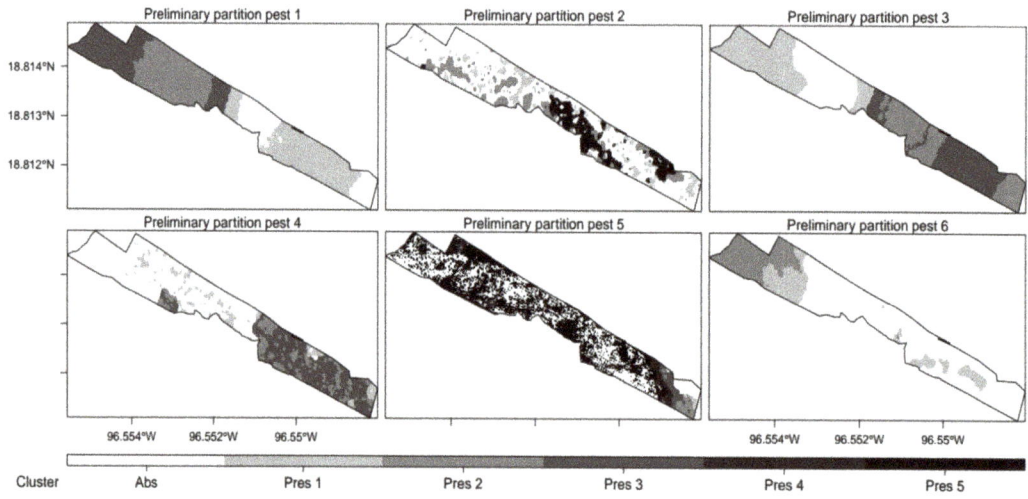

Figure 14. Preliminary models of field partition generated to facilitate management of all six virtual pests within the experimental orchard.

3.5. Validated Management Zones

For all six virtual pests, networks of environmental relationships (Figure 15) showed varying degrees of similarity between the MZ included in their corresponding field partition models, from no environmental relationships at all ($D = 0$) to completely overlapping environments ($D = 1$). However, significance values associated with such between-zone similarities (pD) support only three environmental links that show a maximum D value of 0.02 (thick lines between nodes, pests 2 and 4), which is insignificant in terms of the threshold value used to define strong environmental relationships between MZ ($D \geq 0.1$). Such a condition was interpreted as evidence of environmental independence between the MZ included in preliminary field partition models generated for all six virtual pests. This is important since zones with individual potential to host pest populations are expected to be environmentally independent from the rest [8].

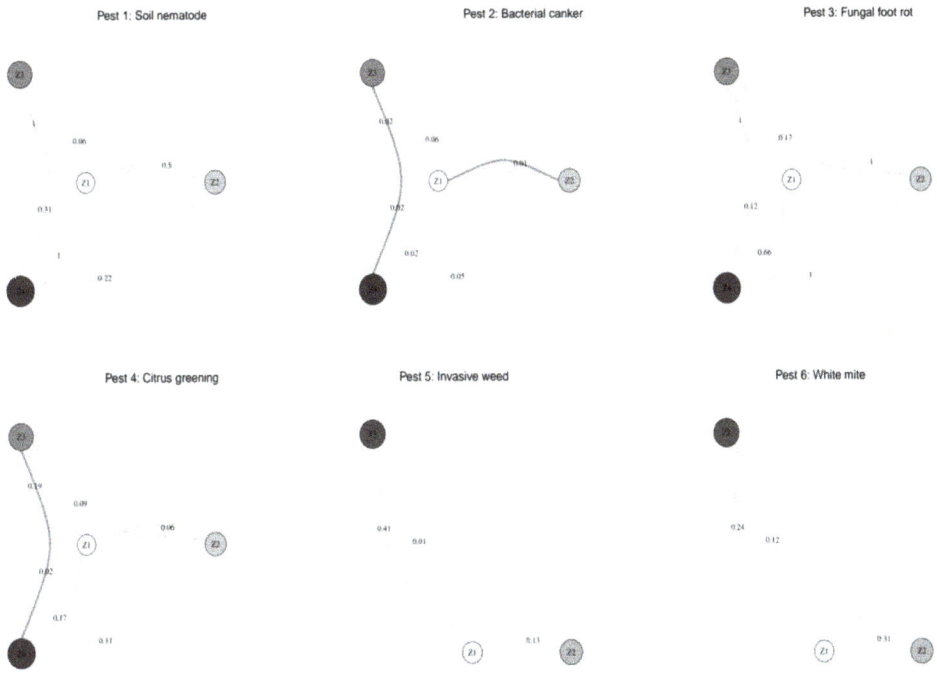

Figure 15. Environmental relations between management zones (by pest, 1–6). Zones are represented by nodes of a different color (corresponding to colors used to represent management zones during previous analyses) and size, with bigger nodes representing higher mean suitability values. Environmental distances between MZ (1-D) are represented by numbers next to network edges (i.e., links). Link width (i.e., slim, normal, thick) corresponds with the three manifestations of environmental relationships between MZ considered here. Slim edges (barely visible) represent statistically insignificant relationships which showed D values below the established threshold for recognition of strong environmental ties. Normal edges (visible but slim) represent statistically insignificant relationships where D values surpassed the established threshold for recognizing environmental bounds between zones. Thick edges represent statistically significant relationships where D values may or may not have surpassed the threshold established for the recognition of environmental bounds between zones.

Although between-zone overlaps observed in the generated networks did not justify the fusion of MZ for any of the analyzed field partition models, such similarities exist and should be considered when using such models to program/implement pest control practices within the experimental orchard. Final field partition models are presented in Figure 16.

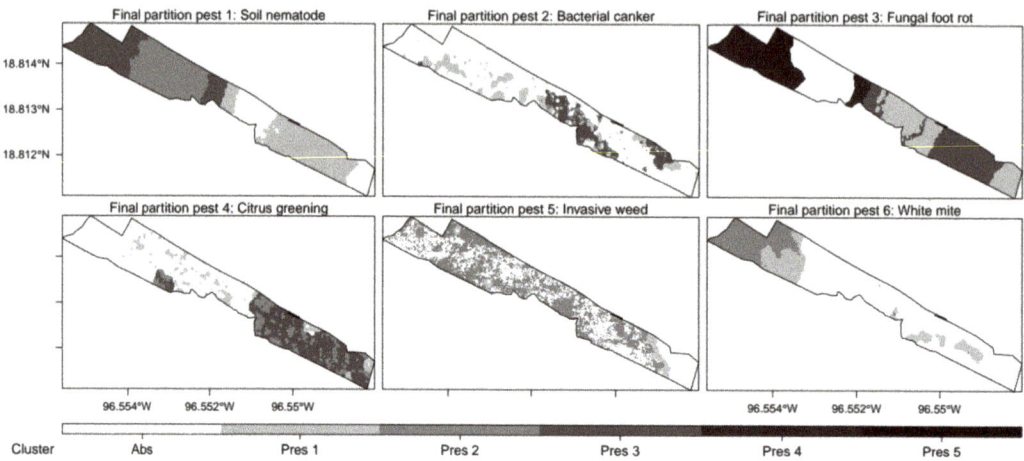

Figure 16. Final partition models generated to optimize management of virtual pests within the experimental orchard.

4. Discussion

4.1. Nested Field Partitioning Essays in eSSPM

Field partitioning strategies currently described in eSSPM are based on one-time implementations of MC algorithms capable of delineating homogeneous and environmentally independent MZ with individual potential to host pest populations [8]. Although this is an efficient approach to partition within-field variability for temporally and spatially stable agricultural phenomena (i.e., yield properties, soil conditions) [61–64], it shows clear limitations when used to partition the spatial variability of agricultural pests which are dynamic in space and time and tend to be present only in specific areas of crop fields (i.e., pest presence zones). Therefore, previous implementations of this method report field partition models with sub-optimal topologies such as presence zones nested within absence clusters, presence zones insensitive to differentiated levels of pest infestation, and more than one pest absence zones [8].

The fact that field partition models generated during the development of this work showed a low propensity to present the inconsistencies mentioned above, was interpreted as evidence that nested partitioning essays are an efficient strategy to minimize the frequency of sub-optimal results during the delineation of MZ with pest control purposes. However, it is necessary to stress the weaknesses that restrain us from claiming universal usefulness for the methods proposed in this paper. In this sense, three types of sub-optimal scenarios were observed: (1) partition models which included nested pest presences in small portions of the crop field (pests 2 and 3), (2) partition models which showed marginal suitability values even in pest presence zones (pest 5), and (3) partition models which showed a pest absence zone considerably larger than expected (pest 4).

For the first scenario, discussions should revolve around the nature of nested field partitioning essays themselves, since even though sub-optimal results in the generated field partition models (i.e., presence zones within absence clusters) were limited to small parts of the crop field, they still exert an influence over the final topologies of such models. This indicates that the two-step process proposed in this paper could be modified to an n-step process which includes as many within-cluster partitioning essays as necessary to isolate nested pest presence zones. In practice, such n-step nested partitioning essays should follow the same logic as their two-step predecessors, with biologically meaningful CVI being used to determine optimal MC algorithms to be used and the appropriate number of MZ to be delineated. However, it should be noticed that this improvement of the zoning method originally proposed will imply more processing time although computational needs will remain the same.

For scenarios two and three, discussions should focus on the type of pests intended to be managed. In these cases, sub-optimal zoning results were observed in field partition models generated for citrus greening (pest 4) and invasive weeds (pests 5) specifically. These two pests share NDVI (pest 4) or NDVI by-products (pest 5) as environmental predictors, which preconditions the crop field to show pest presence sites only where some form of vegetation is also present. However, field partitioning essays for these pests included values of all pixels representative of the experimental orchard. This form of implementation seems to have created a special condition where environmental values from places where no vegetation was present combined with values of complementary pest predictors (i.e., soil PH for pest 4, ambient temperature for pest 5) created false zones of marginal pest presence, which acted as confusion factors during the partitioning of the target crop filed. Based on these observations, we recommend that when the pest to be managed strictly needs the presence of vegetation to manifest, predictors to be used during partitioning essays should include a preprocessing that excludes all environmental values occurring in pixels where such a precondition is not fulfilled.

4.2. Performance of MC Algorithms within the Context of eSSPM

Although most PA zoning essays reported by literature are based on implementations of particular clustering approaches (e.g., fuzzy c-means, FANNY, McQuitty) [10,11], results presented here show that more than one MC algorithms tend to be necessary to delineate MZ with pest control purposes. This is because the spatial nature of data sets to be clustered during field partitioning essays could (and many times do) vary considerably depending on the pest to be managed and on the portion of the crop field being partitioned. Thus, clustering methods of proven efficiency to develop binary field partitions do not necessarily correspond to those that offer the best performance during complementary ones (whether presence-only or absence-only), even when the same pest is being considered. Similarly, any set of MC algorithms used to partition a crop field assuming a pest "A" will rarely perform accurately during partitioning essays developed for a pest "B."

Despite this result, a general pattern was observed where partitioning essays intended to distinguish between pest presence and pest absence sites (i.e., binary, absence-only) were better resolved by hierarchical MC algorithms (i.e., pests 1, 3, 4, 5, and 6), whereas more sophisticated approaches (i.e., SOM, DIA, MCL) were necessary to find differentiated levels of pest presence within general presence clusters (i.e., pests 1, 4, 5 and 6). This trend is consistent with published research that highlights the efficiency of hierarchical MC algorithms (i.e., SL, CL, WL) to partition sets of well-separated binary data (whether biologically meaningful or not) [10,65,66], as well as that of partitioning and model-based algorithms (i.e., MCL, DIA, SOM) to perform this same task with data sets that include observations more closely positioned in statistical space [50,67,68]. It must be noted, however, that plenty of other research works successfully explore the implementation of hierarchical clustering methods to partition spatially close data sets, as well as partitioning and model-based approaches to partition well separated binary data sets. Taking these observations into account, we recommend developing eSSPM field partitioning essays (i.e., binary, complementary) based on a combination of MC algorithms that best suit the particularities of the specific pests and fields to be managed.

4.3. Validation of Field Partition Models Using Biologically Meaningful CVI

A high proportion of field partitioning methods currently described in PA determine optimal numbers of management zones by means of internal and stability CVI [11,37,61,62,64,69], which are specialized algorithms designed to validate unsupervised clustering essays [70]. Internal validation uses intrinsic information in the data to assess the quality of clustering (e.g., compactness, contentedness, fuzziness performance, partition entropy, fuzziness performance), whereas stability measures evaluate the consistency of a clustering topology by comparing results from different iterations where each column (i.e., observation) is removed one at a time during the clustering process (e.g., average proportion of non-overlap, average distance, average distance between means) [50]. It is necessary to consider, however, that these examples do not deal with the delineation of management zones with pest management purposes. Instead, they focus on the management of resources such as fertilizers and water which are inputs needed when soil conditions are not favorable for plant fertility.

The task of enclosing pest populations within environmentally homogeneous zones poses different technical complexities than enclosing qualitative classes of the crop itself. For instance, zoning methods used in fertilization and irrigation PA are based on environmental features that are more stable in time (e.g., soil properties, terrain form) and agricultural phenomena that are driven almost exclusively by such factors (e.g., soil fertility, water deficit). On the other hand, in the case of eSSPM it is necessary to consider that species are living entities constantly evolving to occupy as many environments as possible within the boundaries of their physiological limitations (i.e., adaptation [71]). This means that they will rarely restrict their natural expansion to the limits of a single "climatic component" (i.e., set of environmental conditions) as demonstrated by ecological and agronomic literature [72–75]. On the contrary, species tend to occupy different climatic components simultaneously, depending on factors such as the time of the year and the immediate needs of their populations [76].

Since agroecosystems follow the same physical rules as natural ecosystems (at least in nature), the existence of differentiated microenvironmental conditions can be assumed for most agricultural fields [77,78]. We can also assume that within-field pest distribution will usually converge with more than one microenvironmental component, and that such components are not the only factors governing how pests distribute within the crop field but complementary influences that closely interact with other pest drivers such as the available resources (e.g., food abundance, mating sites) [79]. This is the main reason why we find internal and stability CVI lacking as validators of field partitioning essays in eSSPM, because of their natural tendency to favor partition models that maximize internal coherence of microclimatic components (i.e., cluster) that, although relevant, are not the only drivers of pest within-field dynamics.

BHI compensates for the influence of unknown pest drivers by seeking maximum congruence between microenvironmental components that are somewhat homogeneous in nature (i.e., clusters) and the known spatial distribution of target pests. This way it is possible to evaluate field partitioning essays in function of clusters' capability to enclose individuals of the same class (e.g., levels of pest infestation) rather than their internal structure (which excludes the influence of other factors but microenvironmental). Despite these encouraging conclusions, there are limitations in the use of BHI that need to be mentioned. For instance, when presence-absence thresholds were set too low or too high during binary field partitioning essays, BHI was unable to identify best performing partition models. Under such circumstances, BHI gave higher scores to partition models that were composed of one big cluster containing most observations and a second much smaller cluster that included few isolated values. All these models showed BHI values near to 1.

The reason for the observed phenomenon is that, at least in great measure, original implementations of BHI (and BSI) were designed to be used with genes instead of suitability classes [56]. Gene classes make sense regardless of them forming part of big or small clusters, since they all have a specific function. In our case, however, functional classes were represented by groups of places with similar potential to host pest populations (i.e., MZ). Since this potential is not an intrinsic attribute of the pest itself but of the geographic space occupied by it (i.e., distribution area), its gradation into functional intervals (i.e., pest levels) represents a rather subjective task with more than one equally plausible outcomes. This created scenarios where, even when the same MC algorithm was implemented with the same data sets, using different threshold values to define pest levels resulted in partition models with varying degrees of dissimilarity with ground truth samples (i.e., virtual data).

Although BSI was useful as a secondary validator when different models showed the highest BHI, it did not offer any means to facilitate the validation process in cases where bad model partitions showed high BHI values. To facilitate the recognition of these suboptimal partition models, we recommend the exploration of S^2_T or "total within-zone pest suitability variance". S^2_T was designed to validate field partitioning essays with crop management purposes [36] and recently adapted to the needs of site-specific pest management [8]. Since higher S^2_T values indicate more heterogeneous classes (i.e., MZ), useful field partition models should show low scores for this index. This way, when a model shows extremely high BHI and an extremely low S^2_T values, it should be regarded with caution since suitability thresholds used to define pest levels might still need proper tuning.

4.4. SDM-Based Validation of Management Zones

As mentioned before, once a crop field has been subdivided into individual MZ, it is necessary to corroborate the existence of differences between them in terms of the agricultural phenomenon to be managed. This process, known as validation of management zones, can be developed through different statistical approaches such as clustering validation indexes. CVI represent the most cited approach to the validation of MZ in PA; nevertheless, there are numerous indexes available for such purposes today and no clear agreement in terms of which one offers the best results [10]. Moreover, they are incapable of assessing agronomically meaningful differences between MZ. ANOVA tests, on the other hand, are a straightforward means to corroborate the existence of statistically significant differences between MZ. However, they are limited to the evaluation of individual variables that could or could not reflect the multidimensionality of complex environments. Additionally, they assume independence in the input dataset, a condition that is not met when environmental values are spatially referenced [11]. Finally, although MLM do account for spatial correlation in the data and consider the conjunct effect of different variables over the modeled phenomenon, they demand meticulous parameterization and their implementation tends to be more computationally intensive than other methods [11].

The validation of MZ based on SDM background tests offers different advantages. Since SDM tools (e.g., background tests) were developed to facilitate the study of species geographic distributions, they are spatially explicit in design. This is relevant because they offer *ad hoc* methodologies to minimize the effect of spatial biases usually present in SDM-related processes such as modeling species distributions and comparing environmental preferences between species/populations (e.g., differences in sampling effort between populations, spatially uneven samplings, differences in the habitat available to populations in geographic regions where they do not overlap) [58]. Moreover, SDM tools perform between-zone comparisons in terms of complex multivariate environments and determine environmental similarity based on relative rather than absolute measures (i.e., Schoener's D) [59]. These features make SDM background tests a more realistic and flexible approach to validating MZ than the rest of methodologies discussed above (at least in SSPM). Finally, the use of ecological networks to explain inter-zonal environmental differences is not only easy to set up (only two parameters are needed; i.e, a D threshold and a significance level for such a threshold) but also improves considerably the interpretability of SDM background tests when used to validate site-specific management zones with pest control purposes.

4.5. New Workflow for MZ Delineation in eSSPM

Based on the results and discussions presented during the development of this work, an improved version of current eSSPM zoning methods was described. This new workflow consists of five straightforward steps: (1) description of cause–effect relationships between georeferenced presence–absence data and mapped environmental predictors (not discussed in this paper), (2) single binary partition of a crop field validated through biologically meaningful CVI, (3) n complementary partitions of presence-only and absence-only clusters validated through biologically meaningful CVI, (4) suitability-based classification of MZ, and (5) validation of MZ based on SDM background tests.

5. Conclusions

Results generated in this work support the fact that delineating pest management zones (MZ) based on nested field partitions of environmental features represents an effective means to overcome many of the limitations associated with zoning methods previously described in eSSPM, such as presence zones insensitive to differentiated levels of infestation and nested presences/absence. In general, more than one MC algorithm is necessary to delineate accurate MZ in eSSPM. Hierarchical MC algorithms tended to generate better outputs during binary and absence-only partitioning essays, whereas more sophisticated approaches (i.e., model-based, machine learning) tended to outperform the rest during presence-only complementary partitions.

The validation of partitioning essays based on biologically meaningful CVI (i.e., BHI, BSI) are of great utility during the selection of best-performing algorithms and optimum numbers of MZ to be delineated. Still, due to inaccuracies observed in specific circumstances, we recommend complementing the evaluation of field partitioning essays with $S^2{}_T$. We also conclude that the visual aids used during the classification of MZ (e.g., dendrograms, bubble charts, maps) are important resources that facilitate relevant agronomic decisions such as what zones should be considered individual subfield units and what zones should be fused together. Similarly, SDM background tests displayed as ecological networks represent an efficient and flexible way to corroborate the environmental uniqueness of MZ and corroborate cases of fusion/separation of zones.

A novel workflow to the delineation of MZ within the context of eSSPM was described. It is based on the following sequential steps: (a) selection and mapping of zoning factors, (b) binary partition of the managed crop field, (c) complementary partitions (i.e., presence-only, absence-only) of the managed crop field, (d) classification of zones included in preliminary partition models, and (e) ecological validation of corrected (i.e., final) zones.

Author Contributions: Conceptualization, L.J.M.-V.; writing—original draft preparation, L.J.M.-V. and A.L.-N.; writing—review and editing, L.J.M.-V., A.L.-N., R.L.-C. and S.C.-E.; visualization, L.J.M.-V., A.L.-N., R.L.-C., S.C.-E.; supervision, A.L.-N.; funding acquisition, A.L.-N. and L.J.M.-V. All authors have read and agreed to the published version of the manuscript.

Funding: This research received funding from Fideicomiso o Fondo institucional de Fomento Regional para el Desarrollo Científico, Tecnológico y de Innovación (FORDECyT)-Consejo Nacional de Ciencia y Tecnología (CONACyT) project "Generación de estrategias científico-tecnológicas con un enfoque multidisciplinario e interinstitucional para afrontar la amenaza que representan los complejos ambrosiales en los sectores agrícola y forestal de México" [292399, 2018].

Institutional Review Board Statement: Not applicable.

Informed Consent Statement: Not applicable.

Acknowledgments: We thank Simón del Valle for facilitating unlimited access to his commercial orchard during the development of this research, as well as to the Méndez-Vázquez family for hosting our work team indefinite time during the development of this investigation. We also thank Luis Alberto Sánchez-Tolentino (R.I.P.) and Guadalupe Chávez-Hidalgo for their unvaluable help during fieldwork and laboratory analyses. L.J.M.-V. received scholarship support from the Mexican Consejo Nacional de Ciencia y Tecnología (CONACyT) to develop this project as part of his graduate studies.

Conflicts of Interest: The authors declare no conflict of interest.

Appendix A

A list with the abbreviations and acronyms referred to in this work are presented in Table A1 to facilitate further review.

Table A1. Meanings of abbreviations and acronyms referred to in this work.

Abbreviation	Meaning	Class
AL	average linkage	clustering algorithm
CL	complete linkage	clustering algorithm
CLA	clustering large applications	clustering algorithm
DIA	divisive analysis	clustering algorithm
FNY	fuzzy analysis	clustering algorithm
MCL	model-based clustering	clustering algorithm
PAM	partitioning around medioids	clustering algorithm
SL	single linkage	clustering algorithm
SOM	self-organizing maps	clustering algorithm
WL	Ward's linkage	clustering algorithm
SSPM	site-specific pest management	discipline
eSSPM	ecological site-specific pest management	discipline
IPM	integrated pest management	discipline
PA	precision agriculture	discipline
SDM	species distribution modeling	discipline
SSIPM	site-specific insect pest management	discipline
aFRot	active foot rot	pest driver
cropFVC	fractional vegetation cover of the research orchard	pest driver
cropHeight	height of trees included in the research orchard	pest driver
cropNDVI	normalized differences vegetation index of the research orchard	pest driver
DSM	digital surface model	pest driver
DTM	digital terrain model	pest driver
flowAccum	flow accumulation	pest driver
flowDir	flow direction	pest driver
FVC	fractional vegetation cover	pest driver
iFRot	inactive foot rot	pest driver
maxTemp	maximum ambient temperature	pest driver
minTemp	minimum ambient temperature	pest driver
NDVI	normalized differences vegetation index	pest driver
relHum	relative humidity	pest driver
SI-NDVI	single-image normalized differences vegetation index	pest driver
soilEC	soil electrical conductivity	pest driver
soilPH	soil potential of hydrogen	pest driver
sunRad	sun radiation	pest driver
TDS	total dissolved solids	pest driver
TRI	topographic roughness index	pest driver
VPD	vapor pressure deficit	pest driver
FPS	frames per second	precision agriculture tool
GIS	geographic information system	precision agriculture tool
GPS	global positioning system	precision agriculture tool
MZ	management zones	precision agriculture tool
UAS	unmanned aerial system	precision agriculture tool
ANOVA	analysis of variance	statistical method
BHI	biological homogeneity index	statistical method
BSI	biological stability index	statistical method
CVI	classification validation index	statistical method
D	Schoener's D	statistical method

Table A1. *Cont.*

Abbreviation	Meaning	Class
IDW	inverse distance weights	statistical method
MC	multivariate clustering (algorithm)	statistical method
MLM	mixed linear models	statistical method
PAST	presence–absence suitability threshold	statistical method
pD	probability of D	statistical method
S^2_T	total within-field suitability variance	statistical method

Four types of MC algorithms were used during the development of this work: hierarchical, partitioning, machine learning (based), and model based. Hierarchical clustering algorithms are based on agglomerative methods that yield a dendrogram which can be cut at a chosen height to produce the desired number of clusters [50]. Each observation is initially placed in its own cluster and the clusters are successively joined together in order of their "closeness". The closeness of any two clusters is determined by a dissimilarity matrix and can be based on a variety of agglomeration methods, which in the case on this work were:

1. Average linkage [51], mean distance between observations.
2. Complete linkage [51], maximum distance between observations.
3. Single linkage [51], minimum distance between observations.
4. Ward linkage [55], error sum of squares.

For all cases, the manipulation of the distance function exerts an influence on the combination of any two groups to form a new one [10]. Manipulation of such a distance function can be achieved through the equation:

$$D(G_x, (G_i, G_j)) = \alpha_i D(G_x, G_i) + \alpha_j D(G_x, G_j) + \beta D(G_i, G_j) + \mu |D(G_x, G_i) - D(G_x, G_j)|, \quad (A1)$$

where D is a distance function, α_i, α_j, β and μ are coefficients that have their values determined according to the applied algorithm.

A fifth hierarchical clustering approach was tested in this work:

1. DIANA (Divisive analysis [52]) is an algorithm that initially starts with all observations in a single cluster, and successively divides the clusters until each one contains a single observation; thus, hierarchies are built in $n - 1$ steps. During each step, the cluster C with the largest diameter is selected based on the following equation:

$$diam(C) := max_{i,j \in C} d(i,j) \quad (A2)$$

Assuming $diam(C) > 0$, we then split up C into two clusters A and B, according to a variant of the method of Macnaughton-Smith et al. [80]. At first $A := C$ and $B := \theta$, later one object is moved from A to B and then other objects are moved from A to B.

Partitioning algorithms divide a set of elements into k groups without constructing a hierarchical structure, following the principle that elements in a same group should be more similar than elements belonging to different groups [50]. These algorithms perform a division of the data to identify n natural groups into a certain number of disjoint groups, with a centroid for each group as a reference and employing a distance function. They can perform clustering automatically and seek to achieve the maximum similarity between the elements of the same group and the minimum similarity between different groups [10]. In our case, the following algorithms were used to implement partitioning clustering during the development of this work:

2. PAM (Partitioning around medioids [52]), similar to "k-means", the number of clusters (i.e., k) is fixed in advance and an initial set of cluster centers (i.e., "medioids", in contrast to "means" used in k-means) is required to start the algorithm. PAM is considered more robust than k-means because it admits the use of other dissimilarities

besides Euclidean distance. The implementation of PAM clustering was based on the equation:

$$TD := \sum_{i=1}^{k} \sum_{x_j \in C_i} d(x_j, m_i), \tag{A3}$$

where TD is the total deviation, defined as the sum of dissimilarities of each point $X_j \in C_1$ to medioid m_i of its cluster.

3. CLARA (Clustering large applications [52]), a sampling-based algorithm that implements PAM on a number of sub-datasets, which allows for faster running times when a number of observations is relatively large. CLARA complies with the following algorithm:

   a. Create randomly, from the original dataset, multiple subsets with fixed size (sampsize).
   b. Compute PAM algorithm on each subset and choose the corresponding k representative objects (medioids). Assign each observation of the entire data set to the closest medioid.
   c. Calculate the mean (or the sum) of the dissimilarities of the observations to their closest medioid. This is used as a measure of the goodness of the clustering.
   d. Retain the sub-dataset for which the mean (or sum) is minimal. A further analysis is carried out on the final partition.

4. FANNY (Fuzzy analysis [52]), this algorithm performs fuzzy clustering, where each observation can have partial membership in each cluster. Thus, each observation has a vector that gives the partial membership to each of the clusters. A hard cluster can be produced by assigning each observation to the cluster where it has the highest membership. FANNY clustering is based on the equation:

$$C = \sum_{v=1}^{k} \frac{\sum\sum u_{iv}^r u_{jv}^r d(i,j)}{2 \sum u_{jv}^r}, \tag{A4}$$

where u_{iv} is the membership of element i in relation to group v, n is the number of elements that form the data set, k is the number of groups to be formed, r corresponds to a pertinent exponent, and $d(i, j)$ is the distance between elements i and j.

Machine learning algorithms possess the capability of improving their performance automatically through experience. To do so, they build a mathematical model based on sample data, known as "training data", in order to make predictions or decisions without being explicitly programmed to do so. Although machine learning approaches are commonly associated to modeling and prediction tasks, these tools can also be used to develop clustering essays [81]. In this work, only one machine learning clustering algorithm was used:

5. SOM (Self-organizing maps [54]), an unsupervised learning technique based on neural networks that is popular among computational biologists and machine learning researchers. SOM is a concept of competition network that tries to find the most similar distance between the input vector and neuron with weight vector w_i. SOM always consist of both input vector x and output vector y. At the start of the learning, all the weights (w_i) are initialized to small random numbers. The set of weights forms a vector $w_i = w_{ij}$, $i = 1, 2, \ldots, k_x$, $j = 1, 2 \ldots, k_y$ where k_x is the row number and k_y is the column number. Euclidian distance d between the input vector x and the neuron with weight vector of the given neuron w_c is computed by:

$$d(x, w) = |x(t) - w_c(t)|, \tag{A5}$$

where t is an integer. Next, SOM will search for the winner neuron using the minimum distance (best matching unit, BMU). BMU is calculated as follows:

$$BMU = \operatorname{argmin} |x(t) - w_c(t)| \tag{A6}$$

To increase the similarity with the input vector, weights are adjusted after obtaining the winning neuron. The rule for updating the weight vector is given by:

$$w_i(t+1) = \frac{\sum h_{c(i)}(t) \times x_j}{\sum h_{c(i)}(t)}, \tag{A7}$$

where $w_i(t+1)$ is the updated weight vector, x_j is the input record, $h_c(i)(t)$ is the neighborhood function related to the winning unit c_i at step t, and S is the number of input samples. The neighborhood function (usually assumed as Gaussian) determines the rate of change of the neighborhood around the winner neuron as in equation:

$$h_{c(i)}(t) = e^{\frac{-\left|r_{c(i)} - r_i\right|^2}{2\sigma(t)^2}}, \tag{A8}$$

where $r_{c(i)}$ and r_i are, respectively, the positions on the map of the winning neuron and of the generic unit i; $\sigma(t)$ is the neighborhood radius at the iteration t of the training process and corresponds to the width of the neighborhood function at step t. Initially, $\sigma(t)$ can be as large as the size of the map and then, to guarantee convergence and stability, it decreases linearly with time till one during the process.

Finally, model-based clustering algorithms are those that postulate a generative statistical model for the data and then use a likelihood (or posterior probability) derived from this model as the criterion to be optimized. Model-based clustering has recently gained widespread use both for continuous and discrete domains mainly because it allows one to identify clusters based on their shape and structure rather than on proximity between data points [82]. One model-based clustering algorithm was considered in this work:

6. MCL (Model-based clustering [53]) operates on the assumption that the analyzed data originate from a finite mixture of underlying probability distributions [83]. Each mixture component represents a cluster, and the mixture components and group memberships are estimated using maximum likelihood (EM algorithm). MCL usually assumes a normal or Gaussian mixture model as in the following equation:

$$\prod_{i=1}^{n} \sum_{k=1}^{G} \tau_k \varnothing_k(x_i | \mu_k, \Sigma_k), \tag{A9}$$

where G is the number of components, x represents the data, $\varnothing_k$ are the density and parameters of the kth component in the mixture, μ_k (mean vector) and Σ_k (covariance matrix) are parameters to model each component k by the multivariate distribution, τ_k is the probability that an observation belongs to the kth component, and:

$$\varnothing_k(x_i|\mu_k,\Sigma_k) = (2\pi)^{-p/2}|\Sigma_k|^{-1/2} exp\left\{-\frac{1}{2}(x_i-\mu_k)^T \Sigma_k^1(x_i-\mu_k)\right\}. \tag{A10}$$

References

1. Strickland, R.M.; Ess, D.R.; Parsons, S.D. Precision farming and precision pest management: The power of new crop production technologies. *J. Nematol.* **1998**, *30*, 431–435.
2. Ye, X.; Sakai, K.; Manago, M.; Asada, S.; Sasao, A. Prediction of citrus yield from airborne hyperspectral imagery. *Precis. Agric.* **2007**, *8*, 111–125. [CrossRef]
3. Park, Y.; Krell, R.K.; Carroll, M. Theory, technology, and practice of site-specific insect pest management. *J. Asia Pac. Entomol.* **2007**, *10*, 89–101. [CrossRef]
4. Weisz, R.; Fleischer, S.; Smilowitz, S. Map generation in high-value horticultural integrated pest management: Appropriate interpolation methods for site-specific pest management of Colorado potato beetle (Coleoptera: Chrysomelidae). *J. Econ. Entomol.* **1995**, *88*, 1650–1657. [CrossRef]
5. Weisz, R. Site-specific integrated pest management for high-value crops: Impact on potato pest management. *J. Econ. Entomol.* **1996**, *89*, 501–509. [CrossRef]

6. Park, Y.; Tollefson, J.J. Spatial prediction of corn rootworm (Coleoptera: Chrysomelidae) adult emergence in Iowa cornfields. *J. Econ. Entomol.* **2005**, *98*, 8. [CrossRef] [PubMed]
7. Park, Y.; Perring, T.M.; Farrar, C.A.; Gispert, C. Spatial and temporal distributions of two sympatric *Homalodisca* spp. (Hemiptera: Cicadellidae): Implications for areawide pest management. *Agric. Ecosyst. Environ.* **2006**, *113*, 168–174. [CrossRef]
8. Méndez-Vázquez, L.J.; Lira-Noriega, A.; Lasa-Covarrubias, R.; Cerdeira-Estrada, S. Delineation of site-specific management zones for pest control purposes: Exploring precision agriculture and species distribution modeling approaches. *Comput. Electron. Agric.* **2019**, *167*, 105101. [CrossRef]
9. Elith, J.; Leathwick, J.R. Species distribution models: Ecological explanation and prediction across space and time. *Annu. Rev. Ecol. Evol. Syst.* **2009**, *40*, 677–697. [CrossRef]
10. Gavioli, A.; Godoy de Souza, E.; Bazzi, C.L.; Schenatto, K.; Betzek, N.M. Identification of management zones in precision agriculture: An evaluation of alternative cluster analysis methods. *Biosyst. Eng.* **2019**, *181*, 86–102. [CrossRef]
11. Córdoba, M.A.; Bruno, C.I.; Costa, J.L.; Peralta, N.R.; Balzarini, M.G. Protocol for multivariate homogeneous zone delineation in precision agriculture. *Biosyst. Eng.* **2016**, *143*, 95–107. [CrossRef]
12. Hirzel, A.H.; Helfer, V.; Metral, F. Assessing habitat-suitability models with a virtual species. *Ecol. Model.* **2001**, *145*, 111–121. [CrossRef]
13. Zurell, D.; Berger, U.; Cabral, J.S.; Jeltsch, F.; Meynard, C.N.; Münkemüller, T.; Nehrbass, N.; Pagel, J.; Reineking, B.; Schröder, B.; et al. The virtual ecologist approach: Simulating data and observers. *Oikos* **2010**, *119*, 622–635. [CrossRef]
14. Miller, J.A. Virtual species distribution models: Using simulated data to evaluate aspects of model performance. *Prog. Phys. Geogr. Earth Environ.* **2014**, *38*, 117–128. [CrossRef]
15. Hernández-Landa, L.; López-Collado, J.; Nava-Tablada, M.E.; García-García, C.G. Percepción de la problemática del Huanglongbing por agentes relevantes en zonas urbanas. *Rev. Mex. Cien. Agrícolas* **2017**, *8*, 993–1000. [CrossRef]
16. Bautista-Zúñiga, F.; Solórzano, H.R.; Durán de Bazúa, C. Caracterización y clasificación de suelos con fines productivos en Córdoba, Veracruz, México. *Investig. Geogr.* **1998**, *1*, 21–33. [CrossRef]
17. Moreno-Casasola, P.; Paradowska, K. Especies útiles de la selva baja caducifolia en las dunas costeras del centro de Veracruz. *Madera Bosques* **2009**, *15*, 21–44. [CrossRef]
18. Beisel, N.S.; Callahan, J.B.; Sng, N.J.; Taylor, D.J.; Paul, A.; Ferl, R.J. Utilization of single-image normalized difference vegetation index (SI-NDVI) for early plant stress detection. *Appl. Plant Sci.* **2018**, *6*, e01186. [CrossRef]
19. Malone, M.; Foster, E. A mixed-methods approach to determine how conservation management programs and techniques have affected herbicide use and distribution in the environment over time. *Sci. Total Environ.* **2019**, *660*, 145–157. [CrossRef]
20. Greenspan, S.E.; Morris, W.; Warburton, R.; Edwards, L.; Duffy, R.; Pike, D.A.; Schwarzkopf, L.; Alford, R.A. Low-cost fluctuating-temperature chamber for experimental ecology. *Methods Ecol. Evol.* **2016**, *7*, 1567–1574. [CrossRef]
21. Virk, G.S.; Nagpal, A. Citrus diseases caused by Phytophthora species. *GERF Bull. Biosci.* **2012**, *3*, 18–27.
22. Colauto-Stenzel, N.M.; Janeiro-Neves, C.S.V. Rootstocks for 'Tahiti' lime. *Sci. Agric.* **2004**, *61*, 151–155. [CrossRef]
23. Ouédraogo, M.M.; Degré, A.; Debouche, C.; Lisein, J. The evaluation of unmanned aerial system-based photogrammetry and terrestrial laser scanning to generate DEMs of agricultural watersheds. *Geomorphology* **2014**, *214*, 339–355. [CrossRef]
24. Lum, C.; Mackenzie, M.; Shaw-Feather, C.; Luker, E.; Dunbabin, M. Multi-spectral imaging and elevation mapping from an unmanned aerial system for precision agriculture applications. In Proceedings of the 13th International Conference on Precision Agriculture, St. Louis, MO, USA, 31 July–4 August 2016.
25. Thorp, K.R.; Hunsaker, D.J.; French, A.N. Assimilating leaf area index estimates from remote sensing into the simulations of a cropping systems model. *Trans. ASABE* **2010**, *53*, 251–262. [CrossRef]
26. Wiegand, C.L.; Richardson, A.J.; Kanemasu, E.T. Leaf area index estimates for wheat from LANDSAT and their implications for evapotranspiration and crop modeling. *Agron. J.* **1979**, *71*, 336. [CrossRef]
27. Thorp, K.R.; Wang, G.; West, A.L.; Moran, M.S.; Bronson, K.F.; White, J.W.; Mon, J. Estimating crop biophysical properties from remote sensing data by inverting linked radiative transfer and ecophysiological models. *Remote Sens. Environ.* **2012**, *124*, 224–233. [CrossRef]
28. Arge, L.; Toma, L.; Vitter, J.S. I/O-Efficient Algorithms for Problems on Grid-Based Terrains. *Journal of Experimental Algorithmics. J. Exp. Algorithmics* **2001**, *6*, 1. [CrossRef]
29. Riley, S.J.; DeGloria, S.D.; Elliiot, R.A. Terrain Ruggedness Index That Quantifies Topographic Heterogeneity. *Int. J. Soil Sci.* **1999**, *5*, 23–27.
30. Neteler, M.; Bowman, M.H.; Landa, M.; Metz, M. GRASS GIS: A multi-purpose open source GIS. *Environ. Model. Softw.* **2012**, *31*, 124–130. [CrossRef]
31. Makori, D.; Fombong, A.; Abdel-Rahman, E.; Nkoba, K.; Ongus, J.; Irungu, J.; Mosomtai, G.; Makau, S.; Mutanga, O.; Odindi, J.; et al. Predicting spatial distribution of key honeybee pests in Kenya using remotely sensed and bioclimatic variables: Key honeybee pests distribution models. *ISPRS Int. J. Geo Inf.* **2017**, *6*, 66. [CrossRef]
32. Ficklin, D.L.; Novick, K.A. Historic and projected changes in vapor pressure deficit suggest a continental-scale drying of the United States atmosphere: Increasing U.S. vapor pressure deficit. *J. Geophys. Res. Atmos.* **2017**, *122*, 2061–2079. [CrossRef]
33. Allen, R.G.; Pereira, L.S.; Raes, D.; Smith, M. Crop evapotranspiration: Guidelines for computing crop water requirements. *FAO Irrig. Drain. Pap.* **1998**, *56*, 15.

34. Hartkamp, A.D.; De Beurs, K.; Stein, A.; White, J.W. *Interpolation Techniques for Climate Variables*; CIMMYT: Mexico City, Mexico, 1999.
35. Beckler, A.A.; French, B.W.; Chandler, L.D. Using GIS in areawide pest management: A case study in South Dakota. *Trans. GIS* **2005**, *9*, 109–127. [CrossRef]
36. Zhang, X.; Jiang, L.; Qui, X.; Qui, J.; Wang, J.; Zhu, Y. An improved method of delineating rectangular management zones using a semivariogram-based technique. *Comput. Electron. Agric.* **2016**, *121*, 74–83. [CrossRef]
37. Bazzi, C.L.; Schenatto, K.; Betzek, N.M.; Gavioli, A. A Software for the delineation of crop management zones (SDUM). *Aust. J. Crop Sci.* **2019**, *13*, 26–34. [CrossRef]
38. Shipp, J.L.; Zhang, Y.; Hunt, D.W.A.; Ferguson, G. Influence of humidity and greenhouse microclimate on the efficacy of *Beauveria bassiana* (Balsamo) for control of greenhouse arthropod pests. *Environ. Entomol.* **2003**, *32*, 1154–1163. [CrossRef]
39. Classen, A.T.; Hart, S.C.; Whitman, T.G.; Cobb, N.S.; Koch, G.W. Insect infestations linked to shifts in microclimate: Important climate change implications. *Soil Sci. Soc. Am. J.* **2006**, *70*, 305. [CrossRef]
40. Gogo, E.O.; Saidi, M.; Ochieng, J.M.; Martin, T.; Baird, V.; Ngouajio, M. Microclimate modification and insect pest exclusion using agronet improve pod yield and quality of French bean. *HortScience* **2014**, *49*, 1298–1304. [CrossRef]
41. Corwin, D.L.; Lesch, S.M. Characterizing soil spatial variability with apparent soil electrical conductivity. *Comput. Electron. Agric.* **2005**, *46*, 135–152. [CrossRef]
42. Chaudhari, P.R.; Ahire, D.B.; Chkravarty, M.; Maity, S. Electrical conductivity as a tool for determining the physical properties of Indian soils. *Int. J. Sci. Res. Publ.* **2014**, *4*, 1–4.
43. Chakraborty, K.; Mistri, B. Soil pH as a master variable of agricultural productivity in burdwan-I C.d. block, Barddhaman, West Bengal. *Indian J. Spat. Sci.* **2016**, *11*, 55–64.
44. Philar, U.B. Holistic, Cost Effective Method for Management of Huang Long Bing (HLB), Phytophthora gummosis, Asian Citrus Psyllid and Other Serious Infestations in Citrus and Other Crops. U.S. Patent 10,264,792B2, 23 April 2019.
45. Leroy, B.; Meynard, C.N.; Bellard, C.; Courchamp, F. Virtualspecies, an R package to generate virtual species distributions. *Ecography* **2015**, *39*, 599–607. [CrossRef]
46. O'Bannon, J.H.; Radewald, J.D.; Tomerlin, A.T. Population fluctuation of three parasitic nematodes in Florida citrus. *J. Nematol.* **1972**, *4*, 6.
47. Graham, J.H.; Gottwald, T.R.; Cubero, J.; Achor, D.S. *Xanthomonas axonopodis* pv. *Citri*: Factors affecting successful eradication of citrus canker. *Mol. Plant Pathol.* **2004**, *5*, 1–15. [CrossRef] [PubMed]
48. Hailnu, N.; Fininsa, C.; Mamo, G. Effects of temperature and moisture on growth of common bean and its resistance reaction against common bacterial blight (*Xanthomonas axonopodis* pv. *phaseoli* strains). *J. Plant Pathol. Microbiol.* **2017**, *8*, 419. [CrossRef]
49. Marutani-Hert, M.; Hunter, W.B.; Morgan, J.K. Associated bacteria of Asian citrus psyllid (Hemiptera: Psyllidae: *Diaphorina citri*). *Southwest. Entomol.* **2011**, *36*, 323–330. [CrossRef]
50. Brock, G.; Pihur, V.; Datta, S.; Datta, S. ClValid: An R package for cluster validation. *J. Stat. Softw.* **2008**, *25*, 1–22. [CrossRef]
51. Jain, A.K.; Dubes, R.C. *Algorithms for Clustering Data*; Michigan State University; Prentice Hall: Englewood Cliffs, NJ, USA, 1988.
52. Kaufman, L.; Rousseeuw, P.J. *Finding Groups in Data*; John Wiley & Sons Inc.: Hoboken, NJ, USA, 1990.
53. Tipping, M.E. Deriving cluster analytic distance functions from Gaussian mixture models. In Proceedings of the 9th International Conference on Artificial Neural Networks: ICANN '99, Edinburgh, UK, 7–10 September 1999; pp. 815–820. [CrossRef]
54. Kohonen, T. Exploration of very large databases by self-organizing maps. Proceedings of International Conference on Neural Networks (ICNN'97), Houston, TX, USA, 12 June 1997; Volume 1, pp. PL1–PL6. [CrossRef]
55. Ward, J.H. Hierarchical grouping to optimize an objective function. *J. Am. Stat. Assoc.* **1963**, *58*, 236–244. [CrossRef]
56. Datta, S.; Datta, S. Methods for evaluating clustering algorithms for gene expression data using a reference set of functional classes. *BMC Bioinform.* **2006**, *7*, 397. [CrossRef]
57. Schoener, T.W.; Schoener, A. Distribution of vertebrates on some very small islands. I. Occurrence sequences of individual species. *J. Anim. Ecol.* **1983**, *52*, 209. [CrossRef]
58. Warren, D.L.; Glor, R.E.; Turelli, M. Environmental niche equivalency versus conservatism: Quantitative approaches to niche evolution. *Evolution* **2008**, *62*, 2868–2883. [CrossRef]
59. Warren, D.L. ENMTools: A Toolbox for comparative studies of environmental niche models. *Ecography* **2010**, *33*, 607–611. [CrossRef]
60. Phillips, S.J.; Dudík, M.; Schapire, R.E. A maximum entropy approach to species distribution modeling. In Proceedings of the Twenty-First International Conference on Machine Learning, 83, ICML'04, Banff, AB, Canada, 4–8 July 2004. [CrossRef]
61. Fridgen, J.J.; Fraisse, C.W.; Kitchen, N.R.; Sudduth, K.A. Delineation and analysis of site-specific management zones. In Proceedings of the Second International Conference on Geospatial Information in Agriculture and Forestry, Lake Buena Vista, FL, USA, 10–12 January 2000; Volume 685, pp. 10–12.
62. Hornung, A.; Khosla, R.; Reich, R.; Inman, D.; Westfall, D.G. Comparison of site-specific management zones. *Agron. J.* **2006**, *98*, 407. [CrossRef]
63. Li, Y.; Shi, Z.; Li, F.; Li, H. Delineation of site-specific management zones using fuzzy clustering analysis in a coastal saline land. *Comput. Electron. Agric.* **2007**, *56*, 174–186. [CrossRef]
64. Gavioli, A.; Godoy de Souza, E.; Bazzi, C.L.; Carvalho-Guedes, L.P.; Schenatto, K. Optimization of management zone delineation by using spatial principal components. *Comput. Electron. Agric.* **2016**, *127*, 302–310. [CrossRef]

65. Szeto, L.; Liew, A.W.; Yan, H.; Tang, S. Gene expression data clustering and visualization based on a binary hierarchical clustering framework. *J. Vis. Lang. Comput.* **2003**, *14*, 341–362. [CrossRef]
66. Tamasauskas, D.; Sakalauskas, V.; Kriksciuniene, D. Evaluation framework of hierarchical clustering methods for binary data. In Proceedings of the 12 International Conference on Hybrid Intelligent Systems (HIS), Pune, India, 4–7 December 2012. [CrossRef]
67. Löhr, S.C.; Grigorescu, M.; Hodgkinson, J.H.; Cox, M.E.; Fraser, S.J. Iron occurrence in soils and sediments of a coastal catchment. *Geoderma* **2010**, *156*, 253–266. [CrossRef]
68. Sharma, A.; López, Y.; Tsunoda, T. Divisive hierarchical maximum likelihood clustering. *BMC Bioinform.* **2017**, *18*, 546. [CrossRef]
69. Damian, J.M.; de Castro Pias, O.H.; Cherubin, M.R.; da Fonseca, A.Z.; Fornari, E.Z.; Santi, A.L. Applying the NDVI from satellite images in delimiting management zones for annual crops. *Sci. Agric.* **2020**, *77*, 1–11. [CrossRef]
70. Arbelaitz, O.; Gurrutxaga, I.; Muguerza, J.; Pérez, J.M.; Perona, I. An extensive comparative study of cluster validity indices. *Pattern Recognit.* **2013**, *46*, 243–256. [CrossRef]
71. Wilmer, P.G. Microclimate and the environmental physiology of insects. In *Advances in Insect Physiology*; Berridge, M.J., Treherne, J.E., Wiglessworth, V.B., Eds.; Elsevier: New York, NY, USA, 1982.
72. Sgrò, C.M.; Terblanche, J.F.; Hoffmann, A.A. What can plasticity contribute to insect responses to climate change? *Annu. Rev. Entomol.* **2016**, *61*, 433–451. [CrossRef]
73. Ladányi, M.; Horváth, L. A Review of the potential climate change impact on insect populations: General and agricultural aspects. *Appl. Ecol. Environ. Res.* **2010**, *8*, 143–152. [CrossRef]
74. Kingsolver, J.G.; Woods, H.A.; Buckley, L.B.; Potter, K.A.; MacLean, H.J.; Higgins, J.K. Complex life cycles and the responses of insects to climate change. *Integr. Comp. Biol.* **2011**, *51*, 719–732. [CrossRef] [PubMed]
75. Andrew, N.; Hill, S.J. Effect of climate change on insect pest management. In *Environmental Pest Management: Challenges for Agronomists, Ecologists, Economists and Policymakers*; Coll, M., Wajnberg, E., Eds.; John Wiley & Sons Ltd.: Hoboken, NJ, USA, 2017.
76. Mitchell, M.S.; Powell, R.A. A mechanistic home range model for optimal use of spatially distributed resources. *Ecol. Model.* **2004**, *177*, 209–232. [CrossRef]
77. Johnson, L.F.; Bosch, D.F.; Williams, D.C.; Lobitz, B.M. Remote sensing of vineyard management zones: Implications for wine quality. *Appl. Eng. Agric.* **2001**, *17*, 557–560. [CrossRef]
78. Gardner, A.S.; Maclean, I.M.D.; Gaston, K.G.; Bütikofer, L. Forecasting future crop suitability with microclimate data. *Agric. Syst.* **2001**, *190*, 103084. [CrossRef]
79. Soberón, J. Grinnellian and Eltonian niches and geographic distributions of species. *Ecol. Lett.* **2007**, *10*, 1115–1123. [CrossRef]
80. Macnaughton-Smith, P.; Williams, W.T.; Dale, M.B.; Mockett, L.G. Dissimilarity analysis: A new technique of hierarchical sub-division. *Nature* **1964**, *202*, 1031–1035. [CrossRef]
81. Chollet, F.; Allaire, J.J. *Deep Learning with R*; Manning Publications Co.: Shelter Island, NY, USA, 2017.
82. Meila, M.; Heckerman, D. An experimental comparison of model-based clustering methods. *Mach. Learn.* **2001**, *42*, 9–29. [CrossRef]
83. Scrucca, L.; Fop, M.; Murphy, T.B.; Raftery, A.E. MCL 5: Clustering, classification and density estimation using Gaussian finite mixture models. *R J.* **2016**, *8*, 289. [CrossRef]

Article

TobSet: A New Tobacco Crop and Weeds Image Dataset and Its Utilization for Vision-Based Spraying by Agricultural Robots

Muhammad Shahab Alam [1,*], Mansoor Alam [2], Muhammad Tufail [2,3,*], Muhammad Umer Khan [4], Ahmet Güneş [1], Bashir Salah [5,*], Fazal E. Nasir [2], Waqas Saleem [6] and Muhammad Tahir Khan [2,3]

1. Defense Technologies Institute, Gebze Technical University, Gebze 41400, Turkey; ahmet.gunes@gtu.edu.tr
2. Advanced Robotics and Automation Laboratory, National Center of Robotics and Automation (NCRA), Peshawar 25000, Pakistan; mansooralam129047@gmail.com (M.A.); fazalnasir.uet@gmail.com (F.E.N.); tahir@uetpeshawar.edu.pk (M.T.K.)
3. Department of Mechatronics Engineering, University of Engineering & Technology, Peshawar 25000, Pakistan
4. Department of Mechatronics Engineering, Atilim University, Ankara 06830, Turkey; umer.khan@atilim.edu.tr
5. Department of Industrial Engineering, College of Engineering, King Saud University, Riyadh 11421, Saudi Arabia
6. Department of Mechanical and Manufacturing Engineering, Institute of Technology, F91 YW50 Sligo, Ireland; saleem.waqas@itsligo.ie
* Correspondence: shahab@gtu.edu.tr (M.S.A.); tufail@uetpeshawar.edu.pk (M.T.); bsalah@ksu.edu.sa (B.S.)

Citation: Alam, M.S.; Alam, M.; Tufail, M.; Khan, M.U.; Güneş, A.; Salah, B.; Nasir, F.E.; Saleem, W.; Khan, M.T. TobSet: A New Tobacco Crop and Weeds Image Dataset and Its Utilization for Vision-Based Spraying by Agricultural Robots. *Appl. Sci.* **2022**, *12*, 1308. https://doi.org/10.3390/app12031308

Academic Editors: Anselme Muzirafuti and Dimitrios S. Paraforos

Received: 12 October 2021
Accepted: 22 December 2021
Published: 26 January 2022

Publisher's Note: MDPI stays neutral with regard to jurisdictional claims in published maps and institutional affiliations.

Copyright: © 2022 by the authors. Licensee MDPI, Basel, Switzerland. This article is an open access article distributed under the terms and conditions of the Creative Commons Attribution (CC BY) license (https://creativecommons.org/licenses/by/4.0/).

Abstract: Selective agrochemical spraying is a highly intricate task in precision agriculture. It requires spraying equipment to distinguish between crop (plants) and weeds and perform spray operations in real-time accordingly. The study presented in this paper entails the development of two convolutional neural networks (CNNs)-based vision frameworks, i.e., Faster R-CNN and YOLOv5, for the detection and classification of tobacco crops/weeds in real time. An essential requirement for CNN is to pre-train it well on a large dataset to distinguish crops from weeds, lately the same trained network can be utilized in real fields. We present an open access image dataset (TobSet) of tobacco plants and weeds acquired from local fields at different growth stages and varying lighting conditions. The TobSet comprises 7000 images of tobacco plants and 1000 images of weeds and bare soil, taken manually with digital cameras periodically over two months. Both vision frameworks are trained and then tested using this dataset. The Faster R-CNN-based vision framework manifested supremacy over the YOLOv5-based vision framework in terms of accuracy and robustness, whereas the YOLOv5-based vision framework demonstrated faster inference. Experimental evaluation of the system is performed in tobacco fields via a four-wheeled mobile robot sprayer controlled using a computer equipped with NVIDIA GTX 1650 GPU. The results demonstrate that Faster R-CNN and YOLOv5-based vision systems can analyze plants at 10 and 16 frames per second (fps) with a classification accuracy of 98% and 94%, respectively. Moreover, the precise smart application of pesticides with the proposed system offered a 52% reduction in pesticide usage by spotting the targets only, i.e., tobacco plants.

Keywords: precision agriculture; selective spraying; vision-based crop and weed detection; convolutional neural networks; Faster R-CNN; YOLOv5

1. Introduction

Tobacco is grown in more than 120 countries around the world, covering millions of hectares of land. In Pakistan, it is regarded as an important crop as it generates substantial revenue. According to an estimate, in rural areas of the country, 80k–90k tonnes of Flue-Cured Virginia (*Nicotiana Tabacum*) is produced annually [1]. In addition to being a profitable crop, it is important to highlight that tobacco's leaf is highly susceptible to pests and pathogens, and the crops demand meticulous effort and care in order to protect them from seasonal insects, as shown in Figure 1. Local farmers rely upon the use of conventional agrochemical spray methods for combating these pests and pathogens. Pesticides are applied to tobacco plants usually five to six times in one season (over three months), which makes it a highly pesticide-dependent crop. Two methods are commonly used for pesticide spraying: manual knapsack spraying in which human labor carries the equipment and performs spray on every plant and broadcast spraying via a tractor-mounted sprayer in which the entire field is sprayed indiscriminately. Both methods are imprecise and, therefore, cause serious damage to farmers' health (due to first-hand/direct exposure) and to the environment (due to overdosing) [2–6]. Despite its hazards, agrochemical spraying is still common in practice as it is a viable and economical means to protect tobacco crop from pests and pathogens [7]. The solution, therefore, lies not in eliminating the use of agrochemicals but instead in optimizing their application by embracing advanced techniques and methodologies.

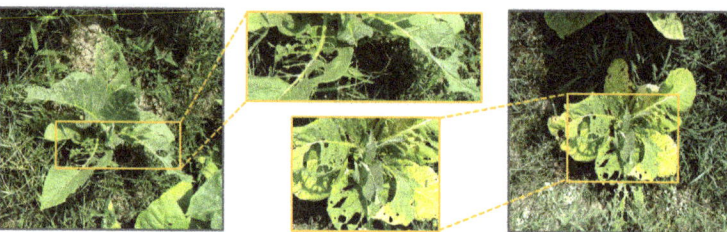

Figure 1. Weeds and tobacco leaf infestation due to pests.

Artificial Intelligence is rapidly bringing a substantial paradigm shift in the agriculture sector. Endowing agricultural spraying systems the cognitive ability of understanding, learning, and responding to different crop conditions greatly improves spraying operations. Precision spraying methods combine techniques from emerging disciplines such as artificial intelligence, robotics, and computer vision, which provides a spraying system the ability to identify plants (crop) and weeds and apply precise doses only on the desired targets [8–13].

Over the last decade, numerous promising attempts have been made by researchers for the development of intelligent spraying systems for different crops [14–23]. Surprisingly, not much work is found in the literature on vision-based site-specific spraying systems for crops. The vision-based system tends to deal with numerous variations, such as varying leaf sizes at different growth stages, varying light intensities, different soil textures, varying leaf colors due to different water levels, high weed densities, and crop plant occlusion by weeds, etc.

Existing methods for vision-based plant/weed detection and precision spraying are mostly based on traditional machine learning-based techniques [24–32]. Although high accuracies have been achieved with these techniques, the hand-crafted features formulation and generation of a decision function over the extracted features make them less robust. Therefore, they are certainly not a preferred choice for tobacco plant and weed detection (keeping in view the factors of variations and complexities involved in tobacco fields) due to poor generalization capabilities. Over the past few years, deep learning-based computer vision algorithms have demonstrated their ability to perform well on complex problems from training examples [33–41]. CNNs are the main architecture of these computer vision algorithms. Deep learning algorithms learn the features and decision functions in an end-to-end fashion. Lopez-Martin et al. [42] proposed a classifier known as gaNet-C for type-of-traffic forecast problem. An additive network model, gaNet, has the capability to forecast k-steps beforehand by utilizing time-series of last computed values for each node. The proposed model demonstrates good performance on two detection forecast problems.

The advantages that deep learning algorithms offer, such as feature learning capabilities, high accuracy, and better performance in intricate problems, make them best suited for complex tasks such as detecting tobacco plants under several variations in outdoor fields. Several studies have been reported with respect to deep learning-based plant and weed detection [43–49]. The latest research on plant and weed detection mainly utilizes computer vision [50–56]. For instance, Costa et al. [57] used deep learning for finding defects in tomatoes by applying Deep ResNet classifiers. According to their finding, ResNet50 with fine-tuned layers was reported as the best model that achieved an average precision of 94.6% and a recall of 86.6%. Moreover, it was observed that fine-tuning outperformed feature extraction process. Santos Ferreira et al. [58] detected weeds in soybean crops using ConVNets and SVM classifiers. ConVNets was able to achieve higher accuracy of more than 97% in weed detection. Yu et al. [59] used deep learning algorithms for detecting multiple weed species in Bermuda grass. The study reported that VGGNet performed well with an F1-score of over 0.95 than compared to GoogleNet. Moreover, F1-scores of over 0.99 were reported for detecting weeds via DetectNet. The authors, based on attained results, concluded the effectiveness of deep convolutional neural networks in the weed detection problem. In another study, Sharpe et al. [60] evaluated three CNNs—DetectNet, VGGNet, and GoogLeNet—for the detection of weeds in strawberry fields. It was observed that the image classification DetectNet model produced the best results for image-based remote sensing of weeds. Le et al. [61] used Faster R-CNN for the detection of weeds in Barley crops using several feature extractors. In the study, mean Average Precision (mAP) with Inception-ResNet-V2 was found better than the mAP for other networks. Moreover, an inference time of 0.38 s per image was also reported. Quan et al. in [62] presented an improved version of the Faster R-CNN vision system for the identification of maize seedlings in tough field environments. The images were taken with a camera at an angle ranging from 0 to 90 degrees. The results reported detection accuracy of 97.71%. In the study performed by [63], the authors reported F1-scores of 88%, 94%, and 94% for SVM, YOLOv3, and Mask R-CNN for detecting weeds in lettuce crops, respectively. The work reported by Wu et al. [64] used YOLOv4-based vision system for detecting apple flowers. The model based on CSPDarkNet-53 framework was simplified with a channel pruning algorithm for detecting the target object in real time. They reported achieving a mAP of 97.31% at a detection speed of 72.33 fps.

Despite the impressive accomplishments in deep learning-based object detection, the performance of these algorithms has yet to be evaluated in the realm of tobacco plants and weeds detection; for instance, the use of region-based methods such as Faster R-CNN or one stage detectors such as YOLOv5. Moreover, published reports also lack experimental validation in actual field environments. The aim of this study includes the replacement of conventional broadcast spraying methods in tobacco fields with a site-specific (drop-on-demand) spraying system. The proposed method detects and classifies tobacco plants and weed automatically, determines their position, i.e., their location in the crop rows, and finally performs agrochemical spray on the detected targets.

This paper focuses on automatic vision-based tobacco plant detection that is considered a vital part of the precision spraying system. The basic frameworks of two off-the-shelf deep-learning algorithms—Faster R-CNN and YOLOv5—are employed for detection and classification models. The robustness and ability of the models are enhanced by fine-tuning detection of tobacco plants in challenging field conditions. Both detection models are tested on a vision-guided mobile robot platform in real tobacco fields. A comparative study is also carried out between both frameworks in terms of robustness, accuracy, and inference/computational speed. The Faster R-CNN-based vision-based model demonstrated higher accuracy but lower real-time detection speed, whereas the YOLOv5-based model produced slightly lower accuracy but higher real-time detection speed. Therefore, YOLOv5-based vision model, based on its performance, is considered best suited for real-time tobacco plant and weed detection. The main contributions of this study are summarized as follows:

1. Development and deployment of a vision-based robotic spraying system for replacement of the conventional broadcast spraying methods with a site-specific selective spraying technique that can detect tobacco plants and weed and classify them in a real time;
2. Building a tobacco image dataset (TobSet) that comprises labeled images of tobacco crop and weed. The dataset is collected under challenging real in-field conditions to

train and evaluate the latest state-of-the-art deep learning algorithms. TobSet is an open-source dataset and is publicly available at https://github.com/mshahabalam/TobSet (accessed on 11 October 2021).

The rest of the paper is organized as follows: Section 2 covers the description about the image dataset and Section 3 briefly explains the materials and methods employed in this study. The workings of Faster R-CNN and YOLOv5 algorithms are discussed in Section 4. The hardware setup for the implementation is explained in Section 5. Evaluation of the proposed approaches is carried out in Section 6 along with discussion and comparative analysis, and a brief concluding remarks are provided in Section 7.

2. Data Description

Due to the unavailability of any image dataset of tobacco plants, we developed an extensive image dataset, TobSet, from the actual fields in Swabi, Khyber Pakhtunkhwa, Pakistan (34°09′07.3″ N 72°21′36.2″ E). The main objective of building this dataset is to provide real-field data for training and evaluating the performance of state-of-the-art algorithms for tobacco crop and weed detection. TobSet comprises (a) 7000 images of tobacco plants and (b) 1000 images of bare soil and weeds (that grow up in tobacco fields), with a resolution of 640 × 480. The images are captured using a 13-megapixels color digital camera possessing a CMOS-image sensor (IMX258 Exmor RS by Sony, Japan), 28 mm focal length, 65.4° horizontal FOV, and 51.4° vertical FOV. A comprehensive dataset was built over a period of 2 months, i.e., starting from the first week of tobacco seedling transplantation from seedbeds to the time when plants gain an approximate height of 1.25 m. All images in the dataset were captured manually by human scouts in the months of June and July 2020. No artificial shading and sources of lightning were used while collecting the images. During image acquisition, the camera's height was adjusted between 1 and 1.5 m. In order to maintain diversity in the dataset, all images in TobSet are captured under several factors of variations: different growth stages, different day timings, varying lighting and weather conditions (i.e., on normal, bright sunny, and cloudy days), and visual occlusions of crop leaves by weeds, etc. The existing literature on vision-based detection of crops and weeds lacks experimental validation on hard real-world datasets such as TobSet. Some sample images from the publicly available TobSet are presented in Figure 2. After data acquisition, the main step involved in crop/weed detection is the annotation of images for ground truth data. All images in the TobSet are manually labeled with the LabelImg tool.

Figure 2. Illustration of factors of variation in the actual tobacco fields.

TobSet is publicly available and offers multi-faceted utilities:

1. It comprises labelled images of tobacco plants and weeds that can be utilized by computer scientists for performance evaluation of their developed computer vision algorithms;
2. Scientists working on agricultural robotics can use it to train their robots for variable rate-spray applications, plant or weed detection, and detection of plant diseases;
3. It can also be used by agriculturists and researchers for studying various aspects of tobacco plant growth, weed management, yield enhancement, leaf diseases, and pest prevention, etc.

3. Materials and Methods

For targeted agrochemical spray, the application equipment must have the following capabilities: (a) discriminating the crop plants from weeds, (b) determining the robot's location in the field, and (c) applying agrochemicals on the targeted plants, i.e., crop or weeds. Considering these aspects, our developed agrochemical spraying robot has three main systems: a vision-based crop or weed identification system, a robot navigation system, and an actuation system for spraying on targeted plants. This paper is focused only on the predominant sensing modalities of developed spraying robot that enables it to identify crop plants and weeds, i.e., a vision-based detection framework.

Due to the nature of the application, i.e., harsh or challenging tobacco field conditions, the vision system essentially must be robust in order to process data and generate accurate results in real-time. Due to excellent performance, deep-learning algorithms are currently state-of-the-art for computer vision applications. This is attributed to the availability of large-sized labeled data, and deeply layered architectures. However, due to increasing depth, the algorithms are computationally very expensive, especially for resource-limited portable machines. The study presented herein aims to develop a deep-learning-based vision framework with low inference cost, thereby it can be used in real-time detection and classification of tobacco crops and weeds. In order to achieve this, two state-of-the-art CNN algorithms, i.e., Faster R-CNN and YOLOv5, are utilized for implementation.

Pesticide application on the tobacco plants begins immediately after the first week when the seedlings are transplanted from the seedbed into the fields and continues periodically until their maturity. As shown in Figure 3, inter-row spacings of approximately 1 m and intra-row spacings of approximately 0.75 m were kept between any two consecutive plants. Therefore, indiscriminate broadcast application of pesticides on the complete tobacco field, particularly at earlier growth stages when the plants' canopy sizes are very small, results in off-the-target pesticide spray on bare soil spots. This unnecessary pesticide application on bare soil or weed patches engenders polluting the environment and leaching of toxic pesticides into the ground.

Figure 3. Inter-row and intra-row plant spacing in tobacco fields.

Moreover, all crop plants across a tobacco field do not necessarily grow homogeneously due to variation in seedling health, size of the plant at the time of transplantation, and water and nutrients variability across the field. Due to these reasons, intra-row and inter-row spacing varies across the entire field according to plant leaf sizes. Our system proposes dividing the camera's field of view into grids. In each grid, the deep-learning-based detector detects plants and assigns a cell to each plant based on its coverage such that it apprehends plant canopies. Since our spray application module comprises flat fan nozzles, the lateral length of the grid is set according to the swath size of each corresponding nozzle. Furthermore, the vertical size of the cell is adjusted based on the detected plant's canopy, as shown in Figure 4, by the green boxes.

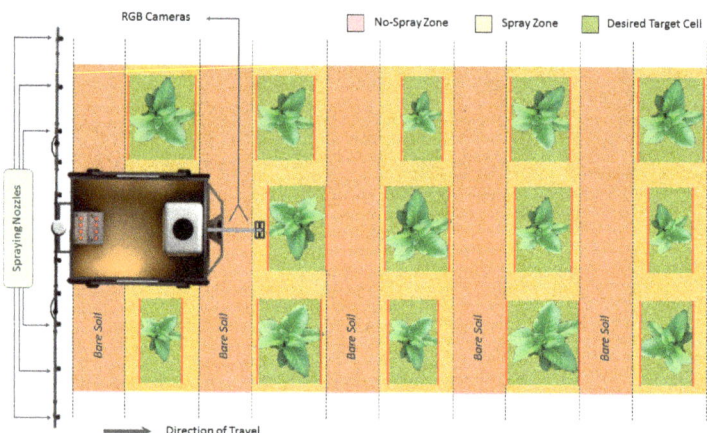

Figure 4. Desired pinpointed spray zones.

Two separate vision systems are employed on the robot. One vision system is for the detection and localization of the tobacco crops and weeds, whereas the other vision system helps with crop row structure detection for guiding the robot along the crop rows. As stated earlier, this paper focuses only on the vision system for crop and weed detection. The tobacco crop or weed detection and spraying processes are performed in the following sequence: (a) acquiring an image with the camera via image grabber; (b) sending the acquired image to the NVIDIA GPU for processing; (c) detection of crop plants and weeds; (d) determining the location of the plant and size of its attributed grid cell based on the plant's coverage; (e) sending the required control signal for spray via USB port to the embedded controller; and (f) actuation of the corresponding nozzles upon reaching the target plant.

4. CNN-Based Detection and Classification Frameworks

The primary objective of this research is to enable an agriculture sprayer robot to identify tobacco plants and weeds in real time using an onboard vision system. Two different deep-learning algorithms are utilized in the detection of tobacco plants and weeds, i.e., Faster R-CNN and YOLOv5. Despite some differences in the overall frameworks of Faster R-CNN and YOLO, both rely upon CNNs as their core working tool. Faster R-CNN processes the entire image using CNN and then divides it for several region proposals in two steps, whereas YOLO splits the image into grid cells and processes it through CNN in one step.

4.1. Faster R-CNN

Faster R-CNN, proposed by Ren et al. [65], is a combination of Fast R-CNN and region proposal network (RPN). The aim behind the introduction of Faster R-CNN was to make the detection process less time consuming and more accurate. Primarily, its structure comprises feature extraction, region proposals, and bounding box regression.

The submodules involved in the algorithm for our tobacco crop and weed detection are explained in the following subsections.

4.1.1. Convolutional Layers

Being a CNN-based detection approach, we use the basic *convolutional*, *relu*, and *pooling* layers for extracting feature maps from tobacco and weeds images. Rather than using the models of Simonyan and Zisserman [66] or Zeiler and Fergus [67], we customized the architecture of the model. The in-depth structure of our model comprised eleven *Conv* layers, eleven *relu* layers, and five *pooling* layers. In each *Conv* layer, the *kernel* size is set to 3 and *padding* and *stride* are set to 1, whereas in the *pooling* layers, the *kernel* size is set to 2, *padding* is set to 0, and *stride* is set to 2. The detection and classification pipeline of the Faster R-CNN-based detection model is shown in Figure 5.

All the convolutions are expanded in the *Conv* layers using *padding* size of 1 to transform the original input image size to $(M+2) \times (N+2)$, and then a *kernel* of size (3×3) is applied to obtain an output image of $(M \times N)$, i.e., (640×480). This helped the input and output matrix sizes to remain unchanged in the *Conv* layers. Moreover, the *pooling* layer, *kernel*, and the *stride* sizes are set to 2 in the *Conv* layers. Thus, every (640×480) matrix that goes past the *pooling* layer is converted to $(640/2) \times (480/2)$. In all of the *Conv* layers, the input and output sizes of the *Conv* and *relu* layers are kept the same. However, the *pooling* layer forces the output length and width to be 1/2 of the input. Next, a matrix with a size of (640×480) is switched to $(640/16) \times (480/16)$ by the *Conv* layers; hence, the feature map produced by *Conv* layers can be associated with the original image. The feature maps are fed to the subsequent RPN and fully connected layers.

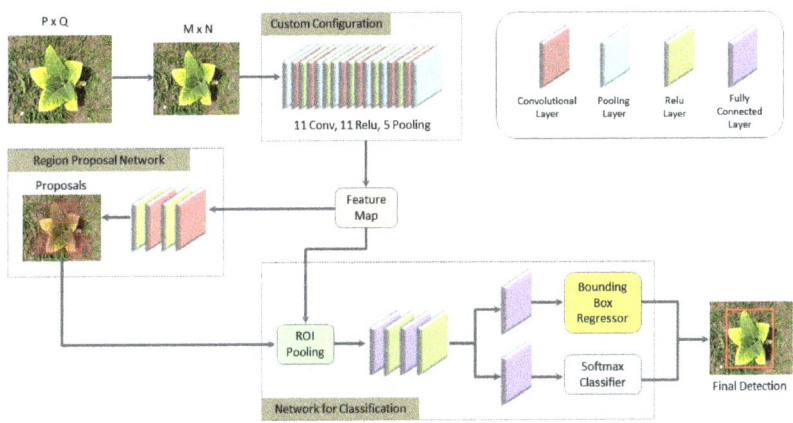

Figure 5. Faster R-CNN-based tobacco crop detection framework.

4.1.2. Region Proposal Networks

The RPN network being small is slid over the feature map for generating regional proposals. RPN classifies the corresponding regions and regresses bounding box locations, simultaneously. To find out that whether the anchors belonged to the foreground or background, we used *softmax* in this layer. Furthermore, the anchors are adjusted with the bounding box regression in order to obtain precise proposals. The classic approach generates a very time-consuming detection framework. Therefore, instead of the traditional sliding window and selective search approaches, the RPN method is used directly for generating detection frames. This served as a plus point of the Faster R-CNN method as compared to classical detection methods in improving the detection frame generation speed to some extent [65].

4.1.3. ROI Pooling

In the ROI *pooling* layer, the region proposals are collected and split into smaller windows. Next, feature maps are extracted from these regions, which are further sent to the subsequent *fullyconnected* layer for determining the target class in this layer. Moreover, our ROI *pooling* layer comprises two inputs:

1. Original feature maps;
2. RPN output proposal boxes of different sizes.

In traditional CNNs such as AlexNet, VGG, etc., the size of the input image essentially should be constant, and the output of the network should also be a fixed-size vector or matrix when the network is trained. Therefore, a remedy is proposed for variable input image sizes: (a) parts of images are cropped, and (b) the images are warped to the desired size. Despite adopting these approaches, either the structure of the entire image is altered after the images are cropped or the shape information of the original image is altered when the images are warped. Similarly to the proposal's generation approach of RPN's bounding box regression on foreground anchors, the image properties achieved in this manner has dissimilar shapes and sizes. To cater with this complexity, ROI pooling is utilized. Since it corresponds to the (640×480) scale, the spatial scale parameter is first used for mapping it back to $(640/16) \times (480/16)$-sized feature maps. Next, horizontal ($pooled_w$) and vertical ($pooled_h$) division of each property is performed. Finally, *maxpooling* is applied to each property. This approach ensured an output of the same size and fixed length.

4.1.4. Classification

Pseudo feature maps are used to compute the proposal's class and, simultaneously, the final position of the detection frame is acquired by the bounding boxes. Since the network deals with $P \times Q$ input size images, they are first scaled down to a constant size of $(M \times N)$, i.e., (640×480), and passed onto the network. The convolution layers contains 11 *Conv* layers, 11 *relu* layers, and 5 *pooling* layers. The RPN network employs 3×3 convolution and then generates foreground or background anchors and the associated bounding box regression offsets. Then, proposals are calculated and ROI pooling is performed, which computes the feature maps and sends them to the subsequent fully connected *softmax* network for classification. The classification section uses the acquired property feature maps for calculating the specific category (i.e., tobacco plants and weeds) that each property belongs to via the *fullyconnected* layer and *softmax*.

Finally the probability for the class is computed, and bounding box regression is once more used for obtaining the position offset for each proposal. The classification section of the proposed network is highlighted by the shaded region of Figure 5. After obtaining the $7 \times 7 = 49$ sized features, feature maps from ROI pooling, and then sending them to the succeeding network, the following two steps were performed:

1. Classification of proposals by *fullyconnected* layer and *softmax*;
2. Bounding box regression on the proposals for acquiring more accurate rectangular boxes.

4.2. You Only Look Once (YOLO)

YOLO is a fast one-stage object detection model that was developed by Redmon et al. [68] in 2015. YOLO as compared to Faster R-CNN is less error-prone to background errors in images as it observes the larger context. The main trait that dignifies YOLO from other similar networks is its capability to detect objects (with bounding boxes) and calculate class probabilities in a single step, i.e., detection and class predictions are performed simultaneously after a single evaluation of the input image. Training is performed on complete images, and the performance of the detection is optimized directly. YOLO, unlike the region proposal and sliding window-based methods, processes the complete image during training and testing phases, which enables it to translate class-specific information and its outlook implicitly.

There are three main elements involved in the YOLO network: (a) backbone, (b) neck, and (c) head. The backbone comprises CNNs that serve the purpose of aggregations and image feature formation from several image granularities. The neck is composed of a series of layers used for mixing and combining the extracted features and subsequently transmitting them to the prediction layer. Finally, the head is used for the features prediction, bounding boxing creation, and class prediction.

The algorithm works by first splitting the input image into a grid of $S \times S$ and then predicting B bounding boxes for each grid cell, as shown in Figure 6. Every bounding box in the grid cell is assigned a confidence score to denote the probability of an object's existence inside the defined box.

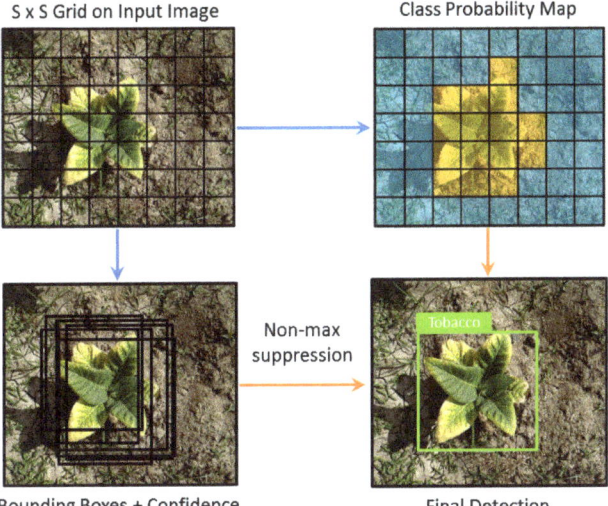

Figure 6. YOLO detection pipeline.

Grid cells are accountable for detecting objects if their centers fall inside a grid cell. If the center of the bounding box (of the same object) is predicted to fall in multiple grid cells, a non-max suppression eliminates redundant bounding boxes and retains the one possessing the highest probability. Each bounding box has four associated predictions that include the (x, y) coordinates of the center of the box, width w, height h, and confidence C. Confidence C can be formulated as follows:

$$C = Pr(Class_i) * IOU_{pred}^{truth} \qquad (1)$$

where IOU is the intersection over union, i.e., the overlapped area between predicted and ground truth bounding boxes. The IOU value of 1 represents a perfect prediction of the bounding box relative to ground truth.

Bounding boxes and conditional class probabilities for each grid cell are computed at the same time. Conditional class probabilities and bounding box confidence predictions during the test phase are multiplied to obtain confidence scores of a particular class of each box as follows.

$$Pr(Class_i|Object) * Pr(Object) * IOU_{pred}^{truth} = Pr(Class_i) * IOU_{pred}^{truth} \qquad (2)$$

Network Architecture

The baseline architecture of YOLOv5 is very similar to YOLOv4, primarily comprising a Backbone, Neck, and Head. The backbone of YOLOv5 can be ResNet-50, VGG16, ResNeXt-

101, EfficientNet-B3, or CSPDarkNet-53. We used the CSPDarkNet-53 neural network as our model's backbone, which encompasses cross-stage partial connections, and it is considered as the most optimal model [57]. CSPDarknet-53 has 53 convolutional layers, and it originates from DenseNet architecture. DenseNet network uses the preceding input, and, prior to stepping into dense layers, it concatenates the previous input with the current one [69]. The robustness of our YOLOv5-based vision framework greatly improved with the CSP application approach, i.e., by applying the CSP1_x to the backbone and CSP2_x to the neck. First, data were fed as input to CSPDarkNet-53 for extracting features. For improving feature extraction from different growth stages of tobacco plants, an additional layer was inserted into the model's backbone, which helped to improve the mAP. Next, the extracted features were fed to PANet (Path Aggregation Network) for fusing features. Finally, output results, i.e., class, score, etc., of detection were provided by the YOLO layer. Our model's head part used an anchor-free one-stage object detector YOLO. The modified YOLOv5 architecture used in this study is illustrated in Figure 7.

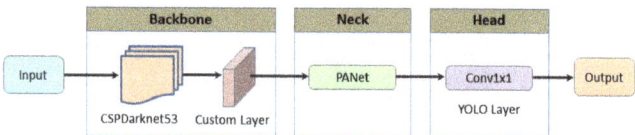

Figure 7. Modified YOLOv5 detection pipeline.

5. Experimental Evaluation

This section deals with the experimental setup that we used for conducting in-field experiments, the dataset used for training both deep learning-based vision models, and the infield real-time results obtained with our vision models.

Hardware Setup

The proposed frameworks are implemented in the tobacco fields with a four-wheeled mobile robot platform. The robot has a track width and wheelbase of 1 and 1.3 m, respectively. In order to protect tobacco plants from the robot, the ground clearance of the platform was carefully chosen as 0.9 m. Moreover, the height of the robot's platform can be adjusted anywhere between 0.9 and 0.4 m depending on different crops.

In order to keep robot design and control simple, a differential drive scheme was chosen with two driving wheels (front) and two passive wheels (rear). The robot is equipped with two DC motors connected to motor controllers for steering and driving the robot along the straight crop rows. Two separate RGB cameras are mounted on the robot: One is used for the crop row detection (for navigation), and the other is used for crop/weed detection (for spraying). The camera for row structure detection is mounted at the front with its face towards the ground and a horizon at an angle of 35° with the horizontal axis, covering three rows simultaneously. The camera for crop and weed detection is mounted at the front of the robot and oriented facing downwards to the ground at a fixed distance of 1.8 m from nozzles on the boom. The distance between the crop and weed detection camera and the boom is kept at the maximum in order to provide the desired time delay between detection and position estimation of the crop plant and the spray application process on every corresponding grid cell.

The vision-based detection system is coupled with spraying equipment and other sensing modules, thereby making a complete precision agricultural robotic spraying system. A 12 V DC diaphragm pump is used to pressurize the fluid system. An electronic pressure regulated valve maintains a constant line pressure when different nozzles on the boom are switched ON and OFF based on feedback from the vision system and other sensing modules. The outflow line from the pump is divided into a bypass line that diverts excess flow back to the tank and a boom line onto which the nozzles are mounted. Two rotary incremental encoders (with resolutions of 1000 pulses per revolution) connected to the embedded controller are mounted on the front wheels' axles to measure the rotation (and thereby speed) of the wheels. The incremental encoders and a GPS module facilitates the robot in determining its position and heading direction for navigation. Moreover, the optical data acquired via cameras are synchronized to the robot's position through incremental encoders and GPS module.

The robot used ROS (Robot Operating System) as the middleware software framework. The cameras were connected to a computer possessing an Intel Core E5-1620, a 3.50 GHz processor, 32 GB RAM, and an 8 GB NVIDIA GTX 1650Ti GPU for processing the images. Moreover, Microsoft Visual Studio and Python were used for program development. The developed agricultural robot sprayer and its overall functional block diagram is shown in Figures 8 and 9, respectively.

Figure 8. Developed prototype of the agricultural robotic sprayer.

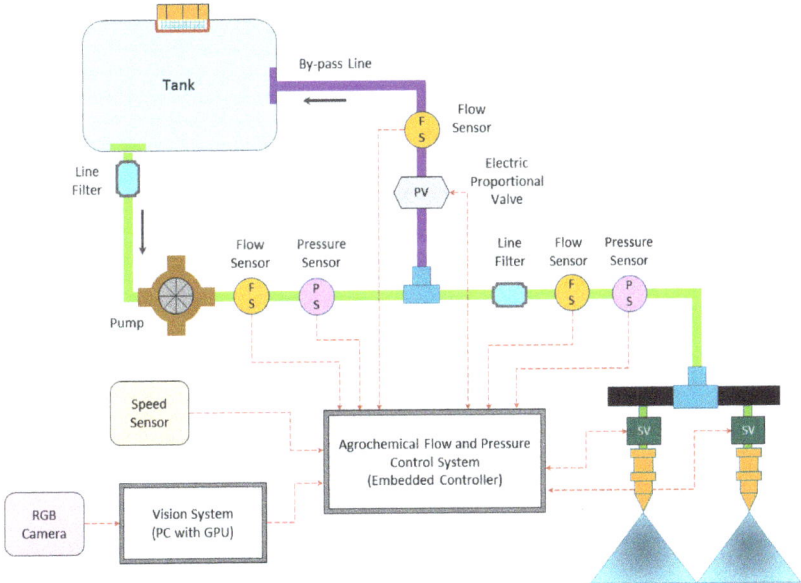

Figure 9. Block diagram illustrating the developed vision and fluid flow control systems.

6. Results and Discussions

In order to validate and demonstrate the effectiveness of both vision-based frameworks for tobacco crop/weed detection and classification, the models are trained and tested on real-field tobacco images from TobSet. The dataset consists of 8000 images; 7000 images are of tobacco plants, and the rest of the images are of weeds. Images from both classes are divided with a 70 to 30 ratio into training and testing sets. The training set comprised a total of 5600 images (4900 tobacco and 700 weeds), whereas the testing set comprised 2400 images (2100 tobacco and 300 weeds).

In the implementation phase, the models are trained using down-sampled images (with a resolution of 640 × 480). A learning rate is initialized as 0.0002 for the training. Google's TensorFlow API is utilized for implementation purposes. Batch sizes of 1 and 10 k epochs are used for training the models. Table 1 lists the hyper-parameters and their corresponding losses (against the epochs) for both models. It can be observed from Table 1 that for obtaining better results with Faster R-CNN-based vision model, the learning rate is kept the same, whereas the other hyper-parameters did change. With an increase in the number of epochs, total loss is reduced. The confusion matrices for Faster R-CNN and YOLOv5-based models, given in Tables 2 and 3 respectively, are used for computing the evaluation measures listed in Table 4.

Table 1. Hyper-parameters for Faster R-CNN and YOLOv5.

S. No.	Learning Rate	Epoch	Loss for Faster R-CNN	Loss for YOLOv5
1	0.0002	2 k	0.046	0.124
2	0.0002	4 k	0.029	0.081
3	0.0002	6 k	0.028	0.066
4	0.0002	8 k	0.025	0.058
5	0.0002	10 k	0.017	0.049

After training the models with the given training set, performance evaluation of both models is conducted on the testing data from TobSet. The accuracy results obtained by using the Faster R-CNN-based vision model show its supremacy over YOLOv5. A total of 635 predictions were produced on unseen test images for each model. Detection results for both models are presented in Figures 10 and 11. The YOLOv5-based model did not perform well on some test samples, as illustrated in Figure 12.

Table 2. Confusion matrix for Faster R-CNN-based model.

		Predicted Class	
		Tobacco	Weeds
True Class	Tobacco	454	6
	Weeds	7	168
		98.48%	96.55%

Table 3. Confusion matrix for YOLOv5-based model.

		Predicted Class	
		Tobacco	Weeds
True Class	Tobacco	437	13
	Weeds	24	161
		94.79%	92.52%

Table 4. Evaluation measures for Faster R-CNN and YOLOv5.

S.No.	Evaluation Measure	Faster R-CNN	YOLOv5
1	Precision	0.9829	0.9481
2	Recall	0.9863	0.9732
3	F1-Score	0.9859	0.9576
4	Accuracy	0.9834	0.9445

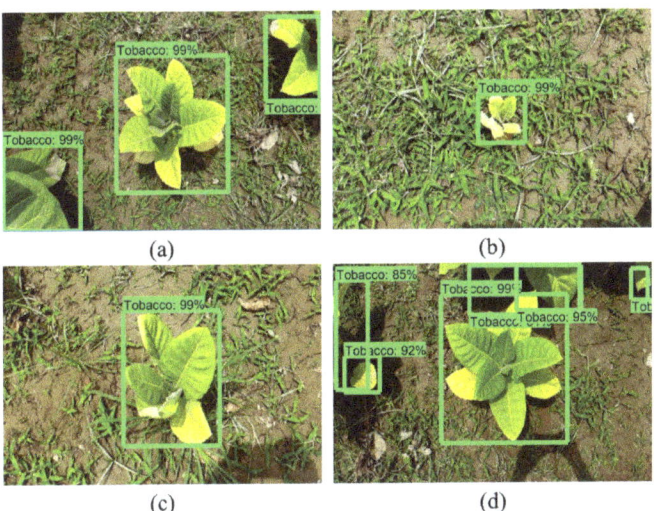

Figure 10. Faster R-CNN detection results of tobacco from testing data in varying scenarios: (a) high intra-row plant distance. (b) high weed density. (c) low weed density. (d) low intra-row plant distance.

Figure 11. YOLOv5 detection results of tobacco from testing data in varying scenarios: (**a**) high intra-row plant distance. (**b**) high weed density. (**c**) low weed density. (**d**) low intra-row plant distance.

Figure 12. YOLOv5 detection results with (**a**) undetected targets, and (**b**) misidentified regions.

Real-Time Inference

The proposed vision models are evaluated in real tobacco fields on a mobile robot spraying platform. For obtaining higher inference in real-time, NVIDIA's optimized library for faster deep-learning inference, i.e., NVIDIA TensorRT, was used. The modified Faster R-CNN and YOLOv5-based vision models identified tobacco plants at 10 and 16 fps, and with classification accuracies of 98% and 94%, respectively, at a robot speed of approximately 3 km/h. The modified YOLOv5-based model can process images at a higher frame rate compared to the Faster R-CNN model, thus making it a better choice for real-time deployment on a spraying robot. Real-time detection results for both models are presented in Figures 13 and 14.

Table 5 presents each model's inference results in real time. YOLOv5 outperformed the Faster R-CNN model in terms of inference speed.

Figure 13. Real-time Faster R-CNN detection of tobacco crop and weeds in scenarios with (**a**) low intra-row plant distance, and (**b**) high intra-row plant distance.

Figure 14. Real-time YOLOv5 detection of tobacco crop and weeds in scenarios with (**a**) low intra-row plant distance, and (**b**) high intra-row plant distance.

Table 5. Inference results of the models in real-time.

Model	Faster R-CNN	YOLOv5
Inference time (ms)	98.5 ± 5	62 ± 5
Frames per second (fps)	10.15	16.12
mAP	0.94	0.91

7. Conclusions

Intelligent precision agriculture robot sprayers for agrochemical application must be robust enough to distinguish crops from weeds to perform targeted spraying to reduce the usage of agrochemicals. In this paper, two different CNN-based approaches, namely, Faster R-CNN and YOLOv5, are explored in order to develop a vision-based framework for the detection and classification of tobacco crop and weeds in the actual fields. Both frameworks are first trained and then tested on a self-developed tobacco plants and weeds dataset, TobSet. The dataset comprises 7000 images of tobacco plants and 1000 images of bare soil and weeds taken manually with digital cameras periodically over 2 months. The Faster R-CNN-based vision framework demonstrated higher accuracy and robustness, whereas the YOLOv5-based vision framework demonstrated lower inference time. Experimental implementation is conducted in the tobacco fields with a four-wheeled mobile robot sprayer with a computer possessing a GPU. Classification accuracies of 98% and 94% and frame rates of 10 and 16 fps were recorded for Faster R-CNN and YOLOv5-based models, respectively. Moreover, the precise smart application of pesticides with the proposed system offered 52% reduction in pesticide usage by pinpointing the targets, i.e., tobacco plants.

Faster R-CNN produces higher accuracy but lower fps on computers (especially without GPUs); high computational cost of training makes it challenging for real-time applications. TobSet demonstrated true assessment of the deep-learning algorithms as it comprises real field images with challenging scenarios possessing different factors of variation, such as dense weed patches, lightening variation, color similarity with weeds, color variation of tobacco plant at different growth stages, and varying growth stages. The classification results of both approaches in real time were slightly lower than the prediction results obtained on the dataset due to higher sunlight intensities. Intended future studies include real-time tobacco plant segmentation for finding canopy size and the desired application flowrate of spray for each tobacco plant.

Author Contributions: Conceptualization, M.S.A., M.T. and M.U.K.; methodology, M.S.A., M.A. and M.T.; software, M.S.A., M.A. and F.E.N.; validation, A.G., B.S., W.S. and M.U.K.; formal analysis, M.T.K., B.S. and W.S.; resources, M.T. and M.T.K.; data curation, M.S.A., M.A. and F.E.N.; writing—original draft preparation, M.S.A. and M.A.; writing—review and editing, M.U.K., M.T. and A.G.; visualization, M.U.K., A.G. and B.S.; supervision, M.T., B.S. and M.T.K.; project administration, M.T., B.S. and M.T.K.; funding acquisition, B.S. All authors have read and agreed to the published version of the manuscript.

Funding: This study received funding from King Saud University, Saudi Arabia, through researcher's supporting project number (RSP-2021/145). The APCs were funded by King Saud University, Saudi Arabia, through researcher's supporting project number (RSP-2021/145).

Institutional Review Board Statement: Not applicable.

Informed Consent Statement: Not applicable.

Data Availability Statement: The dataset used in this study is publicly available on GitHub at https://github.com/mshahabalam/TobSet (accessed on 11 October 2021).

Acknowledgments: The authors extend their appreciation to King Saud University, Saudi Arabia, for funding this study through researcher's supporting project number (RSP-2021/145).

Conflicts of Interest: The authors declare no conflict of interest.

References

1. Iqbal, J.; Rauf, A. Tobacco Revenue and Political Economy of Khyber Pakhtunkhwa. *FWU J. Soc. Sci.* **2021**, *15*, 11–25.
2. Wang, G.; Lan, Y.; Yuan, H.; Qi, H.; Chen, P.; Ouyang, F.; Han, Y. Comparison of spray deposition, control efficacy on wheat aphids and working efficiency in the wheat field of the unmanned aerial vehicle with boom sprayer and two conventional knapsack sprayers. *Appl. Sci.* **2019**, *9*, 218. [CrossRef]
3. Liu, W.; Wu, C.; She, D. Effect of spraying direction on the exposure to handlers with hand-pumped knapsack sprayer in maize field. *Ecotoxicol. Environ. Saf.* **2019**, *170*, 107–111. [CrossRef] [PubMed]
4. Hughes, E.A.; Flores, A.P.; Ramos, L.M.; Zalts, A.; Glass, C.R.; Montserrat, J.M. Potential dermal exposure to deltamethrin and risk assessment for manual sprayers: Influence of crop type. *Sci. Total Environ.* **2008**, *391*, 34–40. [CrossRef]
5. Ellis, M.B.; Lane, A.; O'Sullivan, C.; Miller, P.; Glass, C. Bystander exposure to pesticide spray drift: new data for model development and validation. *Biosyst. Eng.* **2010**, *107*, 162–168. [CrossRef]
6. Kim, K.D.; Lee, H.S.; Hwang, S.J.; Lee, Y.J.; Nam, J.S.; Shin, B.S. Analysis of Spray Characteristics of Tractor-mounted Boom Sprayer for Precise Spraying. *J. Biosyst. Eng.* **2017**, *42*, 258–264.
7. Matthews, G. *Pesticide Application Methods*; John Wiley & Sons: Hoboken, NJ, USA, 2008.
8. Talaviya, T.; Shah, D.; Patel, N.; Yagnik, H.; Shah, M. Implementation of artificial intelligence in agriculture for optimisation of irrigation and application of pesticides and herbicides. *Artif. Intell. Agric.* **2020**, *4*, 58–73. [CrossRef]
9. Mavridou, E.; Vrochidou, E.; Papakostas, G.A.; Pachidis, T.; Kaburlasos, V.G. Machine vision systems in precision agriculture for crop farming. *J. Imaging* **2019**, *5*, 89. [CrossRef]
10. Tian, H.; Wang, T.; Liu, Y.; Qiao, X.; Li, Y. Computer vision technology in agricultural automation—A review. *Inf. Process. Agric.* **2020**, *7*, 1–19. [CrossRef]
11. Kamilaris, A.; Prenafeta-Boldú, F.X. Deep learning in agriculture: A survey. *Comput. Electron. Agric.* **2018**, *147*, 70–90. [CrossRef]
12. Osman, Y.; Dennis, R.; Elgazzar, K. Yield Estimation and Visualization Solution for Precision Agriculture. *Sensors* **2021**, *21*, 6657. [CrossRef]
13. Bechar, A.; Vigneault, C. Agricultural robots for field operations: Concepts and components. *Biosyst. Eng.* **2016**, *149*, 94–111. [CrossRef]
14. Berenstein, R.; Edan, Y. Automatic adjustable spraying device for site-specific agricultural application. *IEEE Trans. Autom. Sci. Eng.* **2017**, *15*, 641–650. [CrossRef]

15. Arakeri, M.P.; Kumar, B.V.; Barsaiya, S.; Sairam, H. Computer vision based robotic weed control system for precision agriculture. In Proceedings of the International Conference on Advances in Computing, Communications and Informatics, Udupi, India, 13–16 September 2017; pp. 1201–1205.
16. Gázquez, J.A.; Castellano, N.N.; Manzano-Agugliaro, F. Intelligent low cost telecontrol system for agricultural vehicles in harmful environments. *J. Clean. Prod.* **2016**, *113*, 204–215. [CrossRef]
17. Faiçal, B.S.; Freitas, H.; Gomes, P.H.; Mano, L.Y.; Pessin, G.; de Carvalho, A.C.; Krishnamachari, B.; Ueyama, J. An adaptive approach for UAV-based pesticide spraying in dynamic environments. *Comput. Electron. Agric.* **2017**, *138*, 210–223. [CrossRef]
18. Zhu, H.; Lan, Y.; Wu, W.; Hoffmann, W.C.; Huang, Y.; Xue, X.; Liang, J.; Fritz, B. Development of a PWM precision spraying controller for unmanned aerial vehicles. *J. Bionic Eng.* **2010**, *7*, 276–283. [CrossRef]
19. Yang, Y.; Hannula, S.P. Development of precision spray forming for rapid tooling. *Mater. Sci. Eng. A* **2008**, *477*, 63–68. [CrossRef]
20. Tellaeche, A.; BurgosArtizzu, X.P.; Pajares, G.; Ribeiro, A.; Fernández-Quintanilla, C. A new vision-based approach to differential spraying in precision agriculture. *Comput. Electron. Agric.* **2008**, *60*, 144–155. [CrossRef]
21. Tewari, V.; Pareek, C.; Lal, G.; Dhruw, L.; Singh, N. Image processing based real-time variable-rate chemical spraying system for disease control in paddy crop. *Artif. Intell. Agric.* **2020**, *4*, 21–30. [CrossRef]
22. Rincón, V.J.; Grella, M.; Marucco, P.; Alcatrão, L.E.; Sanchez-Hermosilla, J.; Balsari, P. Spray performance assessment of a remote-controlled vehicle prototype for pesticide application in greenhouse tomato crops. *Sci. Total Environ.* **2020**, *726*, 138509. [CrossRef]
23. Gil, E.; Llorens, J.; Llop, J.; Fàbregas, X.; Escolà, A.; Rosell-Polo, J. Variable rate sprayer. Part 2–Vineyard prototype: Design, implementation, and validation. *Comput. Electron. Agric.* **2013**, *95*, 136–150. [CrossRef]
24. Alam, M.; Alam, M.S.; Roman, M.; Tufail, M.; Khan, M.U.; Khan, M.T. Real-time machine-learning based crop/weed detection and classification for variable-rate spraying in precision agriculture. In Proceedings of the International Conference on Electrical and Electronics Engineering, Antalya, Turkey, 14–16 April 2020; pp. 273–280.
25. Tufail, M.; Iqbal, J.; Tiwana, M.I.; Alam, M.S.; Khan, Z.A.; Khan, M.T. Identification of Tobacco Crop Based on Machine Learning for a Precision Agricultural Sprayer. *IEEE Access* **2021**, *9*, 23814–23825. [CrossRef]
26. Garcia-Ruiz, F.J.; Wulfsohn, D.; Rasmussen, J. Sugar beet (*Beta vulgaris* L.) and thistle (*Cirsium arvensis* L.) discrimination based on field spectral data. *Biosyst. Eng.* **2015**, *139*, 1–15. [CrossRef]
27. Li, X.; Chen, Z. Weed identification based on shape features and ant colony optimization algorithm. In Proceedings of the International Conference on Computer Application and System Modeling, Taiyuan, China, 22–24 October 2010; Volume 1, p. V1-384.
28. Burgos-Artizzu, X.P.; Ribeiro, A.; Guijarro, M.; Pajares, G. Real-time image processing for crop/weed discrimination in maize fields. *Comput. Electron. Agric.* **2011**, *75*, 337–346. [CrossRef]
29. Cheng, B.; Matson, E.T. A feature-based machine learning agent for automatic rice and weed discrimination. In *International Conference on Artificial Intelligence and Soft Computing*; Springer: Berlin/Heidelberg, Germany, 2015; pp. 517–527.
30. Guru, D.; Mallikarjuna, P.; Manjunath, S.; Shenoi, M. Machine vision based classification of tobacco leaves for automatic harvesting. *Intell. Autom. Soft Comput.* **2012**, *18*, 581–590. [CrossRef]
31. Haug, S.; Michaels, A.; Biber, P.; Ostermann, J. Plant classification system for crop/weed discrimination without segmentation. In Proceedings of the IEEE Winter Conference on Applications of Computer Vision, Steamboat Springs, CO, USA, 24–26 March 2014; pp. 1142–1149.
32. Rumpf, T.; Römer, C.; Weis, M.; Sökefeld, M.; Gerhards, R.; Plümer, L. Sequential support vector machine classification for small-grain weed species discrimination with special regard to Cirsium arvense and Galium aparine. *Comput. Electron. Agric.* **2012**, *80*, 89–96. [CrossRef]
33. Ouyang, W.; Zeng, X.; Wang, X.; Qiu, S.; Luo, P.; Tian, Y.; Li, H.; Yang, S.; Wang, Z.; Li, H.; et al. DeepID-Net: Object detection with deformable part based convolutional neural networks. *IEEE Trans. Pattern Anal. Mach. Intell.* **2016**, *39*, 1320–1334. [CrossRef]
34. Diba, A.; Sharma, V.; Pazandeh, A.; Pirsiavash, H.; Van Gool, L. Weakly supervised cascaded convolutional networks. In Proceedings of the IEEE Conference on Computer Vision and Pattern Recognition, Honolulu, HI, USA, 21–26 July 2017; pp. 914–922.
35. Toshev, A.; Szegedy, C. Deeppose: Human pose estimation via deep neural networks. In Proceedings of the IEEE Conference on Computer Vision and Pattern Recognition, Columbus, OH, USA, 23–28 June 2014; pp. 1653–1660.
36. Chen, X.; Yuille, A. Articulated pose estimation by a graphical model with image dependent pairwise relations. *arXiv* **2014**, arXiv:1407.3399.
37. Noh, H.; Hong, S.; Han, B. Learning deconvolution network for semantic segmentation. In Proceedings of the IEEE International Conference on Computer Vision, Santiago, Chile, 7–13 December 2015; pp. 1520–1528.
38. Long, J.; Shelhamer, E.; Darrell, T. Fully convolutional networks for semantic segmentation. In Proceedings of the IEEE Conference on Computer Vision and Pattern Recognition, Boston, MA, USA, 7–12 June 2015; pp. 3431–3440.
39. Lin, L.; Wang, K.; Zuo, W.; Wang, M.; Luo, J.; Zhang, L. A deep structured model with radius–margin bound for 3d human activity recognition. *Int. J. Comput. Vis.* **2016**, *118*, 256–273. [CrossRef]
40. Cao, S.; Nevatia, R. Exploring deep learning based solutions in fine grained activity recognition in the wild. In Proceedings of the 23rd International Conference on Pattern Recognition, Cancun, Mexico, 4–8 December 2016; pp. 384–389.
41. Doulamis, N. Adaptable deep learning structures for object labeling/tracking under dynamic visual environments. *Multimed. Tools. Appl.* **2018**, *77*, 9651–9689. [CrossRef]

42. Lopez-Martin, M.; Carro, B.; Sanchez-Esguevillas, A. IoT type-of-traffic forecasting method based on gradient boosting neural networks. *Future Gener. Comput. Syst.* **2020**, *105*, 331–345. [CrossRef]
43. Bah, M.D.; Hafiane, A.; Canals, R. Deep learning with unsupervised data labeling for weed detection in line crops in UAV images. *Remote Sens.* **2018**, *10*, 1690. [CrossRef]
44. Yu, J.; Schumann, A.W.; Cao, Z.; Sharpe, S.M.; Boyd, N.S. Weed detection in perennial ryegrass with deep learning convolutional neural network. *Front. Plant Sci.* **2019**, *10*, 1422. [CrossRef] [PubMed]
45. Asad, M.H.; Bais, A. Weed detection in canola fields using maximum likelihood classification and deep convolutional neural network. *Inf. Process. Agric.* **2019**, *7*, 535–545. [CrossRef]
46. Umamaheswari, S.; Arjun, R.; Meganathan, D. Weed detection in farm crops using parallel image processing. In Proceedings of the Conference on Information and Communication Technology, Jabalpur, India, 26–28 October 2018; pp. 1–4.
47. Bah, M.D.; Dericquebourg, E.; Hafiane, A.; Canals, R. Deep learning based classification system for identifying weeds using high-resolution UAV imagery. In *Science and Information Conference*; Springer: Cham, Switzerland, 2018; pp. 176–187.
48. Forero, M.G.; Herrera-Rivera, S.; Ávila-Navarro, J.; Franco, C.A.; Rasmussen, J.; Nielsen, J. Color classification methods for perennial weed detection in cereal crops. In *Iberoamerican Congress on Pattern Recognition*; Springer: Berlin/Heidelberg, Germany, 2018; pp. 117–123.
49. Wang, A.; Zhang, W.; Wei, X. A review on weed detection using ground-based machine vision and image processing techniques. *Comput. Electron. Agric.* **2019**, *158*, 226–240. [CrossRef]
50. Hu, K.; Wang, Z.; Coleman, G.; Bender, A.; Yao, T.; Zeng, S.; Song, D.; Schumann, A.; Walsh, M. Deep Learning Techniques for In-Crop Weed Identification: A Review. *arXiv* **2021**, arXiv:2103.14872.
51. Loey, M.; ElSawy, A.; Afify, M. Deep learning in plant diseases detection for agricultural crops: A survey. *Int. J. Serv. Sci. Manag. Eng. Tech.* **2020**, *11*, 41–58. [CrossRef]
52. Weng, Y.; Zeng, R.; Wu, C.; Wang, M.; Wang, X.; Liu, Y. A survey on deep-learning-based plant phenotype research in agriculture. *Sci. Sin. Vitae* **2019**, *49*, 698–716. [CrossRef]
53. Bu, F.; Gharajeh, M.S. Intelligent and vision-based fire detection systems: A survey. *Image Vis. Comput.* **2019**, *91*, 103803. [CrossRef]
54. Chouhan, S.S.; Singh, U.P.; Jain, S. Applications of computer vision in plant pathology: A survey. *Arch. Comput. Methods Eng.* **2020**, *27*, 611–632. [CrossRef]
55. Bonadies, S.; Gadsden, S.A. An overview of autonomous crop row navigation strategies for unmanned ground vehicles. *Eng. Agric. Environ. Food* **2019**, *12*, 24–31. [CrossRef]
56. Tripathi, M.K.; Maktedar, D.D. A role of computer vision in fruits and vegetables among various horticulture products of agriculture fields: A survey. *Inf. Process. Agric.* **2020**, *7*, 183–203. [CrossRef]
57. da Costa, A.Z.; Figueroa, H.E.; Fracarolli, J.A. Computer vision based detection of external defects on tomatoes using deep learning. *Biosyst. Eng.* **2020**, *190*, 131–144. [CrossRef]
58. dos Santos Ferreira, A.; Freitas, D.M.; da Silva, G.G.; Pistori, H.; Folhes, M.T. Weed detection in soybean crops using ConvNets. *Comput. Electron. Agric.* **2017**, *143*, 314–324. [CrossRef]
59. Yu, J.; Sharpe, S.M.; Schumann, A.W.; Boyd, N.S. Deep learning for image-based weed detection in turfgrass. *Eur. J. Agron.* **2019**, *104*, 78–84. [CrossRef]
60. Sharpe, S.M.; Schumann, A.W.; Boyd, N.S. Detection of Carolina geranium (Geranium carolinianum) growing in competition with strawberry using convolutional neural networks. *Weed Sci.* **2019**, *67*, 239–245. [CrossRef]
61. Le, V.N.T.; Truong, G.; Alameh, K. Detecting weeds from crops under complex field environments based on Faster RCNN. In Proceedings of the IEEE Eighth International Conference on Communications and Electronics, Phu Quoc Island, Vietnam, 13–15 January 2021; pp. 350–355.
62. Quan, L.; Feng, H.; Lv, Y.; Wang, Q.; Zhang, C.; Liu, J.; Yuan, Z. Maize seedling detection under different growth stages and complex field environments based on an improved Faster R–CNN. *Biosyst. Eng.* **2019**, *184*, 1–23. [CrossRef]
63. Osorio, K.; Puerto, A.; Pedraza, C.; Jamaica, D.; Rodríguez, L. A deep learning approach for weed detection in lettuce crops using multispectral images. *AgriEngineering* **2020**, *2*, 471–488. [CrossRef]
64. Wu, D.; Lv, S.; Jiang, M.; Song, H. Using channel pruning-based YOLO v4 deep learning algorithm for the real-time and accurate detection of apple flowers in natural environments. *Comput. Electron. Agric.* **2020**, *178*, 105742. [CrossRef]
65. Ren, S.; He, K.; Girshick, R.; Sun, J. Faster r-cnn: Towards real-time object detection with region proposal networks. *Adv. Neural Inf. Process. Syst.* **2015**, *28*, 91–99. [CrossRef]
66. Simonyan, K.; Zisserman, A. Very deep convolutional networks for large-scale image recognition. *arXiv* **2014**, arXiv:1409.1556.
67. Zeiler, M.D.; Fergus, R. Visualizing and understanding convolutional networks. In *European Conference on Computer Vision*; Springer: Berlin/Heidelberg, Germany, 2014; pp. 818–833.
68. Redmon, J.; Divvala, S.; Girshick, R.; Farhadi, A. You only look once: Unified, real-time object detection. In Proceedings of the IEEE Conference on Computer Vision and Pattern Recognition, Las Vegas, NV, USA, 27–30 June 2016; pp. 779–788.
69. Bochkovskiy, A.; Wang, C.Y.; Liao, H.Y.M. Yolov4: Optimal speed and accuracy of object detection. *arXiv* **2020**, arXiv:2004.10934.

Article

Simulation-Aided Development of a CNN-Based Vision Module for Plant Detection: Effect of Travel Velocity, Inferencing Speed, and Camera Configurations

Paolo Rommel Sanchez [1,2] and Hong Zhang [1,*]

[1] Mechanical Engineering Department, Henry M. Rowan College of Engineering, Rowan University, Glassboro, NJ 08028, USA; sanche45@students.rowan.edu or ppsanchez@up.edu.ph
[2] Agribiosystems Machinery & Power Engineering Division, Institute of Agricultural and Biosystems Engineering, College of Engineering and Agro-Industrial Technology, University of the Philippines Los Baños, Los Baños 4031, Philippines
* Correspondence: zhang@rowan.edu

Citation: Sanchez, P.R.; Zhang, H. Simulation-Aided Development of a CNN-Based Vision Module for Plant Detection: Effect of Travel Velocity, Inferencing Speed, and Camera Configurations. *Appl. Sci.* **2022**, *12*, 1260. https://doi.org/10.3390/app12031260

Academic Editors: Anselme Muzirafuti and Dimitrios S. Paraforos

Received: 15 December 2021
Accepted: 21 January 2022
Published: 25 January 2022

Publisher's Note: MDPI stays neutral with regard to jurisdictional claims in published maps and institutional affiliations.

Copyright: © 2022 by the authors. Licensee MDPI, Basel, Switzerland. This article is an open access article distributed under the terms and conditions of the Creative Commons Attribution (CC BY) license (https://creativecommons.org/licenses/by/4.0/).

Abstract: In recent years, Convolutional Neural Network (CNN) has become an attractive method to recognize and localize plant species in unstructured agricultural environments. However, developed systems suffer from unoptimized combinations of the CNN model, computer hardware, camera configuration, and travel velocity to prevent missed detections. Missed detection occurs if the camera does not capture a plant due to slow inferencing speed or fast travel velocity. Furthermore, modularity was less focused on Machine Vision System (MVS) development. However, having a modular MVS can reduce the effort in development as it will allow scalability and reusability. This study proposes the derived parameter, called overlapping rate (r_o), or the ratio of the camera field of view (S) and inferencing speed (fps) to the travel velocity ($\vec{v}$) to theoretically predict the plant detection rate (r_d) of an MVS and aid in developing a CNN-based vision module. Using performance from existing MVS, the values of r_o at different combinations of inferencing speeds (2.4 to 22 fps) and travel velocity (0.1 to 2.5 m/s) at 0.5 m field of view were calculated. The results showed that missed detections occurred when r_o was less than 1. Comparing the theoretical detection rate ($r_{d,th}$) to the simulated detection rate ($r_{d,sim}$) showed that $r_{d,th}$ had a 20% margin of error in predicting plant detection rate at very low travel distances (<1 m), but there was no margin of error when travel distance was sufficient to complete a detection pattern cycle ($\geq$10 m). The simulation results also showed that increasing S or having multiple vision modules reduced missed detection by increasing the allowable $\vec{v}_{max}$. This number of needed vision modules was equal to rounding up the inverse of r_o. Finally, a vision module that utilized SSD MobileNetV1 with an average effective inferencing speed of 16 fps was simulated, developed, and tested. Results showed that the $r_{d,th}$ and $r_{d,sim}$ had no margin of error in predicting r_{actual} of the vision module at the tested travel velocities (0.1 to 0.3 m/s). Thus, the results of this study showed that r_o can be used to predict r_d and optimize the design of a CNN-based vision-equipped robot for plant detections in agricultural field operations with no margin of error at sufficient travel distance.

Keywords: modeling; simulation; precision agriculture; convolutional neural networks; machine vision; computer vision; modular robot

1. Introduction

The increasing cost and decreasing availability of agricultural labor [1,2] and the need for sustainable farming methods [3–5] led to the development of robots for agricultural field operations. However, despite the success of robots in industrial applications, agricultural robots for field operations remain primarily in the development stage due to the complex characteristics of the farming environment, high cost of development, and high durability, functionality, and reliability requirements [6–8].

As a potential solution to these challenges, field robots with computer vision have been increasing due to the large amount of information that can be extracted from an agricultural scene [9]. Real-time machine vision systems (MVS) are often used to recognize, classify and localize plants accurately for precision spraying [10,11], mechanical weeding [12], solid fertilizer application [13], and harvesting [14,15]. However, using traditional image processing techniques, early machine vision implementations for field operations were difficult due to the vast number of features needed to model and differentiate plant species [13] and work at various farm scenarios [12].

Recently, developments in deep learning allowed Convolutional Neural Networks (CNN) to be used for accurate plant species detection and segmentation [16,17]. However, despite high classification and detection performance, the large computational power requirement of CNN limits its application in real-time operations [18]. As a result, most CNN applications in agriculture were primarily employed in non-real-time scenarios, excluding inferencing speeds in the evaluated parameters among related studies. For example, a study that surveyed CNN-based weed detection and plant species classification reported 86–97% and 48–99% precisions, respectively, but data on inferencing speeds were unreported [19]. Similarly, research on fruit classification and recognition using CNN showed 77–99% precision, but inferencing speeds were also excluded in the measured parameters [19–21].

Few studies have evaluated the real-time performance of CNNs for agricultural applications. For example, in a study by Olsen et al. (2019) [22] on detecting different species of weeds, the real-time performance of ResNet-50 in an NVIDIA Jetson TX2 was only 5.5 fps at 95.1% precision. Optimizing their TensorFlow model using TensorRT increased the inferencing speed to 18.7 fps. The study of Chechliński et al. (2019) [23] using a custom CNN architecture based on U-Net, MobileNets, DenseNet, and ResNet in a Raspberry Pi 3B+ resulted in only 60.2% precision at 10.0 fps. Finally, Partel et al. (2019) [11] developed a mobile robotic sprayer that used YOLOv3 running on an NVIDIA 1070TI at 22 fps and NVIDIA Jetson TX2 at 2.4 fps for real-time crop and weed detection and spraying. The vision system with 1070TI had 85% precision, while the TX2 vision system had 77% precision. Furthermore, the system with TX2 missed 43% of the plants because of its slow inferencing speed. Conversely, the CNN-based sprayer with the faster 1070TI, resulting in higher inferencing speed, only missed 8% of the plants.

However, the travel velocities of surveyed systems were often unevaluated despite operating in real-time. Additionally, theoretical approaches to quantify and account for the effect of travel velocity on the capability of the vision system to sufficiently capture discrete data and effectively represent a continuous field scenario were often not included in the design. Other studies, on the contrary, evaluated the effect of travel velocity on CNN detection performance, but the impact of inferencing speed remained unevaluated. For instance, in the study of Liu et al. (2021) [24], the deep learning-based variable rate agrochemical spraying system for targeted weeds control in strawberry crops equipped with a 1080TI showed increasing missed detections, as travel velocity increases regardless of CNN architecture (VGG-16, GoogleNet or AlexNet). At 1, 3, and 5 km/h, their system missed 5–9%, 6–10%, and 13–17% of the targeted weeds, respectively.

The brief review of the developed systems showed that inferencing speed (fps) and travel velocity ($\vec{v}$) of a CNN-based MVS impact its detection rate (r_d). r_d is the fraction of the number of detected plants to the total number of plants and was often expressed as the recall of the CNN model [11,24,25]. However, a theoretical approach to predict the r_d of a CNN-based MVS as affected by both fps and $\vec{v}$ is yet to be explored.

Current approaches in developing a mechatronic system with CNN-based MVS involve building and testing an actual system to determine r_d [11,24,25]. However, building and testing several CNN-based MVS to determine the effect of $\vec{v}$ and fps on r_d would be very tedious as the process would involve building the MVS hardware, image dataset preparation, training multiple CNN models, developing multiple software frameworks to integrate the different CNN models into the vision system, and testing the MVS. Hence, this study also proposes to use computer simulation, in addition to theoretical modeling, in predicting r_d as a function of the mentioned parameters, to reduce the difficulty of selecting and sizing the components of the CNN-based MVS.

Computer simulations were often used to characterize the effect of design and operating parameters on the overall performance of agricultural robots for field operations. For example, Villette et al. (2021) [26] demonstrated that computer simulations could be used to estimate the required sprayer spatial resolution of vision-equipped boom sprayer, as affected by boom section weeds, nozzle spray patterns, and spatial weed distribution. Wang et al. (2018) [27] used computer modeling and simulation to identify potential problems of a robotic apple picking arm and developed an algorithm to improve the performance by 81%. Finally, Lehnert et al. (2019) [28] used modeling and computer simulation to create a novel multi-perspective visual servoing technique to detect the location of occluded peppers. However, the use of simulation to quantify the effects of fps, $\vec{v}$, and camera configurations on r_d of an MVS remain unexplored.

Furthermore, a survey of review articles showed that almost all robotic systems in agriculture employ fixed configurations and are non-scalable, resulting in less adaptability to complex agricultural environments [7,8,29,30]. Thus, aside from modeling and simulation, this study also proposes a module-based design approach to enable reusability and scalability by minimizing production costs and shortening the lead time of machine development [31]. A module can be defined as a repeatable and reusable machine component that performs partial or full functions and interact with other machine components, resulting in a new machine with new overall functionalities [32].

Therefore, this study proposes theoretical and simulation approaches in predicting combinations of fps, $\vec{v}$, and camera configuration to prevent missed plant detections and aid in developing a modular CNN-based MVS. In particular, based on the brief literature review and identified research gaps, the study specifically aims to make the following contributions:

- The introduction of a dimensionless parameter, called overlapping rate (r_0), which is a quantitative predictor of r_d as a function of $\vec{v}$ and fps, for a specific camera field of view (S);
- A set of Python scripts containing equations and algorithms to model a virtual field, simulate the motion of the virtual camera, and perform plant hill detection;
- An evaluation of r_d based on different combinations of published fps and $\vec{v}$ from existing systems through the proposed theoretical approach and simulation;
- A detailed analysis through simulation of the concept of increasing S or using several adjacent synchronous cameras (n_{vis}) to prevent missed detections in MVS with low fps at high $\vec{v}$; and
- A reusable and scalable CNN-based vision module for plant detection based on Robot Operating System (ROS) and the Jetson Nano platform.

2. Materials and Methods

2.1. Concept

Cameras for plant detection are typically mounted on a boom of a sprayer or fertilizer spreader [10,11,13], as illustrated in Figure 1. They are oriented so that their optical axis is perpendicular to the field and captures top-view images of plants [33].

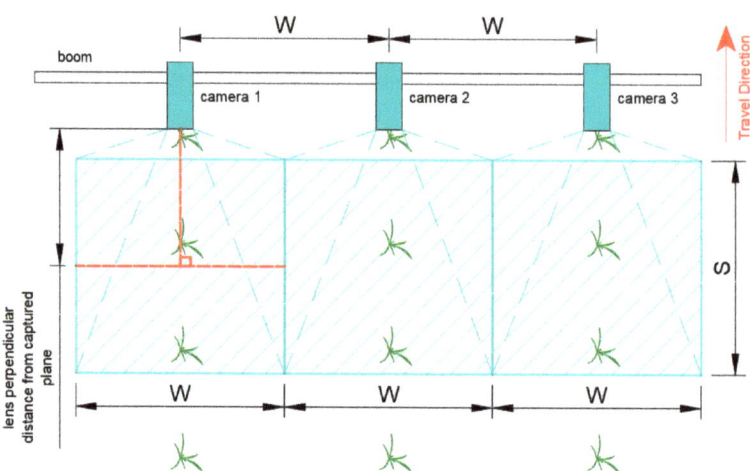

Figure 1. Camera mounting location and orientation in a boom (not drawn in scale).

Depending on the distance between the camera lens and captured plane, lens properties, and sensor size, the frame width or height is equivalent to an actual linear distance. For simplicity, the linear length of the side of a field of view of a frame parallel to the direction of travel shall be denoted as S, in meters per frame. To provide complete visual coverage of the traversed width of the boom, the maximum spacing between adjacent cameras is equal to the length of the side of a field of view of a frame perpendicular to the direction of travel, denoted as W, in meters per frame [10].

During motion, the traverse distance between two consecutive frames of the camera (d_f), in meters, is equal to the product of travel velocity ($\vec{v}$), in m/s, and the time between the frames ($1/fps$), in seconds, as illustrated in Equation (1).

$$d_f = \vec{v} \cdot \frac{1}{fps} = \frac{\vec{v}}{fps} \quad (1)$$

The ratio of S to d_f is proposed as the overlapping rate (r_o), a dimensionless parameter, and is represented by Equation (2).

$$r_o = \frac{S \times fps}{\vec{v}} = \frac{S}{d_f} \quad (2)$$

With a single camera, the value of r_o describes whether there is an overlap or gap between frames. Depending on the value of r_o, certain regions in the traversed field of the machine, with a single camera configuration, will be uniquely captured, captured in multiple frames, or completely missed, as shown in Figure 2 and illustrated by the following cases:

- Case 1: $r_o = 1$. When S and d_f are equal, the extents of each consecutive processed frame are side by side. Hence, both gaps and overlaps are absent. This scenario is ideal since the vehicle velocity and camera frame rate match perfectly.
- Case 2: $r_o > 1$. The vision system accounted for all regions in the traversed field, but there is an overlap between the frames. The length of the current frame that is already accounted for by the previous one is $S - d_f$. The vehicle can run faster if the mechanical capacity allows.
- Case 3: $r_o < 1$. Gaps will occur between each pair of consecutive frames, and the camera will miss certain plants. The length of each gap is $d_f - S$. We can define a gap rate (r_g) as shown in Equation (3) to depict the significance of the gap.

$$r_g = \frac{d_f - S}{d_f} = 1 - r_o \qquad (3)$$

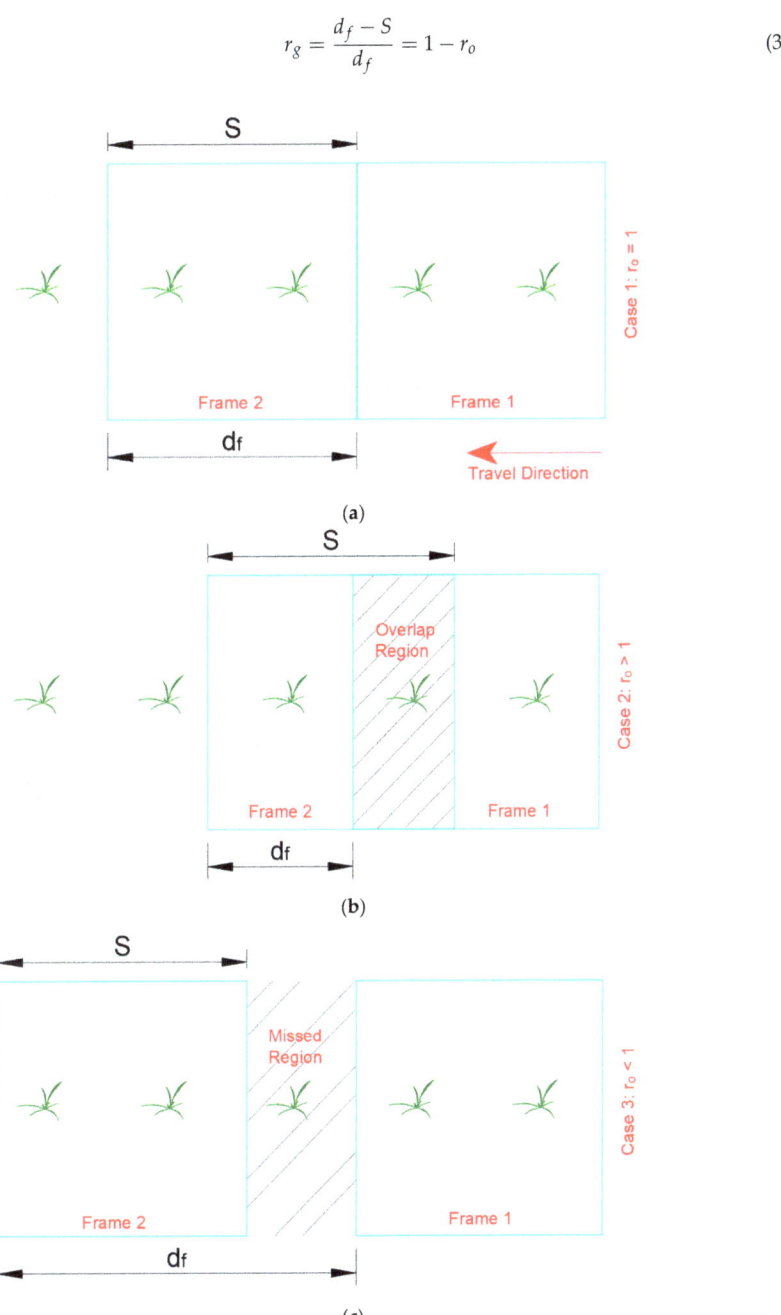

Figure 2. Relative positions of two consecutive processed frames of a vision system: (**a**) $r_o = 1$ since $d_f = S$, resulting in frames with unique bounded regions; (**b**) $r_o > 1$ since $d_f < S$, resulting in overlap region (**b**); and (**c**) $r_o < 1$ since $d_f > S$, resulting in missed region.

2.1.1. Theoretical Detection Rate

The theoretical or maximum detection rate ($r_{d,th}$) can be defined as $\min(1, r_0)$, shown in Equation (4). The $r_{d,th}$ was also a dimensionless parameter.

$$r_{d,th} = \begin{cases} 1, & r_0 \geq 1 \\ r_0, & r_0 < 1 \end{cases} \tag{4}$$

2.1.2. Maximizing Travel Velocity

Setting $r_0 = 1$ in Equation (2) resulted in Equation (5), which is similar to the equation used by Esau et al. (2018) [10] in calculating the maximum travel velocity of a sprayer. However, Equation (5) only describes the maximum forward velocity $\vec{v}_{max}$ that a vision-equipped robot can operate to prevent gaps while traversing the field as a function of S and fps.

$$\vec{v}_{max} = S \times fps \tag{5}$$

2.1.3. Increasing $\vec{v}_{max}$

A consequence of Equation (5) was that increasing the length of the frame S at a constant fps shall increase $\vec{v}_{max}$. Hence, raising the camera mounting height or using multiple adjacent synchronous cameras along a single plant row can increase the effective S. This situation, then, shall increase $\vec{v}_{max}$ without the need for powerful hardware for a faster inferencing speed. When $r_0 < 1$, the number of vision modules (n_{vis}) to prevent missed detection can be calculated using Equation (6). Since r_0 represents the fraction of the field that can be covered by a single camera, the inverse of r_0 represents the number of adjacent cameras that will result in 100% field coverage. The calculated inverse was rounded up to the following number, as cameras are discrete elements.

$$n_{vis} = \left\lceil \frac{1}{r_0} \right\rceil \tag{6}$$

The effective actual ground distance (S_{eff}), in meters, captured side-by-side by identical and synchronous vision modules without gaps and overlaps is equal to the product of n_{vis} and S, as shown in Equation (7). This configuration will then allow the use of less powerful devices while operating at the required $\vec{v}$ of an agricultural field operation such as spraying, as illustrated in Figure 3.

$$S_{eff} = S \times n_{vis} \tag{7}$$

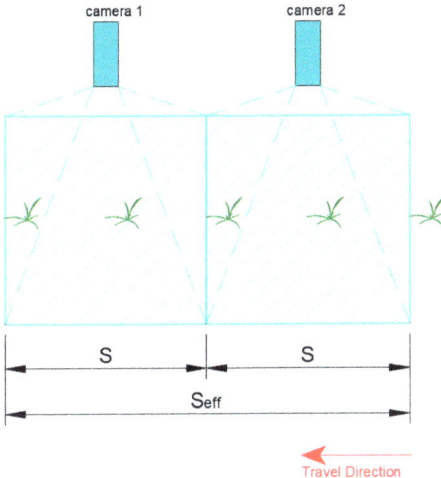

Figure 3. Multiple adjacent cameras for plant detection at $n_{vis} = 2$. Thus, $S_{eff} = 2S$.

2.2. Field Map Modeling

A virtual field was prepared to test the concepts that were presented. A 1000 m field length (d_l) with crops planted in hills at 0.2 m hill spacings (d_h) was used. The number of hills (n_h) and plant hill locations (X_i), in meters, in the entire field length were calculated using Equations (8) and (9), respectively. A section of the virtual field is presented in Figure 4.

$$n_h = \left[\frac{d_l}{d_h}\right] \quad (8)$$

$$X_i = i \times d_h; \ i \in 1, 2, \ldots n_h \quad (9)$$

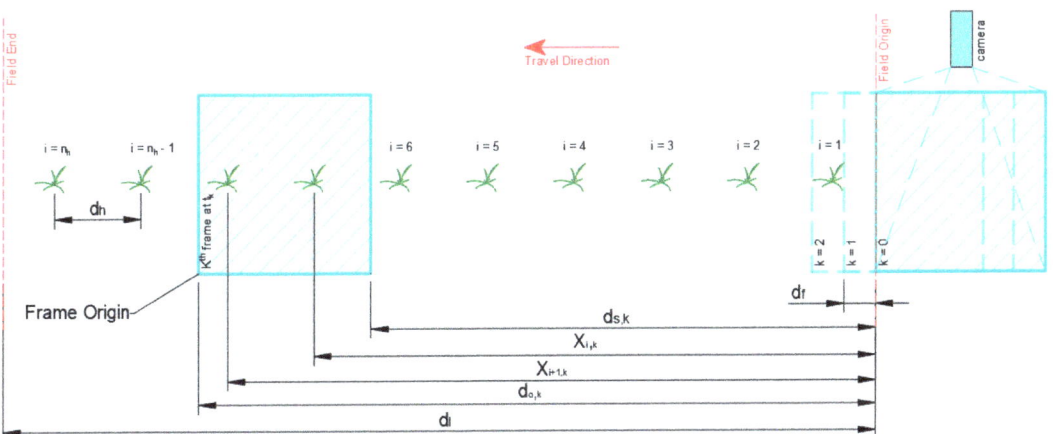

Figure 4. Virtual field with field map and motion modeling parameters. The frame at $k = 0$ represents the frame just outside the virtual field. The frame at $k = 1$ represents the first frame that entered the virtual field.

2.3. Motion Modeling

The robot was assumed to move from right to left of the field during simulation, as shown in Figure 4. Therefore, the right border of the virtual area was the assumed field origin. The total number of frames (K) throughout the motion of the vision system then becomes the number of d_f-sized steps to completely traverse d_l, as shown Equation (10).

$$K = \frac{d_l}{d_f} \qquad (10)$$

The elapsed time after several frame steps (t_k), in seconds, was calculated by dividing the number of elapsed frames (k) by the inferencing speed, as shown in Equation (11). t_k was then used to calculate the distance of the left ($d_{o,k}$) and right ($d_{s,k}$) borders of the virtual camera frame with respect to the field origin, in meters, using kinematic equations as shown in Equations (12) and (13), respectively. In Equation (13), S was subtracted from $d_{o,k}$ due to the assumed right to left motion of the camera.

$$k \in 0, 1, 2, \ldots K$$

$$t_k = k \times \frac{1}{fps} = \frac{k}{fps} \qquad (11)$$

$$d_{o,k} = \vec{v} \times t_k \qquad (12)$$

$$d_{s,k} = d_{o,k} - S \qquad (13)$$

2.4. Detection Algorithm

The simulation was implemented using two Python scripts, which were made publicly available in GitHub. The first script, called "settings.py", was a library that defined the "Settings" object class. This object contained the properties of the virtual field, kinematic motion, and camera parameters for the detection. The second script, "vision-module.py", was a ROS node that published only the horizontal centroid coordinates of the plant hills that would be within the virtual camera frame. The central aspect of ROS was implementing a distributed architecture that allows synchronous or asynchronous communication of nodes [34]. Hence, the ROS software framework was used so that the written simulation scripts for the vision system can be used in simulating the performance and optimizing the code of a precision spot sprayer that was also being developed as part of the future implementation of this study.

When "vision-module.py" was executed, it initially loaded the "Settings" class and fetched the required parameters, including X_i, from "settings.py". The following algorithm was then implemented for the detection:

1. Create an empty NumPy vector of detected hills.
2. For each k frame in K total frames:
   a. t_k, $d_{o,k}$ and $d_{s,k}$ were calculated.
   b. For each i within the number of hills n_h:
      i. All X_i within the left border, $d_{o,k}$, and the right border, $d_{s,k}$, were plant hills within the camera frame
      ii. Append detected hill indices to list
3. The number of detected hills (n_d) was then equal to the number of unique detected hill indices in the list.

In step 1, an empty vector was needed to store the indices of the detected plant hills. In step 2, each k frame represented a camera position as the vision system traversed along the field. Step 2a calculated the elapsed time and the left and right border locations of the frame as described in Section 2.3. The specific detection method was performed in Step 2b, which compared the current distance locations of the left and right bounds of the camera frame to the plant hill locations. The plant hill indices that satisfied Step 2b-i were then appended to the NumPy vector. The duplicates were filtered from the NumPy vector in Step 3, and the remaining elements were counted and stored in the integer variable n_d.

Finally, the simulated detection rate ($r_{d,sim}$) of the vision system was then the quotient of n_d and n_h as shown in Equation (14).

$$r_{d,sim} = \frac{n_d}{n_h} \qquad (14)$$

2.5. Experimental Design

A laptop (Lenovo ThinkPad T15 g Gen 1) with Intel Core i7-10750H, 16 GB DDR4 RAM, and NVIDIA RTX 2080 Super was used in the computer simulation. The script was implemented using Python 2.7 programming language and ROS Melodic Morenia in Ubuntu 18.04 LTS operating system.

The simulation was performed at $S = 0.5$ m, based on the camera configuration of Chechliński et al. (2019) [23]. Sensitivity analysis was performed at d_l values of 1, 10, 100, 1000, and 10,000 m. The literature review showed that 20,000 m was used in the study of Villette et al., 2021 [26]. However, the basis for the d_l used in their study was not explained. Hence, sensitivity analysis was performed in this study to establish the sufficient d_l that would not affect $r_{d,th}$ and $r_{d,sim}$. The resulting values of $r_{d,sim}$ were compared to $r_{d,th}$. Inferencing speed of 2.4 *fps* and travel velocity of 2.5 m/s were used for the sensitivity analysis to have an $r_o < 1$ at $S = 0.5$ m. If a faster inferencing speed or slower travel was used, r_o could be equal to or greater than 1. This result will fall into Case 1 or 2 and could not be used for sensitivity analysis.

The model was then simulated at different values of $\vec{v}$ and fps, as shown in Table 1 to estimate the detection performance of combinations of CNN model, hardware, and $\vec{v}$. Forward walking speeds using a knapsack sprayer typically ranged from 0.1 to 1.78 m/s [35–37]. On the other hand, the travel velocities of boom sprayers ranged from 0.7 to 2.5 m/s [38–41]. Solid fertilizer application using a tractor-mounted spreader operated at 0.89 to 1.68 m/s [13,42]. Finally, a mechanical weeder with rotating mechanisms worked at 0.28–1.67 m/s [43,44]. The literature review showed that 0.1 m/s was the slowest [41] and 2.5 m/s was the highest [40] forward velocities found. The mid-point velocity of 1.3 m/s estimated the typical walking speed using knapsack sprayers [35–37] and forward travel velocities of boom sprayers and fertilizer applicators [38–41].

Table 1. Complete factorial design to determine the detection rate of a CNN-based vision system for agricultural field operation using simulation.

Levels	Parameter	
	$\vec{v}$, m/s	*fps*
Low (−1)	0.1	2.4
Standard (0)	1.3	12.2
High (+1)	2.5	22.0

In addition, 2.4 and 22 *fps* were the inferencing speeds of YOLOv3 running on an NVIDIA TX2 embedded system and a laptop with NVIDIA 1070TI discrete GPU as described in the study of Partel et al. (2019) [11]. Finally, 12.2 *fps* approximated the inferencing time of a custom CNN architecture or SSD MobileNetV1 CNN model optimized in TensorRT and implemented an embedded system [22,23].

The effect of increasing S using multiple camera modules in preventing missed detection was also performed on treatments falling under Case 3.

2.6. Vision Module Development

The development of the vision module was divided into three phases: (1) hardware and software development; (2) dataset preparation and training of the CNN model; and (3) simulation and testing.

2.6.1. Hardware and Software Development

Table 2 summarizes the list and function of the hardware components used to develop the vision module. NVIDIA Jetson Nano with 4 GB RAM was used to perform inferencing on 1280 × 720 at 30 *fps* video from a USB webcam (Logitech StreamCam Plus). Powering the whole system is a power adapter that outputs 5VDC at 4A.

Table 2. Summary of vision module hardware.

Hardware	Model	Function
Webcam	Logitech StreamCam Plus	Realtime video capture
Vision Compute Unit	NVIDIA Jetson Nano 4 GB	Image inferencing
Communication Bus	USB 3.0	Communication with USB devices
Power Adapter	5VDC 4A Power Adapter	Supplies power to the vision compute unit

Table 3 summarizes the software packages used to develop the software framework of the vision module. The software for the vision module was written in Python 2.7. The detectnet object class of Jetson Inference Application Programming Interface (API) was used to develop the major components of the software framework. Detectnet object facilitated connecting to the webcam using gstream, optimizing the PyTorch-based SSD MobileNetV1 model into TensorRT, loading the model, performing inferences on the video stream from the webcam, image processing for drawing bounding boxes onto the processed frame, and displaying the frame. OpenCV is an open-source computer vision library focused on real-time applications. It was used to display the calculated speed of the vision module and convert the detectnet image format from red–green–blue–alpha (RGBA) to blue–green–red (BGR), which was the format needed by ROS for image transmission.

Table 3. Summary of vision module software.

Software Package	Function
NVIDIA Jetson Inference API	Facilitates camera connection, training of object detection model, converting to TensorRT, loading of object detection model, inferencing, and image processing
Python	General programming language to implement the algorithms
OpenCV	Image processing
Robot Operating System (ROS)	Image data, plant coordinate, and processing time transmission
Ubuntu 18.04 ARM	The operating system for Jetson Nano and hosts the other software packages

To enable modularity, the software framework, as illustrated in Figure 5, was also implemented using ROS version Melodic Morenia, which was the version that was compatible with Ubuntu 18.04. A node is a virtual representation of a component that can send or receive messages directly from other nodes. The vision module or node required two inputs: (1) RGB video stream from a video capture device and (2) TensorRT-optimized SSD MobileNetV1 object detection model. It calculates and outputs four parameters, namely: (1) weed coordinates, (2) crop coordinates, (3) processed images, and (4) total delay time. Each parameter was published into its respective topics. Table 4 summarizes the datatype and the function of these outputs.

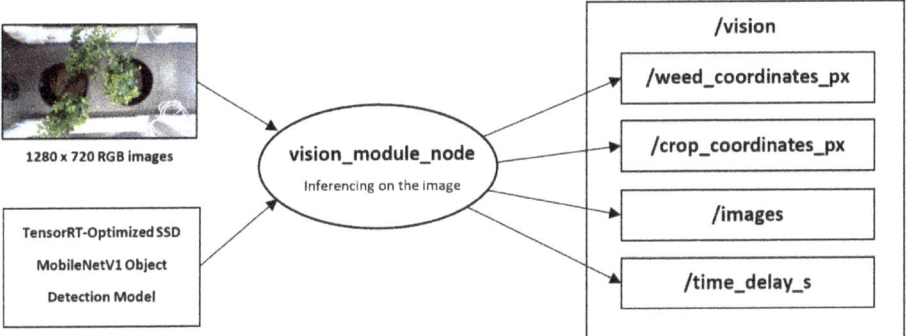

Figure 5. Software framework of the vision module.

Table 4. Output parameters of the vision module with their description.

Parameter	Datatype	Description
Weed coordinates, px	Integer	Array of integers representing the x-coordinate of all detected weed per frame
Crop coordinates, px	Integer	Array of integers representing the x-coordinate of all detected crops per frame
Images	CvBridge	Image data with detections
Time delay, s	Float	Total delay time of the vision module as a result of inferencing, image processing, calculation, and data transmission

2.6.2. Dataset Preparation and Training of the CNN Model

Using the Jetson Inference library, a CNN model for plant detection was trained using SSD MobileNetV1 object detection architecture and PyTorch machine learning framework. A total of 2000 sample images of artificial potted plants at 1280 × 720 composed of 50% weeds and 50% plants were prepared. For CNN model training and validation, 80% and 20% of the datasets were used, respectively. A batch size of 4, base learning rate of 0.001, and momentum of 0.90 were used to train the model for 100 epochs (5000 iterations).

2.6.3. Testing and Simulation

The performance requirement for the vision module was to avoid missed detections for spraying operations at walking speeds, which was 0.1 m/s at minimum [35–37]. The Jetson Nano and webcam were mounted at a height where S was equal to 0.79 m, as shown in Figure 6. S was determined so that the top projections of the potted plants were within the camera frame, and the camera and plants would not collide during motion. A conveyor belt equipped with a variable speed motor was used to reproduce the relative travel velocity of the vision system at 0.1, 0.2, and 0.3 m/s. A maximum of 0.3 m/s was used, since beyond this conveyor speed consistent d_h at 0.2 m was difficult to achieve, even with three people performing the manual loading and unloading, as the potted artificial plants were traveling too fast.

A total of 60 potted plants were loaded onto the conveyor for each conveyor speed setting. Detection was carried out at a minimum conference threshold of 0.5. Detected and correctly classified plants were considered true positives (TP), while detected and incorrectly classified plants were categorized as false positives (FP). Missed detections were classified as false negatives (FN). The precision (p_{actual}) and recall (r_{actual}) of the vision module were then determined using Equations (15) and (16), respectively.

$$p_{actual} = \frac{TP}{TP + FP} \qquad (15)$$

$$r_{actual} = \frac{TP}{TP + FN} \tag{16}$$

Figure 6. Laboratory setup composed of Jetson Nano 4GB, Logitech StreamCam Plus, and variable speed conveyor belt with artificial potted plants.

3. Results and Discussion

The sensitivity of $r_{d,th}$ and $r_{d,sim}$ to the total traversed distance was first determined to establish the d_l that was used in the experimental design. The influence of $\vec{v}$ and fps at specific S on $r_{d,th}$ and $r_{d,sim}$ were then compared and analyzed. Finally, the results of performance testing the vision module were compared to theoretical and simulation results.

3.1. Sensitivity Analysis

As illustrated in Figure 7, the sensitivity analysis results showed that $r_{d,th}$ and $r_{d,sim}$ converged at 10 m traversed distance. The 20% difference of $r_{d,th}$ from $r_{d,sim}$ can be attributed to the different variables considered to determine each parameter. $r_{d,th}$ used inferencing speed, travel velocity, and capture width to theoretically calculate the gaps between consecutive processed frames related to the detection rate, as illustrated in Section 2.1.

On the other hand, $r_{d,sim}$ determined the detection rate based on the number of unique plants within the processed frames, as influenced by traversed distance, hill spacing, inferencing speed, travel velocity, and capture width, as described in Section 2.2, Section 2.3, Section 2.4. Results showed that simulation better approximated the detection rate than theoretical approaches at less than 10 m traversed distance, 0.2 m hill spacing, 2.5 m/s travel velocity, 2.4 *fps*, and 0.5 m frame capture width.

These results infer that at very short distances, $r_{d,sim}$ approximates the detection rate more accurately than $r_{d,sim}$. However, for long traversed distances, the influence of hill spacing on the detection rate was no longer significant and $r_{d,th}$ can simply be used to calculate the detection rate.

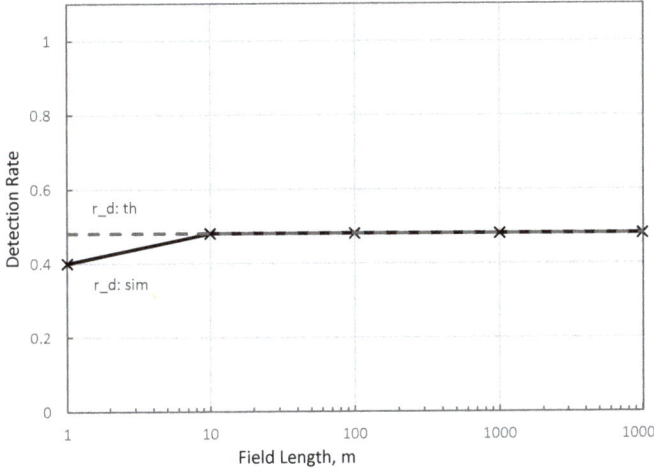

Figure 7. Theoretical (solid) and simulated (broken-line) detection rates at 1, 10, 100, 1000, and 10,000 m field lengths at 0.2 m hill spacing, 2.5 m/s travel velocity, 2.4 fps, and 0.5 m frame capture width.

3.2. Effects of $\vec{v}$ and fps

Table 5 summarizes the theoretical and simulation results on the combinations of $\vec{v}$ and fps at S equal to 0.5-m. Comparing the $r_{d,th}$ to $r_{d,sim}$ for any combinations of the tested parameters showed that detection rates were equal. Results also showed that there were no missed detections at any $\vec{v}$ when the inferencing speeds were at 12.2 and 22 fps (Case 2), as illustrated in Figures 8 and 9. These results infer that one-stage object detection models, such as YOLO and SSD, running on a discrete GPU such as 1070TI, have sufficient inferencing speed to avoid detection gaps in typical ranges of travel velocities for agricultural field operations. The result was also comparable to the 92% precision of the CNN-based MVS with 22 fps inferencing speed in the study of Partel et al. (2019) [11]. Therefore, these results infer that using a one-stage CNN model such as YOLOv3 on a laptop with NVIDIA 1070TI GPU or better can provide sufficient inferencing speed to avoid gaps in different field operations. However, as mentioned in Section 1, the study did not report the travel velocity and field of view length of their setup. Thus, only an estimated performance comparison can be made.

Table 5. Theoretical and simulation performance of a vision system for plant detection at $S = 0.5$ m at three-levels of travel velocity ($\vec{v}$) and inferencing speed (fps).

Treatment No.	$\vec{v}$, m/s	fps	d_f, m/Frame	r_o	Case	r_g	$r_{d,th}$	$r_{d,sim}$
1	0.1	2.4	0.0417	12.00	2	0.00	1.00	1.00
2	0.1	12.2	0.0082	61.00	2	0.00	1.00	1.00
3	0.1	22	0.0045	110.00	2	0.00	1.00	1.00
4	1.3	2.4	0.5417	0.92	3	0.08	0.92	0.92
5	1.3	12.2	0.1066	4.69	2	0.00	1.00	1.00
6	1.3	22	0.0591	8.46	2	0.00	1.00	1.00
7	2.5	2.4	1.0417	0.48	3	0.52	0.48	0.48
8	2.5	12.2	0.2049	2.44	2	0.00	1.00	1.00
9	2.5	22	0.1136	4.40	2	0.00	1.00	1.00

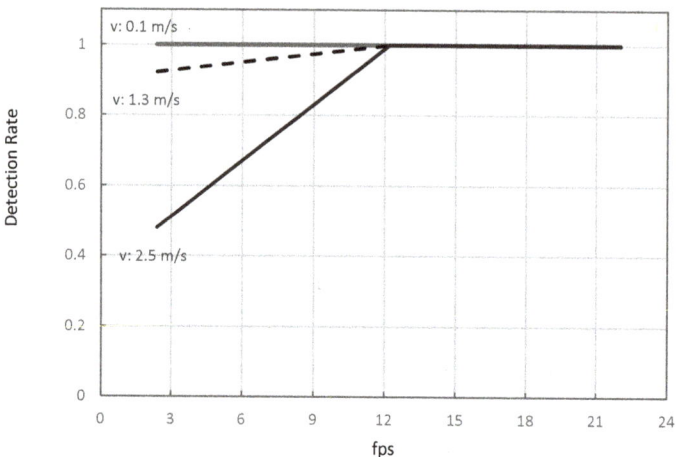

Figure 8. Simulated plant hill detection rates of the vision system moving at 0.1, 1.3, and 2.5 m/s at different inferencing speeds (fps).

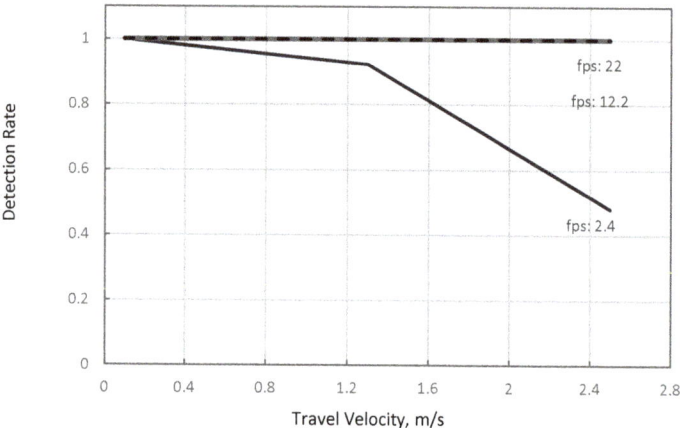

Figure 9. Simulated plant hill detection rates of the vision system moving at 2.4, 12.2, and 22 fps at different travel velocities.

The result of this study also agrees with the results of other studies with known S, $\vec{v}$, and fps. In the study of Chechliński et al. (2019) [23], their CNN-based-vision spraying system had $S = 0.55$ m, $\vec{v} = 1.11$ m/s, and $fps = 10.0$. Applying these values to Equation (2) also yields $r_o > 1$ (Case 2), which correctly predicted their results of full-field coverage. In the study of Esau et al. (2018) [10], their vision-based spraying system had $S = 0.28$ m, $\vec{v} = 1.77$ m/s, and $fps = 6.67$ and also falls under Case 2. Similarly, the vision-based robotic fertilizer application in the study of Chattha et al. (2018) [13] had a $S = 0.31$ m, $\vec{v} = 0.89$ m/s, and $fps = 4.76$. Again, calculating r_o yielded Case 2, which also agrees with their results.

At 2.4 fps, the simulated MVS failed to detect some plant hills when $\vec{v}$ was 1.3 (Treatment 4) or 2.4 m/s (Treatment 7). In contrast, missed detections were absent at 0.1 m/s (Treatment 1). As mentioned in Section 2.5, treatments 1, 4, and 7 represent typical inferencing speeds of CNN models, such as YOLOv3 running in an embedded system, such as NVIDIA TX2 [11]. From these results, it can be inferred that unless CNN object detection models were optimized, such as illustrated in previous studies [23,45], MVS with embedded systems shall only be applicable for agricultural field operations at walking speeds.

Figure 10 illustrates the detected hills per camera frame along the first 10 m traversed distance of treatments simulated at 2.4 *fps* (Treatments 1, 4, and 7). From Figure 10, three information can be obtained: (1) absence of vertical gaps between consecutive frames; (2) horizontal overlaps among consecutive frames; and (3) detection pattern. In Figure 10a, the absence of vertical gaps at 0.1 m/s detections infers that all the hills were detected as the vision moved along the field length. The horizontal overlaps among consecutive frames also illustrate that a plant hill was captured by more than one processed frame. Finally, a detection pattern was observed to repeat every 24 consecutive frames or approximately every 1 m length. The length of the pattern was calculated by multiplying the number of frames to complete a cycle and d_f.

In contrast, the vertical gaps in some consecutive frames at 1.3 m/s, shown in Figure 10b, illustrated the missed detections. Horizontal overlaps were also absent. Hence, the detected plant hills were only represented in the frame once. The vision module traveled too fast and processed the captured frame too slowly at the set capture width, as demonstrated by the detection pattern of one missed plant hill every seven consecutive frames or approximately every 3.8 m traversed distance.

Similar results were also observed at 2.5 m/s travel velocity, as shown in Figure 10c. However, the vertical gaps were more extensive than Figure 10b due to faster travel speed. Observing the detection pattern showed that 14 plant hills were being undetected by the vision system every five frames or approximately every 5.21 m traversed distance. This pattern that forms every 5.21 m further explains the difference in the $r_{d,th}$ and $r_{d,sim}$ in the sensitivity analysis in Section 3.1, when the traversed distance was only 1 m. A complete detection pattern was already formed when the distance was more than 10 m, resulting in better detection rate estimates.

From these results, two vital insights can be drawn. First, at $r_o < 1$, $r_{d,th}$ shall have a margin of error when the length of the detection pattern is less than the traversed distance. Second, concerning future studies, object tracking algorithms, such as Euclidean-distance-based tracking [46], that requires objects should be present in at least two frames, would be not applicable when $r_o \leq 1$. Hence, the importance that $r_o > 1$ in MVS designs is further emphasized.

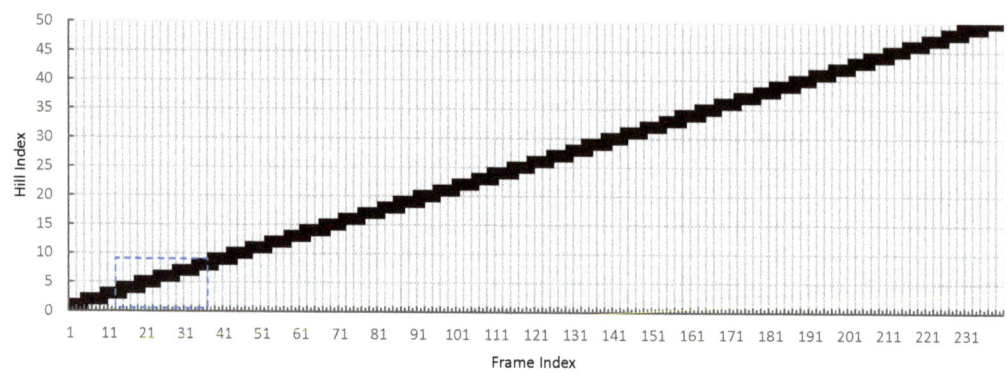

(a)

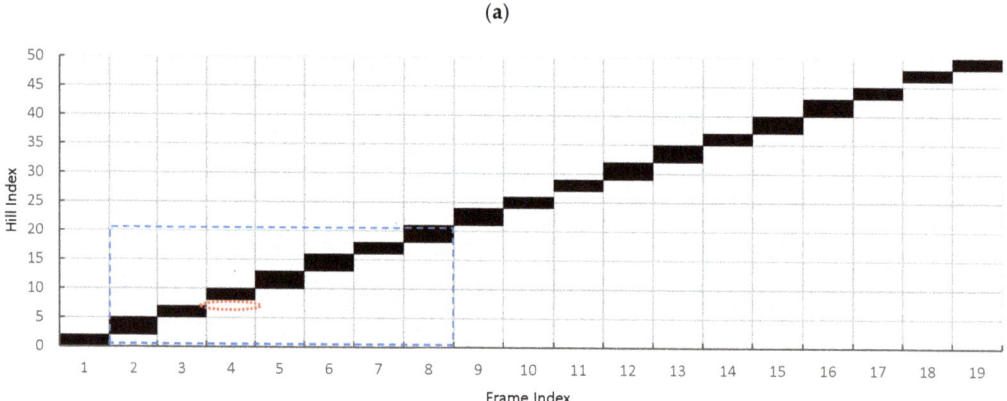

(b)

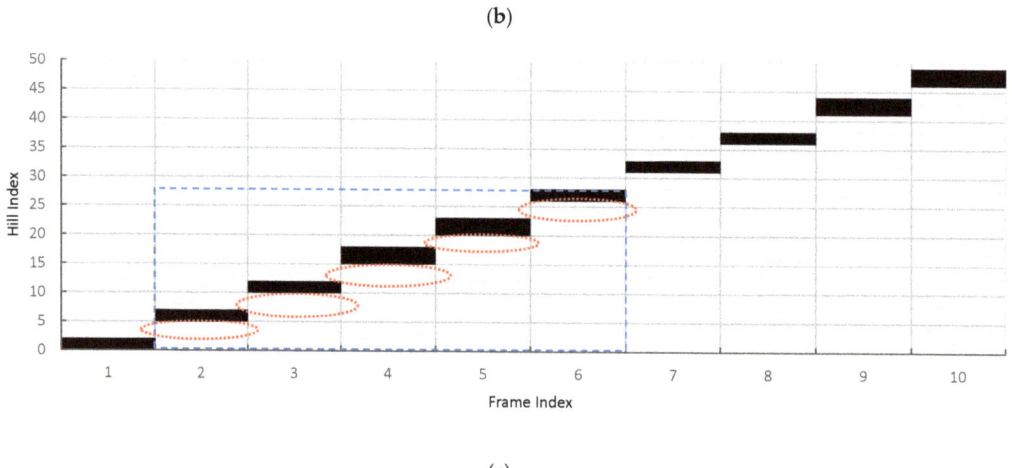

(c)

Figure 10. Range of plant hill indices that were detected per frame along the first 10 m of the simulated field at 2.4 *fps*, 0.5 m capture width, 0.2 m hill spacing at (**a**) 0.1 m/s, (**b**) 1.3 m/s, and (**c**) 2.5 m/s. Blue broken lines enclose a detection pattern, while broken red lines specify the missed plant hills.

3.3. Effect of Increasing S or Multiple Cameras

In cases where $r_o < 1$ (Case 3), a practical solution to increase $\vec{v}_{max}$ is to raise the camera mounting height, which, in effect, shall increase S. However, if raising the camera mounting height is inappropriate as doing so shall also decrease object details, the use of multiple cameras can be a viable solution.

Figure 11 illustrates the effect of increasing the effective S or using multiple cameras on the calculated values of $\vec{v}_{max}$ for the three levels of infencing speeds (2.4, 12.2, and 22 *fps*) simulated at $S = 0.5$ m. The results showed that treatments with missed detections exceeded the allowable $\vec{v}_{max}$. For treatments 4 and 7, the allowable travel velocity was only 1.2 m/s using a single camera module, which was less than the simulated $\vec{v}$ of 1.3 and 2.5 m/s, respectively.

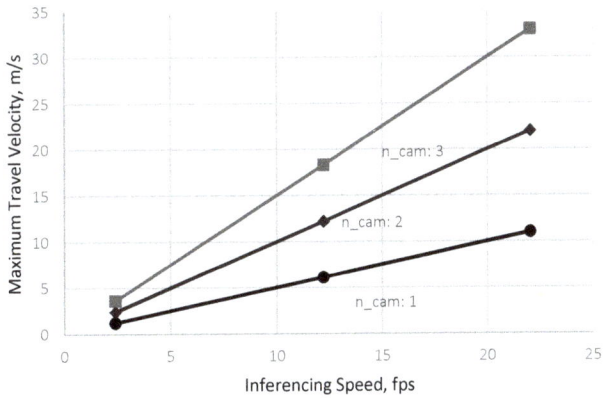

Figure 11. Theoretical maximum travel velocity to prevent missed detections with 1, 2, and 3 cameras at 2.4, 12.2, and 22.0 *fps*.

Calculating n_{vis} using Equation (6) for treatments 4 and 7 showed that 2 and 3 vision modules, respectively, were required to prevent missed detections. Thus, using two vision modules for treatment 4 prevented missed detections, as shown in Figure 12. The 6th, 20th, and 34th frames captured by the second camera detected the plants undetected by the first camera.

As predicted, a two-vision-module configuration for treatment 7 was insufficient in preventing missed detections since the simulated $\vec{v}$ of 2.5 m/s of the vision system was still higher than the increased $\vec{v}_{max}$. As illustrated in Figure 12, without a third camera, the two-camera configuration would still result in an undetected hill on the 16th frame.

Based on these simulated results, the problem of missed detection due to the slow inferencing speed of embedded systems could be potentially solved by using multiple, adjacent, non-overlapping, and colinear cameras along the traversed row when raising the height of the camera was unwanted.

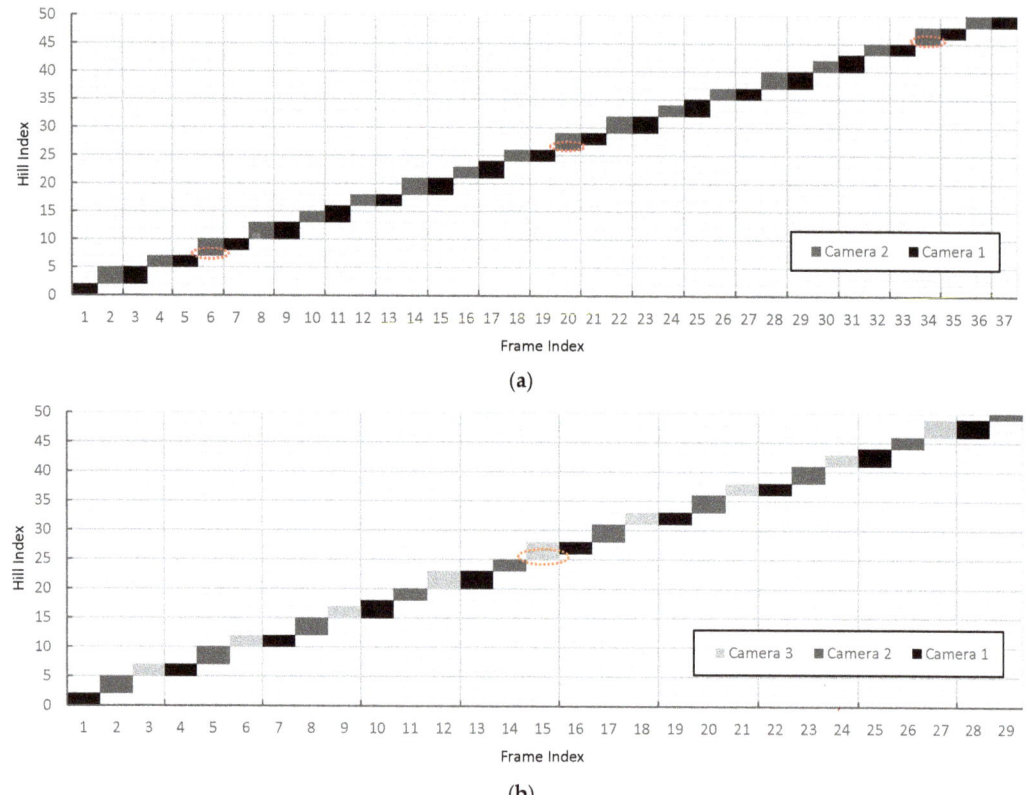

Figure 12. The range of plant hill indices detected per frame along the first 10 m of the simulated field at $fps = 2.4$, $S = 0.5$ m, and $d_h = 0.2$ m: (**a**) detections at 1.3 m/s using two cameras where broken red lines represent plant hills undetected by the first camera but detected by the second camera; and (**b**) detections at 2.5 m/s and three cameras where broken orange lines represent plant hills undetected by the first and second cameras but detected by the third camera.

3.4. Vision Module Simulation and Testing Performance

Figure 13 shows the sample detection of the vision module. Results showed that using a TensorRT-optimized SSD MobileNetV1 to detect plants in 1280 × 720 images on an NVIDIA Jetson Nano 4 GB had an average inferencing speed of 45 *fps*. This average inferencing speed only represented the elapsed time to inference on an already loaded frame. However, due to calculation overheads caused by additional data processing and transmission, the average effective inferencing speed of the vision module was only 16 *fps*, as shown in Figure 13. The effective speed was the average time difference for the vision module to complete a single loop, including grabbing a frame from the camera, inferencing, calculating detection parameters, image processing, and transmitting data.

The results using the theoretical approach and simulation for the vision module are shown in Table 6. Using Equation (2), the configuration of the laboratory setup falls under Case 2 since $r_o > 1$. Then, using Equation (4), $r_{d,th}$ was calculated to be equal to 1.00. Applying Equation (5) yields $\vec{v}_{max} = 12.64$ m/s, which was highly sufficient for the target 0.3 m/s and inferred that multiple vision modules were not required to prevent missed detections. Theoretical prediction of the performance of the vision module showed that the configuration was sufficient to prevent missed detection. Likewise, the theoretical result was confirmed by the simulation results that showed no missed detections ($r_{d,sim} = 1.00$) for both crops and weeds among the simulated $\vec{v}$.

Figure 13. Sample real-time inferencing using trained SSD MobileNetV1 model and optimized in TensorRT. The vision system utilized Jetson Inference API.

Table 6. Theoretical and simulation performance of the CNN-based vision module for plant detection at $S = 0.79$ m, 16 fps, and at three-levels of travel velocity ($\vec{v}$).

$\vec{v}$, m/s	d_f, m/Frame	r_o	Case	r_g	$r_{d,th}$	$r_{d,sim}$
0.1	0.0063	126.40	2	0.00	1.00	1.00
0.2	0.0125	63.20	2	0.00	1.00	1.00
0.3	0.0188	42.13	2	0.00	1.00	1.00

Table 7 summarizes the precision and recall of the trained CNN model in detecting potted plants at different relative travel velocities of the conveyor. Results showed that the combination of an optimized SSD MobileNetV1 in TensorRT running in a Jetson Nano 4 GB have robust detection performance, and incorrect or missed detections were absent despite increasing travel velocity. Comparing the value of r_{actual} to $r_{d,th}$ and $r_{d,sim}$, results showed that the detection rates were equal. The recall was used for comparison instead of precision since the former is the ratio of the correctly detected plants to the total sample plants. This definition of r_{actual} in Equation (16) is equivalent to the definition of $r_{d,sim}$ in Equation (14). Since the r_{actual}, $r_{d,th}$ and $r_{d,sim}$ were equal, these results proved the validity of the theoretical concepts and simulation methods presented in this study. Hence, $r_{d,th}$ and $r_{d,sim}$ can be used to theoretically determine the detection rate of a vision system in capturing plant images as a function of $\vec{v}$ and fps with known S.

Table 7. The detection performance of the CNN-based vision module for detecting a 60 potted plants at different conveyor velocities ($\vec{v}$).

$\vec{v}$, m/s	TP	FP	FN	P_{actual}	r_{actual}
0.1	60	0	0	1.00	1.00
0.2	60	0	0	1.00	1.00
0.3	60	0	0	1.00	1.00

4. Conclusions

This study presented a practical approach to quantify r_d and aid in the development of a CNN-based vision module through the introduction of the dimensionless parameter r_0. The reliability of r_0 in predicting the r_d of an MVS as a function of inferencing speed and travel velocity was successfully demonstrated by having no margin of error compared to simulated and actual MVS at sufficient traversed distance (≥ 10 m). In addition, a set of scripts for simulating the performance of a vision system for plant detection was also developed and showed no margin of error compared to the r_d of actual MVS. This set of scripts was made publicly available to verify the results of this study and provide a practical tool for developers in optimizing design configurations of a vision-based plant detection system.

The mechanism of missed detection was also successfully illustrated by evaluating each of the simulated frames in detail. Using the concept of r_0, simulation, and detailed assessment of each processed frame, the mechanism to prevent missed plant hills by increasing the effective S through synchronous multi-camera vision systems in low-frame processing rate hardware was also successfully presented.

Furthermore, a vision module was also successfully developed and tested. Performance testing showed that the $r_{d,th}$ and $r_{d,sim}$ accurately predicted the r_{actual} of the vision module with no margin of error. The script for the vision module was also made available in a public repository where future improvements shall also be uploaded.

However, despite accomplishing the set objectives in this research, the study encountered limitations that shall be improved in future research. First, the robustness of r_0 in predicting the detection rate was supported mainly by simulation data. The current laboratory tests were only implemented at a maximum travel velocity of 0.3 m/s due to limitations in the manual loading of the test plants. At this time, the study relied on results of other studies to validate the robustness of r_0 at higher travel velocities and different inferencing speeds. Thus, the concepts presented in this shall be further tested to determine the robustness of r_0 at higher travel velocities during the application of the developed vision module on actual field scenarios.

Lastly, the methodology to calculate $r_{d,th}$ and $r_{d,sim}$ assumed that the CNN has 100% precision. In cases less than 100% precision, it can be theorized that $r_{d,th}$ and $r_{d,sim}$ can be multiplied by the precision of the CNN to estimate the effective recall of a CNN-based vision system in evaluating moving objects across the camera frame. However, this concept is yet to be demonstrated and shall also be included in future studies.

Author Contributions: Conceptualization, P.R.S.; methodology, P.R.S. and H.Z.; software, P.R.S.; validation, P.R.S.; formal analysis, P.R.S. and H.Z.; investigation, P.R.S. and H.Z.; resources, P.R.S. and H.Z.; data curation, P.R.S.; writing—original draft preparation, P.R.S.; writing—review and editing, H.Z.; visualization, P.R.S.; supervision, H.Z.; project administration, H.Z.; funding acquisition, P.R.S. and H.Z. All authors have read and agreed to the published version of the manuscript.

Funding: This research received no external funding.

Institutional Review Board Statement: Not applicable.

Informed Consent Statement: Not applicable.

Data Availability Statement: The scripts used in the simulation and vision module were made publicly available in GitHub under BSD license to use, test, and validate the data presented in this study. The scripts for simulation are available at https://github.com/paoap/vision-module-simulation, (accessed on 31 December 2021) while the vision module software framework can be downloaded from https://github.com/paoap/plant-detection-vision-module (accessed on 31 December 2021).

Conflicts of Interest: The authors declare no conflict of interest.

References

1. Pelzom, T.; Katel, O. Youth Perception of Agriculture and Potential for Employment in the Context of Rural Development in Bhutan. *Dev. Environ. Foresight* **2017**, *3*, 2336–6621.
2. Mortan, M.; Baciu, L. A Global Analysis of Agricultural Labor Force. *Manag. Chall. Contemp. Soc.* **2016**, *9*, 57–62.
3. Priyadarshini, P.; Abhilash, P.C. Policy Recommendations for Enabling Transition towards Sustainable Agriculture in India. *Land Use Policy* **2020**, *96*, 104718. [CrossRef]
4. Rose, D.C.; Sutherland, W.J.; Barnes, A.P.; Borthwick, F.; Ffoulkes, C.; Hall, C.; Moorby, J.M.; Nicholas-Davies, P.; Twining, S.; Dicks, L.V. Integrated Farm Management for Sustainable Agriculture: Lessons for Knowledge Exchange and Policy. *Land Use Policy* **2019**, *81*, 834–842. [CrossRef]
5. Lungarska, A.; Chakir, R. Climate-Induced Land Use Change in France: Impacts of Agricultural Adaptation and Climate Change Mitigation. *Ecol. Econ.* **2018**, *147*, 134–154. [CrossRef]
6. Thorp, K.R.; Tian, L.F. A Review on Remote Sensing of Weeds in Agriculture. *Precis. Agric.* **2004**, *5*, 477–508. [CrossRef]
7. Bechar, A.; Vigneault, C. Agricultural Robots for Field Operations: Concepts and Components. *Biosyst. Eng.* **2016**, *149*, 94–111. [CrossRef]
8. Aravind, K.R.; Raja, P.; Pérez-Ruiz, M. Task-Based Agricultural Mobile Robots in Arable Farming: A Review. *Span. J. Agric. Res.* **2017**, *15*, e02R01. [CrossRef]
9. Tian, H.; Wang, T.; Liu, Y.; Qiao, X.; Li, Y. Computer Vision Technology in Agricultural Automation—A Review. *Inf. Process. Agric.* **2020**, *7*, 1–19. [CrossRef]
10. Esau, T.; Zaman, Q.; Groulx, D.; Farooque, A.; Schumann, A.; Chang, Y. Machine Vision Smart Sprayer for Spot-Application of Agrochemical in Wild Blueberry Fields. *Precis. Agric.* **2018**, *19*, 770–788. [CrossRef]
11. Partel, V.; Charan Kakarla, S.; Ampatzidis, Y. Development and Evaluation of a Low-Cost and Smart Technology for Precision Weed Management Utilizing Artificial Intelligence. *Comput. Electron. Agric.* **2019**, *157*, 339–350. [CrossRef]
12. Wang, A.; Zhang, W.; Wei, X. A Review on Weed Detection Using Ground-Based Machine Vision and Image Processing Techniques. *Comput. Electron. Agric.* **2019**, *158*, 226–240. [CrossRef]
13. Chattha, H.S.; Zaman, Q.U.; Chang, Y.K.; Read, S.; Schumann, A.W.; Brewster, G.R.; Farooque, A.A. Variable Rate Spreader for Real-Time Spot-Application of Granular Fertilizer in Wild Blueberry. *Comput. Electron. Agric.* **2014**, *100*, 70–78. [CrossRef]
14. Zujevs, A.; Osadcuks, V.; Ahrendt, P. Trends in Robotic Sensor Technologies for Fruit Harvesting: 2010–2015. *Procedia Comput. Sci.* **2015**, *77*, 227–233. [CrossRef]
15. Tang, Y.; Chen, M.; Wang, C.; Luo, L.; Li, J.; Lian, G.; Zou, X. Recognition and Localization Methods for Vision-Based Fruit Picking Robots: A Review. *Front. Plant Sci.* **2020**, *11*, 510. [CrossRef]
16. Liu, B.; Bruch, R. Weed Detection for Selective Spraying: A Review. *Curr. Robot. Rep.* **2020**, *1*, 19–26. [CrossRef]
17. Jha, K.; Doshi, A.; Patel, P.; Shah, M. A Comprehensive Review on Automation in Agriculture Using Artificial Intelligence. *Artif. Intell. Agric.* **2019**, *2*, 1–12. [CrossRef]
18. Huang, J.; Rathod, V.; Sun, C.; Zhu, M.; Korattikara, A.; Fathi, A.; Fischer, I.; Wojna, Z.; Song, Y.; Guadarrama, S.; et al. Speed/Accuracy Trade-Offs for Modern Convolutional Object Detectors. In Proceedings of the 2017 IEEE Conference on Computer Vision and Pattern Recognition (CVPR), Honolulu, HI, USA, 21–26 July 2017; IEEE: New York, NY, USA, 2017; Volume 84, pp. 3296–3297.
19. Kamilaris, A.; Prenafeta-Boldú, F.X. A Review of the Use of Convolutional Neural Networks in Agriculture. *J. Agric. Sci.* **2018**, *156*, 312–322. [CrossRef]
20. Cecotti, H.; Rivera, A.; Farhadloo, M.; Pedroza, M.A. Grape Detection with Convolutional Neural Networks. *Expert Syst. Appl.* **2020**, *159*, 113588. [CrossRef]
21. Jia, W.; Tian, Y.; Luo, R.; Zhang, Z.; Lian, J.; Zheng, Y. Detection and Segmentation of Overlapped Fruits Based on Optimized Mask R-CNN Application in Apple Harvesting Robot. *Comput. Electron. Agric.* **2020**, *172*, 105380. [CrossRef]
22. Olsen, A.; Konovalov, D.A.; Philippa, B.; Ridd, P.; Wood, J.C.; Johns, J.; Banks, W.; Girgenti, B.; Kenny, O.; Whinney, J.; et al. DeepWeeds: A Multiclass Weed Species Image Dataset for Deep Learning. *Sci. Rep.* **2019**, *9*, 2058. [CrossRef] [PubMed]
23. Chechliński, Ł.; Siemiątkowska, B.; Majewski, M. A System for Weeds and Crops Identification—Reaching over 10 *fps* on Raspberry Pi with the Usage of MobileNets, DenseNet and Custom Modifications. *Sensors* **2019**, *19*, 3787. [CrossRef] [PubMed]
24. Liu, J.; Abbas, I.; Noor, R.S. Development of Deep Learning-Based Variable Rate Agrochemical Spraying System for Targeted Weeds Control in Strawberry Crop. *Agronomy* **2021**, *11*, 1480. [CrossRef]
25. Hussain, N.; Farooque, A.; Schumann, A.; McKenzie-Gopsill, A.; Esau, T.; Abbas, F.; Acharya, B.; Zaman, Q. Design and Development of a Smart Variable Rate Sprayer Using Deep Learning. *Remote Sens.* **2020**, *12*, 4091. [CrossRef]

26. Villette, S.; Maillot, T.; Guillemin, J.P.; Douzals, J.P. Simulation-Aided Study of Herbicide Patch Spraying: Influence of Spraying Features and Weed Spatial Distributions. *Comput. Electron. Agric.* **2021**, *182*, 105981. [CrossRef]
27. Wang, H.; Hohimer, C.J.; Bhusal, S.; Karkee, M.; Mo, C.; Miller, J.H. Simulation as a Tool in Designing and Evaluating a Robotic Apple Harvesting System. *IFAC-PapersOnLine* **2018**, *51*, 135–140. [CrossRef]
28. Lehnert, C.; Tsai, D.; Eriksson, A.; McCool, C. 3D Move to See: Multi-Perspective Visual Servoing towards the next Best View within Unstructured and Occluded Environments. In Proceedings of the 2019 IEEE/RSJ International Conference on Intelligent Robots and Systems (IROS), Macau, China, 3–8 November 2019; IEEE: New York, NY, USA, 2019; pp. 3890–3897.
29. Korres, N.E.; Burgos, N.R.; Travlos, I.; Vurro, M.; Gitsopoulos, T.K.; Varanasi, V.K.; Duke, S.O.; Kudsk, P.; Brabham, C.; Rouse, C.E.; et al. New Directions for Integrated Weed Management: Modern Technologies, Tools and Knowledge Discovery. In *Advances in Agronomy*; Elsevier Inc.: Amsterdam, The Netherlands, 2019; Volume 155, pp. 243–319. ISBN 9780128174081.
30. Hajjaj, S.S.H.; Sahari, K.S.M. Review of Agriculture Robotics: Practicality and Feasibility. In Proceedings of the 2016 IEEE International Symposium on Robotics and Intelligent Sensors (IRIS), Tokyo, Japan, 17–20 December 2016; pp. 194–198.
31. Gauss, L.; Lacerda, D.P.; Sellitto, M.A. Module-Based Machinery Design: A Method to Support the Design of Modular Machine Families for Reconfigurable Manufacturing Systems. *Int. J. Adv. Manuf. Technol.* **2019**, *102*, 3911–3936. [CrossRef]
32. Brunete, A.; Ranganath, A.; Segovia, S.; de Frutos, J.P.; Hernando, M.; Gambao, E. Current Trends in Reconfigurable Modular Robots Design. *Int. J. Adv. Robot. Syst.* **2017**, *14*, 172988141771045. [CrossRef]
33. Lu, Y.; Young, S. A Survey of Public Datasets for Computer Vision Tasks in Precision Agriculture. *Comput. Electron. Agric.* **2020**, *178*, 105760. [CrossRef]
34. Iñigo-Blasco, P.; Diaz-del-Rio, F.; Romero-Ternero, M.C.; Cagigas-Muñiz, D.; Vicente-Diaz, S. Robotics Software Frameworks for Multi-Agent Robotic Systems Development. *Robot. Auton. Syst.* **2012**, *60*, 803–821. [CrossRef]
35. Spencer, J.; Dent, D.R. Walking Speed as a Variable in Knapsack Sprayer Operation: Perception of Speed and the Effect of Training. *Trop. Pest Manag.* **1991**, *37*, 321–323. [CrossRef]
36. Gatot, P.; Anang, R. Liquid Fertilizer Spraying Performance Using A Knapsack Power Sprayer On Soybean Field. *IOP Conf. Ser. Earth Environ. Sci.* **2018**, *147*, 012018. [CrossRef]
37. Cerruto, E.; Emma, G.; Manetto, G. Spray applications to tomato plants in greenhouses. Part 1: Effect of walking direction. *J. Agric. Eng.* **2009**, *40*, 41. [CrossRef]
38. Rasmussen, J.; Azim, S.; Nielsen, J.; Mikkelsen, B.F.; Hørfarter, R.; Christensen, S. A New Method to Estimate the Spatial Correlation between Planned and Actual Patch Spraying of Herbicides. *Precis. Agric.* **2020**, *21*, 713–728. [CrossRef]
39. Arvidsson, T.; Bergström, L.; Kreuger, J. Spray Drift as Influenced by Meteorological and Technical Factors. *Pest Manag. Sci.* **2011**, *67*, 586–598. [CrossRef]
40. Dou, H.; Zhai, C.; Chen, L.; Wang, S.; Wang, X. Field Variation Characteristics of Sprayer Boom Height Using a Newly De-signed Boom Height Detection System. *IEEE Access* **2021**, *9*, 17148–17160. [CrossRef]
41. Holterman, H.J.; van de Zande, J.C.; Porskamp, H.A.J.; Huijsmans, J.F.M. Modelling Spray Drift from Boom Sprayers. *Com-Puter. Electron. Agric.* **1997**, *19*, 1–22. [CrossRef]
42. Yinyan, Y.; Zhichao, H.; Xiaochan, W.; Odhiambo, M.O.; Weimin, D. Motion Analysis and System Response of Fertilizer Feed Apparatus for Paddy Variable-Rate Fertilizer Spreader. *Comput. Electron. Agric.* **2018**, *153*, 239–247. [CrossRef]
43. Machleb, J.; Peteinatos, G.G.; Sökefeld, M.; Gerhards, R. Sensor-Based Intrarow Mechanical Weed Control in Sugar Beets with Motorized Finger Weeders. *Agronomy* **2021**, *11*, 1517. [CrossRef]
44. Fennimore, S.A.; Cutulle, M. Robotic Weeders Can Improve Weed Control Options for Specialty Crops. *Pest Manag. Sci.* **2019**, *75*, 1767–1774. [CrossRef]
45. Pinto de Aguiar, A.S.; Neves dos Santos, F.B.; Feliz dos Santos, L.C.; de Jesus Filipe, V.M.; Miranda de Sousa, A.J. Vineyard Trunk Detection Using Deep Learning—An Experimental Device Benchmark. *Comput. Electron. Agric.* **2020**, *175*, 105535. [CrossRef]
46. Qian, X.; Han, L.; Wang, Y.; Ding, M. Deep Learning Assisted Robust Visual Tracking with Adaptive Particle Filtering. *Signal Processing Image Commun.* **2018**, *60*, 183–192. [CrossRef]

Article

IoT-Ready Temperature Probe for Smart Monitoring of Forest Roads

Gabriel Gaspar [1,2,*], Juraj Dudak [1,2], Maria Behulova [2,*], Maximilian Stremy [2], Roman Budjac [2], Stefan Sedivy [3] and Boris Tomas [4]

1. Research Centre, University of Zilina, Univerzitna 8215/1, 010 26 Zilina, Slovakia; juraj.dudak@uniza.sk
2. Faculty of Materials Science and Technology, Slovak University of Technology, Jana Bottu 2781/25, 917 24 Trnava, Slovakia; maximilian.stremy@stuba.sk (M.S.); roman.budjac@stuba.sk (R.B.)
3. TNtech, s.r.o., Lucna 1014/9, 014 01 Bytča, Slovakia; ssedivy@tntech.eu
4. Faculty of Organization and Informatics, University of Zagreb, Pavlinska 2, 42000 Varazdin, Croatia; boris.tomas@foi.unizg.hr
* Correspondence: gabriel.gaspar@uniza.sk (G.G.); maria.behulova@stuba.sk (M.B.)

Abstract: Currently, we are experiencing an ever-increasing demand for high-quality transportation in the distinctive natural environment of forest roads, which can be characterized by significant weather changes. The need for more effective management of the forest roads environment, a more direct, rapid response to fire interventions and, finally, the endeavor to expand recreational use of the woods in the growth of tourism are among the key factors. A thorough collection of diagnostic activities conducted on a regular basis, as well as a dataset of long-term monitored attributes of chosen sections, are the foundations of successful road infrastructure management. Our main contribution to this problem is the design of a probe for measuring the temperature profile for utilization in standalone systems or as a part of an IoT solution. We have addressed the design of the mechanical and electrical parts with emphasis on the accuracy of the sensor layout in the probe. Based on this design, we developed a simulation model, and compared the simulation results with the experimental results. An experimental installation was carried out which, based on measurements to date, confirmed the proposed probe meets the requirements of practice and will be deployed in a forest road environment.

Keywords: IoT; sensor; probe; temperature profile; forest roads; simulation

Citation: Gaspar, G.; Dudak, J.; Behulova, M.; Stremy, M.; Budjac, R.; Sedivy, S.; Tomas, B. IoT-Ready Temperature Probe for Smart Monitoring of Forest Roads. *Appl. Sci.* **2022**, *12*, 743. https://doi.org/10.3390/app12020743

Academic Editor: Dimitrios S. Paraforos

Received: 8 December 2021
Accepted: 9 January 2022
Published: 12 January 2022

Publisher's Note: MDPI stays neutral with regard to jurisdictional claims in published maps and institutional affiliations.

Copyright: © 2022 by the authors. Licensee MDPI, Basel, Switzerland. This article is an open access article distributed under the terms and conditions of the Creative Commons Attribution (CC BY) license (https://creativecommons.org/licenses/by/4.0/).

1. Introduction

Forest roads are often considered a marginal point of interest within a country's road infrastructure and a considerable number of academic articles are devoted to general transportation routes [1,2]. This fact is not wrong and does not need to be contradicted in the means of value for money. These allegations are based in particular on the volume and nature of the traffic that these roads must bear for a reasonable time period. Currently, there is, on the other hand, an ever-growing demand for high-quality mobility in this specific natural environment, which can be characterized by extreme fluctuations of weather conditions. The list of main reasons includes the requirement for more efficient management of the forest roads environment, a more straightforward, quick approach to fire interventions [3] and the final reason is in the effort for broader recreational utilization of the forests in the development of tourism. The conclusions of some studies are also interesting, such as [4], which points to the direct relationship between the quality of forest infrastructure and the economic value that can be obtained from forests. Not only for these reasons, but the authorities responsible are considerably concerned with ideas of implementing specific management procedures with a long-lasting effect into the process of forest road management. These procedures and systems based on them follow approaches applied within the generally available road networks. The basis of such applied instruments for effective management of road infrastructure, is a comprehensive set of diagnostic tasks regularly performed and a dataset of long-term monitored properties of selected sections [5,6]. The previously mentioned represents the simple starting point for the decision making on early construction interventions in the roads, either through a general routine maintenance, repairs, or reconstruction. At the same time, the information obtained by monitoring existing roads helps us to create new and improve existing standards associated with the comprehensive renewal and construction of new roads, taking into account the geospatial optimization of their situation and application of the innovative material composition as referred to in [7,8]. Within the concept of comprehensive monitoring and diagnostics, several measures need to be put in place. It is mainly a matter of regular data collection application of selected data from the forest road network, such as road bearing capacity, transverse and longitudinal unevenness and roughness. Measurements must be suitably accompanied by a network of stationary meteorological measuring stations at predefined road sections. Accomplishing this task will provide the required inputs for the correct evaluation of the collected data on the condition of the structure and the road surface. Such stations include special sensors evaluating the temperature profile of roads in both the vertical and horizontal directions, soil sensors installed in the immediate vicinity of the road, frost sensors, road pollution sensors and various additional meteorological sensors for use on the road network [9,10]. The data collected using these stations in combination with selected characteristics of the road condition obtained by diagnostics, might bring new results in the form of defining specific dependencies present within the specific road construction in the forest environment. As an application example, we can mention, e.g., determining the relationship between the bearing capacity of the subsoil and the climatic conditions in the forest environment, or determining the relationship between the influence of extreme climatic thermal effects on the road surface roughness. Their aim is, among other issues, to improve existing and conduct new degradation functions related to the life cycle of roads and to determine their residual life. For purposes related to comprehensive diagnostics and monitoring, it is required to design specially modified meteorological sets directly adapted to the tasks associated with the monitoring of forest roads. Several commercial products from reputable manufacturers are available on the market, most of which are intended for precisely defined uses. In our case, we decided to design our own solution that will be suitable for use in a forest environment. However, the proposed device can be used in other areas beyond the originally intended use. A suitable example us smart city solutions [11] and industrial automation solutions [12]. In this article, we focused our efforts on the first part of the joint work, which is the design and implementation of an IoT ready temperature probe for measuring the vertical temperature profile of the soil. Vertical heat transfer in the soil is an indicator of important properties of the soil and thus it is possible to indirectly observe physical events, such as heat transfer transferred by groundwater as an indicator of surface water at different depths [13]. From the hydrology point of view, it is possible to monitor the movement of water and the renewal of the water source through the rain. By creating a vertical temperature soil profile, it is also possible to monitor external atmospheric influences and other interventions in the environment, respectively. In addition to our area of interest in forest roads, it is possible to extend usage to other areas, e.g., agriculture, where the information on heat transfer and form of irrigation in the vertical direction is of significant value [14]. It is interesting to monitor temperature in the depth of plant roots, as plant roots are known to be stimulated to grow in warmer environments [15]. Monitoring of temperature profiles is commonly used, while several solutions of analog and digital temperature sensors are available. In [16], the authors address the issue of monitoring the temperature of the vertical and horizontal profile of the road surface using low-cost temperature sensors DS18B20 [17,18]. The purpose of monitoring the temperature profile [10] and logging the measured data was to investigate the freezing effect on the degradation of the road body in cold areas. The distribution of the measuring points of the vertical profile probe ranged from 30 mm to 4 m. They monitored one cycle of freezing and thawing of the subsoil and the effect of the change in atmospheric temperature on the individual monitored layers. In the article by Liu et al. [19], the authors monitored changes in soil temperature by application using LM35 temperature sensors, which needed to be calibrated for measurement purposes, due to the minor differences in the characteristics of individual sensors. The authors investigated the effect of using a paper cover film on the change and delay of the soil temperature at a depth of 100 mm depending on the change in outdoor temperature for agricultural purposes. However, in [9,13], the authors do not present the design of probes for measuring soil temperature, they only present and interpret the results of measurements. In [20], Izquierdo et al. focus on measuring the vertical temperature profile of a water column through changes in physical properties such as the wavelength of electromagnetic radiation, reflection, and filtering of individual components of the light spectrum due to changes in other physical quantities. Presented technologies consist of optical fibers, FBG sensors (fiber Bragg grating) and an interrogator. Changes in the physical properties of specific light components due to changes in temperature are evaluated and based on changes in the characteristics of the monitored spectrum components, they also allow the determination of the ambient temperature after calibration of the device. This technology is unsuitable for temperature monitoring purposes in an environment other than water due to mechanical vulnerability. The problem with deploying IoT devices in a forest environment is powering them without direct connection to the electricity grid. This represents a serious issue and can be solved utilizing locally available renewable energy sources. Wind energy and solar energy are mainly used [21]. Due to these facts, it is necessary to pay increased attention to the overall energy requirements of the installed equipment [22].

Based on the review of available technologies and our previous experience [23] with 1-wire networks, we decided to base the proposed temperature probe on DS18B20 sensors. Such sensors have a wide variety of utilization in temperature measurements [24,25]. This solution enables unambiguous identification of sensors and thus identification of their position in the probe. The probe is intended for implementation in the area of forest roads for measuring vertical temperature profile of subsoil either as a stand-alone IoT device or in connection with other IoT devices with low energy consumption requirements. Such a design allows utilization in several other areas of interest for monitoring both indoor [26] and outdoor [27] temperature profiles. Its construction allows simple integration into existing devices with 1-wire protocol support. Probe, as such, represents a tiny but essential part of the weather monitoring system for forest roads whose outputs enter into the road management system, as already mentioned above.

2. Materials and Methods

The task of designing a measuring probe for measuring the vertical temperature profile of soil is a set of specific conditions for mechanical and electrical design. Our research team has already designed and implemented meteorological measurement sets as part of previous studies [28,29]. This real expertise makes proposing our temperature probe design easier. DBAR is the name of our unique temperature probe design for monitoring the vertical temperature profile of the subsoil. The following components must be present in the sensor, according to our design:

- study housing for use in severe settings with mechanical, chemical, and environmental stress;
- sensor elements at predetermined distances;
- internal wiring with minimized influence on measurements;
- connecting cabling with minimal impact on measurements; and
- hermetically sealed mechanical elements.

In this chapter, we will present the process of development of individual components and final prototypes based on them.

2.1. Probe Electrical Design

Our design utilizes 1-wire technology, which allows connecting multiple temperature sensors on a single twisted pair cable under certain conditions. The basis of 1-wire technology is a serial protocol using a data line and a ground reference for communication. Additional wire for device power is optional. The 1-wire master initiates and controls communication with one or more 1-wire slave devices on the 1-wire bus. Each 1-wire slave device has a unique, unchangeable, factory-programmed 64-bit identification number that serves as the device address on the 1-wire bus. The 8-bit family code, as a subset of the 64-bit ID, identifies the type and functionality of the device. Most 1-wire devices allow obtaining power for their own operation from the 1-wire bus, which is referred to as a parasitic power mode. In this case, the device does not need a third wire and just a single wire is shared for both communication and power to the device. 1-wire is widely supported by many IoT devices [30–32], which was another reason for our choice of sensor type DS18B20. Furthermore, in the case of a custom solution, the implementation of 1-wire protocol on a selected microcontroller is trivial. The sensors are available in packages with the dimensions of common transistors (TO-92) and IC (TSOC, TDFN, SOT23). We decided to use the variant in the TO-92 package, as it is more economically advantageous and allows easier work in laboratory conditions and the connection of sensors will use parasitic connection, as depicted in Figure 1.

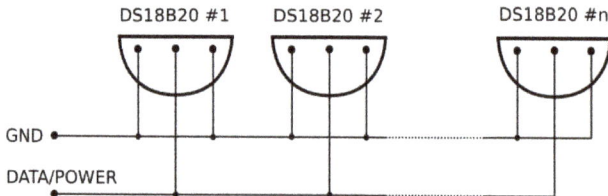

Figure 1. Parasitic connection of sensors.

2.2. Probe Mechanical Design

For the probe mechanical parts, it is necessary to choose a material that will protect the sensory element from environmental influences, both mechanical and chemical, and others depending on the type of environment. Furthermore, concerning the parameters influencing the heat transfer through the casing, two material alternatives were chosen:

- polypropylene as a thermoplastic polymer, which is one of the commonly used plastics in many areas of industry. It is suitable for our intended use as a probe housing due to its temperature, mechanical, and chemical resistance. A range of pipes of various diameters is available for use in plumbing installations. Various fittings are also available and the principle of joining them is with the use of a plastic welding machine,
- polyethylene is a commonly used thermoplastic polymer. It is characterized by high resistance to acids, alkalis, and some other chemicals. It is suitable as a probe housing for measuring the temperature profile, also due to its temperature resistance. Fittings and accessories are mostly available, joining is realized by thermal fusion or by using an appropriate adhesive.

Considering the availability of materials, we decided to produce two experimental prototypes marked according to the material used as PPR for polypropylene and HDPE for polyethylene. PPR alternative—commonly available PPR used for the implementation of plumbing installations were used to produce the probe cover. A series with a dimension of 20 mm was chosen for the housing.

Figure 2 shows a cross-section of the proposed probes interconnection pieces. We used commonly available tubing and couplings for PPR material, and we used custom made couplings for HDPE material.

2.3. Probe Internal Connection Proposal

The basic requirement is that the sensors in the probe are placed so that they are mounted at distances precisely defined for temperature profile measurements. To achieve this goal, it is necessary to consider the need to connect the sensors to a common bus, which is physically implemented as a twisted pair cable.

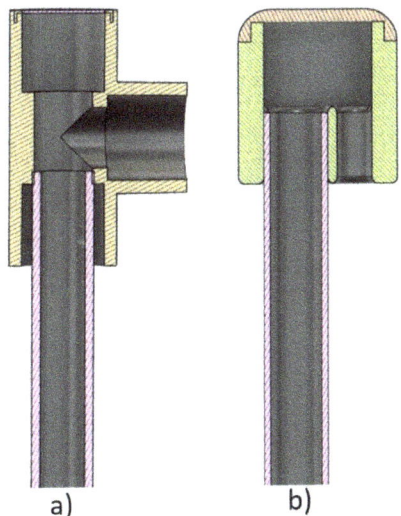

Figure 2. Cross-sections of proposed probes: (**a**) PPR and (**b**) HDPE.

Identification of individual sensors in the probe is ensured using their unique identification numbers. Connecting the sensors directly to the cable is the simplest method of implementing a temperature probe. The problem is the way to achieve the required distances between the sensors. In this case, it is necessary to add some reserve on the cable to realize bending and insulation of joints. The sensors must be attached to an auxiliary profile inserted into the housing, as shown in Figure 3.

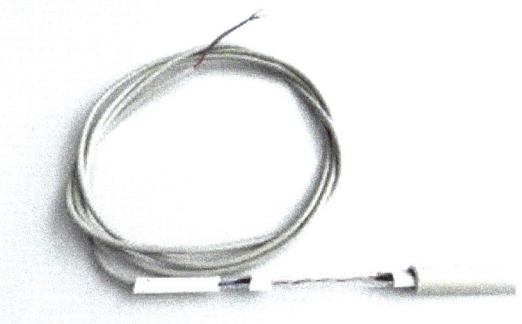

Figure 3. Sensors connected on a cable.

Although this method is simple, it did not work entirely as expected. It was too laborious, and the precise placement of the sensors could not be easily ensured. Due to this fact, we proposed a method of implementing a temperature probe by connecting sensors utilizing printed circuit boards. The printed circuit boards were designed with a 50 mm grid and a module length of 250 mm in Figure 4. This modular system makes it possible to create a probe of the required length according to the immediate needs. In practice, we expect probes with lengths between 1 m and 1.5 m.

Figure 4. PCB modules for sensors.

Custom dimensions of the probe can be achieved by a simple technique of cutting the PCB profile at the selected location as can be seen in Figure 5. With this method, no installation on the auxiliary profile or cable reserve is required, as the implementation only contains connecting cables to the last PCB module. The grid of 50 mm is suitable for purposes of our probe but can be modified in the future if different probe parameters are required.

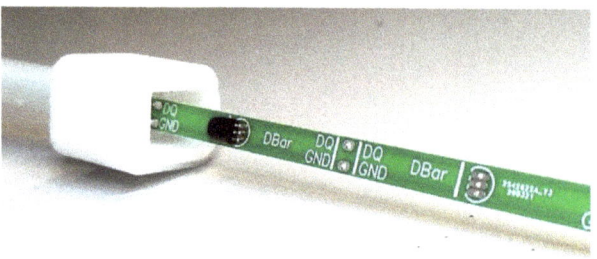

Figure 5. Sensors installed on a PCB module.

When using 1-wire technology, the required cabling is reduced only to a simple twisted pair cable with copper cores. The 2×0.8 mm² cable was chosen, which is commonly available in several versions for installations in both indoor and outdoor environments. An alternative is also available for placement directly in the ground or building structures (concrete, mortar, etc.). After connecting the sensors and preparing the wiring for connection to the measuring board, such a system is inserted into the prepared probe housing. Many sealing compounds are available in the market for the sealing of electronic components. For the purposes of this device, a two-component polyurethane sealant Elantron PU 310/PH 27 [33] with a recommended operating temperature of a maximum of 100 °C was chosen.

2.4. Individual Sensors Position Identification

In addition to the commonly used 64-bit sensor ID, we decided to use 2 free bytes of the non-volatile memory of the DS18B20 sensor. These are utilized to set the upper and lower limits for alarms that are used in temperature-critical applications shown in Table 1. In our design, they carry information about the position of the individual sensor in the probe, with the designation "0" for the sensor located at the bottom of the probe.

Stored value is in format: value = $UB_H UB_L$. The stored value has width of 16 bits, and the meaning of this value is position of the temperature sensor in the temperature probe.

Table 1. Memory map of DS1820 temperature sensor.

	Scratchpad	EEPROM
Byte 0–1	Temperature registers	Not Available
Byte 2	TH Register or User Byte 1	User Byte 1 (UB_H)
Byte 3	TL Register or User Byte 1	User Byte 2 (UB_L)
Byte 4	Configuration register	Configuration register
Byte 5–8	Reserved	Not Available

2.5. Probe System Architecture and Design for Forest Road Measurements

The probe design for measuring the vertical temperature profile of the subsoil considers the two primary applications shown in the block diagram in Figure 6. The first method is the connection of the probe to the IoT device or to the measurement control panel, which enables the online transfer of measurement data to a server solution, where the data is processed and prepared for user evaluation. The second method is collecting data into a local standalone logger, while the transfer of data to the server solution is provided by downloading data from the logger and then uploading them to the database. This solution is designed for long-term measurements, where it is not required to transmit the measured data frequently, rather it is vital to download them at specified interval as a batch.

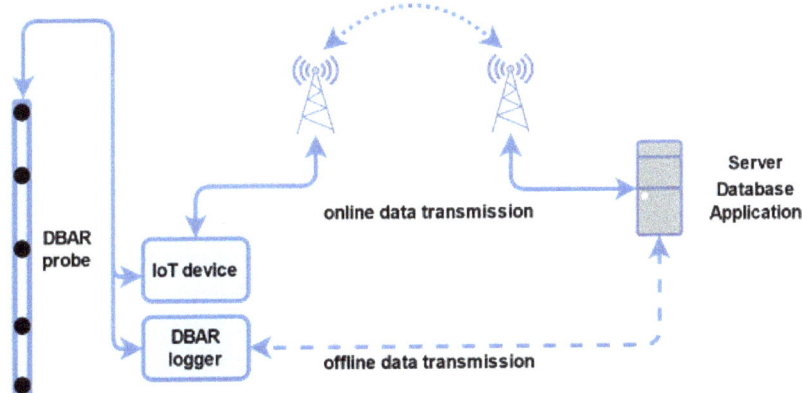

Figure 6. System architecture block diagram.

3. Simulation Model and Experimental Measurement Proposal

For numerical simulation of the heat transfer from the environment and experimental verification of obtained results, two experimental probes were proposed. Both probes are 250 mm long and the sensors are placed on the printed circuit boards with a 50 mm distance. The experimental probes in Figure 7 differ in the applied housing material, dimensions and the number of sensors used for proposed experiments. The decision regarding the housing material was made based on its availability on the market considering mainly its environmental resistance. Considering this, PPR and HDPE plastic materials were selected. These plastic materials offer high abrasion and corrosion resistance to soil chemicals. Both materials are also suitable in terms of operating temperature, as the experimental setup is supposed to be installed in the forest, not in a laboratory environment. Another reason is the appropriate thickness of the available plastic tubes. This property affects the heat transfer from the environment to probes attached inside the housing.

In case of probe 1 (PPR probe in Figure 7a), the probe housing is a polypropylene plastic tube with a diameter of 20 mm and a wall thickness of 3.2 mm. Sensors marked 100, 110, 120 are placed in the body of the probe with an offset of 4.5 mm from the printed circuit board, with a location corresponding to the thermometers 00, 10, 20. Sensors marked 01, 02, 03 are placed on the body of the probe with the location corresponding to the thermometers 00, 10, 20. The sensor marked 04 is, as in the case of probe 1, an auxiliary thermometer for measuring the ambient temperature.

The base construction of the probe 2 (HDPE probe in Figure 7b) uses thermoplastic polymer (HDPE material) as a tube housing material with a diameter of 10 mm and a wall thickness of 1 mm. Sensors DS18B20 marked 00, 05, 10, 15, 20, 25 are in the body of the probe with a polyurethane sealant. Sensors marked 01, 02, 03 are placed on the body surface of the probe with the location corresponding to the thermometers 00, 10, 20. The sensor marked 04 is an auxiliary sensor for measuring the ambient temperature.

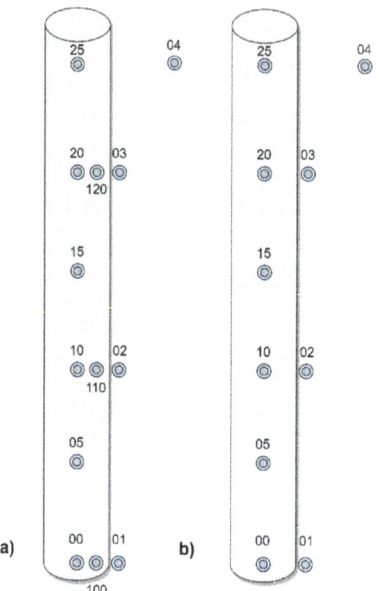

Figure 7. Experimental probes: (**a**) PPR and (**b**) HDPE.

There are sensors inside the tube, as we have already mentioned. These probes are encapsulated using two components consisting of a resin and curing agent. PUR sealant is used to isolate electric components from the surrounding environment to prevent damage caused by water, air humidity. Equally important, dissipation factor using PUR sealant is acceptable.

The proposed experiments covered two basic types of measurements:

(A) Measurement in a water bath to verify the measurement procedure during a step temperature change.
1. The measurement starts at ambient temperature.
2. The probe is immersed in a water bath at a temperature approximately 55 °C (domestic hot water).
3. The measurement is completed after approaching the temperatures measured by sensors 00, 05, 10, 15, 20, 25 and the auxiliary thermometer 04.

The measurement is repeated at different values of the water bath temperature (30 °C and 10 °C). The time resolution of the measurement is assumed to be ~1 s.

(B) Measurement in a climate chamber to verify the measurement procedure in case of gradual temperature change.
1. The probe is placed into the climate chamber and the measurement starts.
2. The temperature increases by 10 °C comparing the initial temperature is adjusted in the climate chamber, followed by a dwell time of approximately 6–8 min at maximum temperature. Then the climate chamber is switched off [34].
3. The measurement is completed after approaching the temperatures measured by sensors 00, 05, 10, 15, 20, 25 and auxiliary thermometer 04.

The measurement is repeated at different maximum temperatures of the climate chamber (plus 15 °C and 25 °C). The time resolution of the measurement is assumed to be ~1 s.

3.1. Simulation Model Proposal

Numerical simulation of temperature fields during the probe heating and cooling is based on the solution of Fourier–Kirchhoff partial differential equation in the form [35]

$$\rho c_p \frac{\partial T}{\partial t} = \lambda \nabla^2 T + \dot{q}_v \qquad (1)$$

in which T is the temperature, ρ is the density, c_p is the specific heat, λ is the thermal conductivity, and $\dot{q}_v$ is the volumetric heat source density, i.e., the heat generated per unit time in a unit volume. Geometrical, thermal, initial, and boundary conditions are necessary to define to accomplish solution of the Equation (1). Analysis of temperature fields was performed by the system ANSYS 18.1 [36] using the implemented finite element method.

Because the temperature changes along the height of the probes are negligible for the investigated heat conduction processes, it is possible to use a simplified axially symmetric model of the probe 1 and 2 with the dimensions according to Figure 8. This assumption was also confirmed by experimental measurements (see Section 4). The maximum standard deviations of the temperatures measured by the sensors in the axis of probe 1 (Figure 7a—with the exception of sensor 25) are 0.52 °C, which is at the level of accuracy of the sensors specified by the manufacturer [17]. The finite element mesh was generated using the PLANE 77 element.

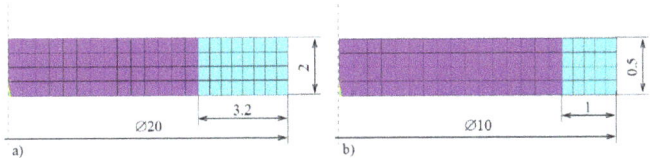

Figure 8. Axis-symmetric finite element model for the analysis of temperature fields (**a**) probe 1 (PPR) and (**b**) probe 2 (HDPE).

The thermophysical properties of PPR and HDPE pipes as well as PUR sealant are summarized in Table 2.

Table 2. Used materials properties [33,37–39].

Property	PPR	HDPE	PU310/PH27
Thermal conductivity [W·m^{-1}·K^{-1}]	0.24	0.4	0.375
Density [kg·m^{-3}]	898	956	1290
Specific heat [J·kg^{-1}·K^{-1}]	2000	1840	1900

The initial temperature of probes was proposed to be equal to the ambient temperature, generally 20 °C. The boundary conditions were defined in accordance with the described experimental program using the boundary condition of the 3rd type. The imperfect thermal contact between the housing wall and the PUR sealant was modeled by contact thermal resistance.

3.2. Simulation Results

Figures 9 and 10 illustrate the temperature distribution in probes 1 and 2 in the time of 30 s, 60 s, 90 s, and 300 s, respectively, after immersing the probe in a water bath at approximately 55 °C. The temperature differences in the probe 2 are smaller in comparison with the probe 1. In time of 300 s, the temperatures in probe 2 are from 52 °C to 56 °C. Probe 1 has temperatures from 48 °C to 56 °C, while the temperatures of PUR sealant are from 48 °C to 52 °C. The response of the probe 1 to the sudden temperature change is delayed more.

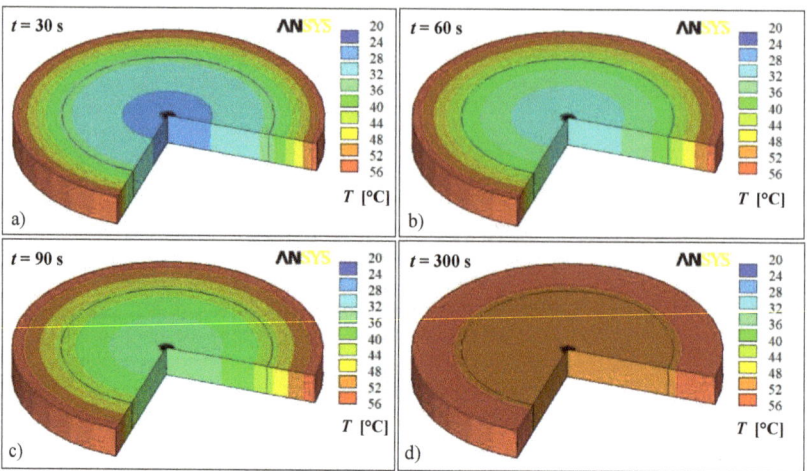

Figure 9. Temperature distribution in the probe 1 in the time of (**a**) 30 s, (**b**) 60 s, (**c**) 90 s, and (**d**) 300 s after immersing the probe in a water bath at approximately 55 °C.

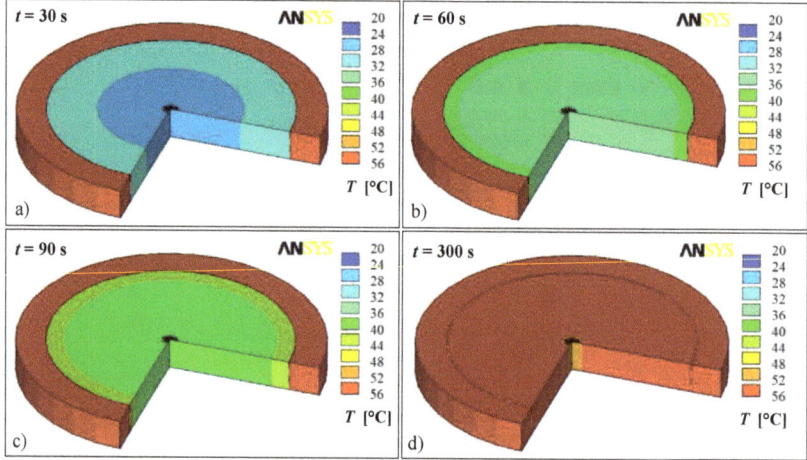

Figure 10. Temperature distribution in the probe 2 in the time of (**a**) 30 s, (**b**) 60 s, (**c**) 90 s, and (**d**) 300 s after immersing the probe in a water bath at approximately 55 °C.

Figures 11 and 12 illustrate the temperature fields in the probe 1 and probe 2, respectively, in chosen times during probe heating and cooling in the climate chamber. Comparing the temperature distribution during the period of heating in the time of 1200 s (Figures 11a and 12a) and 2000 s (Figures 11b and 12b), there can be seen that the temperature rises in the probe 2 is slightly faster than in the probe 1. Minimal temperature differences of 0.1 °C in the probe 2 are reached approximately in the time of 2480 s, i.e., at the end of dwell time (Figure 12c). As it follows from Figure 11c, the minimal temperature differences in the probe 1 are during the phase of probe cooling in the time of 2770 s.

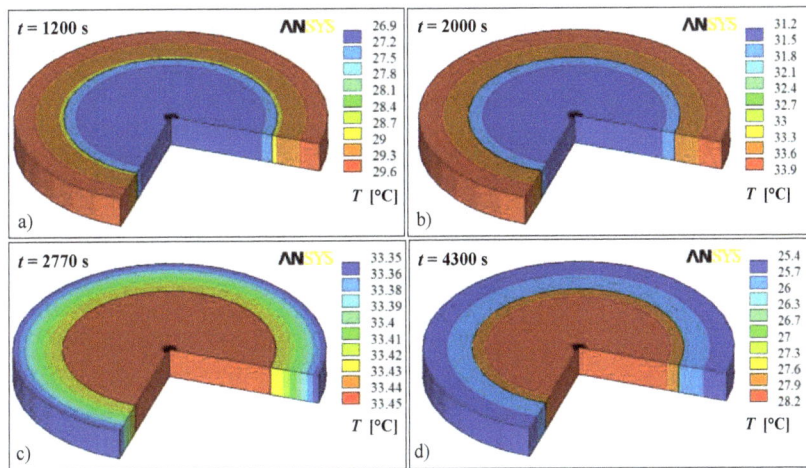

Figure 11. Temperature distribution in probe 1 at chosen times during its heating by 10 °C and subsequent cooling in the climate chamber, (**a**) $t = 1200$ s, (**b**) $t = 2000$ s, (**c**) $t = 2770$ s and (**d**) $t = 4300$ s.

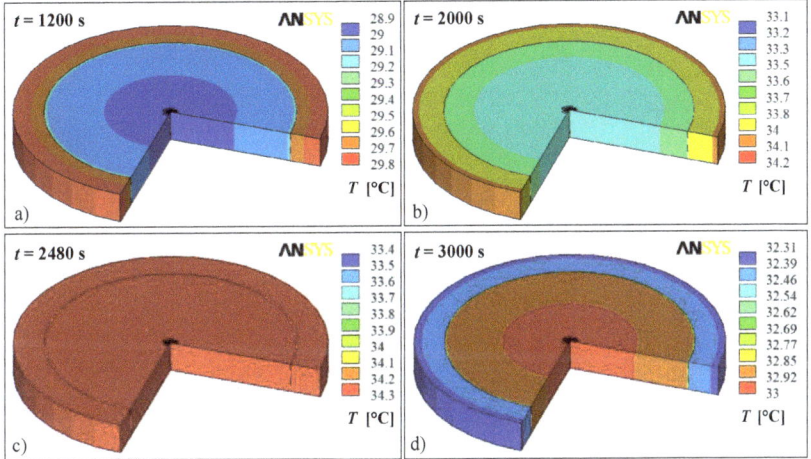

Figure 12. Temperature distribution in probe 2 at chosen times during its heating by 10 °C and subsequent cooling in the climate chamber, (**a**) $t = 1200$ s, (**b**) $t = 2000$ s, (**c**) $t = 2480$ s and (**d**) $t = 3000$ s.

4. Simulation and Experimental Result Comparison and Discussion

To verify the simulation model, the computed and experimentally obtained results were compared. In case of measurement A (immersing the probe in a water bath), the time dependences of the temperatures measured by the sensors and the temperatures calculated in the probe axis were used in the comparison. Evaluation of time histories measured by sensors and computed on the probe surface and in the probe axis was taken into account for the measurement B (in a climate chamber).

In all the figures in this section, the time records of the temperatures measured by the sensors and thermometers marked according to Figure 7 are plotted by lines as follows: the auxiliary thermometer 04—black thick solid line; thermometers on the probe surface 01—red dashed line; 02—green dashed line; 03—blue dashed line, sensors in the probe body 00, 05, 10, 15, 20, 25; and 100, 110, 120—solid line of a specific color. The computed time histories of temperatures in the probe axis and on the probe surface are presented by black dashed lines and black dotted lines, respectively.

Figures 13–15 illustrate the dependence of measured and computed temperatures on the time after immersing the probe 1 to a water bath at a temperature of 55 °C, 30 °C, and 10 °C, respectively. The temperatures measured by the sensor 25 are apparently affected by the top cover of the probe, where we used a substitute made of different material instead of a regular coupling. This problem was fixed in the experimental installation probe by using appropriate coupling cap. Otherwise, in general, a suitable match was demonstrated among the measured and calculated results. The temperature in the probe axis equalizes to the surface temperature for about 800 s after a sudden change in surrounding temperature. This time can be considered as a response time of the probe 1 to the sudden temperature change.

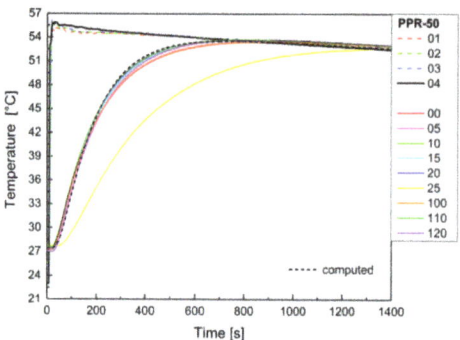

Figure 13. Time-history of measured temperatures and the temperatures computed in the probe axis (measurement A—water bath temperature of 55 °C/probe 1).

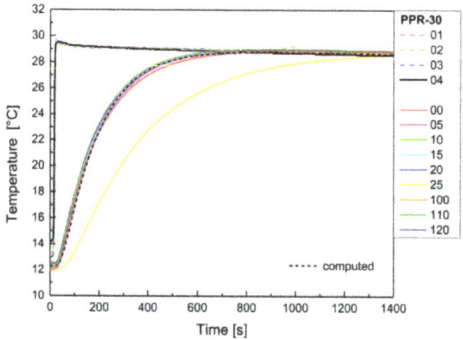

Figure 14. Time-history of measured temperatures and the temperatures computed in the probe axis (measurement A—water bath temperature of 30 °C/probe 1).

In Figures 16–18, the time histories of measured and computed temperatures are shown for the probe 2 and measurement A. The response time in the case of probe 2 is considerably shorter. The axis temperature reaches the surface temperature approximately in the time of 250 s.

The dependences of measured and computed temperatures on the time during the measurements in a climate chamber (measurement B) are shown in Figures 19–24. The temperature differences between the surroundings and surface temperatures, and the surface and axial temperatures for the probe 1 (Figures 19–21) are larger compared to that for the probe 2 (Figures 22–24). Considering the design and dimensions of probes 1 and 2, this result was expected. However, the performed measurements and calculations provide a possibility of feasible assessment of the response time of the designed probes to a gradual change of surrounding temperature. Finally, it can be concluded, that as well as in the case of measurement B, a sufficient correlation was achieved between the measured and calculated results.

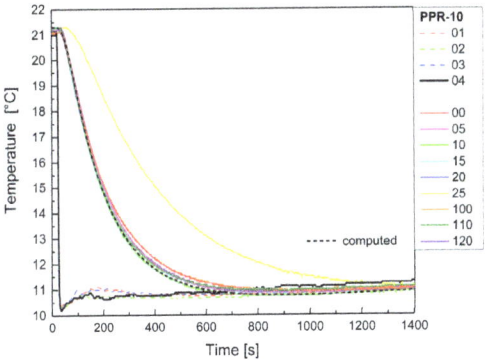

Figure 15. Time-history of measured temperatures and the temperatures computed in the probe axis (measurement A—water bath temperature of 10 °C/probe 1).

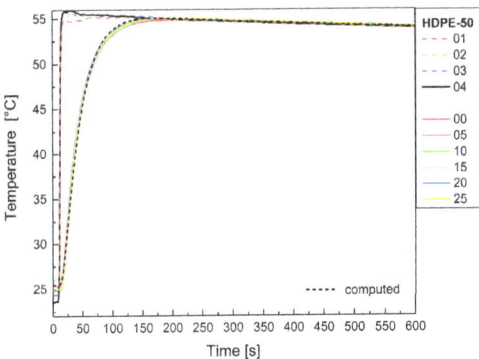

Figure 16. Time-history of measured temperatures and the temperatures computed in the probe axis (measurement A—water bath temperature of 55 °C/probe 2).

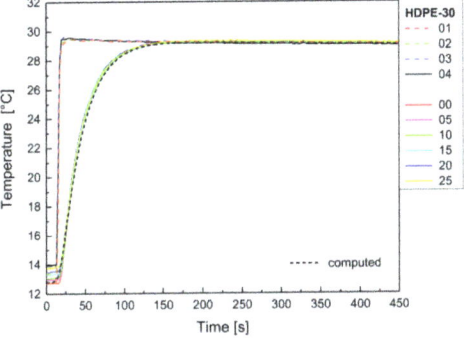

Figure 17. Time-history of measured temperatures and the temperatures computed in the probe axis (measurement A—water bath temperature of 30 °C/probe 2).

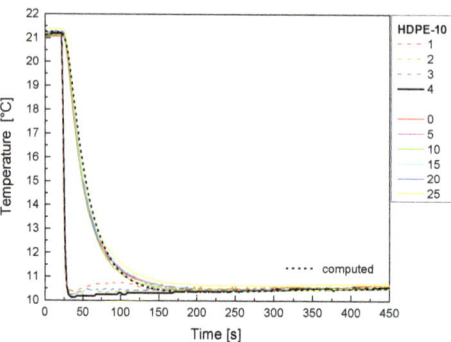

Figure 18. Time-history of measured temperatures and the temperatures computed in the probe axis (measurement A—water bath temperature of 10 °C/probe 2).

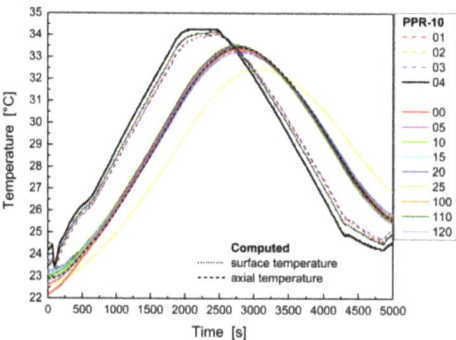

Figure 19. Time-history of measured temperatures and computed surface and axial temperatures (measurement B—temperature increase in climate chamber by 10 °C/probe 1).

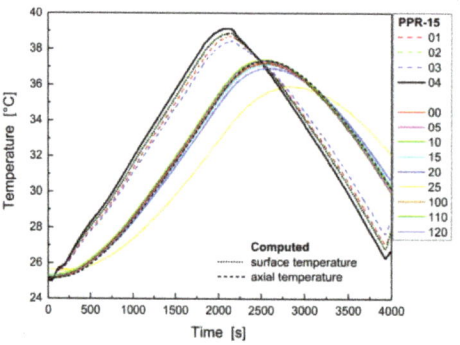

Figure 20. Time-history of measured temperatures and computed surface and axial temperatures (measurement B—temperature increase in climate chamber by 15 °C/probe 1).

For the practical verification of the probe, we carried out a pilot installation in the premises of the University of Žilina in the vicinity of other experimental installations in the field of meteorology. The probe contains nine pieces of DS18B20 thermometers spaced with 0.1 m grid. The installation was completed at the end of October 2021. The implementation is as follows: Sensor 9 is placed 0.1 m above the road surface, Sensor 8 is placed on the road surface, and Sensors 7–1 are placed 0.1 m to 0.7 m below the road surface. Figure 25 shows a graph of the temperature profile measurements from the start of the measurements in October 2021 to December 2021. Sensor 9 varies the most with respect to direct weather exposure at 0.1 m above the road surface. The effect of ambient temperature on the remaining sensors depends on their location below the road surface.

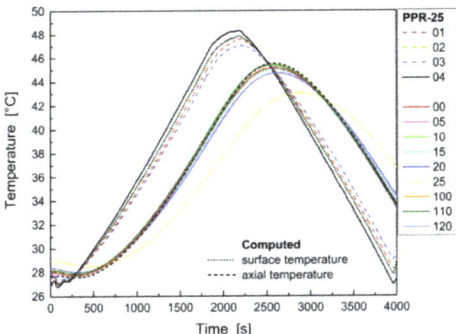

Figure 21. Time-history of measured temperatures and computed surface and axial temperatures (measurement B—temperature increase in climate chamber by 25 °C/probe 1).

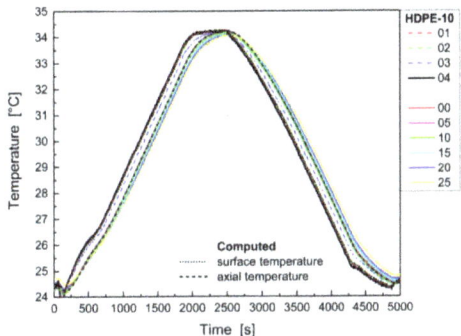

Figure 22. Time-history of measured temperatures and computed surface and axial temperatures (measurement B—temperature increase in climate chamber by 10 °C/probe 2).

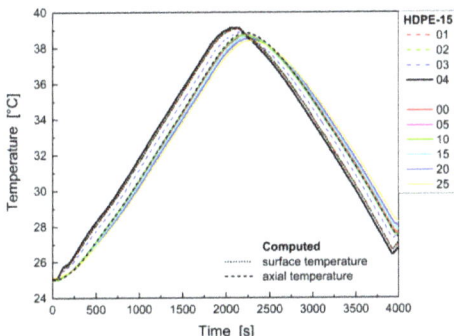

Figure 23. Time-history of measured temperatures and computed surface and axial temperatures (measurement B—temperature increase in climate chamber by 15 °C/probe 2).

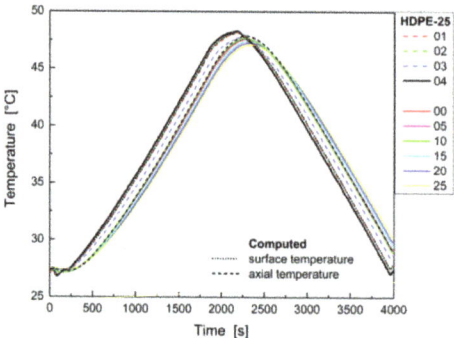

Figure 24. Time-history of measured temperatures and computed surface and axial temperatures (measurement B—temperature increase in climate chamber by 25 °C/probe 2).

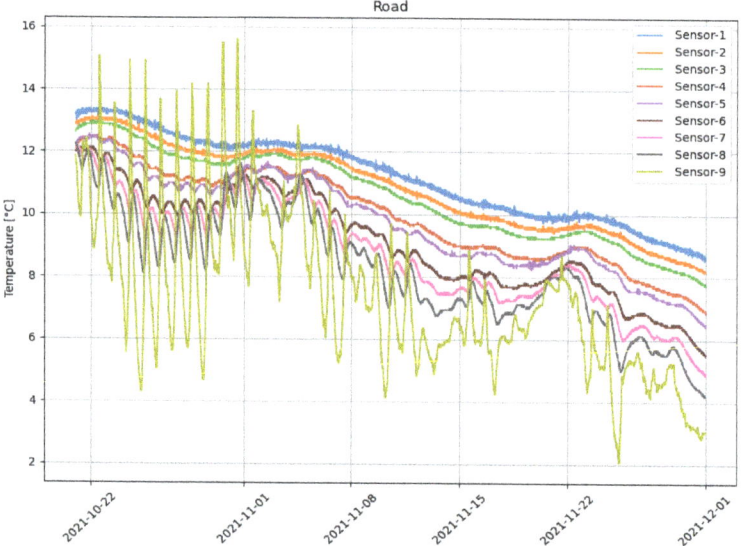

Figure 25. Temperature measurements October–December 2021.

The course of measurements for one week is shown in Figure 26. It can be seen that there is a transport delay in the effect of the exterior temperature on the individual thermometers depending on their location in the probe. This phenomenon is expected and arises primarily due to the parameters of the probe installation environment, e.g., subsoil. Analyzing the results of the temperature probe 1 (Figures 19–21) and probe 2 (Figures 22–24) sensors, it is clear that the transport delay between the individual sensors is approximately 120 s (Probe 1) and 480 s (Probe 2). Considering the heat transfer rate in the ground (asphalt road, construction materials, subsoil) in Figure 26, where the delay among sensors is from 4 to 6 h, the own transport delay of the temperature probe is negligible.

Novelty of Proposed Solution and Future Research

The novelty of the solution can be in our opinion divided into several categories. The first category is the design of the probe itself. Commonly available solutions using, for example, thermocouples or PT sensors require extensive cabling. Typically, this involves the utilization of 2-, 3-, or 4-wire connections. Such solutions increase in physical size as the number of sensors used in the probe grows due to the need for individual cabling for each installed sensor. Cabling then represents a significant parameter affecting the probe measurements—due to the amount of wire metal used. There is a considerable degree of influence on the installed sensors and subsequent measurements. Presented solution uses a custom PCB design with glass-laminate backing and conductive copper connections. Due to the usage of DS18B20 digital thermometers, the copper used is optimized for only 2 conductive paths. Therefore, we were able to reduce the metal used significantly and thus reduce the parasite influences.

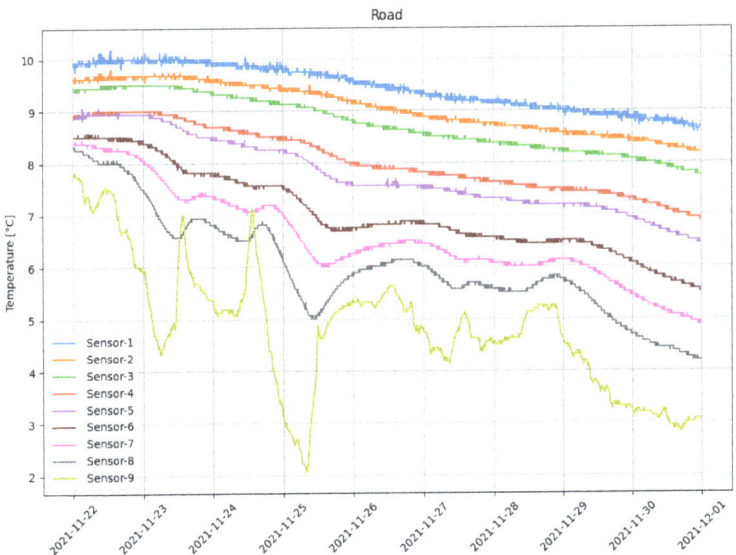

Figure 26. Vertical temperature measurements during one week.

Another benefit is the possibility of deploying sensors at precisely defined distances. This is ensured by the PCB design, where in the case of our experimental probe we used a 50 mm pitch, which defines the maximum density of sensors placement. This pitch can be modified to other values such as 10 mm by simply changing the PCB design. The modularity of the probe is ensured by the usage of several PCB modules, which can be connected by jumpers and, if necessary, shortened at designated locations, which are indicated on the PCB silkscreen in front of the installed sensor.

We have designed the probe housing in two versions using two different materials. Both designs are applicable to forest road environments. The difference is perceived mainly in terms of practical installation in the forest road, taking into account the need to drill a hole with the least possible diameter in order not to disturb the subsoil and thus affect the measurements significantly. The probe uses 1-wire technology, which is widely known and often directly used in a number of available IoT solutions. This solution therefore allows the probe to be connected to data loggers, measurement control panels, stand-alone data acquisition solutions as well as online connectivity via communication converters or IoT solutions, making the probe a rather versatile device.

Furthermore, in our opinion, this is a new solution for forest roads, as the forest road area, although similar in nature to the general road area, has specific requirements for the equipment used. In our case, we have started to design a solution for the measurement of the vertical temperature profile of forest roads as part of a research project to gather information on subsoil freezing progress. In addition, the solution for the measurement of the horizontal temperature profile has already been validated from previous activities [40,41]. The said solution also uses 1-wire technology.

In additional support of our novelty claims, we note that the novelty of our solution may also be supported by the granted utility model number SK122021U1 [42].

The measured data, once processed, will enter the forest road management system, where it should be part of a predictive model for preventive maintenance and repair planning. The predictive model will have a higher relevance after the addition of other parts of the measurement system and the measurement of more annual cycles. However, this activity is not part of our temperature probe design and is beyond the scope of this manuscript.

Although the proposed probe is suitable for its intended use in the management of forest road environments, we see further opportunities for modifications and improvements in order to maximize the range of its applications. We also claimed intellectual property rights on our probe design [42]. Future research activities will aim at subtle details, which may seem to be negligible, although in overall may affect measurement characteristics of the proposed probe, for example:

1. analysis of the effect of PUR sealant aging on the response time and measurement accuracy. It is assumed that due to aging, the PUR sealant will shrink and thus the contact between the individual components of the probe will be lost. From a thermal point of view, this will increase the thermal resistance, e.g., between the probe housing and the PUR sealant, thus extending the response time;
2. investigation of temperature fields, mainly the temperature gradients along the height of the probe, which occur due to variation in the surrounding temperature depending on the time and depth in the soil layer. This research will require the preparation and solution of a 3D simulation model, including the definition of the temperature profile in the soil depending on the depth below the earth's surface in different seasons, as well as a description of temperature changes during the day;
3. optimization of the internal probe layout with respect to self-heating of the measuring elements, space requirements and dimensions; and
4. analysis of several available sealants to design specific probe solutions for environments with special requirements.

5. Conclusions

The growing need for higher-quality transportation in a specific natural setting, typically characterized by significant temperature changes, puts pressure on local civil engineers to improve quality. The interest in more efficient forest environment management, a more pleasant approach to fire interventions, and finally, the attempt to expand recreational use of the forest in the development of tourism are the major causes for forest management. The installation of a network of stationary measuring sites on designated road sections is also part of the criteria, with the goal of providing valuable inputs for the accurate evaluation of the gathered data on the structure and road surface. This includes, for example, unique sensors that assess the temperature profile of roads in both the vertical and horizontal directions, soil sensors put along the road, frost sensors, road pollution sensors, and different meteorological sets for use on the forest road network. Our approach opens the potential of fusing data from such professional equipment with data from our suggested temperature probe for measuring the temperature profile of the soil to create a measurement set with additional value that contributes to the achievement of the above purposes.

Our proposal represents a comprehensive solution with possibilities for future extensions. We have addressed the design of the mechanical solution in two variants for use in several types of environments, considering the requirements for environments with mechanical, chemical, and environmental stresses. The electrical wiring of the sensors used is straightforward, but it was necessary to design their arrangement in the probe so that their positions were guaranteed. We proposed two solutions, of which the solution using PCB elements that can be adapted to the desired length, currently with a 50 mm grid, proved to be the most suitable. Such wiring allows data collection via a standalone logger, or the probe can be connected to an IoT device supporting a 1-wire bus. Based on the proposed solution, we have created a simulation model, which then enters the verification process. We designed and implemented a set of experiments in a water bath and a climate chamber. Subsequently, we discussed and compared the results of the experiments and simulation in Chapter 4. The presented figures clearly confirm the substantial agreement between the experimental and simulation data. Our proposed probe will primarily be used to measure the temperature profile of forest roads as input to models used in their management. However, its application is much broader in different areas of the economy, such as agriculture, energy, technological processes, automation, smart city solutions and more with requisite for measuring temperature profiles.

6. Patents

SK122021U1 Temperature probe with adjustable position of the installed sensors.

Author Contributions: Conceptualization, G.G., J.D., S.S., M.B., M.S., R.B. and B.T.; Data curation, G.G., M.B. and M.S.; Formal analysis, G.G., M.B. and M.S.; Funding acquisition, G.G. and S.S.; Investigation, J.D., S.S., R.B. and B.T.; Methodology, G.G., J.D., S.S. and M.B.; Project administration, G.G., M.S. and R.B.; Resources, G.G., J.D., S.S., M.B. and R.B.; Supervision, G.G. and M.S.; Validation, S.S., R.B. and B.T.; Writing—original draft, G.G., M.B. and R.B.; Writing—review and editing, G.G., M.B., R.B. and B.T. All authors have read and agreed to the published version of the manuscript.

Funding: This paper was supported with project of basic research: "Expanding the base of theoretical hypotheses and initiating assumptions to ensure scientific progress in methods of monitoring hydrometeors in the lower troposphere", which is funded by the R&D Incentives contract. This paper was supported under the project of Operational Programme Integrated Infrastructure: Research and development of contactless methods for obtaining geospatial data for forest monitoring to improve forest management and enhance forest protection, ITMS code 313011V465. The project is co-funding by European Regional Development Fund.

Institutional Review Board Statement: Not applicable.

Informed Consent Statement: Not applicable.

Data Availability Statement: Not applicable.

Conflicts of Interest: The authors declare no conflict of interest.

References

1. Roberts, R.; Inzerillo, L.; di Mino, G. Using UAV Based 3D Modelling to Provide Smart Monitoring of Road Pavement Conditions. *Information* **2020**, *11*, 568. [CrossRef]
2. Trubia, S.; Severino, A.; Curto, S.; Arena, F.; Pau, G. Smart Roads: An Overview of What Future Mobility Will Look Like. *Infrastructures* **2020**, *5*, 107. [CrossRef]
3. Akay, A.; Serin, H.; Sessions, J.; Bilici, E.; Pak, M. Evaluating the Effects of Improving Forest Road Standards on Economic Value of Forest Products. *Croat. J. For. Eng.* **2021**, *42*, 245–258. [CrossRef]
4. Zhang, F.; Dong, Y.; Xu, S.; Yang, X.; Lin, H. An approach for improving firefighting ability of forest road network. *Scand. J. For. Res.* **2020**, *35*, 547–561. [CrossRef]
5. Florková, Z.; Pepucha, L. Microtexture diagnostics of asphalt pavement surfaces. *IOP Conf. Ser. Mater. Sci. Eng.* **2017**, *236*, 12–25. [CrossRef]
6. Kneib, G. Non-destructive Pavement Testing for Sustainable Road Management. In *Proceedings of the 9th International Conference on Maintenance and Rehabilitation of Pavements—Mairepav9*; Springer: Cham, Switzerland, 2020; pp. 675–685.
7. Fidelus-Orzechowska, J.; Strzyżowski, D.; Cebulski, J.; Wrońska-Wałach, D. A Quantitative Analysis of Surface Changes on an Abandoned Forest Road in the Lejowa Valley (Tatra Mountains, Poland). *Remote Sens.* **2020**, *12*, 3467. [CrossRef]
8. Kruchinin, I.N.; Pobedinsky, V.V.; Kovalev, R.N. Fuzzy simulation of forest road surface parameters. *IOP Conf. Ser. Earth Environ. Sci.* **2019**, *316*, 012026. [CrossRef]
9. Stepanova, D.; Sukuvaara, T.; Karsisto, V. Intelligent Transport Systems–Road weather information and forecast system for vehicles. In Proceedings of the 2020 IEEE 91st Vehicular Technology Conference (VTC2020-Spring), Antwerp, Belgium, 25–28 May 2020; pp. 1–5. [CrossRef]
10. Erhan, L.; di Mauro, M.; Anjum, A.; Bagdasar, O.; Song, W.; Liotta, A. Embedded Data Imputation for Environmental Intelligent Sensing: A Case Study. *Sensors* **2021**, *21*, 7774. [CrossRef]
11. Wong, M.; Wang, T.; Ho, H.; Kwok, C.; Lu, K.; Abbas, S. Towards a Smart City: Development and Application of an Improved Integrated Environmental Monitoring System. *Sustainability* **2018**, *10*, 623. [CrossRef]

12. Baire, M.; Melis, A.; Lodi, M.B.; Tuveri, P.; Dachena, C.; Simone, M.; Fanti, A.; Fumera, G.; Pisanu, T.; Mazzarella, G. A Wireless Sensors Network for Monitoring the Carasau Bread Manufacturing Process. *Electronics* **2019**, *8*, 1541. [CrossRef]
13. Anderson, M.P. Heat as a Ground Water Tracer. *Groundwater* **2005**, *43*, 951–968. [CrossRef]
14. García, L.; Parra, L.; Jimenez, J.M.; Parra, M.; Lloret, J.; Mauri, P.V.; Lorenz, P. Deployment Strategies of Soil Monitoring WSN for Precision Agriculture Irrigation Scheduling in Rural Areas. *Sensors* **2021**, *21*, 1693. [CrossRef]
15. Júnior, R.D.N.; Amaral, G.C.D.; Pezzopane, J.E.M.; Toledo, J.V.; Xavier, T.M.T. Ecophysiology of C3 and C4 plants in terms of responses to extreme soil temperatures. *Theor. Exp. Plant Physiol.* **2018**, *30*, 261–274. [CrossRef]
16. Chunpeng, H.; Yanmin, J.; Peifeng, C.; Guibo, R.; Dongpo, H. Automatic Measurement of Highway Subgrade Temperature Fields in Cold Areas. In Proceedings of the 2010 International Conference on Intelligent System Design and Engineering Application, Changsha, China, 13–14 October 2010; Volume 1, pp. 409–412. [CrossRef]
17. *Programmable Resolution 1-Wire Digital Thermometer DS18B20*; Maxim Integrated Products, Inc.: San Jose, CA, USA, 2019. Available online: https://datasheets.maximintegrated.com/en/ds/DS18B20.pdf (accessed on 5 June 2021).
18. Morais, R.; Mendes, J.; Silva, R.; Silva, N.; Sousa, J.J.; Peres, E. A Versatile, Low-Power and Low-Cost IoT Device for Field Data Gathering in Precision Agriculture Practices. *Agriculture* **2021**, *11*, 619. [CrossRef]
19. Liu, C.; Ren, W.; Zhang, B.; Lv, C. The application of soil temperature measurement by LM35 temperature sensors. In Proceedings of the 2011 International Conference on Electronic & Mechanical Engineering and Information Technology, Harbin, China, 12–14 August 2011; Volume 4, pp. 1825–1828. [CrossRef]
20. Izquierdo, C.G.; Garcia-Benadí, A.; Corredera, P.; Hernandez, S.; Calvo, A.G.; Fernandez, J.D.R.; Nogueres-Cervera, M.; De Torres, C.P.; Del Campo, D. Traceable sea water temperature measurements performed by optical fibers. *Measurement* **2018**, *127*, 124–133. [CrossRef]
21. Othman, A.; Maga, D. Indoor Photovoltaic Energy Harvester with Rechargeable Battery for Wireless Sensor Node. In Proceedings of the 2018 18th International Conference on Mechatronics-Mechatronika (ME), Brno, Czech Republic, 5–7 December 2018; pp. 1–6.
22. Andreadis, A.; Giambene, G.; Zambon, R. Monitoring Illegal Tree Cutting through Ultra-Low-Power Smart IoT Devices. *Sensors* **2021**, *21*, 7593. [CrossRef]
23. Gaspar, G.; Sedivy, S.; Dudak, J.; Skovajsa, M. Meteorological support in research of forest roads conditions monitoring. In Proceedings of the 2021 2nd International Conference on Smart Electronics and Communication (ICOSEC), Trichy, India, 7–9 October 2021; pp. 1265–1269. [CrossRef]
24. Saavedra, E.; del Campo, G.; Santamaria, A. Smart Metering for Challenging Scenarios: A Low-Cost, Self-Powered and Non-Intrusive IoT Device. *Sensors* **2020**, *20*, 7133. [CrossRef]
25. Kampczyk, A.; Dybeł, K. The Fundamental Approach of the Digital Twin Application in Railway Turnouts with Innovative Monitoring of Weather Conditions. *Sensors* **2021**, *21*, 5757. [CrossRef]
26. Rosero-Montalvo, P.D.; Erazo-Chamorro, V.C.; López-Batista, V.F.; Moreno-García, M.N.; Peluffo-Ordóñez, D.H. Environment Monitoring of Rose Crops Greenhouse Based on Autonomous Vehicles with a WSN and Data Analysis. *Sensors* **2020**, *20*, 5905. [CrossRef]
27. Croce, S.; Tondini, S. Urban microclimate monitoring and modelling through an open-source distributed network of wireless low-cost sensors and numerical simulations. In Proceedings of the 7th International Electronic Conference on Sensors and Applications, Basel, Switzerland, 15–30 November 2020; p. 8270. [CrossRef]
28. Fabo, P.; Sedivy, S.; Kuba, M.; Buchholcerova, A.; Dudak, J.; Gaspar, G. PLC based weather station for experimental measurements. In Proceedings of the 2020 19th International Conference on Mechatronics-Mechatronika (ME), Prague, Czech Republic, 2–4 December 2020; pp. 1–4. [CrossRef]
29. Gaspar, G.; Sedivy, S.; Mikulova, L.; Dudak, J.; Fabo, P. Local Weather Station for Decision Making in Civil Engineering. In *Informatics and Cybernetics in Intelligent Systems*; Springer: Cham, Switzerland, 2021; pp. 199–206.
30. Blázquez, C.S.; Piedelobo, L.; Fernández-Hernández, J.; Nieto, I.M.; Martín, A.F.; Lagüela, S.; González-Aguilera, D. Novel Experimental Device to Monitor the Ground Thermal Exchange in a Borehole Heat Exchanger. *Energies* **2020**, *13*, 1270. [CrossRef]
31. Lewis, G.D.; Merken, P.; Vandewal, M. Enhanced Accuracy of CMOS Smart Temperature Sensors by Nonlinear Curvature Correction. *Sensors* **2018**, *18*, 4087. [CrossRef]
32. Koritsoglou, K.; Christou, V.; Ntritsos, G.; Tsoumanis, G.; Tsipouras, M.G.; Giannakeas, N.; Tzallas, A.T. Improving the Accuracy of Low-Cost Sensor Measurements for Freezer Automation. *Sensors* **2020**, *20*, 6389. [CrossRef]
33. Elan-tron PU 310/PH 27. ELANTAS Group. Available online: https://www.elantas.com (accessed on 16 August 2021).
34. Vötsch Industrietechnik GmbH, "Climate Test Chambers." March 2021. Available online: https://www.weiss-technik.com/fileadmin/Redakteur/Mediathek/Broschueren/WeissTechnik/Umweltsimulation/Voetsch/Voetsch-Technik-VC3-VCS3-EN.pdf (accessed on 15 May 2021).
35. Incropera, F.P.; DeWitt, D.P. *Fundamentals of Heat and Mass Transfer*; Wiley: Hoboken, NJ, USA, 1996. Available online: https://books.google.sk/books?id=UAZRAAAAMAAJ (accessed on 16 August 2021).
36. ANSYS, Inc. *Ansys Mechanical | Structural FEA Analysis Software*; Ansys: Canonsburg, PA, USA. Available online: https://www.ansys.com/products/structures/ansys-mechanical (accessed on 5 September 2021).
37. "HDPE Data Sheet.pdf". Available online: https://www.directplastics.co.uk/pub/pdf/datasheets/HDPE%20Data%20Sheet.pdf (accessed on 16 August 2021).

38. "Raw Material/PPR". Available online: http://ppr.almatherm.de/index.php?option=com_content&view=article&id=%208&Itemid=139 (accessed on 16 August 2021).
39. Yazdi, A.; Zebarjad, S.; Sajjadi, S.; Esfahani, J. On the sensitivity of dimensional stability of high densitypolyethylene on heating rate. *Express Polym. Lett.* **2007**, *1*, 92–97. [CrossRef]
40. Dudak, J.; Fabo, P.; Skovajsa, M.; Dvorský, L.; Kurota, P. Acta Logistica Moravica. *Syst. Monit. Environ. Parameters Road Infrastruct.* **2014**, *4*, 1804–8315.
41. Dudak, J.; Gaspar, G.; Sedivy, S.; Pepucha, L.; Florkova, Z. Road structural elements temperature trends diagnostics using sensory system of own design. *IOP Conf. Ser. Mater. Sci. Eng.* **2017**, *236*, 012036. [CrossRef]
42. Šedivý, Š.; Gašpar, G.; Fabo, P.; Farbák, M. Temperature Probe with Adjustable Position of the Installed Sensors. SK122021U1. Available online: https://worldwide.espacenet.com/patent/search/family/075979890/publication/SK122021U1?q=Gabriel%20Ga%C5%A1par (accessed on 26 May 2021).

Article

Cucumber Leaf Diseases Recognition Using Multi Level Deep Entropy-ELM Feature Selection

Muhammad Attique Khan [1,*], Abdullah Alqahtani [2], Aimal Khan [3], Shtwai Alsubai [2], Adel Binbusayyis [2], M Munawwar Iqbal Ch [4], Hwan-Seung Yong [5] and Jaehyuk Cha [6]

Citation: Khan, M.A.; Alqahtani, A.; Khan, A.; Alsubai, S.; Binbusayyis, A.; Ch, M.M.I.; Yong, H.-S.; Cha, J. Cucumber Leaf Diseases Recognition Using Multi Level Deep Entropy-ELM Feature Selection. *Appl. Sci.* **2022**, *12*, 593. https://doi.org/10.3390/app12020593

Academic Editors: Anselme Muzirafuti and Dimitrios S. Paraforos

Received: 19 December 2021
Accepted: 5 January 2022
Published: 7 January 2022

Publisher's Note: MDPI stays neutral with regard to jurisdictional claims in published maps and institutional affiliations.

Copyright: © 2022 by the authors. Licensee MDPI, Basel, Switzerland. This article is an open access article distributed under the terms and conditions of the Creative Commons Attribution (CC BY) license (https://creativecommons.org/licenses/by/4.0/).

1 Department of Computer Science, HITEC University Taxila, Taxila 47080, Pakistan
2 College of Computer Engineering and Sciences, Prince Sattam bin Abdulaziz University, Al-Kharj 16273, Saudi Arabia; Aq.alqahtani@psau.edu.sa (A.A.); Sa.alsubai@psau.edu.sa (S.A.); a.binbusayyis@psau.edu.sa (A.B.)
3 Department of Computer & Software Engineering, CEME NUST Rawalpindi, Rawalpindi 46000, Pakistan; aimalkhan.eme@gmail.com
4 Institute of Information Technology, Quaid-i-Azam University, Islamabad 44000, Pakistan; mmic@qau.edu.pk
5 Department of Computer Science & Engineering, Ewha Womans University, Seoul 03760, Korea; hsyong@ewha.ac.kr
6 Department of Computer Science, Hanyang University, Seoul 04763, Korea; chajh@hanyang.ac.kr
* Correspondence: attique.khan@hitecuni.edu.pk

Abstract: Agriculture has becomes an immense area of research and is ascertained as a key element in the area of computer vision. In the agriculture field, image processing acts as a primary part. Cucumber is an important vegetable and its production in Pakistan is higher as compared to the other vegetables because of its use in salads. However, the diseases of cucumber such as Angular leaf spot, Anthracnose, blight, Downy mildew, and powdery mildew widely decrease the quality and quantity. Lately, numerous methods have been proposed for the identification and classification of diseases. Early detection and then treatment of the diseases in plants is important to prevent the crop from a disastrous decrease in yields. Many classification techniques have been proposed but still, they are facing some challenges such as noise, redundant features, and extraction of relevant features. In this work, an automated framework is proposed using deep learning and best feature selection for cucumber leaf diseases classification. In the proposed framework, initially, an augmentation technique is applied to the original images by creating more training data from existing samples and handling the problem of the imbalanced dataset. Then two different phases are utilized. In the first phase, fine-tuned four pre-trained models and select the best of them based on the accuracy. Features are extracted from the selected fine-tuned model and refined through the Entropy-ELM technique. In the second phase, fused the features of all four fine-tuned models and apply the Entropy-ELM technique, and finally fused with phase 1 selected feature. Finally, the fused features are recognized using machine learning classifiers for the final classification. The experimental process is conducted on five different datasets. On these datasets, the best-achieved accuracy is 98.4%. The proposed framework is evaluated on each step and also compared with some recent techniques. The comparison with some recent techniques showed that the proposed method obtained an improved performance.

Keywords: crops diseases; data augmentation; deep learning; entropy; features fusion; machine learning

1. Introduction

Agriculture is one of the most important research topics globally nowadays [1]. Agriculture is a significant source of income and the economy of a country is based on the quality and yields of crops [2]. Cucumber is an important vegetable and during the year 2020, the global cucumber planting area was around 2.25 million hectares and the global yield was 90.35 million tons [3]. The production of crops is highly threatened by diseases and failure to identify and prevent cucumber diseases causes a reduction of cucumber vegetable yield and quality. The failure in early diagnosis causes significant economic losses to growers. Therefore, the rapid diagnosis of crops diseases helps to increase the quality and yield and also increases the national economy [4].

Mostly, identification is accomplished using typical methods like seeing through naked eyes or through a microscope [5]. The results of the manual visual estimation are generally unreliable while the microscopic assessments are generally time-consuming and costly. Most of the agriculturists of underdeveloped countries are illiterate [6]. They are compelled to return those charges along with other expenses like pesticides and fertilizer. The cucumber diseases like anthracnose, powdery mildew, downy mildew, and cucumber mosaic can destroy a large number of crops and the result will be a huge loss and vegetable deficiency [7]. Significant work has been done to accomplish a method that can boost the fastness and accuracy of the process. The methods necessarily contained some sort of computerization [8]. A large number of techniques presented until now are based on digital image processing and machine learning to identify the crops' diseases and achieve the desired output [9].

Image processing has many applications in the domain of computer vision such as medical imaging [10], agriculture [11], and named a few more. Agriculture is a hot application of image processing for the identification and classification of crops and plant diseases [12]. Although detection of cucumber abnormalities and then classifying them using image processing techniques is a critical task due to some sequence of steps [13]. A computerized method consists of some important steps such as preprocessing of original leaf images, detection of the infected region, feature extraction using handcrafted methods, and finally reduction and classification. Recognition of diseased portions in images is the key factor as it can influence the design and performance of the classification algorithms [14]. However, the error in the detection of the infected region extracted the irrelevant features that reduces the recognition accuracy.

Deep learning (DL) [15] is a hot research topic nowadays [16] and is employed everywhere for the detection and recognition tasks for several applications [17] such as biometric [18], image classification [19], surveillance [20], medical [21], and agriculture [22]. The researcher of computer vision introduced many techniques using machine learning and deep learning for plants diseases recognition [23]. Hussain et al. [24] introduced a deep learning technique for the identification of multiple cucumber leaf diseases. They extract deep learning features through two fine-tuned deep models including VGG19 and Inception V3. Both fine-tuned models were trained on the selected dataset using the transfer learning approach. The main advantage of training through TL is to save memory and time. The extracted features were fused by implementing the parallel maximum fusion technique to get the maximum information of each trained image. In the end, a whale optimization algorithm (WOA) was applied to select the robust features and perform classification. The purpose of feature selection is to get the best features because, in the fusion process, a few redundant features were also added. They achieved a maximum of 96.5% accuracy on the selected leaf dataset. In [13], researchers built an automated detection classification model for cucumber leaf diseases. In the first phase, pre-processing was performed to enhance the local contrast of images and to make the infected region more visible. This step makes the infected region clearer that later helped in the accurate segmentation using a novel Sharif saliency-based (SHSB) technique. Then researchers fused the proposed saliency method with active contour segmentation to improve the segmentation accuracy that later extracts the relevant features. In the feature extraction phase, they utilized VGG-19 and VGG-M pre-trained models. The extracted features were refined through three parameters including local entropy, local interquartile range, and local standard deviation. In the final classification, the best accuracy of 98.08% was achieved on multi-class SVM. The strength of this work was less computational time that can be useful for a real-time computerized system. Wang et al. [3] introduced a deep learning-based technique for the recognition of cucumber leaf diseases under complex backgrounds. They fused DeepLabV3+ and U-Net models instead of a single network. In the first step, DeepLabV3+ was used to segment the leaves from the images. Then the diseased area was segmented using U-Net. The fused models give better accuracy than the accuracy reported by the individual models. Researchers in [25], introduced a model for the identification of crop diseases in real-world images. The proposed trilinear convolutional neural network utilized bilinear pooling. In the laboratory environment, the proposed technique achieved 99.99% accuracy and in the real-world environment, the obtained accuracy is 84.11%. Kianat et al. [7] proposed a hybrid system for the recognition of cucumber diseases. In the pre-processing step, the data augmentation was applied using different angles to increase the image count in the dataset. In this step, contrast stretching was also performed to visually improve the images. The features were extracted from binary robust invariant scalable keypoints (BRISK), histogram of gradient (HOG), and features from the accelerated segmented test (FAST). Initially, the irrelevant features were eliminated by utilizing the probability distribution-based entropy (PDbE) technique. Then features were fused using the serial-based method and implemented Manhattan distance-controlled entropy (MDcE) method was to select the robust features. The proposed model achieved maximum accuracy of 93.5%. These techniques faced a major challenge of irrelevant feature extraction that were tried to be resolved through feature selection techniques [26].

Visual inspection of crops was carried out by farmers and agriculture experts. This evaluation process is exhausting, time-consuming, and highly subjective. The development of computer vision systems to identify, recognize, and classify disease-affected crops will keep humans out of the equation, allowing for unbiased, accurate disease-infection decisions [1]. An automatic classification system consists of various steps as mentioned above. Preprocessing is an important step, the aim is to remove noise and improve the quality of original images that later helps in important feature extraction. The extracted features from the refined images are used for the training of deep learning models that are further employed for feature extraction and classification. The key problems which are considered in this work are (i) training a deep learning model on an imbalanced dataset gives the high priority in the prediction to higher numbers of sample class; (ii) disease spots and background objects differ in appearance; (iii) changes in the shape, color, texture, and origin of the disease; (iv) irrelevant and redundant features extraction, and (v) choosing the superlative features for the classification.

In this article, our major focus is to design an automated computerized method for cucumber leaf diseases recognition using deep learning and Entropy-ELM-based best feature selection. The recent methods focused on the infected region identification and then employed for feature extraction; however, the error in the identification step misleads the irrelevant feature extraction that later reduces the classification accuracy. Our major contributions are:

(i) Four mathematical functions such as horizontal flip, vertical flip, rotate 45, and rotation 60 are implemented for the sake of data augmentation. Later, four deep learning models are fine-tuned and trained on the augmented dataset.
(ii) Deep learning features are extracted from the average pooling layer instead of the fully connected layer. The extracted deep features are passed to the Softmax classifier and compared the accuracy. Based on the accuracy value, the Densenet201 fine-tuned model is selected for the rest of the process. Moreover, all fine-tuned model features are fused using a new parallel approach.
(iii) An Entropy-ELM based best feature selection technique is proposed. The proposed technique is applied on both the Densenet201 feature vector and fused vector, that later serially fused for the final classification.
(iv) To determine which step of the proposed framework is better performed, a comparison is made between all hidden steps.

The rest of the manuscript is organized as follows: a proposed methodology that includes augmentation of the dataset, deep learning-based feature extraction, and Entropy-

ELM-based best feature selection, is presented in Section 2. Results are discussed in Section 3 with the help of tables and graphs. Finally, the conclusion of the manuscript is given in Section 4.

2. Proposed Methodology

In this work, an automated framework is proposed for cucumber leaf diseases recognition using deep learning and Entropy-ELM-based best feature selection. The proposed framework is illustrated in Figure 1. In this figure, it is shown that the initial augmentation step is applied to the original images by creating more training data. Then two different phases are utilized. In the first phase, four pre-trained deep models are fine-tuned and selected the best of them based on the accuracy. Features are extracted from the selected fine-tuned model and refined through the Entropy-ELM technique. In the second phase, fused the features of all four fine-tuned models and apply the Entropy-ELM technique, and finally fused with phase 1 selected feature. Finally, the fused features are classified using machine learning classifiers for the final output.

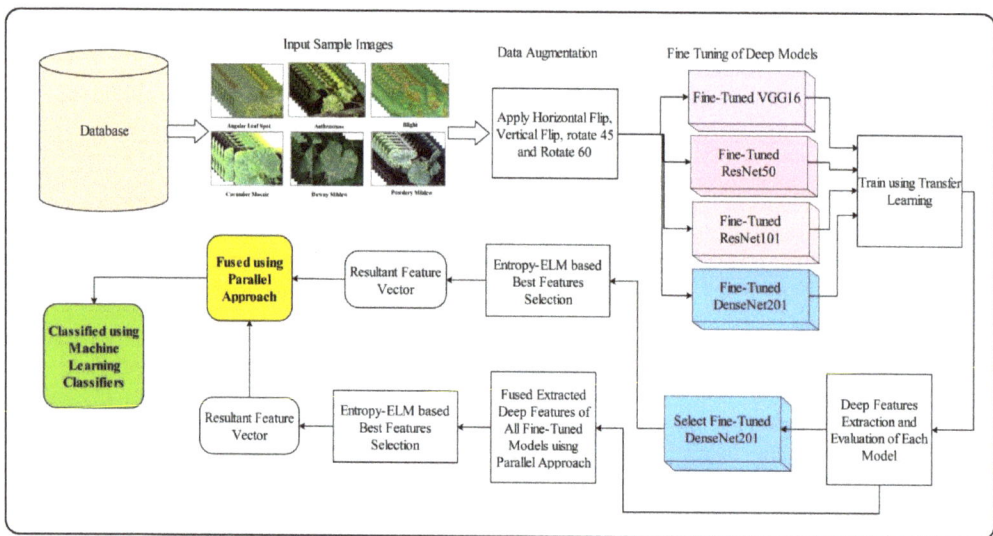

Figure 1. Proposed framework for cucumber leaf diseases recognition using deep learning and Entropy-ELM.

2.1. Dataset Collection and Augmentation

The experiments were performed on the publically available dataset named the Cucumber leaf diseases scan dataset [27]. This dataset consists of six different diseases such as anthracnose, powdery mildew, downy mildew, angular spot, mosaic, and blight. A sample of images are illustrated in Figure 2. Each class has 100 to 150 images originally that are not enough to train a deep learning model. Therefore, we design a simple algorithm (Algorithm 1) for data augmentation that includes four operations such as horizontal flip, vertical flip, rotate 45, and rotate 60. This algorithm is applied to each cucumber disease class and increases the number of images to 2000 in each class. In the later steps, this augmented dataset is utilized for the training of deep models.

Figure 2. Sample images of Cucumber leaf diseases.

Algorithm 1: (Data Augmentation)
Step 1: Input Original Database
Step 2: Consider First Disease Class
Step 3: Count Images of Step 2 (Selected Disease Class)
Step 4: For i = 1 to Total Images of each Class
- Horizontal Flip and Image Write
- Vertical Flip and Image Write
- Rotate 45 and Image Write
- Rotate 60 and Image Write
Step 5: Repeat Step 2, 3, and 4 for the Rest of the Disease Classes
End

2.2. Deep Learning Architecture

Four deep learning pre-trained models are employed in this work for feature extraction. The selected models are—VGG16, ResNet50, ResNet101, and DenseNet201. As mentioned in Figure 1, all selected models are initially fine-tuned and then trained through transfer learning using an augmented dataset. A brief description of each deep model is given below.

VGG16 [28] is a pre-trained model that was created by the Visual Geometry Group. This group is a combination of students and teachers focused on Computer Vision at Oxford University. This model is reflected to be one of the best computer vision models in the world. A unique feature of VGG16 is that rather than having numerous hyper-parameters it concentrates on having used identical PL and MPL of 2 × 2 filters of stride 2 and CL of 3 × 3 filters with a stride 1. VGG16 continues the same organization containing Convolutional and Maxpool Layers continuously during the course of the entire structural design. In the end, VGG has 2 Fully Connected Layers afterward a Softmax to output. Due to the fact that the VGG16 has 16 layers with weights, it has the name VGG16. This model was originally trained on an ImageNet dataset having 1000 object classes. The prediction of this model was done by the Softmax layer, defined as:

$$\Theta = w0x0 + w1x1 + \ldots + wkxk = \sum_{i=0}^{k} w_i x_i = w^T x \qquad (1)$$

ResNet [29] also known as a Deep Residual Network, have proved to perform with great accuracy and efficiency with a Deep Framework and to create an extra straight pathway for the transmission of data through the network. Within such Deep Systems, the deprivation issue arises because of the rise of Network Layers and the precision begins to dilute which results in its reduction quickly. Backpropagation does not come across the Vanishing Gradient problem when working with RESNET. There are some "Shortcut Connections" that a Residual Network has which are to be equivalent to a regular Convolutional Layer which aids the network to comprehend the Global Features. Then an input x has to be added to the output layer by adding the Shortcut connection, afterward some weight layers below. After the application of these Shortcut Connections, they permitted the network by avoiding the layers which were not beneficial while training. Hence, the output came in an ideal modification of the number of layers to perform rapid training. Mathematical, the output of $H(x)$ can be expressed as

$$H(x) = F(x) - x \qquad (2)$$

A type of Residual Mapping is used to train the weight layers which is expressed as,

$$F(x) = H(x) - x \qquad (3)$$

The above-mentioned function $F(x)$ signifies stacked nonlinear weight layers. Several properties of ResNet50 include the fact that it has 64 kernels including 7×7 Convolutional layers. It also includes 16 residual blocks. There are 23 million trainable parameters.

ResNet101 model utilizes Residual links that the angles can stream straightforwardly over to hinder the slopes to get 0 after the utilization of Chain Rule. There are 104 convolutional layers altogether in ResNet101. Alongside, it comprises 33 squares of layers altogether and 29 of these squares utilize past squares yield straightforwardly which is characterized as leftover associations above. Hence the above-mentioned residuals were using such main Operand of Summation (OOS) administrator towards the termination of every square to obtain the contribution of the accompanying squares. Leftover 4 squares get the past square's yield and apply it to a CL with a channel size of 1×1 and a step of 1 after a clump standardization layer, which performs standardization activity and the resultant yield is shipped off the summation administrator at the yield of that block. Mathematically, this model working is defined as follows:

$$u(x, t+1) = u(x,t) + w(x,t) * u(x,t) \qquad (4)$$

$$\widehat{T}_x = \frac{1}{2}\sigma^2 \frac{\partial^2}{\partial x^2} + b\frac{\partial}{\partial x} + c \Leftrightarrow \widehat{T}_p = -\frac{1}{2}\sigma^2 p^2 + ibp + c \qquad (5)$$

$$\widehat{T}_p \widetilde{u}(p,t) = \frac{d}{dt}\widetilde{u}(p,t) \qquad (6)$$

$$\widetilde{u}(p,t) = e^{\widehat{T}_p t}\widetilde{u}(p, 0) \qquad (7)$$

$$\widetilde{u}(p,t) \approx (1 + \widehat{T}_p t)\widetilde{u}(p, 0) \qquad (8)$$

Densenet-201 [30] is a convolutional neural network that is 201 layers deep. In this model, each layer gets feature maps from all preceding layers, the network can be thinner and more compact, resulting in fewer channels. The extra number of channels for each layer is the growth rate k. As a result, it has better computational and memory efficiency. The transition layers between two contiguous dense blocks are 11 Conv followed by 22 average pooling. Within the dense block, feature map sizes are uniform, allowing them to be readily concatenated. A global average pooling is done after the last dense block, and then a softmax classifier is added. The error signal can be transmitted more directly to earlier levels. As previous layers can get direct supervision from the final classification layer, this is a form of implicit deep supervision.

2.3. Transfer Learning Based Feature Extraction

Transfer learning (TL) is a process of reusing a pre-trained model for a new task [31], as illustrated in Figure 3. The ImageNet dataset was used as a source dataset of the pre-trained model. The pre-trained model is fine-tuned and transfer knowledge through the TL concept. In the last, the new fine-tuned model is trained on the augmented cucumber dataset that is utilized for further feature extraction. The features are extracted from the deep layers like FC7 for VGG, Average Pool for ResNet50, ResNet101, and Densenet201. Several hyperparameters are employed during the training process such as 0.0001 learning rate, max epochs are 200, the mini-batch size is 16, and the activation function is sigmoid.

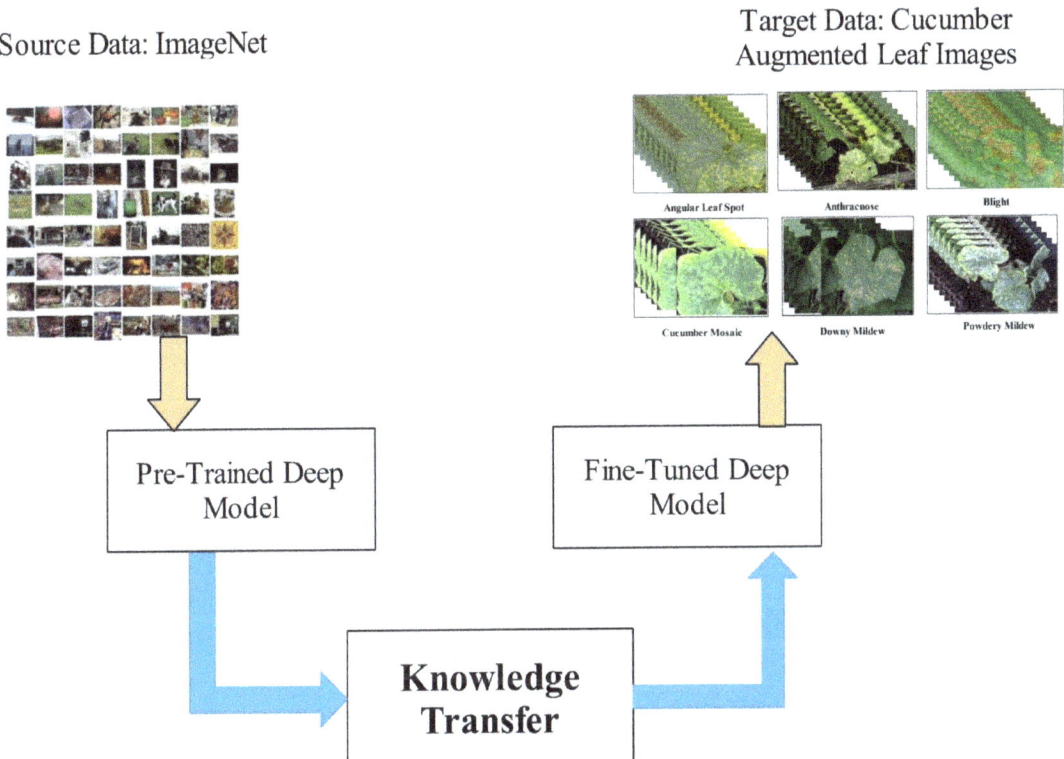

Figure 3. Transfer learning-based training of deep models for cucumber leaf diseases recognition.

2.4. Entropy-ELM Based Features Selection and Parallel Fusion

Feature selection is an important and hot research topic nowadays [32]. The main purpose of feature selection is to increase the system accuracy and minimize the computational time by focusing on the selection of the most important features [33]. In this work, a new technique is proposed named Entropy-ELM for the best feature selection. This proposed technique worked in the following steps: (i) compute the entropy of input vector; (ii) based on the entropy value, a threshold function is employed that return two vectors—fulfill the threshold value (selected) and not-selected; (iii) ELM [34] employed as a fitness function and selected threshold passed features are utilized as an input. Mathematically, the entropy formulation is defined as follows:

$$H_1 = -\sum_{k=1}^{G} P_k Id(P_k) \qquad (9)$$

$$H_{diff,1} = -\int_0^1 h_1(w) 1 d[h_1(w)] dw. \tag{10}$$

$$H_1 = H_{diff,1} + \text{Id}(G) = H_{diff,1} + H_{1,max} \tag{11}$$

$$\iint_{whole\ image} [\Delta w(w,y)]^k dxdy \propto \int_{-1}^1 (\Delta w)^k h_d(\Delta w) d\Delta w = M_k \tag{12}$$

$$T = \begin{cases} Sel(k) & for\ Features(i) \geq H_1 \\ NotSelec(l) & for\ Features(i) < H_1 \end{cases} \tag{13}$$

The detail of this selection process is given in Algorithm 2.

Algorithm 2: (Entropy-ELM)
Step 1: Input Feature Vector $N \times K$ // K is the length of features
Step 2: For i = 1 to N
Step 3: Computer Entropy through Equations (9)–(12)
Step 4: Define Threshold Function as Equation (13)
Step 5: Check Fitness through ELM
Step 6: Evaluate the Accuracy
Step 7: Repeat Step 2–6, until accuracy on the top side
End
Selected Feature Vector

Finally, the parallel fusion approach is opted to get the fused feature vector. This approach is based on the following three steps. In the first step, get the maximum length feature vector. As we have two feature vectors X and X_1, where the length of vectors is $N \times K$ and $N \times K_1$, respectively. In the second step, compute the entropy value and perform padding for the lower size feature vector. In the third step, correlation is computed among K and K_1 features for the final fusion. The fused vector is finally utilized for the classification through supervised learning classifiers.

$$Fusion = \psi(K, K_1) \tag{14}$$

where K and $K_1 \in X$ and X_1

3. Experimental Results

The proposed framework is evaluated on the selected cucumber dataset having a ratio of 70:15:15 which means that 70% of the images are utilized to train the model, whereas the 15% for testing and 15% for validation. We combined the testing and validation images and performed testing (30%). All the experimental results are computed with K-Fold cross-validation, whereas the value of K is 10. Several classifiers are implemented as discussed in Table 1. The performance of each classifier is computed through several measures such as recall rate, precision rate, F1-Score, accuracy, and time. The entire framework simulations are conducted on Simulink MATLAB2021a using a Personal Desktop.

Table 1. Brief description of selected classifiers.

Classifiers	Details
LSVM	Kernel scale: Automatic, Box constraint level: 1 Multiclass method: One-vs-One
QSVM	Kernel scale: Automatic, Box constraint level: 1 Multiclass method: One-vs-One
CSVM	Kernel scale: Automatic, Box constraint level: 1 Multiclass method: One-vs-One
MGSVM	Kernel scale: 45, Box constraint level: 1 Multiclass method: One-vs-One
FKNN	No of neighbor: 10, Distance matric: Euclidean Distance weight: Equal
Subspace_KNN	Learner type: Nearest neighbors, No of learners: 30 Subspace dimensions: 1024
Weighted_KNN	No of neighbor: 10, Distance matric: Euclidean Distance weight: Squared inverse
Cosine_KNN	No of neighbor: 10, Distance matric: cosine Distance weight: Equal
Cubic_KNN	No of neighbor: 10, Distance matric: Minkowsi (cubic) Distance weight: Equal
Medium_KNN	No of neighbor: 10, Distance matric: Euclidean Distance weight: Equal

3.1. Results

The detailed experimental process of the proposed framework is conducted in this section. The results are computed using the following steps: (i) classification using originally collected dataset on fine-tuned pre-trained models; (ii) classification using augmented dataset on fine-tuned deep models and select the best deep model for the further processing; (iii) best deep model features are refined using a new technique name Entropy-ELM; (iv) fusion of fine-tuned deep model features (augmented dataset), and (v) fused both step features using a parallel approach

3.2. Results on Original Cucumber Dataset

The results of the proposed method on the original cucumber dataset are given in Table 2. In this table, accuracy is computed for each fine-tuned deep model using the original dataset. Fine-tuned VGG16 (F-VGG16) obtained the maximum accuracy of 56.9% on the MG SVM classifier. The fine-tuned ResNet50 and ResNet101 obtained the best accuracy of 58.7 and 55.1% on Cubic SVM and Quadratic SVM, respectively. The fine-tuned Densenet201 deep model obtained an accuracy of 61.9% on Quadratic SVM. Based on these results, it is noticed that the originally collected dataset have several issues like imbalancing and short training data. Using these data, the fine-tuned Densenet201 gives better results for all classifiers.

Table 2. Classification results on originally selected cucumber dataset without data augmentation step for several fine-tuned deep learning models.

Classifier	F-VGG16	F-ResNet50	F-ResNet101	F-DenseNet201
Cubic SVM	55.6	58.7	54.2	61.8
Quadratic SVM	55.1	54.2	55.1	61.9
MG SVM	56.9	48	50.7	60.4
Fine KNN	50.7	48.9	49.8	52
*Linear SVM	51.6	53.3	51.1	58.4
ESD	24.9	47.1	46.7	56.4
ES KNN	50.7	55.1	53.8	51
WKNN	51.6	47.6	40.9	49.5
Cosine KNN	47.6	50.2	43.6	49
Medium KNN	45.3	45.3	36.6	49.5

3.3. Results on Augmented Cucumber Dataset

Experimental results of fine-tuned VGG16 pre-trained model after augmentation are given in Table 3. The best-obtained accuracy is 93.8% on Cubic SVM, whereas the recall rate and precision rates are 93.84 and 93.92%, respectively. The second best-obtained accuracy is 93.6%, which was accomplished on Quadratic SVM, whereas the recall rate and precision rates are 93.66 and 93.72%, correspondingly. The execution time of Linear SVM is better than the rest of the classifiers.

Table 3. Classification results of fine-tuned VGG16 deep model after data augmentation.

Classifier	Recall Rate (%)	Precision Rate (%)	Accuracy (%)	FNR (%)	F1 Score (%)	Time (Sec)
Cubic SVM	93.84	93.92	93.8	6.16	93.88	300
Quadratic SVM	93.66	93.72	93.6	6.34	93.69	242
MG SVM	91.78	92.04	91.8	8.22	91.91	463
Fine KNN	88.54	88.54	88.5	11.46	88.54	562
Linear SVM	90.56	90.88	90.6	9.44	90.72	188
ESD	92.98	93	93	7.02	92.99	1526
ES KNN	88.72	88.74	88.7	11.28	88.73	1550
WKNN	86.22	86.46	86.2	13.78	86.34	1038
Cosine KNN	80.8	81.36	80.8	19.20	81.08	648
Medium KNN	79.44	80.9	79.5	20.56	80.16	589

The classification accuracy of fine-tuned ResNet50 on the augmented dataset is given in Table 4. This table presents the highest obtained accuracy on Cubic SVM of 94.6%, whereas the recall and precision rates are 94.36 and 94.46%, correspondingly. The second top accuracy is 94.4% obtained on Quadratic SVM, whereas the recall and precision rates are 94.26 and 94.36%, respectively. Similar to fine-tuned VGG16, the Quadratic SVM executed fast than the rest of the classifiers.

Table 4. Classification results of fine-tuned ResNet50 deep model after data augmentation.

Classifier	Recall Rate (%)	Precision Rate (%)	Accuracy (%)	FNR (%)	F1 Score (%)	Time (Sec)
Cubic SVM	94.36	94.46	94.6	5.64	94.41	964
Quadratic SVM	94.26	94.36	94.4	5.44	94.61	387
MG SVM	91.38	91.7	91.5	8.62	91.54	1169
Fine KNN	86.6	86.96	86.6	13.40	86.78	681
Linear SVM	91.12	91.48	91	8.88	91.30	657
ESD	90.22	90.5	90.5	9.78	90.36	655
ES KNN	91.86	91.6	91.7	8.14	91.73	600
WKNN	79.66	81.44	78	20.34	80.54	748
Cosine KNN	82.38	82.72	82.4	17.62	82.55	539
Medium KNN	72.64	76.6	72.6	27.36	74.57	392

Experimental results of fine-tuned ResNet101 pre-trained model are given in Table 5. The best-obtained accuracy of 97.7% was accomplished on Cubic SVM. The recall and precision rates are 97.7 and 97.7%, correspondingly. The second best-obtained accuracy is 97.2% on Quadratic SVM. The recall and precision rates are 97.24 and 97.32%, correspondingly. In this experiment, the Linear SVM was executed fast than the rest of the selected classifiers.

Table 5. Classification results of fine-tuned ResNet101 deep model after data augmentation.

Classifier	Recall Rate (%)	Precision Rate (%)	Accuracy (%)	FNR (%)	F1 Score (%)	Time (Sec)
Cubic SVM	97.7	97.76	97.7	2.30	97.7	608
Quadratic SVM	97.24	97.32	97.2	2.76	97.2	574
MG SVM	94.64	94.7	94.6	5.36	94.6	1086
Fine KNN	94.42	94.44	94.4	5.58	94.4	1285
Linear SVM	94.32	94.62	94.3	5.68	94.4	513
ESD	95.82	96.68	95.8	4.18	96.2	2394
ES KNN	96.36	96.26	96.3	3.64	96.3	4072
WKNN	92.16	92.48	92.3	7.84	92.3	1501
Cosine KNN	86.46	86.84	86.5	13.54	86.6	1404
Medium KNN	83.44	84.44	83.5	16.56	83.93	1342

The classification results of fine-tuned Densenet201 pre-trained model are given in Table 6. In this table, the obtained best accuracy is 98.4% on Cubic SVM. Moreover, the recall and precision rates are 98.44 and 98.5%, correspondingly. Figure 4 illustrated the confusion matrix of Cubic SVM that was utilized for the verification of recall rate. The second best-obtained accuracy is 97.4%, which was accomplished on Quadratic SVM. The computation time of each classifier is also noted and the minimum time is 302 (sec) for LSVM. At the first step comparison among without augmented and augmented datasets, it is noted that the accuracy obtained on the augmented dataset is significantly better. In the second step comparison, it is noted that the fine-tuned DenseNet201 model achieved better results than VGG16, ResNet50, and ResNet101. Based on this analysis, the fine-tuned DenseNet201 is selected for the rest of the experiments.

Table 6. Classification results of fine-tuned Densenet201 deep model after data augmentation.

Classifier	Recall Rate (%)	Precision Rate (%)	Accuracy (%)	FNR (%)	F1 Score (%)	Time (Sec)
Cubic SVM	98.44	98.5	98.4	1.56	98.47	355
Quadratic SVM	97.4	97.46	97.4	2.60	97.43	330
MG SVM	95.32	95.62	95.4	4.68	95.47	623
Fine KNN	93.42	93.42	93.4	6.58	93.42	734
Linear SVM	92.62	93.16	92.7	7.38	92.89	302
ESD	96.62	96.6	96.6	3.38	96.61	1495
ES KNN	94.1	94.08	94.1	5.90	94.09	1923
WKNN	92.26	92.4	92.3	7.74	92.33	927
Cosine KNN	85.44	85.94	85.3	14.56	85.69	807
Medium KNN	85.9	86.9	85.8	14.10	86.40	764

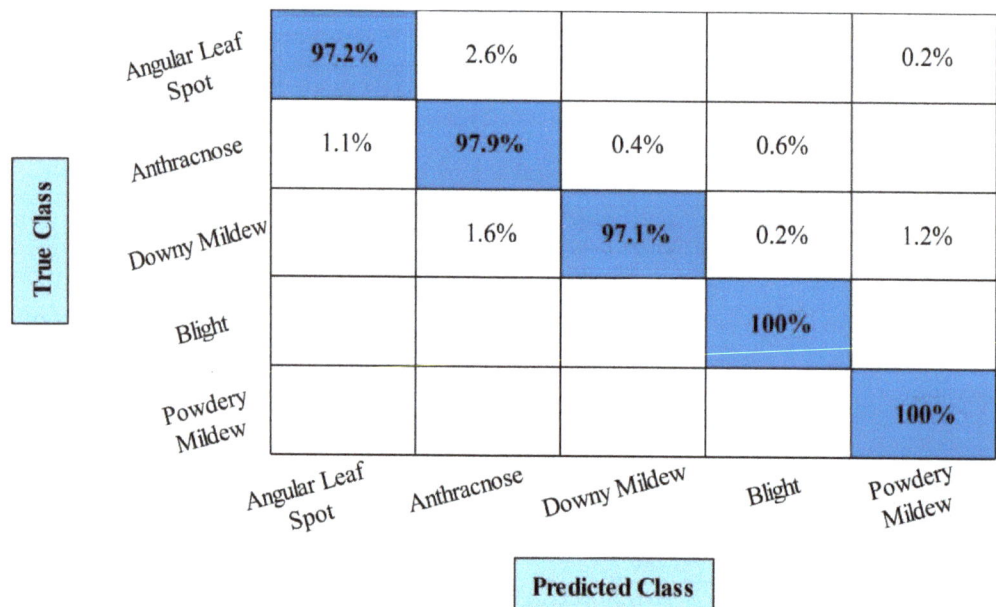

Figure 4. Confusion matrix-based representation of Cubic SVM accuracy.

The fine-tuned deep learning model is selected based on the better accuracy and applied proposed Entropy-ELM feature selection technique. The results are given in Table 7. This presents the best accuracy of 98% on Cubic SVM. The other computed measures are the recall rate which is 98.02, the precision rate at 97.98, and the F1-Score at 98%. The recall rate of Cubic SVM can be also verified through a confusion matrix, illustrated in Figure 5. Compared to the results given in Table 6, it is noted that the accuracy is a bit reduced but on the other side, a huge change occurred in the computation time. The time is also plotted in Figure 6 (FDenseNet201 and Dense Entropy-ELM).

Table 7. Classification results using proposed Entropy-ELM selection approach.

Classifier	Recall Rate (%)	Precision Rate (%)	Accuracy (%)	FNR (%)	F1 Score (%)	Time (Sec)
Cubic SVM	98.02	97.98	98.0	1.98	98.00	116
Quadratic SVM	96.54	96.62	96.5	3.46	96.58	135
MG SVM	84.2	81.4	84.0	15.80	82.78	197
Fine KNN	94.82	94.84	94.8	5.18	94.83	217
Linear SVM	82.82	83.14	82.8	17.18	82.98	133
ESD	93.3	93.78	93.0	6.70	93.54	201
ES KNN	93.16	92.2	93.0	6.84	92.68	405
WKNN	94.22	94.28	94.2	5.78	94.25	384
Cosine KNN	88.16	88.92	88.4	11.84	88.54	93
Medium KNN	85.78	86.8	85.8	14.22	86.29	204

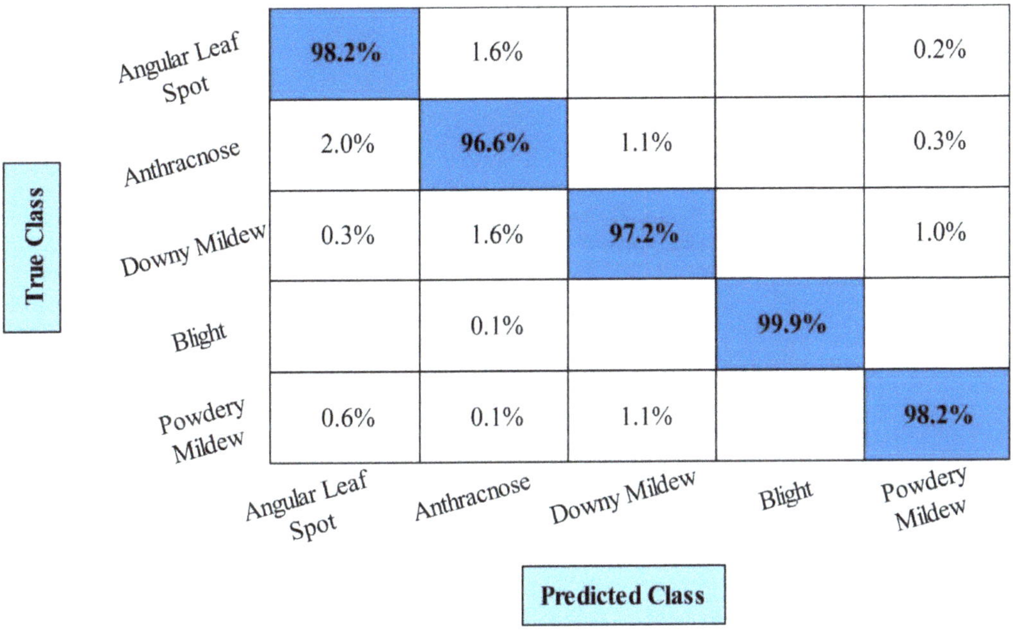

Figure 5. Confusion matrix of Cubic SVM after employing feature selection technique.

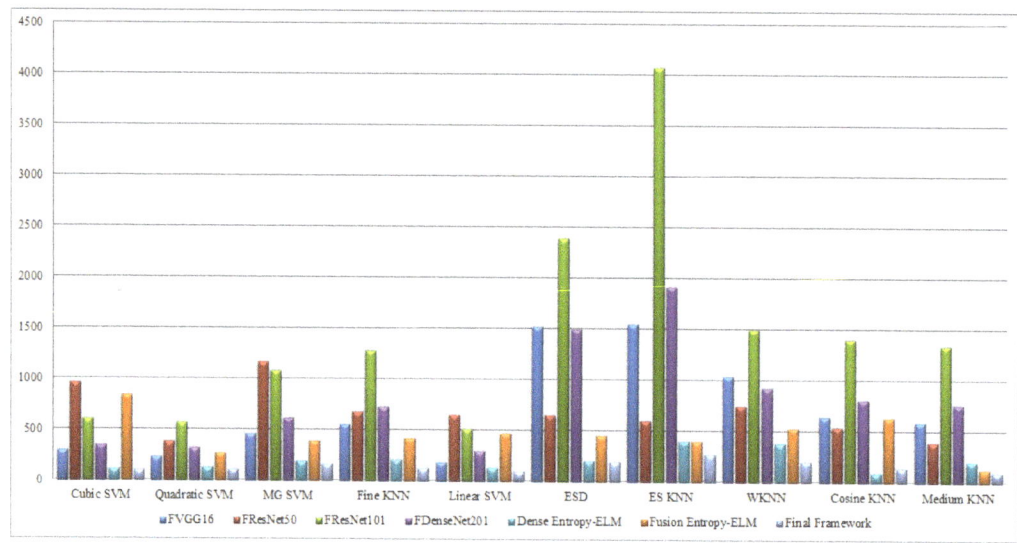

Figure 6. Comparison of all experiments in terms of testing time.

After the selection of the best dense features, in the next step all fine-tuned deep model features are fused using the proposed parallel approach. The results of this experiment are given in Table 8. The best-noted accuracy in this table is 98.2% on Cubic SVM. The recall and precision rates are 97.92 and 98.12%, respectively. Figure 7 illustrated the confusion matrix that can be utilized for the verification of the recall rate. The time of each classifier is also noted and plotted in Figure 6 (Fusion Entropy-ELM). In comparison with the results of Tables 6 and 7, it is noted that the overall accuracy is improved but the time is more increased than in the Dense Entropy-ELM step.

Table 8. Classification results using proposed parallel features fusion and Entropy-ELM selection of all pre-trained deep models using augmented dataset.

Classifier	Recall Rate (%)	Precision Rate (%)	Accuracy (%)	FNR (%)	F1 Score (%)	Time (Sec)
Cubic SVM	97.92	98.12	98.2	2.08	97.02	847
Quadratic SVM	97.74	97.38	97.7	2.26	97.56	277
MG SVM	94.84	95.06	94.8	5.16	94.95	392
Fine KNN	94.88	94.9	94.9	5.12	94.89	422
Linear SVM	94.76	95	94.8	5.24	94.88	475
ESD	96.26	96.28	96.3	3.74	96.27	454
ES KNN	96.66	96.66	96.7	3.34	96.66	400
WKNN	92.78	92.86	92.8	7.22	92.82	534
Cosine KNN	86.78	81.32	86.8	13.22	83.96	635
Medium KNN	84.82	85.6	84.8	15.18	85.21	129

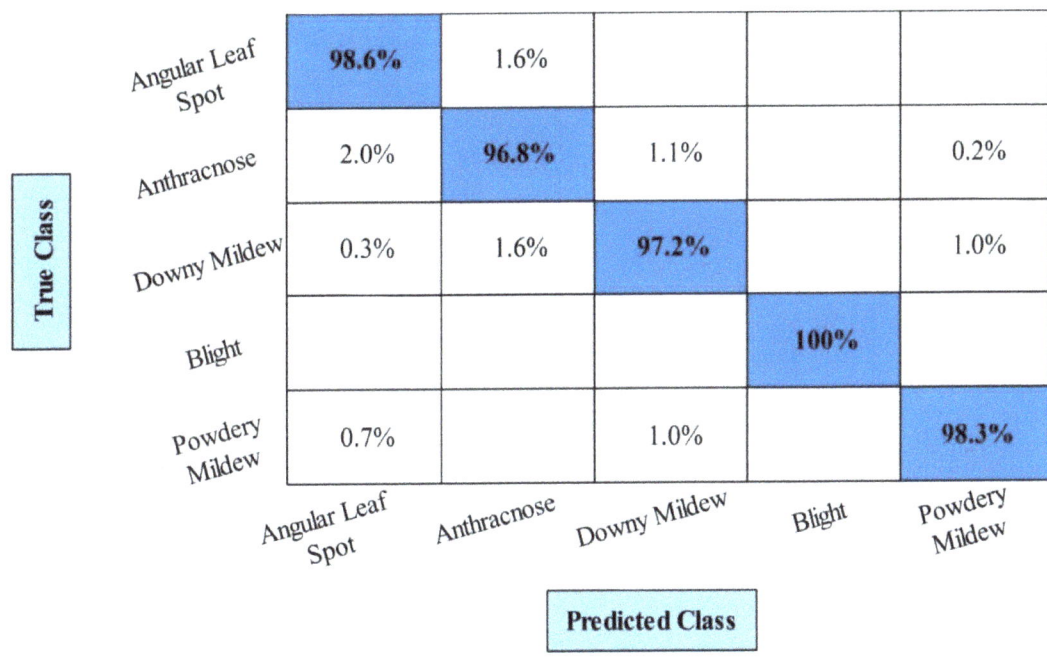

Figure 7. Confusion matrix of Cubic SVM after parallel fusion of all selected pre-trained deep features and Entropy-ELM selection.

Finally, the features of Dense Entropy-ELM and Fusion Entropy-ELM are fused using the proposed parallel approach, and the results are given in Table 9. This table presents the best-obtained accuracy of 98.50% on Cubic SVM. The noted precision rate is 98.30, recall rate is 98.36 and F1-Score is 98.48%, respectively. The second best-noted accuracy is 97.5% on Quadratic SVM. The recall rate of Cubic SVM can be verified through a confusion matrix plotted in Figure 8. This figure shows the correct prediction rate of each class in the diagonal. Compared to the results of this experiment with all previous experiments, it is clearly noted that the accuracy is improved and computational time is significantly reduced.

Table 9. Proposed framework classification results using augmented dataset.

Classifier	Recall Rate (%)	Precision Rate (%)	Accuracy (%)	FNR (%)	F1 Score (%)	Time (Sec)
Cubic SVM	98.36	98.3	98.5	1.74	98.48	111
Quadratic SVM	98.1	97.5	97.5	1.90	97.80	117
MG SVM	95.74	96.06	95.8	4.26	95.90	175
Fine KNN	94.42	94.42	94.4	5.58	94.42	130
Linear SVM	93.06	93.68	93.2	6.94	93.37	103
ESD	96.36	96.42	96.4	3.64	96.39	196
ES KNN	94.82	94.8	94.8	5.18	94.81	277
WKNN	92.2	92.56	92.2	7.80	92.38	201
Cosine KNN	85.48	85.78	85.4	14.52	85.63	142
Medium KNN	85.58	84.36	85.5	14.42	84.97	104

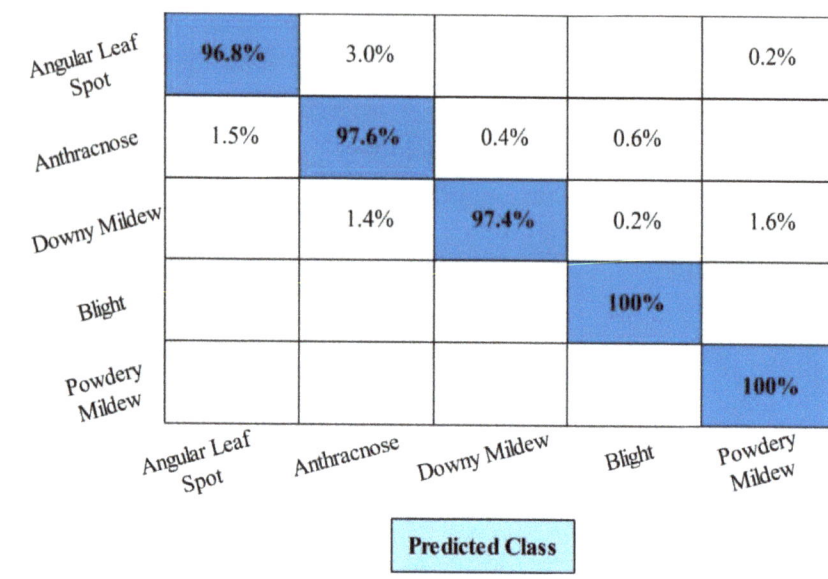

Figure 8. Confusion matrix of proposed framework of cucumber leaf diseases for Cubic SVM.

3.4. Discussion

Figure 1 showed the proposed framework that includes a few important steps. This figure illustrated the importance of the data augmentation step. The results without data augmentation having less accuracy than the results obtained after the data augmentation. Moreover, the selection of important features improves the accuracy that is later fused through a parallel approach. This step not only improves the classification accuracy but also reduced the computational time, as plotted in Figure 6. This figure clearly shows that the final fusion step significantly reduced the computational time than the rest of the steps on all classifiers.

In the last, the proposed framework accuracy is compared with recent SOTA techniques, as given in Table 10. The methods mentioned in this table are from the year 2017–2022. Moreover, all the methods mentioned in this table used the same leaf dataset. The recent best accuracy was 98.08% and 96.50% achieved by Khan et al. [13] and Hussain et al. [24]. The other methods such as Lin et al. [35] achieved an accuracy of 96.08% on the same dataset. The proposed framework achieved an accuracy of 98.48% that is improved than the SOTA techniques.

Table 10. Comparison with SOTA for cucumber leaf diseases recognition.

Methods	Year	Accuracy (%)
Zhang et al. [27]	2017	85.7
Ma et al. [36]	2018	93.4
Lin et al. [35]	2019	96.08
Khan et al. [13]	2020	98.08
Zhang et al. [37]	2021	90.67
Jaweria et al. [7]	2021	93.50
Hussain et al. [24]	2022	96.50
Proposed		98.48

4. Conclusions

Agriculture is a hot topic of research nowadays. In agriculture, deep learning showed significant success from the last decade for the recognition of plant diseases. In this article, a deep learning and Entropy-ELM based framework is proposed for the recognition of cucumber leaf diseases. In the proposed framework, four pre-trained deep models are trained and selected one of them based on the accuracy that is later employed for the selection of best features using the proposed Entropy-Elm technique. In the opposite step, features of all pre-trained models are fused and apply the feature selection technique. In the last, features of both steps are fused and perform classification. The proposed framework is tested on an augmented cucumber leaf dataset and achieved an accuracy of 98.48%. Comparison with the existing techniques showed the proposed framework obtained improved results. From the results, it is concluded that the augmentation process improves the recognition accuracy but also increases the time that was the first limitation of this framework; therefore a feature selection technique is proposed to maintain the accuracy and reduce the computational time. Through feature selection and fusion process, important information is obtained that later improves the classification accuracy. Another limitation of this work was the reduction of a few features that were ignored during the selection process. In the future, EfficientNet deep model will be implemented and features will be refined through the Butterfly metaheuristic algorithm instead of the heuristic search approach [20]. Moreover, reinforcement learning and Graph CNN shall be applied and refined through feature selection algorithms for the better results [38–42].

Author Contributions: Conceptualization, M.A.K., A.A., A.K., S.A., A.B., M.M.I.C., H.-S.Y. and J.C.; methodology, M.A.K., A.A., A.K., S.A., A.B., M.M.I.C., H.-S.Y. and J.C.; software, M.A.K., A.A., A.K., S.A., A.B., M.M.I.C., H.-S.Y. and J.C.; validation, M.A.K., A.A., A.K., S.A., A.B., M.M.I.C., H.-S.Y. and J.C.; formal analysis, M.A.K., A.A., A.K., S.A., A.B., M.M.I.C., H.-S.Y. and J.C.; investigation, M.A.K., A.A., A.K., S.A., A.B., M.M.I.C., H.-S.Y. and J.C.; resources, M.A.K., A.A., A.K., S.A., A.B., M.M.I.C., H.-S.Y. and J.C.; data curation, M.A.K., A.A., A.K., S.A.,A.B., M.M.I.C., H.-S.Y. and J.C.; writing and original draft preparation, M.A.K., A.A., A.K., S.A., A.B., M.M.I.C., H.-S.Y. and J.C.; writing, review and editing, M.A.K., A.A., A.K., S.A., A.B., M.M.I.C., H.-S.Y. and J.C.; visualization, M.A.K., A.A., A.K., S.A., A.B., M.M.I.C., H.-S.Y. and J.C.; supervision, M.A.K., A.A., A.K., S.A., A.B., M.M.I.C., H.-S.Y. and J.C.; project administration, M.A.K., A.A., A.K., S.A., A.B., M.M.I.C., H.-S.Y. and J.C.; funding acquisition, M.A.K., A.A., A.K., S.A., A.B., M.M.I.C., H.-S.Y. and J.C. All authors have read and agreed to the published version of the manuscript.

Funding: This work was supported by the National Research Foundation of Korea (NRF) grant funded by the Korean government (Ministry of Science and ICT; MSIT) under Grant RF-2018R1A5A7059549.

Institutional Review Board Statement: Not applicable.

Informed Consent Statement: Not applicable.

Data Availability Statement: Not applicable.

Conflicts of Interest: The authors declare no conflict of interest.

References

1. Iqbal, Z.; Sharif, M.; Shah, J.H.; ur Rehman, M.H.; Javed, K. An automated detection and classification of citrus plant diseases using image processing techniques: A review. *Comput. Electron. Agric.* **2018**, *153*, 12–32. [CrossRef]
2. Agarwal, M.; Gupta, S.; Biswas, K. A new conv2d model with modified relu activation function for identification of disease type and severity in cucumber plant. *Sustain. Comput. Inform. Syst.* **2021**, *30*, 100473. [CrossRef]
3. Wang, C.; Du, P.; Wu, H.; Li, J.; Zhao, C.; Zhu, H. A cucumber leaf disease severity classification method based on the fusion of DeepLabV3+ and U-Net. *Comput. Electron. Agric.* **2021**, *189*, 106373. [CrossRef]
4. Sanju, S.K.; VelammaL, D.B. An Automated Detection and Classification of Plant Diseases from the Leaves Using Image processing and Machine Learning Techniques: A State-of-the-Art Review. *Ann. Rom. Soc. Cell Biol.* **2021**, *25*, 15933–15950.
5. Pourdarbani, R.; Sabzi, S.; Rohban, M.H.; Hernández-Hernández, J.L.; Gallardo-Bernal, I.; Herrera-Miranda, I.; García-Mateos, G. One-Dimensional Convolutional Neural Networks for Hyperspectral Analysis of Nitrogen in Plant Leaves. *Appl. Sci.* **2021**, *11*, 11853. [CrossRef]

6. Tahir, M.B.; Javed, K.; Kadry, S.; Zhang, Y.-D.; Akram, T.; Nazir, M. Recognition of apple leaf diseases using deep learning and variances-controlled features reduction. *Microprocess. Microsyst.* **2021**, 104027. [CrossRef]
7. Kianat, J.; Sharif, M.; Akram, T.; Rehman, A.; Saba, T. A joint framework of feature reduction and robust feature selection for cucumber leaf diseases recognition. *Optik* **2021**, *240*, 166566. [CrossRef]
8. Tariq, U.; Hussain, N.; Nam, Y.; Kadry, S. An Integrated Deep Learning Framework for Fruits Diseases Classification. *Comput. Mater. Contin.* **2022**. [CrossRef]
9. Khirade, S.D.; Patil, A. Plant disease detection using image processing. In Proceedings of the 2015 International Conference on Computing Communication Control and Automation, Pune, India, 26–27 February 2015; pp. 768–771.
10. Rehman, A.; Naz, S.; Khan, A.; Zaib, A.; Razzak, I. Improving coronavirus (COVID-19) diagnosis using deep transfer learning. *MedRxiv* **2020**. [CrossRef]
11. Yasmeen, U.; Tariq, U.; Khan, J.A.; Yar, M.A.E.; Hanif, C.A.; Mey, S.; Nam, Y. Citrus Diseases Recognition Using Deep Improved Genetic Algorithm. *Comput. Mater. Contin.* **2021**, *73*, 1–15. [CrossRef]
12. Ramesh, S.; Hebbar, R.; Niveditha, M.; Pooja, R.; Shashank, N.; Vinod, P. Plant disease detection using machine learning. In Proceedings of the 2018 International Conference on Design Innovations for 3Cs Compute Communicate Control (ICDI3C), Bangalore, India, 25–28 April 2018; pp. 41–45.
13. Akram, T.; Sharif, M.; Javed, K.; Raza, M.; Saba, T. An automated system for cucumber leaf diseased spot detection and classification using improved saliency method and deep features selection. *Multimed. Tools Appl.* **2020**, *79*, 18627–18656.
14. ur Rehman, Z.; Ahmed, F.; Damaševičius, R.; Naqvi, S.R.; Nisar, W.; Javed, K. Recognizing apple leaf diseases using a novel parallel real-time processing framework based on MASK RCNN and transfer learning: An application for smart agriculture. *IET Image Process.* **2021**, *15*, 2157–2168. [CrossRef]
15. Razzak, M.I.; Imran, M.; Xu, G. Efficient brain tumor segmentation with multiscale two-pathway-group conventional neural networks. *IEEE J. Biomed. Health Inform.* **2018**, *23*, 1911–1919. [CrossRef]
16. Feizizadeh, B.; Mohammadzade Alajujeh, K.; Lakes, T.; Blaschke, T.; Omarzadeh, D. A comparison of the integrated fuzzy object-based deep learning approach and three machine learning techniques for land use/cover change monitoring and environmental impacts assessment. *GISci. Remote Sens.* **2021**, *58*, 1–28. [CrossRef]
17. Feizizadeh, B.; Omarzadeh, D.; Kazemi Garajeh, M.; Lakes, T.; Blaschke, T. Machine learning data-driven approaches for land use/cover mapping and trend analysis using Google Earth Engine. *J. Environ. Plan. Manag.* **2021**, 1–33. [CrossRef]
18. Saleem, F.; Khan, M.A.; Alhaisoni, M.; Tariq, U.; Armghan, A.; Alenezi, F.; Choi, J.-I.; Kadry, S. Human gait recognition: A single stream optimal deep learning features fusion. *Sensors* **2021**, *21*, 7584. [CrossRef]
19. Hussain, N.; Kadry, S.; Tariq, U.; Mostafa, R.; Choi, J.-I.; Nam, Y. Intelligent Deep Learning and Improved Whale Optimization Algorithm Based Framework for Object Recognition. *Hum. Cent. Comput. Inf. Sci* **2021**, *11*, 34.
20. Khan, S.; Alhaisoni, M.; Tariq, U.; Yong, H.-S.; Armghan, A.; Alenezi, F. Human Action Recognition: A Paradigm of Best Deep Learning Features Selection and Serial Based Extended Fusion. *Sensors* **2021**, *21*, 7941. [CrossRef]
21. Muhammad, K.; Sharif, M.; Akram, T.; Kadry, S. Intelligent fusion-assisted skin lesion localization and classification for smart healthcare. *Neural Comput. Appl.* **2021**, 1–16. [CrossRef]
22. Akram, T.; Sharif, M.; Alhaisoni, M.; Saba, T.; Nawaz, N. A probabilistic segmentation and entropy-rank correlation-based feature selection approach for the recognition of fruit diseases. *EURASIP J. Image Video Process.* **2021**, *2021*, 1–28.
23. Zhang, S.; Wang, Z. Cucumber disease recognition based on Global-Local Singular value decomposition. *Neurocomputing* **2016**, *205*, 341–348. [CrossRef]
24. Hussain, N.; Khan, M.A.; Tariq, U.; Kadry, S.; Yar, M.A.E.; Mostafa, A.M.; Alnuaim, A.A.; Ahmad, S. Multiclass Cucumber Leaf Diseases Recognition Using Best Feature Selection. *Comput. Mater. Contin.* **2022**, *2022*, 3281–3294. [CrossRef]
25. Wang, D.; Wang, J.; Li, W.; Guan, P. T-CNN: Trilinear convolutional neural networks model for visual detection of plant diseases. *Comput. Electron. Agric.* **2021**, *190*, 106468. [CrossRef]
26. Razzak, I.; Saris, R.A.; Blumenstein, M.; Xu, G. Integrating joint feature selection into subspace learning: A formulation of 2DPCA for outliers robust feature selection. *Neural Netw.* **2020**, *121*, 441–451. [CrossRef]
27. Zhang, S.; Wu, X.; You, Z.; Zhang, L. Leaf image based cucumber disease recognition using sparse representation classification. *Comput. Electron. Agric.* **2017**, *134*, 135–141. [CrossRef]
28. Qassim, H.; Verma, A.; Feinzimer, D. Compressed residual-VGG16 CNN model for big data places image recognition. In Proceedings of the 2018 IEEE 8th Annual Computing and Communication Workshop and Conference (CCWC), Las Vegas, NV, USA, 8–10 January 2018; pp. 169–175.
29. Theckedath, D.; Sedamkar, R. Detecting affect states using VGG16, ResNet50 and SE-ResNet50 networks. *SN Comput. Sci.* **2020**, *1*, 1–7. [CrossRef]
30. Wang, S.-H.; Zhang, Y.-D. DenseNet-201-based deep neural network with composite learning factor and precomputation for multiple sclerosis classification. *ACM Trans. Multimed. Comput. Commun. Appl.* **2020**, *16*, 1–19. [CrossRef]
31. Brusa, E.; Delprete, C.; Di Maggio, L.G. Deep Transfer Learning for Machine Diagnosis: From Sound and Music Recognition to Bearing Fault Detection. *Appl. Sci.* **2021**, *11*, 11663. [CrossRef]
32. Attique Khan, M.; Sharif, M.; Akram, T.; Kadry, S.; Hsu, C.H. A two-stream deep neural network-based intelligent system for complex skin cancer types classification. *Int. J. Intell. Syst.* **2021**. [CrossRef]

33. Khan, M.A.; Alhaisoni, M.; Tariq, U.; Hussain, N.; Majid, A.; Damaševičius, R.; Maskeliūnas, R. COVID-19 Case Recognition from Chest CT Images by Deep Learning, Entropy-Controlled Firefly Optimization, and Parallel Feature Fusion. *Sensors* **2021**, *21*, 7286. [CrossRef]
34. Huang, G.-B.; Zhu, Q.-Y.; Siew, C.-K. Extreme learning machine: Theory and applications. *Neurocomputing* **2006**, *70*, 489–501. [CrossRef]
35. Lin, K.; Gong, L.; Huang, Y.; Liu, C.; Pan, J. Deep learning-based segmentation and quantification of cucumber powdery mildew using convolutional neural network. *Front. Plant Sci.* **2019**, *10*, 155. [CrossRef]
36. Ma, J.; Du, K.; Zheng, F.; Zhang, L.; Gong, Z.; Sun, Z. A recognition method for cucumber diseases using leaf symptom images based on deep convolutional neural network. *Comput. Electron. Agric.* **2018**, *154*, 18–24. [CrossRef]
37. Zhang, J.; Rao, Y.; Man, C.; Jiang, Z.; Li, S. Identification of cucumber leaf diseases using deep learning and small sample size for agricultural Internet of Things. *Int. J. Distrib. Sens. Netw.* **2021**, *17*, 15501477211007407. [CrossRef]
38. Yuan, H.; Tang, G.; Guo, D.; Wu, K.; Shao, X.; Yu, K.; Wei, W. BESS Aided Renewable Energy Supply using Deep Reinforcement Learning for 5G and Beyond. *IEEE Trans. Green Commun. Netw.* **2021**, *11*, 1–12. [CrossRef]
39. Yu, K.; Tan, L.; Mumtaz, S.; Al-Rubaye, S.; Al-Dulaimi, A.; Bashir, A.K.; Khan, F.A. Securing Critical Infrastructures: Deep-Learning-Based Threat Detection in IIoT. *IEEE Commun. Mag.* **2021**, *59*, 76–82. [CrossRef]
40. Zhang, Q.; Yu, K.; Guo, Z.; Garg, S.; Rodrigues, J.; Hassan, M.M.; Guizani, M. Graph Neural Networks-driven Traffic Forecasting for Connected Internet of Vehicles. *IEEE Trans. Netw. Sci. Eng.* **2021**, *4*, 1–9. [CrossRef]
41. Afza, F.; Sharif, M.; Kadry, S.; Manogaran, G.; Saba, T.; Ashraf, I.; Damaševičius, R. A framework of human action recognition using length control features fusion and weighted entropy-variances based feature selection. *Image Vis. Comput.* **2021**, *106*, 104090. [CrossRef]
42. Rashid, M.; Sharif, M.; Javed, K.; Akram, T. Classification of gastrointestinal diseases of stomach from WCE using improved saliency-based method and discriminant features selection. *Multimed. Tools Appl.* **2019**, *78*, 27743–27770.

Article

Guava Disease Detection Using Deep Convolutional Neural Networks: A Case Study of Guava Plants

Almetwally M. Mostafa [1,*], Swarn Avinash Kumar [2], Talha Meraj [3], Hafiz Tayyab Rauf [4], Abeer Ali Alnuaim [5] and Maram Abdullah Alkhayyal [1]

1. Department of Information Systems, College of Computer and Information Sciences, King Saud University, P.O. Box 51178, Riyadh 11543, Saudi Arabia; 442204299@student.ksu.edu.sa
2. Indian Institute of Information Technology, Allahabad 211015, Uttar Pradesh, India; swarnavinashkumar@ieee.org
3. Department of Computer Science, COMSATS University Islamabad—Wah Campus, Wah Cantt 47040, Pakistan; talha_cui@ciitwah.edu.pk
4. Department of Computer Science, Faculty of Engineering & Informatics, University of Bradford, Bradford BD7 1DP, UK; h.rauf4@bradford.ac.uk or hafiztayyabrauf093@gmail.com
5. Department of Computer Science and Engineering, College of Applied Studies and Community Services, King Saud University, P.O. Box 22459, Riyadh 11495, Saudi Arabia; abalnuaim@ksu.edu.sa
* Correspondence: almetwaly@ksu.edu.sa

Abstract: Food production is a growing challenge with the increasing global population. To increase the yield of food production, we need to adopt new biotechnology-based fertilization techniques. Furthermore, we need to improve early prevention steps against plant disease. Guava is an essential fruit in Asian countries such as Pakistan, which is fourth in its production. Several pathological and fungal diseases attack guava plants. Furthermore, postharvest infections might result in significant output losses. A professional opinion is essential for disease analysis due to minor variances in various guava disease symptoms. Farmers' poor usage of pesticides may result in financial losses due to incorrect diagnosis. Computer-vision-based monitoring is required with developing field guava plants. This research uses a deep convolutional neural network (DCNN)-based data enhancement using color-histogram equalization and the unsharp masking technique to identify different guava plant species. Nine angles from 360° were applied to increase the number of transformed plant images. These augmented data were then fed as input into state-of-the-art classification networks. The proposed method was first normalized and preprocessed. A locally collected guava disease dataset from Pakistan was used for the experimental evaluation. The proposed study uses five neural network structures, AlexNet, SqueezeNet, GoogLeNet, ResNet-50, and ResNet-101, to identify different guava plant species. The experimental results proved that ResNet-101 obtained the highest classification results, with 97.74% accuracy.

Keywords: data augmentation; deep learning; guava disease; plant disease detection

1. Introduction

Food production is currently one of the greatest challenges with the growing global population. It is estimated that food consumption will double by 2050. Therefore, food production needs a more high-yielding and sustainable environment to increase the plant yield [1,2]. Guava is an important plant that belongs to the Myrtaceae plant family. It was initially allocated in the American tropics; guava was discovered in Portugal in the early 17th Century [3]. It is popular in tropical and nontropical countries such as Bangladesh, India, Pakistan, Brazil, and Cuba [4]. Guava contains phosphorus, calcium, nicotinic acid, and many other essential food components [5]. Furthermore, it normalizes blood pressure, has benefits for diabetes, provides immunity against dysentery, and eliminates diarrhea [6]. Regarding guava's growing environment, it can grow in a variety of soils with a wide range of pH (4.4 to 4.9), where it can also sustain intensive and extensive climate change [7].

The delightful aroma of the generally spherical guava fruit makes it attractive [8]. The growing age of fruits and vegetables may change and pass from various stages, making it challenging to recognize various factors that make them behave differently during different stages. Therefore, image acquisition of vegetables and fruits is the first important step to effectively analyze quality attributes such as color and texture. Illumination also affects receiving these features from the sensor in fruit image collection [9]. Computer-vision-based fruit and vegetable disease detection can lead to large-scale automatic vegetable and fruit monitoring [10]. This helps in taking earlier steps to take care of specific hazards that disturb actual yields, such as the need for fertilizers to be applied to increase the growth rate [11]. Various diseases affect the production of guava fruits, such as anthracnose [5], canker, dot, mummification, and rust. Farmers are very knowledgeable about these diseases, but they mostly do not know of early prevention methods to protect them against further loss. This ultimately leads to significant loss in guava production [12]. These different diseases are caused by different factors of guava plants; for example, canker is caused by algae and was first discovered by Ruehle [13].

Similarly, Dastur is another guava disease that is caused by dry rot [14]. These types of diseases affect guava production, which leads to economic and environmental loss [15]. Environmental loss is any kind of loss, including energy, water, clean air, and land loss, where as far as the economic loss is concerned, this results in financial loss in production. Pakistan is a country in the Asian Pacific whose economy is mostly based on agricultural production. The agricultural significance of Pakistan can be analyzed from its gross domestic product (GDP), with agriculture being 25% of its annual GDP [16]. Many agricultural countries produce guava as a domestic product, and Pakistan is globally fourth in guava production, as it annually produces 1,784,300 t [17]. To diagnose guava diseases in a timely manner, accurate detection is necessary, as false detection may lead to the poor production of guava species. Manual observation may be time consuming and lead to the wrong interpretations.

This led us to produce an automatic system for guava disease detection [18], as the production of guava fruits creates severe issues in developed and underdeveloped countries [19,20]. The automation of disease detection is currently the fastest, least expensive, and most accurate solution [21]. It could cost more, but it can lead to a colossal time reduction by automating the disease detection process [22]. For prediction models, the RGB color channel images are primarily used, which are visually distinguishable by color. The color features could be strong descriptors to distinguish different diseases. However, obtaining deep feature-based models could be more robust, as this covers many other aspects such as geometry, pattern, texture, and other local features. For this, a local guava disease-based RGB image dataset was collected by a high-display-quality camera here. It contains four types of disease, namely canker, dost, rust, and mummification, with the fifth category as the healthy class. Further details of the dataset are discussed in Section 3. The proposed study was inspired by deep learning (DL), and a comparative analysis of various pretrained models is proposed. The main contributions are as follows:

1. Augmented data cover different aspects of view to provide more real-time data visualization and big data usage for DL;
2. The first local Pakistani guava disease detection dataset using DL;
3. State-of-the-art DL models used to validate Pakistani guava disease detection.

The rest of the manuscript is divided into three sections. Section 2 presents the related work. Section 3 is the methodology. Section 4 outlines the results and discussion.

2. Related Work

Plant disease detection is becoming increasingly automated. However, both machine learning and deep learning methods are used [23] in order to provide intelligent automated solutions, with a few recent studies on both categories are discussed below.

2.1. Machine-Learning-Based Plant Disease Detection

Some experts confirmed the labeling to identify unhealthy from healthy guava fruits. Handcrafted features named local binary patterns (LBPs) are extracted and further reduced using principal component analysis (PCA). The multiple types of machine-learning (ML) classifiers are used, where a cubic support vector machine performed the best among the various methods in [24]. Edge- and threshold-based segmentation is performed for plant disease detection using images of leaves. Multiple features of color, texture, and shape are extracted, which are fed into a neural network classifier that classifies different plant diseases [25].

Shadows are removed from the background by enhancing, resizing, and isolating the region of interest (ROI), with clustering performed by using K-means for pomegranate fruit disease detection in [26]. Guava disease detection was performed using basic color transformation functions in image processing to detect the actual diseased parts of plant leaves. Classification was performed using support vector machine (SVM) and K-nearest neighbor (KNN) in [27]. Apple disease detection was performed using the spot segmentation method, where feature extraction and fusion were also performed. The decorrelation method was used for the fusion of extracted features in [28]. A soil-based analysis to recognize the soil indicator that plays an important role in plant yield was also used in [29]. Similarly, the weather forecasting history could play an important role in plant monitoring systems to avoid any natural hazards, as in [30]. The hue–saturation–intensity (HSI) color space was initially used, where unhealthy areas were detected using textures. Multiple features were extracted after the color conversion of the data. Features such as homogeneity, energy, and other cluster-based features were extracted. Lastly, SVM was used for classification in [31].

2.2. Deep-Learning-Based Plant Disease Detection

Demand for deep-learning (DL)-based studies is increasing due to their promising results. Big data are used for DL model training for the prediction of automated detection. Therefore, a similar study used more than 54,000 images of 14 crop diseases with 26 different diseases types. A deep convolutional neural network (DCNN) was proposed with a 99.35% accuracy achieved on the held-out test dataset. Lastly, smartphone-assisted automatic crop disease detection was proposed and an app was suggested for development in [32]. A similar big data dataset was used for plant disease detection. The open dataset of more than 87,000 images was used with 25 different plant categories. Multiple DCNN architectures were used, and the best-performing network achieved a 99.53% accuracy. The reported results showed that this tool model can be used for real-time plant disease identification [33]. Symptom-based gaps found by researchers that cover it by proposing their own CNN with a visualization technique were also missing in previous architectures.

Modified networks for plant disease identification were applied that improved the results of [34]. In-depth features and transfer learning using famous architectures were applied on famous models' architectures. Deep-feature-based classification using SVM and other ML classifiers showed better results than those of the transfer-learning method. Moreover, the fully connected layer of state-of-the-art architectures such as VGG-16, VGG-19, and AlexNet showed better accuracy than that of other fully connected layers [35]. The images of specific conditions and various symptoms were acquired in real time, and these were missed in public datasets. To tackle this limitation, data augmentation was performed, which took a single input leaf image from multiple views that covered certain conditions on the same leaf input image. It also covered multiple diseases affecting leaves. Augmentation-based predictions increased the accuracy by 12%. Furthermore, the data limitation suggested using data augmentation in [36]: the Plant–Village dataset contained differently annotated apple black rot images. Fine-tuned DL models were trained, and the best accuracy was achieved by VGG-16, at 90.4% [37]. Different ML- and DL-based methods are shown in Table 1.

Table 1. Summary of recent studies on the detection of guava diseases.

References	Year	Species	Domain	Method	Results
[38]	2021	Plant–Village	Deep learning	C-GAN and DenseNet121-based transfer learning	Accuracy (5 classes) = 99.51% Accuracy (7 classes) = 98.65% Accuracy (10 classes) = 97.11%
[4]	2021	Guava plant disease	Machine learning	HSV, RGB, LBP, and classical machine learning methods	99% accuracy
[39]	2020	Cotton plant disease	Deep learning	Proposed CNN	~
[40]	2019	Plant disease dataset	Deep learning	Traditional augmentation and GAN for data generation and neural network	Accuracy = 93.67%
[41]	2019	Tomato and brinjal	Deep learning	GLCM, adaptive neuro-fuzzy system	Tomato accuracy = 90.7%, brinjal accuracy = 98.0%
[42]	2019	Potato tuber	Deep learning	CNN	90–10 split training–testing accuracy = 96%
[33]	2018	Open plant dataset	Deep learning	AlexNet, VGG, and other CNN	Best accuracy = 99.53%
[43]	2018	Papaya leaves	Machine learning	HOG features, random forest classifier	Accuracy = 70.14%

Although DL has shown excellent results in plant disease detection, it still faces some challenges. The big data challenge compromises previous studies because they used limited data. The data limitation can be reduced by using various strategies that also produce a more confident model by covering a different aspect of a specific input sample of plant disease [44].

3. Methodology

The automation of plant disease monitoring is taking the place of manual monitoring. Many researchers have used real-time experimentation of plant disease monitoring and achieved satisfying results. A local Pakistani dataset was collected for guava plant and fruit disease detection in the proposed study. Data augmentation was used to meet the big data usage challenge, where it was also used to cover model overfitting problems. Data augmentation was performed using the affine transformation method; to enhance the region of interest (ROI), unsharp masking and the histogram equalization method were performed, better sharpening the ROI and removing any existing noise in the augmented data. The final augmented and enhanced data were fed into various fine-tuned state-of-the-art classification methods. All the steps are shown in Figure 1.

In the framework, augmented and enhanced images were given to 5 different predefined architectures by replacing their last layer according to the given data classes. AlexNet was the first model in the ImageNet competition that changed the image classification and object detection using deep-learning models. SqueezeNet, GoogLeNet, and ResNet followed with many others. Famous ones with different kinds of architectures were used to check the effectiveness of a given local guava dataset. The proposed method showed an initial step on a newly collected local Pakistani dataset, where more methods can be adopted using a different ML and DL technology. The details of the dataset before and after augmentation and the details of other used CNN architectures are shown in Section 3.1 by discussing their fine-tuned parameters, with the weights in Sections 3.2–3.4. The achieved results on the validation data are shown in Section 4.

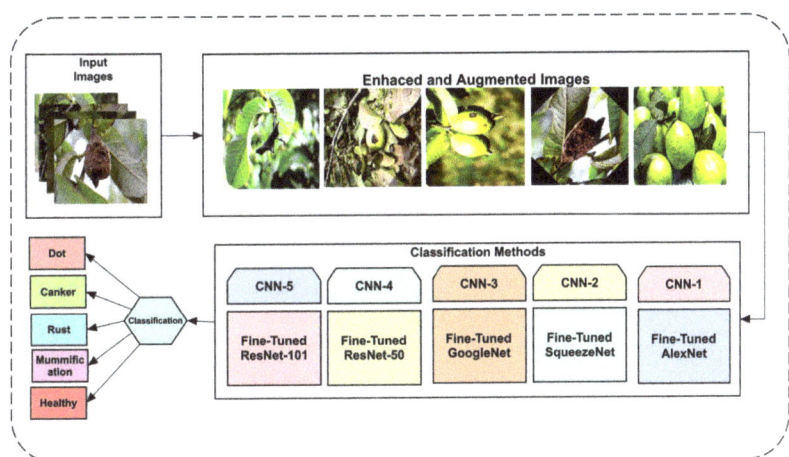

Figure 1. Proposed framework for guava plant disease detection.

3.1. Dataset Normalization

The dataset was initially collected using a high-definition camera with different resolutions. Images were of different angles and orientations, which may lead to the misguidance of the prediction model due to the illusion factor, spatial resolution changes, camera settings, background changes, and many other real-time factors. Therefore, the data were first resized to be equal in size using the bicubic interpolation method. This uses 4 by 4 neighborhood pixels to interpolate the 16 nearest pixels, primarily used in many image-editing tools. This improved the results as compared to those of the bilinear and nearest-neighbor methods. Interpolation was used to resize the image. The resized image was again augmented and enhanced; its histogram-based representation is shown in Figure 2.

Figure 2. (**top**, **left**) Original and (**top**, **right**) enhanced. (**bottom**, **left**) Original and (**bottom**, **right**) enhanced image histogram.

The histogram shows that the data intensity levels were equalized after the preprocessing of data resizing and enhancement.

3.2. Data Augmentation and Enhancement

Resized images were rotated or transformed using the affine transformation method. Different angles with a 360 rotation were used with an angle difference of 45°. There were 9 angles in total in each sample instance applied for all classes, namely 0°, 45°, 90°, 135°, 180°, 225°, 270°, 315°, and 360°. Affine transformation was calculated as given in Equation (1), and the applied augmentation sample for each category is shown in Figure 3.

$$Rotation = \begin{bmatrix} cos(a) & sin(a) & 0 \\ -sin(a) & cos(a) & 0 \\ 0 & 0 & 1 \end{bmatrix} \quad (1)$$

The angle of rotation according to an affine rotation is shown in Equation (1). A is the angle value that was changed nine times for an image to obtain the new rotated image.

Data enhancement was applied with a combination of unsharp masking and color histogram equalization, and both of these methods were applied and calculated using Equations (2) and (3).

$$f(I) = \alpha f - \beta f_l \quad (2)$$

The output enhanced image was calculated in $f(I)$, where α and β are constant, to be the input image that is multiplied where the original image is processed and subtracted via low-pass filter process mask f_l. Histogram equalization was used in the image processing to enhance a given RGB image, and the three channels were individually evaluated using Equation (3).

$$T_k = (L-1)cdf(P) \quad (3)$$

The cumulative distribution of the given intensity was calculated over the probability of occurrences, as calculated in Equation (4)

$$Cdf(P) = \sum_{k=-\infty}^{P} Prob(k) \quad (4)$$

This calculated accumulative distributive value was then multiplied with maximal value intensity, and the newly calculated transformed intensity was calculated and mapped to the corresponding pixels throughout the given image.

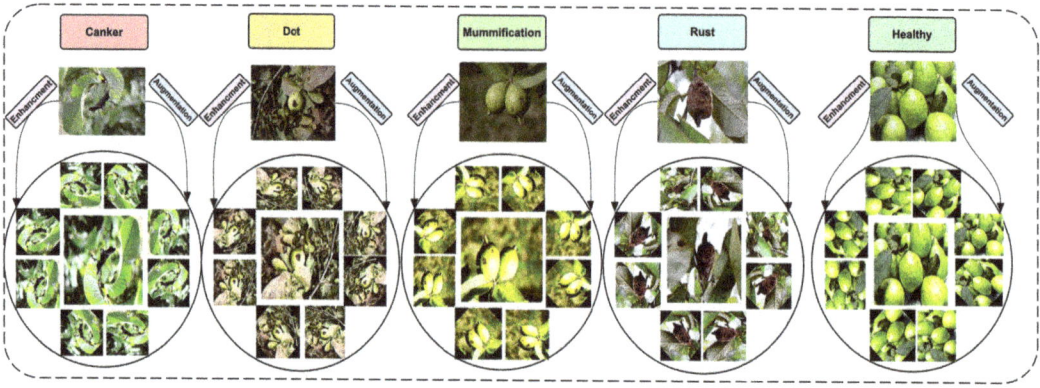

Figure 3. Five types of guava species image samples with their enhanced and rotated 9 angle images in circular view in 4 circles.

Rotated images covered a different aspect of the actual time occurrence, which could be any of the orientations for the user. The training and predictions of rotated augmented data offered promising results.

3.3. Basics of Convolutional Neural Networks

There are many proposed CNN architectures used in various aspects of intelligent classification and object-detection systems. These architectures have slight differences in their networks, and the analyzed primary layers and components are discussed here. We discuss these basics before explaining state-of-the-art models of classification that were also used.

3.3.1. Convolutional Layer

The convolutional layer is called as such when at least one convolutional operation is used in the input layers of an architecture. The convolutional operation uses various parameters such as kernels, where the kernel size is specified for parameter initialization. Similarly, padding and stride size are also initialized and used in convolutional operations. The convolutional operation is summarized in Equation (5).

$$Conv_i^l = Bias_i^l + \sum_{j=1}^{a_i^{(l-1)}} w_{i,j}^{(l-1)} * C_i^l \tag{5}$$

In Equation (5), $Conv_i^l$ is the output of a convolved operation in which $Bias_i^l$ is the bias matrix, with the ith iterative region of operation on which convolved window w is evolving, and i, j represents the window size of the rows and columns. Iterated convolved window C_i^l is multiplied with the corresponding pixels of the given image, where the selected area is defined by window size $w_{i,j}$.

3.3.2. Batch Normalization

Batch normalization is a normalization operation, such as the min–max data normalization performed in data cleaning. Batch normalization is a normalization in which a batch of input data is normalized, and it can be written as in Equation (6).

$$x_i' = \frac{x_i - \mu_B}{\sigma_B^2} \tag{6}$$

It normalizes data, where the data transformation has taken place, such as a mean output close to 0, and the standard deviation output remains close to 1. In Equation (6), input x of a particular instance is subtracted from the mean (μ) of batch b, where after subtraction, a ratio is calculated over the square of the standard deviation (σ) of that particular batch (B) where instance x belongs, and a normalized value of x_i' is returned as output.

3.3.3. Pooling Layer

Pooling pools over a specific item from some scenarios, where pooling in CNN is used to calculate a max, min, and average pool to take a single output value from a defined kernel window. The stride is also used as a parameter to define the ongoing or iterating step for a pooling value. The pooling value is calculated as in Equation (7).

$$D_{output} = x_h * x_w * x_d \tag{7}$$

The output dimension after performing pooling is represented as D_{output}, where x is the input instance and instance height, the width represented as h, w, and the color channel dimension is represented as d, for instance x.

3.3.4. Rectified Linear Unit

ReLU is an activation unit where other activation units such as tanh and sigmoid are also used, and it is used in various studies. ReLU is also called the piecewise linear function, and it simply outputs an identical input variable to the input if it is >0; otherwise, it is 0. Lastly, it maximally excludes the misguiding value in calculating an output class by a prediction model of artificial intelligence. We can simply write the ReLU as in Equation (8).

$$ReLU = \max(0, x) \tag{8}$$

The linear behavior of this activation function makes it a commonly used activation function. In Equation (8), the output is shown as ReLU, where input x is to be taken as the max, which is calculated very straightforwardly if the input value is positive >0; then, it outputs the simple input as it is where <0, or negative values are only taken as 0.

3.3.5. Softmax

The softmax function returns probability values in the range of 0–1, where maximum likelihood returns a higher probability value. It is somehow matched to multilinear regression, where multiple classes are predicted using internal values. The softmax function can be calculated as Equation (9).

$$\sigma(\vec{z})_i = \frac{e^{z_i}}{\sum_{j=1}^{k} e^{z_j}} \tag{9}$$

In Equation (5), the softmax operation is calculated as input vector $\vec{z}$ where z_i are all the input values of vector z. Exponential function e is applied over each value that gives a positive value greater than 0. The denominator value confirms all values sum up to give one value. The final K is the output class number that changes from application to application.

3.4. Classification Using the AlexNet Architecture

AlexNet was the first model in deep learning to change the trend of image identification and classification tasks. It was initially proposed for detecting and classifying objects using a benchmark dataset from ImageNet. Using the AlexNet architecture, the image size for the input layer is taken to be 227 × 227 × 3. In this architecture, there are only 5 convolutional layers and 3 fully connected layers, giving 25 layers in total. The last layers were altered in the proposed framework, and then, we used fine-tuned network parameters. The modifications are shown in Table 2.

In Table 2, all layers above remain the same as in AlexNet, where Fc-8 is first altered with four layers, and the two layers of the softmax activation class output correspondingly give the output for the five categories of guava species.

Table 2. AlexNet for guava disease detection.

Layers	Categories	Activations	Weights
Data Layer	Image Input	227 × 227 × 3	-
Convolve-1	Convolution	55 × 55 × 96	11 × 11 × 3 × 96
ReLU-1	ReLU	55 × 55 × 96	-
Normalization-1	Cross-Channel Normalization	55 × 55 × 96	-
Pool-1	Max-Pooling	27 × 27 × 96	-
Convolve-2	Grouped Convolution	27 × 27 × 256	5 × 5 × 48 × 128
ReLU-2	ReLU	27 × 27 × 256	-
Normalization-2	Cross-Channel Normalization	27 × 27 × 256	-
Pool-2	Max-Pooling	13 × 13 × 256	-
Convolve-3	Convolution	13 × 13 × 384	3 × 3 × 256 × 384
ReLU-3	ReLU-3	13 × 13 × 384	-

Table 2. *Cont.*

Layers	Categories	Activations	Weights
Convolve-4	Grouped Convolution	$13 \times 13 \times 384$	$3 \times 3 \times 192 \times 192$
ReLU-4	ReLU	$13 \times 13 \times 384$	-
Convolve-5	Grouped Convolution	$13 \times 13 \times 256$	$3 \times 3 \times 192 \times 128$
ReLU-5	ReLU	$13 \times 13 \times 256$	-
Pool-5	Max-Pooling	$6 \times 6 \times 256$	-
FC-6	Fully Connected	$1 \times 1 \times 4096$	4096×9216
ReLU-6	ReLU	$1 \times 1 \times 4096$	-
Drop-6	Dropout	$1 \times 1 \times 4096$	-
Fc-7	Fully Connected	1×14096	4096×4096
ReLU-7	ReLU	$1 \times 1 \times 4096$	-
Drop-7	Dropout	$1 \times 1 \times 4096$	-
FC-8	Fully Connected	$1 \times 1 \times 5$	5×4096
Softmax	Softmax	$1 \times 1 \times 5$	-
Class Output	Classification	-	-

3.5. Classification Using GoogLeNet Architecture

Google developers focused on the proposed AlexNet model and then introduced the inception module and changed it sequentially by stacking up layers. This introduced different and smaller kernel size windows with more layers in them. It became the winner of the 2014 ILSVRC competition. The inception modules that were the fundamental contribution by GoogLeNet are shown for the trained architecture on the guava dataset, and the layer-by-layer parameters are shown in Table 3.

Table 3. GoogLeNet network used for guava disease detection.

Layers	Categories	Activations	Weights
Inception-3a-1×1	Convolution	$28 \times 28 \times 64$	$1 \times 1 \times 192 \times 64$
Inception-3a-3×3	Convolution	$28 \times 28 \times 128$	$3 \times 3 \times 196 \times 128$
Inception-3a-5×5	Convolution	$28 \times 28 \times 32$	$5 \times 5 \times 16 \times 132$
Inception-4a-1×1	Convolution	$14 \times 14 \times 192$	$1 \times 1 \times 480 \times 192$
Inception-4a-3×3	Convolution	$14 \times 14 \times 208$	$3 \times 3 \times 96 \times 208$
Inception-4a-5×5	Convolution	$14 \times 14 \times 48$	$5 \times 5 \times 16 \times 48$
Inception-5a-1×1	Convolution	$7 \times 7 \times 256$	$1 \times 1 \times 832 \times 256$
Inception-5a-3×3	Convolution	$7 \times 7 \times 320$	$3 \times 3 \times 160 \times 320$
Inception-5a-5×5	Convolution	$7 \times 7 \times 128$	$5 \times 5 \times 32 \times 128$
FC	Fully Connected	$1 \times 1 \times 5$	5×1024
Softmax	Softmax	$1 \times 1 \times 5$	-
Classification	Classification Output	-	-

The overall architecture remains similar, where the last three layers are altered, and the upper-layer connections remains connected.

3.6. Classification Using the SqueezeNet Architecture

SqueezeNet was introduced with five modules. It is claimed that the 3×3 kernel size should be reduced to 1×1, reducing the size of the overall parameter. Downsampling is also reduced into layers. More feature maps are thus learned by the layers. The introduced fire module contains the squeeze layer, and it has 1×1 filters. They are fed an expanding layer that is a mixture of 1 × 1 and 3 × 3 kernels. The SqueezeNet layer architecture and parameters with altered layers are shown in Table 4.

Table 4. SqueezeNet network used for guava disease detection.

Layers	Categories	Activations	Weights
Fire3-Squeeze-1×1	Convolution	56 × 56 × 16	1 × 1 × 128 × 16
Fire4-Squeeze-1×1	Convolution	28 × 28 × 32	1 × 1 × 128 × 32
Fire5-Squeeze-1×1	Convolution	28 × 28 × 32	1 × 1 × 256 × 32
Fire6-Squeeze-1×1	Convolution	14 × 14 × 48	1 × 1 × 256 × 48
Fire7-Squeeze-1×1	Convolution	14 × 14 × 48	1 × 1 × 384 × 48
Fire8-Squeeze-1×1	Convolution	14 × 14 × 64	1 × 1 × 384 × 64
Fire9-Squeeze-1×1	Convolution	14 × 14 × 64	1 × 1 × 512 × 64
Last-convolve	Convolution	14 × 14 × 5	1 × 1 × 512 × 5
ReLU-Convolve	ReLU	14 × 14 × 5	1 × 1 × 512 × 5
Pool-4	Global Average Pooling	1 × 1 × 5	-
Softmax	Softmax	1 × 1 × 5	-
Classification	Classification Output	-	-

The fire modules in the hyperparameter continuation produce three tunable parameters, namely s1 × 1, e1 × 1, and e3 × 3. AlexNet's level of accuracy was achieved by the actual SqueezeNet with 50× fewer parameters, and the model size was reduced to just 0.5 MB because of the decrease in the kernel sizes and the fire modules used in this architecture.

3.7. Classification Using the ResNet-50 Architecture

ResNet was introduced with the residual block concept mainly to answer the overfitting issue created in DL models. It uses a considerable number of layers, such as 50, 101, and 152. As ti is suggested by GoogLeNet to use a small kernel size, it uses small convolutional kernels where denser or more layers are used to meet or improve the validity of the data. The introduced residual block uses a 1 × 1 layer that reduces the dimension, a 3 × 3 layer, and a 1 × 1 layer used to restore the dimensions of the given input. The layer-based ResNet was used, so we used a 50- and 101-layer architecture; the 50-layer architecture of ResNet is shown in Table 5.

Table 5. ResNet-50 network used for guava disease detection.

Layers	Categories	Activations	Weights
Res-2a	Convolution	56 × 56 × 256	1 × 1 × 64 × 256
Res-3a	Convolution	28 × 28 × 512	1 × 1 × 256 × 512
Res-4a	Convolution	14 × 14 × 1024	1 × 1 × 512 × 1024
Res-5a	Convolution	7 × 7 × 2048	1 × 1 × 1024 × 2048
FC	Fully Connected	1 × 1 × 5	5 × 2048
Softmax	Softmax	1 × 1 × 5	-
Class-Output	Classification	-	-

The basic residual branches of ResNet-50 with their learned parameters in guava disease detection are shown in Table 5. The residual block re-concatenates spatial information from the previous block to preserve information in each calculated feature map of the residual block. For the proposed framework, the last layers are altered with five categories to classify them on the basis of previous learning on the augmented guava data of the proposed study.

3.8. Classification Using ResNet-101 Architecture

The ResNet-based study produced many variants of its introduced residual blocks, such as 18, 19, 34, 50, and 101, and the densest of 152. Making it increasingly denser did not improve the accuracy after a certain point. This may be due to many factors, such as learning saturation, loopholes in the proposed architecture, and hyperparameter optimization. Therefore, the mainly used networks were ResNet-50 and 101. The learned-weight-based architecture of the residual blocks using the ResNet-101 architecture for guava disease detection are described in Table 5; the difference between 50 and 101 is in their architecture. There are 347 layers in total in ResNet-101 and 177 layers in the ResNet-50 model. Learning mainly changed after Res-branch 4a, as hundreds of layers are added after it learns in different ways, as described in the ResNet-101 architecture.

4. Results and Discussion

The proposed study used augmented data of actual given locally collected data in Pakistan for guava disease detection. The data had enough images to train the DL model. The dataset details for before and after augmentation are shown in Table 6.

Table 6. Dataset description with and without augmentation.

Categories	Number of Images (without Augmentation)	Number of Images (with Augmentation)
Canker	77	693
Dot	76	684
Mummification	83	747
Rust	70	630
Healthy	15	135
Total	321	2889

4.1. Evaluation Measure

There are mainly four types of prediction instances, which we can consider in the formulation of these above-mentioned evaluation measures: true positive (*TP*), false positive (*FP*), true negative (*TN*), and false negative (*FN*). These are described in detail below.

4.1.1. Accuracy

Accuracy is the most commonly used measure in the ML and DL domains for classification. It can briefly be described as truly predicted instances over total instances, including wrong and right predictions. In terms of the four types used above, the equation for accuracy can be written as follows:

$$Accuracy = \frac{(TP + TN)}{(TP + TN + FP + FN)} \quad (10)$$

The equation can be described as the ratio of the summation of *TP* and *TN* over the summation of *TP*, *TN*, *FP*, and *FN*.

4.1.2. Specificity

This is the measure among the right predictions over the total that were from both the positive and negative classes. Briefly, negatively labeled objects are measured over the total of true- and false-negative instances. It can be written as:

$$Specificity = \frac{TN}{(TN + FP)} \quad (11)$$

The specificity equation is defined as the ratio over *TN* and the summation of *TN* and *FP*.

4.1.3. F1 Score

The F1 score is an essential measure, as it is calculated for both the essential measures of recall and precision. Recall is also known as sensitivity and is the measure to detect positive class predictions among true positives and false negatives. For the F1 score, we need to calculate the sensitivity as follows:

$$Recall\ (Sensitivity) = \frac{TP}{(TP+FN)} \quad (12)$$

The second measure, precision, is also separately calculated and used in the F1 score measurement. Precision is calculated to obtain a truly predicted class over true and false positives. It is represented as:

$$Precision = \frac{TP}{(TP+FP)} \quad (13)$$

After obtaining the precision and recall, the F1 score can be calculated. After having TP over TP and FN by recall and by obtaining TP over TP and FP, we can obtain more precise measurements for true-positive predictions. The final F1 score can be calculated as:

$$F1 - score = 2 \times \frac{Precision * Recall}{Recall + Precision} \quad (14)$$

Therefore, the F1 score equation can be defined as two multiplied by the ratio of the product and summation of precision and recall.

4.1.4. Kappa–Cohen Index

The last measure is to have confidence about the statistical analysis of the results, as statistical analysis is broadly used in many aspects of scientific work. Therefore, a statistical measure that gives confidence over confusion matrix values was calculated for evaluation. The kappa index gives the confidence over a certain range of confusion-matrix-based calculated values. If its value range lies in the range of 0–20, it promises that 0–4% of the data are reliable to use for prediction. If its index value lies in the range of 21–39, it promises a 4–15% data reliability. If it lies between 40 and 59, it promises 15–35% data reliability. If it lies between 60 and 79, it promises a 35% to 63% reliability. If it lies between 80 and 90, then it promises strong data reliability, and if it lies at more than 90, that means 82% to 100% reliability. It can be calculated as:

$$Agreement = \frac{\left(\frac{cm_1 * rm_1}{n}\right) + \left(\frac{cm_2 x rm_2}{n}\right)}{n} \quad (15)$$

The agreement type was calculated using $cm_1, cm_2, rm_1,$ and rm_2, where cm_1 represents Column 1 and cm_2 represents Column 2. rm_1 and rm_2 represent Rows 1 and 2 of any two-class confusion matrix. This formulation is the general form of a two-class confusion matrix; in the case of our proposed methodology, there are five columns and rows that extend the formulation to up to five rows and columns.

The total augmented data were later split into a 70/30 ratio for training and testing data, where the fine-tuned parameters for each of the five training models remained different and showed different results. The training and testing data number of instances became 2023 and 866, respectively.

There were five different kinds of architectures applied to classify guava diseases using the augmented image data. The individual class testing data prediction results are discussed with their overall results. The evaluation measures accuracy, sensitivity, specificity, precision, recall, and the kappa index were also used as statistical measures. The results are shown in Table 7.

Table 7. Results obtained on the basis of individual class testing data.

Models	Categories	Accuracy	Specificity	F1-Score	Precision	Kappa
AlexNet	Canker	0.98077	0.99088	0.97608	0.97143	0.52096
	Dot	0.98049	0.99697	0.98529	0.99015	0.5309
	Healthy	1	1	1	1	0.90762
	Mummification	0.98661	0.99533	0.98661	0.99533	0.48509
	Rust	0.98942	0.99705	0.98942	0.98942	0.5648
	Overall	0.9850	0.9960	0.9875	0.9875	0.9531
GoogLeNet	Canker	0.96635	0.99696	0.9781	0.99015	0.52859
	Dot	0.97561	0.98638	0.96618	0.95694	0.52695
	Healthy	1	1	1	1	0.90762
	Mummification	0.99429	0.99377	0.97297	0.98182	0.49202
	Rust	0.99471	0.99114	0.98172	0.96907	0.56
	Overall	0.9758	0.9937	0.9798	0.9796	0.9242
SqueezeNet	Canker	0.97596	0.99392	0.97831	0.98068	0.52404
	Dot	1	1	0.96698	1	0.51541
	Healthy	1	1	1	1	0.90762
	Mummification	0.98214	0.9891	0.97561	0.96916	0.48364
	Rust	0.91534	1	0.9558	1	0.5866
	Overall	0.9711	0.9924	0.9753	0.9772	0.9098
ResNet-50	Canker	0.99519	0.99696	0.99281	0.99043	0.51957
	Dot	1	0.99697	0.99515	0.99034	0.52957
	Healthy	1	1	1	1	0.90762
	Mummification	0.98661	1	1	1	0.48733
	Rust	1	1	1	1	0.56351
	Overall	0.9954	0.9988	0.9962	0.9962	0.9856
ResNet-101	Canker	0.99519	0.98784	0.97872	0.96279	0.51484
	Dot	0.98049	0.99849	0.98772	0.99505	0.53178
	Healthy	1	1	1	1	0.90762
	Mummification	0.98214	1	1	1	0.48887
	Rust	0.98413	0.99557	0.98413	0.98413	0.56544
	Overall	0.9861	0.9964	0.9883	0.9884	0.9567

The individual class and overall results for each model were evaluated. Several evaluation measures were used: accuracy, specificity, F1 score, precision, and kappa.

The first model, AlexNet, showed 98% accuracy, 99% specificity, a 97.60% F1 score, 97.14% precision, and a 0.5296 value for kappa as the canker class prediction results of the testing data. Accuracy is a general measure over all positive and negative instances that are either wrongly or correctly predicted. The 98% accuracy of the canker class showed accurate predictions of positive and negative classes and mainly showed excellent and satisfactory results in TN predictions over TN and FP; specificity showed that true negatives were mostly predicted right among TN and FP. Precision showed TP over TP and FP, with a 97.14% value; this means that it had less accurate predictions. For a positive class to see the combined effect of TP and TN, the F1 score measure was used. The F1 score showed a 97% value, which summarizes both of the above measures. The kappa index showed a weak level of agreement for the canker class. The dot class showed 98.04% accuracy, 99.69% specificity, 98.52% F1 score, 99% precision, and a 0.53 kappa value. The dot class showed less accuracy than that of the canker class, but it needs to be discussed with another evaluation measure to analyze the predictions of positive and negative instances.

Specificity, which was 99.7% in the case of the dot class, represented TN among TN and FP where the precision was 99% for TP over TP and FP. The F1 score of this measure showed 98.52%, which summarizes the recall and precision, which could be considered to be more promising factors compared to sensitivity, specificity, and precision. The last statistical measure showed a weak level of agreement for this class. The third and healthy class showed accurate results in all networks where the data agreement value was also 0.97, showing an extraordinary level of agreement on the given data. The next class, mummification, showed 98.66% accuracy, 99.53% specificity, a 98.66% F1 score, a 99.53% precision value, and a 0.485 value for kappa. The accuracy for the mummification class was slightly better than that for the canker and dot classes. Other values such as specificity were lower than those for canker and dot, but with large differences. The precision value was higher than that of the canker class and lower than that of the dot class. To summarize both precision and specificity, the F1 score was used, which was nearer to that of the dot class and higher than that of the canker class. The kappa value is a decision-maker index, with a 0.48 value, which was less than that of the canker and dot classes, but also had weak agreement with the data reliability. The kappa value for the last class was 0.56 due to the one value for both precision and specificity, but 56 also lies in the weak agreement class. Lastly, the overall or mean results were evaluated, showing 98.50% accuracy, 99.60% specificity, a 98.75% F1 score, 98.75% precision, and a 0.9531 value for the kappa index. The mean or overall value was an actual representation of a model that produced good results; it was either about the accuracy, specificity specifically for TN, and precision for the TP value, and the F1 score represents the recall and precision. A kappa value of more than 90% is a strong agreement level, and the data reliability is also high when kappa is more than 90. Therefore, AlexNet overall showed satisfactory results.

GoogLeNet showed testing results on the canker class as follows: 96.635% accuracy, 99.69% specificity, 97.81% F1 score, 99.01% precision, and 0.5285 kappa value. Accuracy showed promising results where, if examining the specificity value over TN values, it showed a value of 99.69%. Similarly, the precision value over the TP values showed 99%, and the F1 score over precision and recall showed a value of 97.81%, which showed more confidence than precision and specificity did. The last important measure is the kappa index, 0.5285, which showed weak promise as it was of only one class over other classes. The mean value showed the actual effect of the kappa stat. The other dot class showed values of 97.56% accuracy, 98.638% specificity, a 96.618% F1 score, 95.69% precision, and a 0.5269 kappa value. The accuracy value as compared to that of the canker class was lower. In the case of the dot class, where the specificity value was also slightly lower, its precision value was lower than that of the canker class. The F1 score based on precision and recall showed a slightly higher score in the canker class, where the last kappa Cohen index was similar and lied on weak agreement of data reliability. The mummification showed a 99.42% value of the accuracy, a 99.37% value of the specificity, a 97.29% value for the F1 score, a 98.18% precision value, and a 0.49 kappa value. For an accuracy value higher than those of the canker and dot classes, where the specificity value was lower than that of the canker, and slightly lower than that of the dot class, this means that some TN instances had variation in these cases. The F1 score showed a lower score than that of the dot class and higher than that of the canker class. This means the combined effect of TP and FP was more promising for mummification compared to that for the dot class. The last class of rust showed 99.47% accuracy, 99.114% specificity, a 98.172% value for the F1 score, 96.907% for precision, and 0.56 for the Cohen index. The accuracy value was higher than that of the canker, dot, and mummification classes. The F1 score showed a 98.17% value that was nearest the canker, mummification, and dot classes. Therefore, the accuracy value was the measure to analyze the test prediction results where other values such as the F1 score also mattered and made the results distinguishable.

SqueezeNet testing showed results for the canker class of 97.596% accuracy, 99.392% specificity, a 97.838% F1 score, 98% precision, and a 0.52 kappa value. The accuracy value, a general assumption of model performance, had a lower value than that of specificity, precision, and the F1 score. The specificity value was much higher, which means that true negatives over the TN and FP had higher rates for the canker class. TP cases were also predicted with a 98% value over the FP and TP as the precision scores. The recall- and precision-based F1 score showed 97.83%, which is intermediate between the precision and specificity. The kappa value was 0.52 and lied on the weak agreement of the data. The second dot-class-based results showed 100% accuracy, 100% specificity, a 96.68% F1 score, 100% precision, and a 0.51 kappa value. Although the accuracy was good for this class, specificity showed a 100% result. The F1 score that covered recall and precision validity had a 96.68% score and 0.52 kappa value. The mummification class showed a predictivity of 98.214% accuracy, 98.91% specificity, a 97.56% F1-score, a 96.916% precision value, and 0.48 for the kappa index. The last class of rust showed predictivity measures of 91.53% accuracy, 100% specificity, a 95.58% F1 score, and 100% precision. The accuracy value was lower than that of the other three classes where precision and specificity were 100% in this case. The global or overall results for the all classes had a primary issue. Accuracy was 97.11%, lower than that of AlexNet and GoogLeNet, where specificity was lower than that of both AlexNet and GoogLeNet, and the F1 score, precision, and kappa were lower for this model testing the data predictions. The kappa index showed promise for these data.

ResNet-50 and -101 had much more improved results than those of AlexNet, SqueezeNet, and GoogLeNet. ResNet-50 showed canker class results of 99.51 accuracy, 99.68% specificity, a 99.28% of F1 score, a 99.03% precision value, and 0.51 for precision value of the kappa index. Compared to the previous cases of AlexNet, GoogLeNet, and SqueezeNet, the results were overall improved for all measures. Similarly, for the dot class, the accuracy value was 100%, 99.697% specificity, a 99.515% F1 score, a 99.034% precision value, and a 0.52 kappa index. The canker class results were not better than those of the dot class if we look at the accuracy measures, with only a slight difference in the F1 scores. Mummification results showed 98.661% accuracy, 100% specificity, a 100% F1 score, 98.25% precision, and a 0.48 kappa value. Although it had 100% accuracy as an individual class, the specificity and precision values were also higher than those of the canker and dot classes. The rust class showed 100% accuracy, 100% specificity, a 100% F1 score, and a 100% precision value. The overall results were improved as compared to the above models' mean results. The global mean results were 99.54% accuracy, which was better than that of SqueezeNet, GoogLeNet, and AlexNet. Specificity was also better than that in the three models. The F1 score and precision were both better than those in the previously discussed three models. Although it had a higher value than that of the previous models, the last kappa had the same class of confidence. The data reliability was higher for all models.

The last model of the proposed study also achieved good results as compared to the other models. The canker class showed predictivity values of 99.519% accuracy, 98.784% specificity, an F1 score of 97.87%, a precision of 96.27%, and a 0.51 kappa index. ResNet-50's accuracy had a dominant result compared to the previous class results of the other models, while other evaluation values also showed more improvement in this model. If the accuracy value was improved, other values were not so improved, but different cases showed overall improvement of the results. In ResNet-101, the dot class showed 98.049% accuracy, 99.84% specificity, 99.505% precision, and a 0.53 kappa value. Accuracy, specificity, precision, and the F1 score were overall improved for each class, which did not happen in the previous models' results for any class. The mummification class again showed 100% accuracy in this model testing. The mummification class showed 100% accurate results in other models where other values were not improved to such an extent: 98.41% accuracy, 99.84% specificity, 98.93% F1 score, and 99.47% precision. The last class showed consistency in the improvement of the results for each class by also showing promising results here. Lastly, the overall mean results of ResNet-50 proved it to be more accurate than the four other models. The accuracy was also better than that of the others. Similarly, other measures also showed excellent performance. The graphical illustration of all five models' mean testing results is shown in Figure 4.

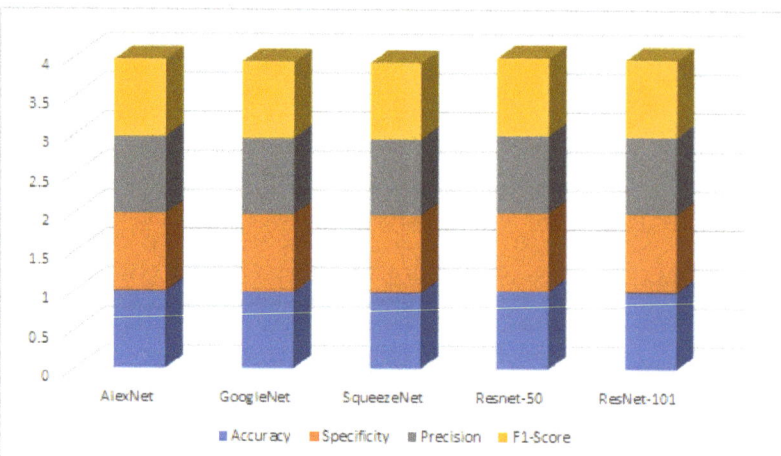

Figure 4. Guava disease classification result visualization.

The overall results showed that the kappa value had overall excellent data reliability for all models. The mummification, the healthy and dot classes were more distinct classes to distinguish them from the other two classes, as they showed 100 true results many times. The densest model with more residual connections showed more accurate results, which means that using a small kernel size with an increasingly denser network improved the classification results.

For individual cases or data-based testing analysis, the confusion matrices were designed and evaluated, and they are shown in Table 8.

Table 8. Confusion matrix obtained using several state-of-the-art networks.

Networks	AlexNet					GoogLeNet					SqueezeNet					ResNet-50					ResNet-101				
Classes	C	D	H	M	R	C	D	H	M	R	C	D	H	M	R	C	D	H	M	R	C	D	H	M	R
C	204	0	0	3	1	201	3	0	2	2	203	0	0	5	0	207	1	0	0	0	207	0	0	0	1
D	3	201	0	0	1	0	200	0	2	3	0	205	0	0	0	0	205	0	0	0	2	201	0	0	2
H	0	0	40	0	0	0	0	40	0	0	0	0	40	0	0	0	0	40	0	0	0	0	40	0	0
M	3	0	0	221	0	2	5	0	216	1	1	3	0	220	0	2	1	0	221	0	3	1	0	220	0
R	0	2	0	0	187	0	1	0	0	188	3	11	0	2	173	0	0	0	0	189	3	0	0	0	186

The confusion matrices of all architectures showed that the rust class was the most distinguishable among other guava diseases. If we discuss the AlexNet model results, four wrong cases were predicted as wrong in the rust and mummification classes, four wrong predictions were found for the dot class, where the wrong predictions lied on canker and rust. There were three wrong predictions for mummification, in the class of canker, and there were two wrong predictions in the rust class; these wrong predictions were two for dot. The second GoogLeNet architecture made 201 correct predictions in the canker class, and seven wrong predictions lied in the dot, mummification, and rust class, while no prediction lied in the healthy class. The five wrong predictions for the dot class were two predicted as mummification and three was rust, whereas 200 were predicted as right. This makes it one case less accurate than AlexNet, as that made four wrong predictions in the dot class. Mummification was predicted as nine wrong classes in the canker, dot, mummification, and rust categories, where two-hundred sixteen cases were rightly predicted. In rust, 188 cases were rightly predicted. One was predicted wrongly in the dot class. It was highly more efficient than AlexNet was in this category, as that predicted two wrong cases in the dot class, and GoogLeNet predicted only one wrong. The SqueezeNet architecture model made five wrong predictions, with two-hundred and three correct predictions of the canker class.

The five wrong predictions were in the mummification class. In the dot class, there was no wrong prediction, and all 205 test instances were rightly predicted. In mummification, there were four wrongly predicted cases and two-hundred and twenty rightly predicted cases. Most were predicted in the dot class with three instances, with one predicted in the canker class. Rust had 16 wrong cases, and most were in the dot class—11 out of 16. ResNet-50 was similar to ResNet-101, where the difference was mainly of several layers and its parameters. ResNet-50 predicted one wrong cases for the canker class with one wrong prediction in dot, and two-hundred and seven were correctly predicted. All dot class predictions (205) were correctly predicted in the dot class. The mummification class in this model's predictions had 221 correct predictions in dot. It has one wrong and two wrong predictions in the canker class. ResNet-101 also had higher accuracy results as compared to those of all other models. According to its confusion matrix, it made one wrong prediction for the canker class into the rust class. As compared to ResNet-50, it had one wrong case prediction, but in a different class. Regarding the dot class, ResNet-50 made no wrong predictions, and ResNet-101 made four wrong predictions. In the third case of the mummification class, there were four wrong predictions, and ResNet-50 made three wrong predictions. There were three wrong predictions in the last class, rust. There were 186 correct predictions for ResNet-101.

The above analysis of the five models shows that ResNet-50 had an overall high rate of correct predictions; for wrong predictions in the dot class data, it was highly difficult for each model, as it was predicted as the wrong class in most cases. The other classes also misled the models, where the most challenging and least robust class was dot. Therefore, the dot class may need more confident and robust approaches to classify it from other classes. The mummification class had no wrong predictions in ResNet-50 and -101, where it only had a higher rate of wrong predictions in the cases of SqueezeNet with four, where the two remaining models also did not make very many accurate predictions for this class. The healthy class overall in all models remained accurate with no wrong prediction by any model. It made the normal class easily distinguishable by any model. However, the challenge was differentiating the guava disease categories.

5. Conclusions

Guava is an important plant to monitor with the growing population; its production demand is also increasing. Pakistan is a leading global guava producer. Hence, for automatic monitoring, the study proposed a DL-based guava disease detection system. Data were preprocessed and enhanced using a color histogram and unsharp masking method. Enhanced data were then augmented over the nine angles using the affine transformation method—augmented enhanced data used by five DL networks by altering their last layers. The AlexNet, GoogLeNet, SqueezeNet, ResNet-50, and ResNet-101 architectures were used. The results of all networks showed adequate measurements, and ResNet-101 was the most accurate model. Future work should use more data augmentation methods such as generative adversarial networks. Other federated-learning-based DL architectures can be applied for classification to obtain more robust and confident results for this Pakistani guava disease dataset.

Author Contributions: Conceptualization, A.M.M., S.A.K., T.M., H.T.R., A.A.A. and M.A.A.; funding acquisition, A.M.M., A.A.A. and M.A.A.; methodology, A.M.M., S.A.K., T.M., H.T.R., A.A.A. and M.A.A.; software, A.M.M. and S.A.K.; visualization, H.T.R.; writing—original draft, A.M.M., S.A.K., T.M., H.T.R., A.A.A. and M.A.A.; supervision, H.T.R. All authors have read and agreed to the published version of the manuscript.

Funding: The authors extend their appreciation to the Deanship of Scientific Research at King Saud University for funding this work through Research Group No. RG-1441-425.

Institutional Review Board Statement: Not applicable.

Informed Consent Statement: Not applicable.

Data Availability Statement: Not applicable.

Conflicts of Interest: The authors declare that there is no conflict of interest.

References

1. Hunter, M.C.; Smith, R.G.; Schipanski, M.E.; Atwood, L.W.; Mortensen, D.A. Agriculture in 2050: Recalibrating targets for sustainable intensification. *Bioscience* **2017**, *67*, 386–391. [CrossRef]
2. Mirvakhabova, L.; Pukalchik, M.; Matveev, S.; Tregubova, P.; Oseledets, I. Field heterogeneity detection based on the modified FastICA RGB-image processing. *J. Phys. Conf. Ser.* **2018**, *1117*, 012009. [CrossRef]
3. Bose, T.; Mitra, S. Guava. In *Fruits Tropical and Subtropical*; Bose, T.K., Ed.; Naya Udyog: Calcutta, India, 1990; pp. 280–303.
4. Almadhor, A.; Rauf, H.T.; Lali, M.I.U.; Damaševičius, R.; Alouffi, B.; Alharbi, A. AI-Driven Framework for Recognition of Guava Plant Diseases through Machine Learning from DSLR Camera Sensor Based High Resolution Imagery. *Sensors* **2021**, *21*, 3830. [CrossRef] [PubMed]
5. Amusa, N.; Ashaye, O.; Oladapo, M.; Oni, M. Guava fruit anthracnose and the effects on its nutritional and market values in Ibadan, Nigeria. *World J. Agric. Sci.* **2005**, *1*, 169–172. [CrossRef]
6. Al Haque, A.F.; Hafiz, R.; Hakim, M.A.; Islam, G.R. A Computer Vision System for Guava Disease Detection and Recommend Curative Solution Using Deep Learning Approach. In Proceedings of the 2019 22nd International Conference on Computer and Information Technology (ICCIT), Dhaka, Bangladesh, 18–20 December 2019; pp. 1–6.
7. Keith, L.M.; Velasquez, M.E.; Zee, F.T. Identification and characterization of *Pestalotiopsis* spp. causing scab disease of guava, Psidium guajava, in Hawaii. *Plant Dis.* **2006**, *90*, 16–23. [CrossRef]
8. Pachanawan, A.; Phumkhachorn, P.; Rattanachaikunsopon, P. Potential of Psidium guajava supplemented fish diets in controlling Aeromonas hydrophila infection in tilapia (*Oreochromis niloticus*). *J. Biosci. Bioeng.* **2008**, *106*, 419–424. [CrossRef]
9. Sonka, M.; Hlavac, V.; Boyle, R. *Image Processing, Analysis, and Machine Vision*; Cengage Learning: Boston, MA, USA, 2014.
10. Somov, A.; Shadrin, D.; Fastovets, I.; Nikitin, A.; Matveev, S.; Hrinchuk, O. Pervasive agriculture: IoT-enabled greenhouse for plant growth control. *IEEE Pervas. Comput.* **2018**, *17*, 65–75. [CrossRef]
11. Bharathi Nirujogi, D.M.M.; Naidu, L.N.; Reddy, V.K.; Sunnetha, D.S.; Devi, P.R. Influence of carrier based and liquid biofertilizers on yield attributing characters and yield of guava cv. Taiwan White. *Pharma Innov. J.* **2021**, *10*, 1157–1160.
12. Sain, S.K. *AESA BASED IPM Package for Guava*. National Institute of Plant Health Management: Telangana, India, 2014.
13. Ruehle, G.; Brewer, C. *The FDA Method. Official Method of US Food and Drug Administration, US Department of Agriculture, and National Assn. of Insecticide and Disinfectant Manufacturers for Determination of Phenol Coefficients of Disinfectants*; MacNair-Dorland Co.: New York, NY, USA, 1941; pp. 189–201.
14. Misra, A. Guava diseases—their symptoms, causes and management. In *Diseases of Fruits and Vegetables: Volume II*; Springer: Berlin/Heidelberg, Germany, 2004; pp. 81–119.
15. Shadrin, D.; Pukalchik, M.; Uryasheva, A.; Tsykunov, E.; Yashin, G.; Rodichenko, N.; Tsetserukou, D. Hyper-spectral NIR and MIR data and optimal wavebands for detection of apple tree diseases. *arXiv* **2020**, arXiv:2004.02325.
16. Raza, S.A.; Ali, Y.; Mehboob, F. Role of agriculture in economic growth of Pakistan. *Int. Res. J. Financ. Econ.* **2012**, 180–186.
17. Pariona, A. Top Guava Producing Countries in the World. *Worldatlas*. 2017. Available online: https://www.worldatlas.com/articles/top-guava-producingcountries-in-the-world.html (accessed on 17 July 2018).
18. Pujari, J.D.; Yakkundimath, R.; Byadgi, A.S. Grading and classification of anthracnose fungal disease of fruits based on statistical texture features. *Int. J. Adv. Sci. Technol.* **2013**, *52*, 121–132.
19. Gilligan, C.A. Sustainable agriculture and plant diseases: An epidemiological perspective. *Philos. Trans. R. Soc. B Biol. Sci.* **2008**, *363*, 741–759. [CrossRef]
20. Adenugba, F.; Misra, S.; Maskeliūnas, R.; Damaševičius, R.; Kazanavičius, E. Smart irrigation system for environmental sustainability in Africa: An Internet of Everything (IoE) approach. *Math. Biosci. Eng.* **2019**, *16*, 5490–5503. [CrossRef]
21. Zhou, R.; Kaneko, S.; Tanaka, F.; Kayamori, M.; Shimizu, M. Disease detection of Cercospora Leaf Spot in sugar beet by robust template matching. *Comput. Electron. Agric.* **2014**, *108*, 58–70. [CrossRef]
22. Ali, H.; Lali, M.; Nawaz, M.Z.; Sharif, M.; Saleem, B. Symptom based automated detection of citrus diseases using color histogram and textural descriptors. *Comput. Electron. Agric.* **2017**, *138*, 92–104. [CrossRef]
23. Stasenko, N.; Chernova, E.; Shadrin, D.; Ovchinnikov, G.; Krivolapov, I.; Pukalchik, M. Deep Learning for improving the storage process: Accurate and automatic segmentation of spoiled areas on apples. In Proceedings of the 2021 IEEE International Instrumentation and Measurement Technology Conference (I2MTC), Glasgow, UK, 17–20 May 2021; pp. 1–6.
24. Almutiry, O.; Ayaz, M.; Sadad, T.; Lali, I.U.; Mahmood, A.; Hassan, N.U.; Dhahri, H. A Novel Framework for Multi-Classification of Guava Disease. *CMC—Comput. Mater. Continua* **2021**, *69*, 1915–1926. [CrossRef]
25. Gavhale, K.R.; Gawande, U. An overview of the research on plant leaves disease detection using image processing techniques. *IOSR J. Comput. Eng. (IOSR-JCE)* **2014**, *16*, 10–16. [CrossRef]
26. Deshpande, T.; Sengupta, S.; Raghuvanshi, K. Grading & identification of disease in pomegranate leaf and fruit. *Int. J. Comput. Sci. Inf. Technol.* **2014**, *5*, 4638–4645.
27. Thilagavathi, M.; Abirami, S. Application of image processing in diagnosing guava leaf diseases. *Int. J. Sci. Res. Manag. (IJSRM)* **2017**, *5*, 5927–5933.

28. Khan, M.A.; Lali, M.I.U.; Sharif, M.; Javed, K.; Aurangzeb, K.; Haider, S.I.; Altamrah, A.S.; Akram, T. An optimized method for segmentation and classification of apple diseases based on strong correlation and genetic algorithm based feature selection. *IEEE Access* **2019**, *7*, 46261–46277. [CrossRef]
29. Gasanov, M.; Petrovskaia, A.; Nikitin, A.; Matveev, S.; Tregubova, P.; Pukalchik, M.; Oseledets, I. Sensitivity analysis of soil parameters in crop model supported with high-throughput computing. In *International Conference on Computational Science*; Springer: Berlin/Heidelberg, Germany, 2020; pp. 731–741.
30. Gasanov, M.; Merkulov, D.; Nikitin, A.; Matveev, S.; Stasenko, N.; Petrovskaia, A.; Pukalchik, M.; Oseledets, I. A New Multi-objective Approach to Optimize Irrigation Using a Crop Simulation Model and Weather History. In *International Conference on Computational Science*; Springer: Berlin/Heidelberg, Germany, 2021; pp. 75–88.
31. Arivazhagan, S.; Shebiah, R.N.; Ananthi, S.; Varthini, S.V. Detection of unhealthy region of plant leaves and classification of plant leaf diseases using texture features. *Agric. Eng. Int. CIGR J.* **2013**, *15*, 211–217.
32. Mohanty, S.P.; Hughes, D.P.; Salathé, M. Using deep learning for image-based plant disease detection. *Front. Plant Sci.* **2016**, *7*, 1419. [CrossRef]
33. Ferentinos, K.P. Deep learning models for plant disease detection and diagnosis. *Comput. Electron. Agric.* **2018**, *145*, 311–318. [CrossRef]
34. Saleem, M.H.; Potgieter, J.; Arif, K.M. Plant disease detection and classification by deep learning. *Plants* **2019**, *8*, 468. [CrossRef]
35. Türkoğlu, M.; Hanbay, D. Plant disease and pest detection using deep learning-based features. *Turk. J. Electr. Eng. Comput. Sci.* **2019**, *27*, 1636–1651. [CrossRef]
36. Barbedo, J.G.A. Plant disease identification from individual lesions and spots using deep learning. *Biosyst. Eng.* **2019**, *180*, 96–107. [CrossRef]
37. Wang, G.; Sun, Y.; Wang, J. Automatic image-based plant disease severity estimation using deep learning. *Comput. Intell. Neurosci.* **2017**, *2017*, 2917536.
38. Abbas, A.; Jain, S.; Gour, M.; Vankudothu, S. Tomato plant disease detection using transfer learning with C-GAN synthetic images. *Comput. Electron. Agric.* **2021**, *187*, 106279. [CrossRef]
39. Patil, B.V.; Patil, P.S. Computational method for Cotton Plant disease detection of crop management using deep learning and internet of things platforms. In *Evolutionary Computing and Mobile Sustainable Networks*; Springer: Berlin/Heidelberg, Germany, 2021; pp. 875–885.
40. Arsenovic, M.; Karanovic, M.; Sladojevic, S.; Anderla, A.; Stefanovic, D. Solving current limitations of deep learning based approaches for plant disease detection. *Symmetry* **2019**, *11*, 939. [CrossRef]
41. Sabrol, H.; Kumar, S. Plant leaf disease detection using adaptive neuro-fuzzy classification. In *Science and Information Conference*; Springer: Berlin/Heidelberg, Germany, 2019; pp. 434–443.
42. Oppenheim, D.; Shani, G.; Erlich, O.; Tsror, L. Using deep learning for image-based potato tuber disease detection. *Phytopathology* **2019**, *109*, 1083–1087. [CrossRef]
43. Ramesh, S.; Hebbar, R.; Niveditha, M.; Pooja, R.; Shashank, N.; Vinod, P. Plant disease detection using machine learning. In Proceedings of the 2018 International Conference on Design Innovations for 3Cs Compute Communicate Control (ICDI3C), Bangalore, India, 25–28 April 2018; pp. 41–45.
44. Nagaraju, M.; Chawla, P. Systematic review of deep learning techniques in plant disease detection. *Int. J. Syst. Assur. Eng. Manag.* **2020**, *11*, 547–560. [CrossRef]

Article

Mango Leaf Disease Recognition and Classification Using Novel Segmentation and Vein Pattern Technique

Rabia Saleem [1], Jamal Hussain Shah [1,*], Muhammad Sharif [1], Mussarat Yasmin [1], Hwan-Seung Yong [2] and Jaehyuk Cha [3]

1. Department of Computer Science, COMSATS University Islamabad, Wah Campus, Islamabad 45550, Pakistan; rabia_278@hotmail.com (R.S.); sharif@ciitwah.edu.pk (M.S.); mussaratabdullah@gmail.com (M.Y.)
2. Department of Computer Science & Engineering, Ewha Womans University, Seoul 120-750, Korea; hsyong@ewha.ac.kr
3. Department of computer Science, Hanyang University, Seoul 04763, Korea; chajh@hanyang.ac.kr
* Correspondence: jhshah@ciitwah.edu.pk

Abstract: Mango fruit is in high demand. So, the timely control of mango plant diseases is necessary to gain high returns. Automated recognition of mango plant leaf diseases is still a challenge as manual disease detection is not a feasible choice in this computerized era due to its high cost and the non-availability of mango experts and the variations in the symptoms. Amongst all the challenges, the segmentation of diseased parts is a big issue, being the pre-requisite for correct recognition and identification. For this purpose, a novel segmentation approach is proposed in this study to segment the diseased part by considering the vein pattern of the leaf. This leaf vein-seg approach segments the vein pattern of the leaf. Afterward, features are extracted and fused using canonical correlation analysis (CCA)-based fusion. As a final identification step, a cubic support vector machine (SVM) is implemented to validate the results. The highest accuracy achieved by this proposed model is 95.5%, which proves that the proposed model is very helpful to mango plant growers for the timely recognition and identification of diseases.

Keywords: mango leaf; CCA; vein pattern; leaf disease; cubic SVM

1. Introduction

Countries dependent upon agriculture are facing a terrible threat and great loss due to plant diseases, which cause a decline in the quality and quantity of fruits and yields [1]. Pakistan is among those countries where a large amount of its income is earned by importing and producing a variety of crops, vegetables, and fruits cultivated in different areas of the country [2]. Therefore, it is necessary to identify diseased plants by implementing computer vision and image processing techniques [3,4]. Recently, deep learning (DL) techniques, specifically, convolutional neural networks (CNNs), have achieved extraordinary results in many applications, including the classification of plant diseases [5,6]. The mango is a highly popular fruit and is available in summer [7]. It is important in the agricultural industry of Pakistan due to its huge production volume. Several approaches to the detection and identification of mango plant leaf diseases have been proposed in the literature. Although a large number of diseases affect mango orchards, only some of them are causing great loss to the economy of the country. A few of the more common diseases [8] include powdery mildew, sooty mold, anthracnose and apical necrosis, as shown in Figure 1. In the current era, computer scientists are aiming to devise computer-based solutions to identify the diseases in their initial phase. This will aid farmers to safeguard the crop until it is harvested, resulting in the reduction in economic loss [9]. According to agricultural experts, naked eye observation is the traditional method used to recognize plant diseases. This is very expensive and time-consuming as it requires continuous monitoring [3,10]. Hence, it is almost impossible to accurately recognize the diseases of plant at an initial stage. Unfortunately, very few techniques that address the diseases of the mango, also called the king of fruits, have been reported before now [11,12] due to the complicated and complex structure and pattern of the plant. Hence, there is a need for efficient and robust techniques to identify mango diseases automatically, accurately, and efficiently [4,13–15]. For this purpose, the images used as a baseline can be captured by digital and mobile cameras [16,17].

Machine learning (ML) plays an important role in the identification of diseases [16,18]. ML is a sub-branch of artificial intelligence (AI) [19]. It enables computer-based systems to provide accurate and precise results. Real-world objects are the main objects to inspire ML techniques [20]. A few computerized techniques, for instance, segmentation by K-means and classification using SVM to identify the diseased area, are reported in [21].

Hence, as per the available and above discussion, there is a strong need for the automatic detection, identification, and classification the diseases of the mango plant. Keeping this in mind, this article covers the following issues: data augmentation to increase the dataset; tracking the color, size, and texture of the diseased part of leaf; handling background variation and the diseased part; the proper segmentation of the unhealthy part of the leaf; and, at a later stage, robust feature extraction and fusion to classify the disease. The following key contributions were performed to achieve this task:

1. Image resizing and augmentation in order to set query images.
2. A method for the segmentation of the diseased part.
3. Fusion of color and LBP features by performing canonical correlation analysis (CCA).
4. Using classifiers of ten different types to perform identification and recognition.

2. Literature Review

In current times, various types of techniques and methods have been established for the detection of plant leaf diseases. These are generally characterized into disease finding or disease detection methods and disease sorting or disease classification methods [22]. Many techniques use segmentation, feature fusion, and image classification implemented on cotton, strawberry, mango, tomato, rice, sugarcane, and citrus. Similarly, these methods are appropriate for leaf, flower, and fruit diseases because they use proper segmentation, feature extraction, and classification. To make a computer-based system work efficiently, ML techniques are generally used to enhance the visualization of the disease symptoms and to segment the diseased part for classification purposes. Mango fruit is important in the agricultural sector due to its massive production volume in Pakistan. Therefore, several approaches for the detection and identification of mango plant leaf diseases have been proposed to prevent a loss of harvest.

Iqbal et al. [23] considered segmentation, recognition, and identification techniques. They found that almost all the techniques are in the initial stage. Moreover, they discussed almost all existing methods along with their advantages, limitations, challenges, and models of ML (image processing) for the recognition and identification of diseases.

Shin et al. [24], in their study on powdery mildew of strawberry leaves, achieved an accuracy of 94.34% by combining an artificial neural network (ANN) and speeded-up robust features (SURF). However, by using an SVM and GLCM, their highest classification accuracy was 88.98%. They used HOG, SURF, GLCM, and two supervised ML methods (ANN and SVM).

Pane et al. [25] adopted an ML technique using a wavelength between 403 and 446nm for the detection of the blue color. They precisely distinguished unhealthy and healthy leaves of wild rocket, also called salad leaves. Bhatia et al. [26] used the Friedman test to rank multiple classifiers and post hoc analysis was also performed using the Nemenyi test. In their study, they found the MGSVM to be the superior classifier with an accuracy of 94.74%. Lin et al. [17] proved their results to be 3.15% more accurate than the traditional method used on pumpkin leaves. They used PCA in order to obtain 97.3% accurate results.

Shah [27] extracted the color features to detect diseases on cotton leaves. Kahlout et al. [28] developed an expert system for the detection of diseases including powdery mildew and sooty mold on all members of the citrus family. Sharif et al. [29] recommended a computerized system to segment and classify the diseases of citrus plants. In the first part of their suggested system, they used an optimized weight technique to recognize unhealthy parts of the leaf. Secondly, color, geometric, and texture descriptors were combined. Lastly, the best features were nominated by a hybrid feature selection technique consisting of the PCA approach called entropy and they obtained 90% accuracy. Udayet et al. [30] proposed a method to classify anthracnose (diseased) leaves of mango plants. They used a multiple layer convolutional neural network (CNN) for this task. The method was applied to 1070 images collected by their own cameras and gadgets. Consequently, the classification accuracy was raised to 97.13%. Kestur et al. [31] presented a segmentation method in 2019 based on deep learning called Mango net. The results using this method are 73.6% accurate. Arivazhagan et al. [11] used a CNN and showed 96.6% accurate outcomes. No preprocessing or feature extraction was performed in this proposed technique. Srunitha [32] detected unhealthy regions for mango diseases, including red rust, anthracnose, powdery mildew, and sooty mold.

Sooty mold and powdery mildew both create a layer on the leaf by disturbing its vein pattern, as the vein is a vital part of the plant. No techniques have been proposed to make this specific diagnosis, which is a major weakness in the available literature. So, a novel segmentation technique was constructed in this work that segments the diseased part by considering the vein pattern of mango leaves. After segmentation, two features (color and texture) are extracted and used for fusion and classification purposes. We used a self-collected dataset to accomplish this task. The dataset was collected using mobile cameras and digital gadgets. Details about this work are given in Section 3. A concise and precise description of the experimental results and their interpretation is given in Section 4, while Section 5 presents the conclusion drawn from those results.

3. Material and Method

The primary step of this work was the preparation of a dataset. The dataset used for this work was a collection of self-collected images that were captured using different types of image capturing gadgets. These images were collected from different mango growing regions in Pakistan, including Multan, Lahore, and Faisalabad, in the form of RGB images. The collected images were resized to 256 × 256 after being annotated by expert pathologists, as they showed dissimilar sizes. Figure 1 presents the workflow of the proposed technique, which comprises the following steps: (1) the preprocessing of images consisting of data augmentation followed by image resizing; (2) the use of the proposed model to segment the images obtained as a result of the resizing operation (codebook); (3) color and texture (LBP) feature extraction. Finally, the images were classified by using 10 different types of classifiers.

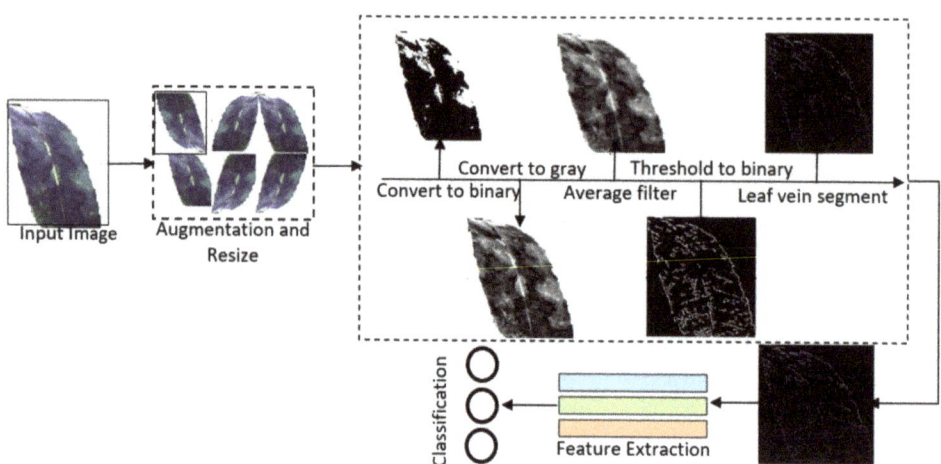

Figure 1. Framework of the proposed computerized system.

3.1. Preprocessing

The purpose of preprocessing is to improve the segmentation and classification accuracy by enhancing the quality of the image. The detailed sketch of each phase implemented for this purpose is as follows:

Resizing and Data Augmentation

A total of 29 images of healthy and unhealthy mango plant leaves were collected. Some of the images (2 out of 29) were distorted upon applying the resize operation. The distorted images were discarded, and the remaining 27 images were augmented by flipping and rotating [33] them horizontally, vertically, and both horizontally and vertically, as well as by using power-law transformations with gamma = 0.5 and c = 1, as shown in Figure 2. As such, 135 images were made available for tuning the proposed algorithm, as shown in Table 1.

Original Image   Horizontal Flip   Horizontal Flip   Horizontal + Vertical Flip   Horizontal Flip

Figure 2. Data augmentation steps.

Table 1. Distribution of the data.

Sooty Mold	Powdery Mildew	Healthy	Total
45	45	45	135

An equal image size of 256 × 256 was utilized for the current study. From the whole dataset, only diseased images were used for segmentation, while the other 45 images of healthy leaves were used for classification.

3.2. Proposed Leaf Vein-Seg Architecture

The second step after preprocessing is the proposed leaf vein-seg architecture. The powder-like, purplish-white fungi growing mainly on the leaves that makes the plants dry and brown is called powdery mildew [34]. The honey-dew-like insect secretions that form a brownish layer on the leaf of mango plants are called sooty mold.

The proposed architecture is a stepwise process that extracts the veins of the leaf. The extracted veins are helpful in further processing and in the classification of diseases of the mango plant. First, the RGB input image is converted to a binary image. The binary image is then converted to a gray-scale image to extract a single channel from the image. CLAHE is applied to it, which improves the quality of the image by improving its contrast. CLAHE operates on small regions of an image called tiles and takes care of the over amplification of contrast in an image [35]. Bilinear interpolation is used to remove artificial boundaries by combining the neighboring tiles.

An average filter of 9×9 is then applied to the output gray-scale image to exclude its background. A mean or average filter smooths the image by reducing the intensity variation among the neighbor pixels. This filter works by replacing the original value of a pixel with the average value of its own and neighboring pixels. It moves pixel-by-pixel through the whole image. An average 9×9 filter is shown in matrix form below. Mathematically, it can be represented by Equation (1).

$$I_{new(x,y)} = \sum_{j=-1}^{1} \sum_{i=-1}^{1} 1 \times I_{old}(x+i, y+j) \tag{1}$$

The output is then normalized to make the image pixel values between 0 and 255 as given in Equation (2).

$$I_{new(x,y)}^{normalized} = \frac{1}{\sum_{j=-1}^{1} \sum_{i=-1}^{1} 1} \sum_{j=-1}^{1} \sum_{i=-1}^{1} 1 \times I_{old}(x+i, y+j) \tag{2}$$

The edges of the image are detected, and noise is removed by sharpening the image. In the next step, the difference is calculated from the images obtained in the previous two steps: the gray-scale images and the images obtained after the application of the average filer along with its normalization, as the images should be the same size. This step is performed to compare the images and correct uneven luminance. Its mathematical representation is shown in Equation (3):

$$D(x,y) = I_1(x,y) - I_{new(x,y)}^{normalized} \tag{3}$$

where the output of the gray-scale image is $I_1(x,y)$ and $I_{new(x,y)}^{normalized}$ is the output after the application of the average filter and its normalization. The threshold is then applied to the obtained images as $D(x,y)$. A method was designed to perform this task. The global image threshold is computed by setting a threshold level to $D(x,y)$, obtained in Equation (3). The computed global threshold level is used to convert the intensity image to a binary image. Standardized intensity values lie between 0 and 1. The histogram is segmented into two parts to normalize the image by using a starting threshold value that should be half of the maximum dynamic range.

$$H = 2B - 1 \tag{4}$$

Foreground and background values are computed by using the sample mean values. Mean values associated with the foreground are (mf, 0), and the gray values associated with the background are (mb, 0). In this way, threshold value 1 is computed and this process is repeated until the threshold value does not change any more. The image obtained is converted to a binary image. The next step is to take the complement of binary image D. All zeros became one and all ones became zero. Mathematically, it can be represented in Equation (5).

$$D = D' \tag{5}$$

Another method was designed to obtain the final segmented output in the form of edges and veins that is used to detect the diseased leaf of the mango plant. It takes as input the binary image obtained in Equation (5) to produce a mask of indicated colors for the output image while using a 1×3 size vector with values between 0 and 1 for the color. The mask must be represented in a logical two-dimensional matrix where [0 0 0] shows white and [1 1 1] shows black. The output of this step is a segmented RGB image. The processed images, including both diseased and healthy leaves of a mango plant, are presented in Figure 3.

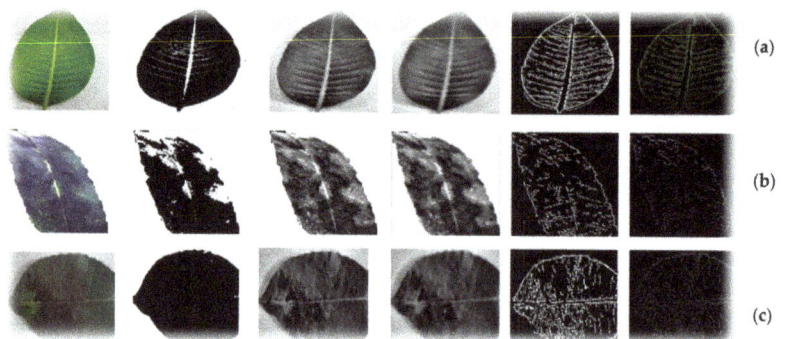

Figure 3. Stepwise output of leaves: (**a**) healthy, (**b**) powdery mildew, and (**c**) sooty mold.

3.3. Features Extraction and Fusion

This is a helpful phase of an artificially intelligent, automated system based on ML. Color and texture features are helpful descriptors. Information about the color is obtained by using color features, whereas texture features provide texture analysis for the diseased leaves of the mango plant. In order to conduct the classification of diseases (powdery mildew, sooty mold) of the mango plant, color, shape, and texture features are extracted in this proposed method. The structure of the feature (color, shape, and texture) extraction is shown in Figure 4.

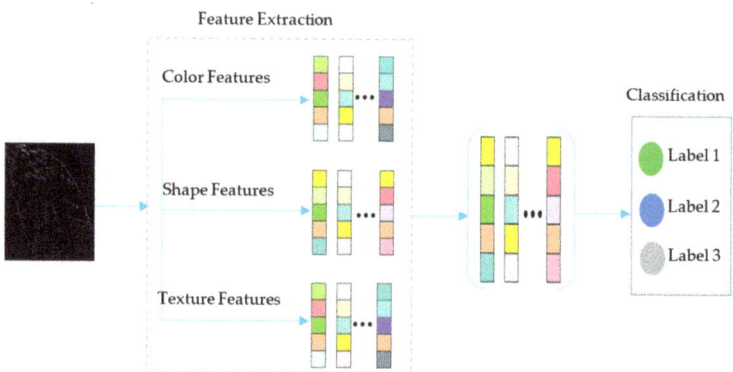

Figure 4. Structure of feature extraction.

The comprehensive explanation of each phase is as follows: preprocessed images are passed through the codebook and a segmented image is achieved in the training phase; the segmented image is used for feature extraction; and then color, shape, and texture features are extracted.

Color Features: These are a significant resource for the detection and recognition of diseased/unhealthy parts of mango plants, as every mango plant leaf disease has a different color and shade. In this paper, the diseases on mango plant leaves are recognized or identified by the extraction of color features. As stated earlier, each disease has its own pattern and shading, so four types of color spaces are utilized to obtain the extracted colors. Features including RGB; hue, saturation, and variance (HSV); luminance, a, b component (LAB); hue, intensity, and saturation (HIS) are obtained from the mango leaf images. Different information is obtained along every channel. Therefore, the maximum information about the defective part of the mango leaf is obtained by applying color spaces. Color features are obtained by using the six types of different statistical metrics mentioned below from Equations (6)–(11). One vector, sized 1×3000, is used to combine these features for each channel. The statistical metrics are calculated by the following formulae:

$$\overline{A} = (\sum a_i)/n, \qquad (6)$$

$$\sigma = \frac{1}{n}\sqrt{\sum_{i=1}^{n}(a_i - a')^2} \qquad (7)$$

$$V = \frac{1}{n}\sum_{i=1}^{n}(a_i - a')^2 \qquad (8)$$

$$Entropy = \sum_{i=1}^{c} -p_i \log_2 p_i \qquad (9)$$

$$KR = n\frac{\sum_{i=1}^{n}(A_i - A_{avg})^4}{(\sum_{i=1}^{n}(A_i - A_{avg})^2)^2} \qquad (10)$$

$$Skewness = \frac{\frac{1}{n}\sum_{i=1}^{n}(a_i - a')^3}{(\frac{1}{n}\sum_{i=1}^{n}(a_i - a')^2)^{\frac{3}{2}}} \qquad (11)$$

where $\overline{A}$ denotes the mean feature, σ denotes the standard deviation feature, V represents the variance feature, $Entropy$ describes the entropy, KR represents the kurtosis feature, and $Skewness$ indicates the skewness feature.

Texture Features: Texture features alone cannot find identical images. Other features, such as color, work with texture features to segregate texture and non-texture features. To handle complications in image texture, an ancient but easy technique, local binary pattern (LBP), is implemented [36]. Hence, LBP was used in this research to extract texture features. Its description is mentioned below:

$$LBP_{A,B} = \sum_{A=0}^{A-1} F(g_A - g_c)2^A, F(x) = \begin{cases} 1 \ if \ x \geq 0; \\ 0 \ otherwise \end{cases} \qquad (12)$$

In Equation (12), the value of a pixel is denoted by A, the value of the radius by B, the neighborhood point by g_a, the center point by g_c, and the binomial factor is denoted by 2^A. As a result, a 1×800 size vector was generated after extracting features from different channels and color spaces.

3.4. Features Fusion and Classification

Finally, a CCA-based feature reduction is applied to the extracted features. A serial-based fusion approach is used to fuse the resultant reduced vectors. A vector of dimensions $n \times 2000$ is obtained simply by concatenating the features. This is used for classification and fed into the classifiers. Ten different types of classification techniques were implemented for the analysis of classification accuracy. In the past, agricultural applications suffered from the unavailability of data due to its complex structure and collection cost, especially for mango plants because these plants are available only in select regions. The cost of labeling for data acquisition is also very high [37]. Hence, this issue encouraged us to collect data by ourselves from mango growing areas in Pakistan. We adopted 2 strategies in this research: (1) data augmentation, and (2) a segmentation technique to segment the diseased parts and veins of mango leaves. This is a unique technique as it has not been performed in earlier agricultural applications, especially with the mango plant. As a final point, the segmented images were entered into a computer-based system for feature fusion, and then for identification. The computation was about 45 min as the segmentation of one image took almost 7.5 s. All simulations were performed on a personal computer with the following specifications: 64-bit Windows operating system with MATLAB version 2018, 32 GB RAM with an Intel® Xeon® processor and central processing unit of 2.2GHz, GPU GeForce GT×1080.

4. Experimental Results and Analysis

In this section, the results of the proposed algorithm are discussed in both graphic and tabular form. For validation, out of the total 135 images used, 45 images were of sooty mold, 45 images were of powdery mildew, and 45 were of healthy mango plants. As a primary step, the proposed segmentation technique was used. Second, the classification results for this segmentation technique were tabulated by implementing different standardized classifiers. The detailed results, with their descriptions, are discussed in the following section. The identification of each disease is analyzed with the images of the healthy leaves of the mango plant. Then, the classification accuracy of all diseased leaves is compared with the classification of the healthy leaves of the mango plant. The proposed technique was tested with 10 of the most ideal classifiers with 1- fold cross-validation.

4.1. Test 1: Powdery Mildew vs. Healthy

In this test, 45 powdery mildew and 45 healthy mango leaf images were classified. Table 2 shows that an accuracy of 96.6% was attained through cubic SVM. It proved to be the highest amongst all the other competing classifiers. Furthermore, 0.16, 0.97, 0.97, and 3.4 were the obtained values of sensitivity, specificity, AUC, and FNR, respectively. Sensitivity and specificity mathematically describe the accuracy of a test that reports the presence or absence of a condition. The confusion matrix of this test is also given in Table 3.

Table 2. Powdery mildew vs. healthy mango leaves.

Methods	Sensitivity	Specificity	AUC	FNR (%)	Accuracy (%)
Linear discriminant	0.24	0.89	0.82	17.8	82.2
Linear SVM	0.22	0.91	0.85	5.6	94.4
Quadratic SVM	0.22	0.91	0.86	6.7	93.3
Cubic SVM	0.16	0.97	0.97	3.4	96.6
Fine KNN	0.22	0.91	0.84	11.2	88.8
Medium KNN	0.18	0.84	0.86	16.7	83.3
Cubic KNN	0.20	0.71	0.84	18.9	81.1
Weighted KNN	0.18	0.82	0.86	11.2	88.8
Subspace discriminant	0.2	0.82	0.86	12.3	87.7
Subspace KNN	0.18	0.84	0.86	16.7	83.3

Table 3. Confusion matrix of powdery mildew vs. healthy mango leaves.

Classification-Class	Classification-Class	
	Powdery Mildew	Healthy
Powdery mildew	97.7%	<1%
Healthy	<1%	95.6%

4.2. Test 2: Sooty Mold vs. Healthy

In this test, 45 images of sooty mold and 45 healthy mango leaf images were classified. Table 4 shows that an accuracy of 95.5% was attained by using linear SVM. It proved to be the highest amongst all the other competing classifiers. Furthermore, 0.05, 0.95, 0.95, and 4.5 are the obtained values of sensitivity, specificity, AUC, and FNR, respectively. The confusion matrix of this test is also given in Table 5.

Table 4. Sooty mold vs. healthy mango leaves.

Methods	Sensitivity	Specificity	AUC	FNR (%)	Accuracy (%)
Linear discriminant	0.47	0.74	0.63	36.7	63.3
Linear SVM	0.05	0.95	0.95	4.5	95.5
Quadratic SVM	0.22	0.91	0.86	6.7	93.3
Cubic SVM	0.22	0.91	0.85	5.6	94.4
Fine KNN	0.24	0.67	0.71	28.9	71.1
Medium KNN	0.2	0.82	0.86	12.3	87.7
Cubic KNN	0.23	0.82	0.86	11.2	88.8
Weighted KNN	0.20	0.87	0.93	12.5	83.3
Subspace discriminant	0.29	0.91	0.82	18.9	81.1
Subspace KNN	0.24	0.6	0.75	32.2	67.8

Table 5. Confusion matrix of sooty mold vs. healthy mango leaves.

Classification Class	Classification Class	
	Sooty Mold	Healthy
Sooty mold	95.5%	<1%
Healthy	<1%	95.5%

4.3. Test 3: Diseased vs. Healthy

This section presents the findings of the classification all unhealthy and healthy images of mango plant leaves. Table 6 shows the classification results for all the diseases obtained after feature fusion based on CCA. This test was performed on all 135 images. The accuracy of 95.5% was attained using a cubic SVM, which is the highest among all the other competing classifiers. Moreover, 0.03, 0.93, 0.99, and 4.5 were the values of the sensitivity, specificity, AUC, and FNR, respectively. The confusion matrix of this test is also given in Table 7.

Table 6. Diseased vs. healthy mango leaves.

Methods	Sensitivity	Specificity	AUC	FNR (%)	Accuracy (%)
Linear discriminant	0.13	0.69	0.78	27.4	72.6
Linear SVM	0.03	0.88	0.98	6.7	93.3
Quadratic SVM	0.14	0.76	0.88	25.2	74.8
Cubic SVM	0.03	0.93	0.99	4.5	95.5
Fine KNN	0.10	0.56	0.73	33.3	66.7
Medium KNN	0.06	0.83	0.88	11.2	88.8
Cubic KNN	0.03	0.89	0.98	7.5	92.5
Weighted KNN	0.11	0.79	0.86	20	80
Subspace discriminant	0.13	0.71	0.87	28.1	71.9
Subspace KNN	0.08	0.47	0.77	35.6	64.4

Table 7. Confusion matrix of diseased vs. healthy mango leaves.

Classification Class	Classification Class		
	Healthy	Powdery Mildew	Sooty Mold
Healthy	97.8%	<1%	-
Powdery mildew	-	97.8%	<1%
Sooty mold	-	-	91.1%

4.4. Discussion

The achieved results are more efficient than the results presented by Srunitha et al. [32] in 2018 who introduced k-means for the segmentation of the diseased part and a multiclass SVM for classification purposes and obtained an accuracy of 96%, as shown in Table 8. However, the list of diseases they detected did not contain powdery mildew, whereas our proposed method showed an accuracy of 95.5% while detecting powdery mildew as well. None of the studies available have yet checked the vein pattern of plant leaves. As it is the most important part of the plant as concerns food and water transportation, the proposed technique is far superior to the available techniques. Furthermore, the mobile-based system proposed by Anantrasirichai et al. [38] in 2019 uses classification techniques and obtained an accuracy of 80% when detecting the diseases of mango plants. In comparison, the achieved accuracy of the proposed model is much better, at 95.5%.

Table 8. Comparison of different segmentation and classification techniques with proposed model.

Methods	Year	Technique	Accuracy (%)
K means by Srunitha et al. [32]	2018	Multiclass SVM	96%
Mobile phone based by Anantrasirichai et al. [38]	2019	Classification	80%
Proposed		Leaf vein-seg	95.5%

5. Conclusions

A novel segmentation technique was introduced in this paper. Two types of mango leaf disease, powdery mildew and sooty mold, were recognized. A self-collected dataset was used to perform this task. A leaf vein segmentation technique that detects the vein pattern of the leaf was proposed. The leaf's features were extracted on the basis of color and texture after performing the segmentation. Ten different classifiers were used to obtain the results. The overall performance of the proposed method is much improved compared to already available methods. However, the following improvements will be considered in the future: (1) increase the number of images in the dataset, (2) minimize the identification time through feature optimization algorithms [39–41] to implement it in real time, and (3) implement some latest deep learning models [42–45].

Author Contributions: Conceptualization, R.S. and J.H.S.; methodology, R.S., J.H.S., and M.S.; software, R.S. and M.S.; validation, J.H.S., M.Y., and M.S.; formal analysis, M.Y. and J.C.; investigation, H.-S.Y. and J.C.; resources, M.Y. and J.C.; data curation, H.-S.Y. and J.C.; writing—original draft preparation, R.S. and M.S.; writing—review and editing, J.C. and M.Y.; visualization, M.S. and M.Y.; supervision, J.H.S. and J.C.; project administration, H.-S.Y.; funding acquisition, H.-S.Y. All authors have read and agreed to the published version of the manuscript.

Funding: This work was supported by the National Research Foundation of Korea (NRF) grant funded by the Korea government (Ministry of Science and ICT; MSIT) under Grant RF-2018R1A5A7059549.

Institutional Review Board Statement: Not applicable.

Informed Consent Statement: Not applicable.

Conflicts of Interest: The authors declare no conflict of interest.

References

1. Tran, T.-T.; Choi, J.-W.; Le, T.-T.H.; Kim, J.-W. A comparative study of deep CNN in forecasting and classifying the macronutrient deficiencies on development of tomato plant. *Appl. Sci.* **2019**, *9*, 1601. [CrossRef]
2. Shah, F.A.; Khan, M.A.; Sharif, M.; Tariq, U.; Khan, A.; Kadry, S.; Thinnukool, O. A Cascaded Design of Best Features Selection for Fruit Diseases Recognition. *Comput. Mater. Contin.* **2021**, *70*, 1491–1507. [CrossRef]
3. Akram, T.; Sharif, M.; Saba, T. Fruits diseases classification: Exploiting a hierarchical framework for deep features fusion and selection. *Multimed. Tools Appl.* **2020**, *79*, 25763–25783.
4. Rehman, M.Z.U.; Ahmed, F.; Khan, M.A.; Tariq, U.; Jamal, S.S.; Ahmad, J.; Hussain, I. Classification of Citrus Plant Diseases Using Deep Transfer Learning. *CMC Comput. Mater. Contin.* **2022**, *70*, 1401–1417.

5. Maeda-Gutierrez, V.; Galvan-Tejada, C.E.; Zanella-Calzada, L.A.; Celaya-Padilla, J.M.; Galván-Tejada, J.I.; Gamboa-Rosales, H.; Luna-Garcia, H.; Magallanes-Quintanar, R.; Guerrero Mendez, C.A.; Olvera-Olvera, C.A. Comparison of convolutional neural network architectures for classification of tomato plant diseases. *Appl. Sci.* **2020**, *10*, 1245. [CrossRef]
6. Hussain, N.; Khan, M.A.; Tariq, U.; Kadry, S.; Yar, M.A.E.; Mostafa, A.M.; Alnuaim, A.A.; Ahmad, S. Multiclass Cucumber Leaf Diseases Recognition Using Best Feature Selection. *CMC Comput. Mater. Contin.* **2022**, *70*, 3281–3294. [CrossRef]
7. Khan, M.A.; Sharif, M.I.; Raza, M.; Anjum, A.; Saba, T.; Shad, S.A. Skin lesion segmentation and classification: A unified framework of deep neural network features fusion and selection. *Expert Syst.* **2019**, *14*, 1–23. [CrossRef]
8. Rosman, N.F.; Asli, N.A.; Abdullah, S.; Rusop, M. Some common disease in mango. In *AIP Conference Proceedings*; AIP Publishing LLC: Melville, NY, USA, 2019; p. 020019.
9. Jogekar, R.N.; Tiwari, N. A review of deep learning techniques for identification and diagnosis of plant leaf disease. In *Smart Trends in Computing and Communications: Proceedings of SmartCom 2020*; Springer: New York, NY, USA, 2021; pp. 435–441.
10. Runno-Paurson, E.; Lääniste, P.; Nassar, H.; Hansen, M.; Eremeev, V.; Metspalu, L.; Edesi, L.; Kännaste, A.; Niinemets, Ü. Alternaria Black Spot (*Alternaria brassicae*) Infection Severity on Cruciferous Oilseed Crops. *Appl. Sci.* **2021**, *11*, 8507. [CrossRef]
11. Arivazhagan, S.; Ligi, S.V. Mango leaf diseases identification using convolutional neural network. *Int. J. Pure Appl. Math.* **2018**, *120*, 11067–11079.
12. Saleem, R.; Shah, J.H.; Sharif, M.; Ansari, G.J. Mango Leaf Disease Identification Using Fully Resolution Convolutional Network. *Comput. Mater. Contin.* **2021**, *69*, 3581–3601. [CrossRef]
13. Latif, M.R.; Khan, M.A.; Javed, M.Y.; Masood, H.; Tariq, U.; Nam, Y.; Kadry, S. Cotton Leaf Diseases Recognition Using Deep Learning and Genetic Algorithm. *Comput. Mater. Contin.* **2021**, *69*, 2917–2932. [CrossRef]
14. Adeel, A.; Khan, M.A.; Akram, T.; Sharif, A.; Yasmin, M.; Saba, T.; Javed, K. Entropy-controlled deep features selection framework for grape leaf diseases recognition. *Expert Syst.* **2020**. [CrossRef]
15. Aurangzeb, K.; Akmal, F.; Khan, M.A.; Sharif, M.; Javed, M.Y. Advanced machine learning algorithm based system for crops leaf diseases recognition. In Proceedings of the 6th Conference on Data Science and Machine Learning Applications (CDMA), Riyadh, Saudi Arabia, 4–5 March 2020; pp. 146–151.
16. Saeed, F.; Khan, M.A.; Sharif, M.; Mittal, M.; Goyal, L.M.; Roy, S. Deep neural network features fusion and selection based on PLS regression with an application for crops diseases classification. *Appl. Soft Comput.* **2021**, *103*, 107164. [CrossRef]
17. Tariq, U.; Hussain, N.; Nam, Y.; Kadry, S. An Integrated Deep Learning Framework for Fruits Diseases Classification. *Comput. Mater. Contin.* **2021**, *71*, 1387–1402.
18. Khan, M.A.; Akram, T.; Sharif, M.; Javed, K.; Raza, M.; Saba, T. An automated system for cucumber leaf diseased spot detection and classification using improved saliency method and deep features selection Multimed. *Tools Appl.* **2020**, *79*, 18627–18656.
19. Webster, C.; Ivanov, S. Robotics, artificial intelligence, and the evolving nature of work. In *Digital Transformation in Business and Society*; Springer: New York, NY, USA, 2020; pp. 127–143.
20. Adeel, A.; Khan, M.A.; Sharif, M.; Azam, F.; Shah, J.H.; Umer, T.; Wan, S. Diagnosis and recognition of grape leaf diseases: An automated system based on a novel saliency approach and canonical correlation analysis based multiple features fusion. *Sustain. Comput. Inform. Syst.* **2019**, *24*, 100349. [CrossRef]
21. Febrinanto, F.G.; Dewi, C.; Triwiratno, A. The implementation of k-means algorithm as image segmenting method in identifying the citrus leaves disease. In Proceedings of the IOP Conference Series: Earth and Environmental Science, East Java, Indonesia, 17–18 November 2019; p. 012024.
22. Khan, M.A.; Akram, T.; Sharif, M.; Alhaisoni, M.; Saba, T.; Nawaz, N. A probabilistic segmentation and entropy-rank correlation-based feature selection approach for the recognition of fruit diseases. *EURASIP J. Image Video Process.* **2021**, *2021*, 1–28. [CrossRef]
23. Iqbal, Z.; Khan, M.A.; Sharif, M.; Shah, J.H.; ur Rehman, M.H.; Javed, K. An automated detection and classification of citrus plant diseases using image processing techniques: A review. *Comput. Electr. Agric.* **2018**, *153*, 12–32. [CrossRef]
24. Shin, J.; Chang, Y.K.; Heung, B.; Nguyen-Quang, T.; Price, G.W.; Al-Mallahi, A. Effect of directional augmentation using supervised machine learning technologies: A case study of strawberry powdery mildew detection. *Biosyst. Eng.* **2020**, *194*, 49–60. [CrossRef]
25. Pane, C.; Manganiello, G.; Nicastro, N.; Cardi, T.; Carotenuto, F. Powdery Mildew Caused by Erysiphe cruciferarum on Wild Rocket (*Diplotaxis tenuifolia*): Hyperspectral Imaging and Machine Learning Modeling for Non-Destructive Disease Detection. *Agriculture* **2021**, *11*, 337. [CrossRef]
26. Bhatia, A.; Chug, A.; Singh, A.P. Statistical analysis of machine learning techniques for predicting powdery mildew disease in tomato plants. *Int. J. Intell. Eng. Inform.* **2021**, *9*, 24–58. [CrossRef]
27. Shah, N.; Jain, S. Detection of disease in cotton leaf using artificial neural network. In Proceedings of the 2019 Amity International Conference on Artificial Intelligence (AICAI), Dubai, United Arab Emirates, 4–6 February 2019; pp. 473–476.
28. El Kahlout, M.I.; Abu-Naser, S.S. An Expert System for Citrus Diseases Diagnosis. *Int. J. Acad. Eng. Res. (IJAER)* **2019**, *3*, 1–7.
29. Sharif, M.; Khan, M.A.; Iqbal, Z.; Azam, M.F.; Lali, M.I.U.; Javed, M.Y. Detection and classification of citrus diseases in agriculture based on optimized weighted segmentation and feature selection. *Comput. Electr. Agric.* **2018**, *150*, 220–234. [CrossRef]
30. Singh, U.P.; Chouhan, S.S.; Jain, S.; Jain, S. Multilayer convolution neural network for the classification of mango leaves infected by anthracnose disease. *IEEE Access* **2019**, *7*, 43721–43729. [CrossRef]
31. Kestur, R.; Meduri, A.; Narasipura, O. MangoNet: A deep semantic segmentation architecture for a method to detect and count mangoes in an open orchard. *Eng. Appl. Artif. Intell.* **2019**, *77*, 59–69. [CrossRef]

32. Srunitha, K.; Bharathi, D. Mango leaf unhealthy region detection and classification. In *Computational Vision and Bio Inspired Computing*; Springer: New York, NY, USA, 2018; pp. 422–436.
33. Hussain, Z.; Gimenez, F.; Yi, D.; Rubin, D. Differential data augmentation techniques for medical imaging classification tasks. In Proceedings of the AMIA Annual Symposyum Proceedings, Washington, DC, USA, 6–8 November 2017; pp. 979–984.
34. Ajitomi, A.; Takushi, T.; Sato, Y.; Arasaki, C.; Ooshiro, A. First report of powdery mildew of mango caused by Erysiphe quercicola in Japan. *J. Gen. Plant Pathol.* **2020**, *86*, 316–321. [CrossRef]
35. Sepasian, M.; Balachandran, W.; Mares, C. Image enhancement for fingerprint minutiae-based algorithms using CLAHE, standard deviation analysis and sliding neighborhood. In Proceedings of the World Congress on Engineering and Computer Science, San Francisco, CA, USA, 22–24 October 2008; pp. 22–24.
36. Yosinski, J.; Clune, J.; Bengio, Y.; Lipson, H. How transferable are features in deep neural networks? *arXiv* **2014**, arXiv:1411.1792.
37. Shin, H.C.; Roth, H.R.; Gao, M.; Lu, L.; Xu, Z.; Nogues, I.; Yao, J.; Mollura, D.; Summers, R.M. Deep convolutional neural networks for computer-aided detection: CNN architectures, dataset characteristics and transfer learning. *IEEE Trans. Med. Imaging* **2016**, *35*, 1285–1298. [CrossRef] [PubMed]
38. Anantrasirichai, N.; Hannuna, S.; Canagarajah, N. Towards automated mobile-phone-based plant pathology management. *arXiv* **2019**, arXiv:1912.09239.
39. Khan, S.; Alhaisoni, M.; Tariq, U.; Yong, H.-S.; Armghan, A.; Alenezi, F. Human Action Recognition: A Paradigm of Best Deep Learning Features Selection and Serial Based Extended Fusion. *Sensors* **2021**, *21*, 7941. [CrossRef]
40. Saleem, F.; Alhaisoni, M.; Tariq, U.; Armghan, A.; Alenezi, F.; Choi, J.-I.; Kadry, S. Human gait recognition: A single stream optimal deep learning features fusion. *Sensors* **2021**, *21*, 7584. [CrossRef]
41. Arshad, M.; Khan, M.A.; Tariq, U.; Armghan, A.; Alenezi, F.; Younus Javed, M.; Aslam, S.M.; Kadry, S. A Computer-Aided Diagnosis System Using Deep Learning for Multiclass Skin Lesion Classification. *Comput. Intell. Neurosci.* **2021**, *21*, 1–27. [CrossRef]
42. Nasir, M.; Sharif, M.; Javed, M.Y.; Saba, T.; Ali, H.; Tariq, J. Melanoma detection and classification using computerized analysis of dermoscopic systems: A review. *Curr. Med Imaging* **2020**, *16*, 794–822. [CrossRef]
43. Muhammad, K.; Sharif, M.; Akram, T.; Kadry, S. Intelligent fusion-assisted skin lesion localization and classification for smart healthcare. *Neural Comput. Appl.* **2021**, *12*, 1–16.
44. Khan, M.; Sharif, M.; Akram, T.; Kadry, S.; Hsu, C.H. A two-stream deep neural network-based intelligent system for complex skin cancer types classification. *Int. J. Intell. Syst.* **2021**, *14*, 1–28.
45. Akram, T.; Sharif, M.; Kadry, S.; Nam, Y. Computer decision support system for skin cancer localization and classification. *Comput. Mater. Contin.* **2021**, *70*, 1–15.

Article

One-Dimensional Convolutional Neural Networks for Hyperspectral Analysis of Nitrogen in Plant Leaves

Razieh Pourdarbani [1,*], Sajad Sabzi [2], Mohammad H. Rohban [2], José Luis Hernández-Hernández [3], Iván Gallardo-Bernal [4], Israel Herrera-Miranda [4,*] and Ginés García-Mateos [5,*]

1 Department of Biosystems Engineering, University of Mohaghegh Ardabili, Ardabil 56199-11367, Iran
2 Department of Computer Engineering, Sharif University of Technology, Tehran 11155-1639, Iran; s.sabzi@sharif.edu (S.S.); rohban@sharif.edu (M.H.R.)
3 National Technological of México/Campus Chilpancingo, Chilpancingo 39070, Guerrero, Mexico; joseluis.hernandez@itchilpancingo.edu.mx
4 Government and Public Management Faculty, Autonomous University of Guerrero, Chilpancingo 39087, Guerrero, Mexico; Igallardo@uagro.mx
5 Computer Science and Systems Department, University of Murcia, 30100 Murcia, Spain
* Correspondence: r_pourdarbani@uma.ac.ir (R.P.); israelhm@uagrovirtual.mx (I.H.-M.); ginesgm@um.es (G.G.-M.)

Featured Application: The proposed methodology is able to estimate the amount of nitrogen in plant leaves, using spectral information in the visible (Vis) and near infrared (NIR) ranges, obtaining a mean relative error below 1%. Thus, it will enable the development of portable devices to detect overuse of nitrogen fertilizers in the crops in a fast and non-destructive way. Although it has been tested in cucumber plants, the proposed method can be applied to other types of horticultural crops, repeating the training of the neural network when the new datasets of spectral data and measured nitrogen is available.

Abstract: Accurately determining the nutritional status of plants can prevent many diseases caused by fertilizer disorders. Leaf analysis is one of the most used methods for this purpose. However, in order to get a more accurate result, disorders must be identified before symptoms appear. Therefore, this study aims to identify leaves with excessive nitrogen using one-dimensional convolutional neural networks (1D-CNN) on a dataset of spectral data using the Keras library. Seeds of cucumber were planted in several pots and, after growing the plants, they were divided into different classes of control (without excess nitrogen), $N_{30\%}$ (excess application of nitrogen fertilizer by 30%), $N_{60\%}$ (60% overdose), and $N_{90\%}$ (90% overdose). Hyperspectral data of the samples in the 400–1100 nm range were captured using a hyperspectral camera. The actual amount of nitrogen for each leaf was measured using the Kjeldahl method. Since there were statistically significant differences between the classes, an individual prediction model was designed for each class based on the 1D-CNN algorithm. The main innovation of the present research resides in the application of separate prediction models for each class, and the design of the proposed 1D-CNN regression model. The results showed that the coefficient of determination and the mean squared error for the classes $N_{30\%}$, $N_{60\%}$ and $N_{90\%}$ were 0.962, 0.0005; 0.968, 0.0003; and 0.967, 0.0007, respectively. Therefore, the proposed method can be effectively used to detect over-application of nitrogen fertilizers in plants.

Keywords: nitrogen prediction; 1D convolution neural networks; cucumber; crop yield improvement

1. Introduction

Proper management of inputs, that is, the application of fertilizer in accordance with the nutritional needs of each plant, is necessary to achieve a healthy crop and optimal yield, which is a new concept in the so-called field of precision agriculture [1]. Among the nutrients needed by plants, nitrogen is one of the most important because it is a main component of chlorophyll. However, nitrogen deficiency/overdose in plants is one of the major nutritional problems that is affecting current agricultural practices. The first sign of nitrogen deficiency is the paleness of old leaves, because this element can move from the lower old leaves to the upper young leaves. The leaves are usually yellowish green and light yellow due to the lack of chlorophyll formation. Late red and reddish-purple growth is observed as a result of anthocyanin dye formation; also, the leaves, stems and branches become thin in nitrogen deficiency status [2–4].

Unfortunately, in order to increase yield, farmers have resorted to excessive use of nitrogen fertilizers and are not very willing to use biological, organic and micronutrient fertilizers [5]. However, studies indicate that nitrate accumulation in agricultural products does not play a very positive role in sustainable production and public health [6,7].

Excessive consumption of nitrogen makes the plant susceptible to diseases and pests. In fruit crops, it causes damage to flowers and blossoms, and reduces the quality of the product. If the main source of nitrogen supply is ammonium, it can poison the plant, as a result of which the vascular tissues are damaged and the leaves appear as cups. Therefore, at the present stage, scientific management of production and consumption of fertilizers is inevitable to improve the structure of production and the level of public health by improving the optimal use of fertilizers.

Many farmers perform leaf analysis when they observe signs of food disturbance or unknown symptoms on the leaves. Leaf analysis helps growers better manage nutrient inputs by determining the type and amount of fertilizer needed and identifying problems that may lead to poor crop performance; but, at that time, it is evident that the plant has been damaged. Therefore, in order to get a better result, disorders must be identified before the onset of symptoms. Leaf analysis is time consuming and requires special methods to interpret the data correctly [8,9]. On the other hand, this method is destructive. Thus, the application of non-destructive and fast methods is very important.

In recent years, spectroscopy and hyperspectral imaging have been recognized as useful tools for classifying the internal and external properties of biological products. For example, Sabzi et al. [10] detected early the nitrogen in cucumber plants using hyperspectral images. They captured hyperspectral images of the leaves with excess and standard amount of nitrogen for several consecutive days in laboratory. The results showed that the visible and near-infrared (Vis/NIR) hyperspectral imaging technique was able to early detect plants with excess of nitrogen at a rate of 96.11%. Wang et al. [11] analyzed hyperspectral data from mature leaves of tea plants by partial least squares-discriminant analysis (PLS-DA) and least squares-support vector machines (LS-SVM) to classify different nitrogen status. The results showed that the LS-SVM model was more able to predict different rates of nitrogen with an accuracy of 82%.

Currently, deep learning (DL) models are an extremely active area of research and they are applied in many domains, such as is in this study. They are taught using large labeled datasets and neural network structures that directly characterize features. So, they are trained without the need to extract the features manually. One of the most popular types of DL models are convolutional neural networks (CNN or ConvNet). Automatic feature extraction in CNNs makes them practical and attractive for computer vision tasks such as object detection and segmentation. A CNN learns to recognize different features of an image using up to hundreds of hidden layers. They were introduced in 1990, inspired by experiments performed by Hubel and Wiesel [12] on the visual cortex. The uses of CNN have grown so rapidly that, in a short time, they have revolutionized many areas of computer vision, such as human action detection, object detection, face detection and tracking. In this regard, Zhu et al. [13] conducted a review of traditional machine learning and deep learning methods, including AlexNet, VGG Net and fully convolutional networks (FCN). According to their declaration, despite the success of the machine vision methods, they will have low accuracy if the background is too noisy, or the lighting conditions are poor, since they will not be able to detect small changes in the food.

Watchareeruetai et al. [14] used a CNN for studying deficiencies on plants considering different nutrients, such as Ca, Fe, K, Mg and N. A dataset consisting of 3000 leaf images was analyzed. The results indicated that the proposed method is superior to trained humans in the detection of nutrient deficiency. The CNN classifier had an accuracy of 94%. Espejo-García et al. [15] aimed to use fine-tuning, instead of ImageNet, in pre-trained CNN in order to improve the obtained performance. Experimental results showed that the overall performance can be increased by the proposed method. Some structures, such as Xception and Inception-Resnet, improved it by 0.51% and 1.89%, respectively. Comparing machine learning with DL, Sharma et al. [16] stated that DL has been analyzed and implemented in various applications and has shown remarkable results, thus they need to be explored more broadly, since they can also be useful in most fields. Liu et al. [17] categorized hyperspectral images using long short-term memory (LSTM). Specifically, for each pixel, they fed spectral values one by one in different LSTM channels to train spectral properties. Principal component analysis (PCA) was used first to extract the first components of a hyperspectral image, and then local patches were selected. The row vectors of each patch are then transferred to the LSTM to determine the spatial properties of the central pixel. Then, in the classifier step, the spectral and spatial properties of the pixels are fed into soft-max classifiers, to obtain two different results. A strategy for decision fusion was used to obtain more spatial-spectral results. Tian et al. [18] estimated soluble solids in apples using spectral data and DL. In their proposed model, the spectral data of apples were investigated and determined using a random frog algorithm; and DL was used to train and test the detection of geographical origin with spectral data as input. Partial least squares (PLS) were used to create individual calibration models, and then to estimate soluble solids. Competitive adaptive reweighted sampling (CARS) was used to select the optimal wavelengths. Compared to the individual source model, the proposed multi-source model obtained more accurate results for predicting soluble solid content of apples from multiple geographical origins, obtain an RP of 990 and RMSEP 0.274. Cai et al. [19] estimated soil nutrients also using spectroscopy and DL. The simulation results indicated that the proposed model was able to improve the efficiency of obtaining the features with high reliability. This solves the problem of traditional models.

Cucumber is a fruit that is rich in useful nutrients, containing some compounds and antioxidants, which may help improve, and even prevent, some diseases in humans [20]. Reliable diagnosis of the nutritional status of agricultural products is an important aspect of farm production, since both excess and deficiencies of nutrients can lead to damages and reduced yields. According to the literature, most of the previous research works used statistical methods, simple perceptron neural networks and DL neural networks with predefined structures, such as ImageNet and LSTM [21]. Although all these methods were successful on their own, more accurate methods are needed for everyday use.

Consequently, the present paper describes a new one-dimensional (1D) convolutional neural network to estimate the nitrogen content of cucumber leaves. The innovation of this paper resides in the application of nitrogen overdose at 3 different levels, in order to investigate whether it is possible to detect nitrogen-rich cucumber on site and in real-time. Hyperspectral analysis of fruits and plants has been a very active area of research in the literature, and particularly in our research group, as previously seen. The main novelty of the present work resides in the proposal of a new 1D-CNN architecture with 12 layers. It contains six convolutional layers and three max-pooling layers, finishing with a dense layer, which produces the final estimation of nitrogen content. Another important element is the addition of a dropout layer that, by randomly removing some weights of the previous layer, avoids problems of overfitting.

2. Materials and Methods

The constituent stages of the proposed methodology for non-destructive estimation of nitrogen in cucumber plants can be seen in Figure 1. The steps of this methodology are described in detail in the following subsections.

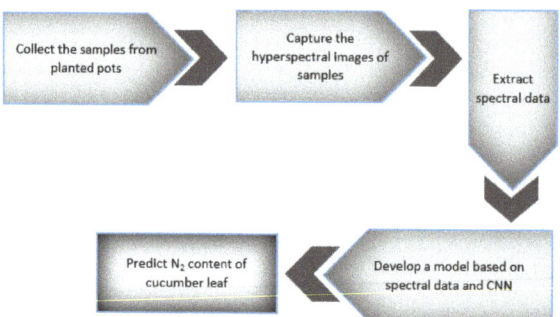

Figure 1. Main stages of the non-destructive estimation of nitrogen in cucumber plants.

2.1. Cultivation and Collection of Samples to Extract Spectral Data

According to the stages of the methodology in Figure 1, the first step is to prepare the samples with standard nitrogen fertilizer and with nitrogen over-dose. Hence, several seeds of cucumber of Super Arshiya'F1 variety were planted into 40 pots, and they were grown under laboratory conditions. All pots received the same inputs until the leaves grew; then the pots were divided into 4 groups of 10 pots. The first group with standard nitrogen was considered as the control treatment. A sample view of the pots in this category is shown in Figure 2. The second, third and fourth groups received nitrogen fertilizer overdoses by 30%, 60% and 90%, respectively.

Figure 2. Cucumber pots prepared for data collection.

After 24 h of applying the treatments, sample leaves were obtained from each group and their spectral signatures were extracted by a hyperspectral camera. The process for obtaining the samples of the dataset was as follows. First, 10 leaves were picked from each category, making a total of 40 cucumber leaves. For each leaf, the spectral images were captured, as described in Section 2.3, giving 327 images per leaf (one for each spectral band considered). This makes a total of 13,080 images. Then, the leaves were transferred to laboratory for the measurement of the nitrogen content by chemical analysis, as presented in Section 2.4. After manually analyzing the images, 10 patches were selected for each leaf in the 327 images, calculating the mean of each patch. Thus, 100 samples are available for each category, which consist of a tuple of 327 values, and the corresponding nitrogen content measured.

Although the process for obtaining this data is expensive (in terms of volume of images and laboratory analysis), the number of 100 samples per class is clearly insufficient for deep neural network analysis. For this reason, data augmentation was performed to obtain more synthetic samples. This was carried out by random weighted averaging of the real samples (i.e., computing a weighted average of two random tuples, and the same weighted average of the corresponding measured nitrogen content). Using this procedure, 900 synthetic samples per class were produced, totaling 1000 samples per class in the dataset (i.e., 4000 samples in the four classes).

2.2. Extraction of Spectral Data from Cucumber Leaves and Statistical Analysis

For the extraction of the spectral information of the cucumber leaves, a hyperspectral camera (Noor Iman Tajhiz Co., Kashan, Isfahan, Iran) at the range of 400–1100 nm was used. More specifically, 327 spectral images uniformly distributed in this range were obtained for each leaf, with a wavelength increment of 3.37 nm. To block out ambient light, the camera was located in an illumination chamber and exposed to two 10-watt tungsten halogen lamps (SLI-CAL, StellarNet Inc., Tampa, FL, USA). The spectral information in the Vis-NIR region was extracted and stored in a laptop. This laptop was a model DELL (DELL Co., Round Rock, TX, USA) with Intel Core i5, 2430 M at 2.40 GHz, 4 GB of RAM, and Windows 10. The original wavelengths were corrected by two methods: multiplicative scatter correction (MSC) in ParLeS software (Raphael Viscarra Rossel, Curtin University, Bentley, Australia), shown in Figure 3; and the smoothing operation by the Savitzky–Golay (SG) filtering algorithm [22].

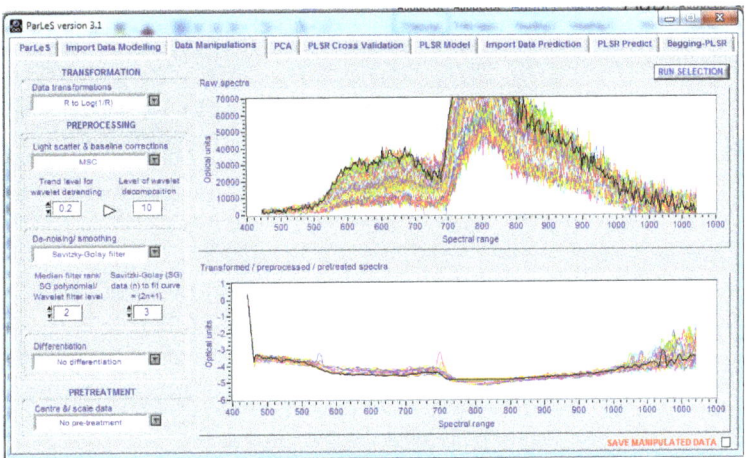

Figure 3. Sample of the application of spectral preprocessing using ParLeS software.

Before creating the regression models, we checked whether the spectral information of the treatments are statistically significantly different or not. The spectral data of the four treatments (control, $N_{30\%}$, $N_{60\%}$ and $N_{90\%}$) were examined with two statistical tests: ANOVA test [23] and Duncan test [24]. The null hypothesis is that the mean values of the spectral wavelengths are the same for all the categories, and the alternative hypothesis is that they are different. That is, the input to the statistical tests are the 4000 samples, each of them a tuple of 327 values.

As it is well known, ANOVA test analyzes the null hypothesis for all the classes, while the Duncan test is a multiple comparison procedure between the pairs of classes. The results obtained for the ANOVA test are presented in Table 1, and those corresponding to Duncan test are contained in Table 2.

The results of these tests indicate that the differences between the classes are statistically significant, both in the ANOVA test (Table 1) and in the Duncan test (Table 2), with a high level of significance. That is, the spectral wavelengths that reflect the presence of nitrogen are affected by the amount of nitrogen contained in the leaves. Thus, this justifies the development of different regression models for each class, since each of them produces distinct spectral features.

Table 1. ANOVA analysis for spectral data of the four categories of treatments: control, $N_{30\%}$, $N_{60\%}$ and $N_{90\%}$. The rows indicate the sum of squares between and within the classes, the number of degrees of freedom, the means of the squares between and within classes, the F-score (mean square between classes divided by mean square within classes), and the p-value of significance for an alpha of 0.05.

Category	Sum of Squares	Degrees of Freedom (df)	Mean Square	F-Score	Significance
Between Groups	8.15×10^{11}	3	2.72×10^{11}	5.69×10^{3}	0
Within Groups	1.91×10^{11}	3996	4.78×10^{7}		
Total	1.01×10^{12}	3999			

Table 2. Duncan test for spectral data related to the four categories of treatments, control, $N_{30\%}$, $N_{60\%}$ and $N_{90\%}$, at a significance level of 5%. Mean values of the classes are sorted in increasing order.

Treatment	Number	Subset for Alpha = 0.05			
		Control	$N_{30\%}$	$N_{60\%}$	$N_{90\%}$
$N_{60\%}$	1000	9.2864×10^{2}			
$N_{30\%}$	1000		1.7571×10^{3}		
Control	1000			2.8540×10^{3}	
$N_{90\%}$	1000				3.4779×10^{4}

These tests do not prove any evidence about the linearity or non-linearity between the spectral information and the nitrogen content of the leaves. They only indicate that each class produces different spectral data. This is the reason for creating separate regression models. Although we could create a single regression model for all the classes, the statistical tests involve that, since the spectral data are different, more precise results can be obtained with individual models for each category.

2.3. Extraction of Nitrogen in Cucumber Leaves Using Laboratory Destructive Method

The Kjeldahl method was used to measure the total nitrogen content on the leaves. This method [25] includes three steps of digestion, distillation and titration. First, the leaves were dried in an oven and then powdered. Then, they are digested with sulfuric acid, so the nitrogen in the sample could be converted to ammonium sulfate. The nitrogen in ammonium sulfate was released in the form of ammonia and converted to ammonium borate with boric acid and titrated using 1% normal sulfuric acid; and then, the total nitrogen content of the sample can be obtained by calculating the consumed acid, as indicated in Equation (1):

$$N_{total} (\%) = \frac{(vs - vb)}{md} \times N_{H2SO4} \times 0.014_{meqN} \times 100 \qquad (1)$$

where:

vs: Volume consumed by the samples (mL),
vb: Volume consumed by the control treatment (mL),
N_{H2SO4}: Normality of sulfuric acid (equation/L),
md: Dry weight of the sample (g).

2.4. Non-Destructive Estimation of Nitrogen Using Convolutional Neural Networks

Deep learning (DL) is a category of machine learning algorithms that uses multiple layers to extract high-level features from raw input [26]. These kind of networks use multiple layers of information processing, especially nonlinear information, to perform the conversion or extraction of supervised or unsupervised features, generally for the purpose of pattern analysis or recognition, classification and clustering. Most deep learning methods use neural network structures, which is why DL models are often referred to as deep neural networks. Among them, convolutional neural networks (CNN) are one of the most popular techniques, since they do not require manual feature extraction. CNNs learn to recognize features from the samples using a large number of hidden layers. Each hidden layer increases the complexity of the features analyzed. Moreover, CNN networks do not change the structure of the input and pay attention to the connection between neighboring values.

Until now, most of the research has been conducted by known CNN structures such as AlexNet, VGG Net, ZF Net, GoogLeNet and fully convolutional networks (FCN) [27]. But in this paper, we proposed our personalized structure. In fact, it has been developed by examining several personalized structures by trial and error. Finally, the structure with the highest prediction rate was selected and proposed in this work.

More specifically, the proposed structure of the convolutional neural network used in this study is shown in Figure 4. The input vector of the algorithm contains the spectral data of a sample, and the output is the actual amount of nitrogen. Since, in our case, the input data is a one-dimensional (1D) spectrum, we used 1D convolutions. As shown, the architecture is similar to the well-known funnel design, where the size of the input is progressively reduced while the number of features increases. This is carried out by a succession of 1D convolutions and max-pooling layers. All the convolutions use the rectified linear (ReLu) activation function. The first part of the network contains 3 consecutive convolutional layers, then there are three steps of max-pooling and convolution. The second of these steps also includes a dropout layer, a method which is commonly used to avoid overfitting, since it randomly removes some weights in the previous convolutions, thus making the system more robust. The dropout rate was set to 0.1. Finally, the values of the last layer are flattened (producing 20,160 values) and a dense layer is responsible for producing the final estimation of nitrogen content.

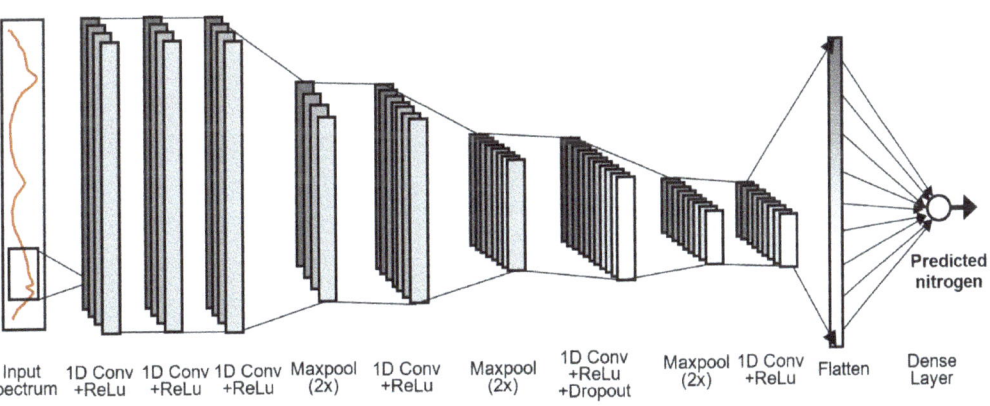

Figure 4. Architecture of the proposed 1D convolutional neural network for the estimation of nitrogen content.

It should be noted that among the 1000 samples in each category, 70% were used as the training data, and the remaining 30% as the test data, both sets disjoint. Moreover, the synthetic samples of the augmentation process were obtained after partitioning the real samples, so there is no mix of the training and test samples. The number of total epochs, batch-size, verbose and validation split were 200, 12, 1 and 0.1, respectively. The exact parameters of the proposed 1D-CNN are presented in Table 3.

Table 3. Hyperparameters of the 1D convolutional neural network (1D-CNN) structure proposed in this paper for the estimation of nitrogen amount in cucumber leaves. Output shape refers to the size × number of features.

Layer (Type)	Output Shape	Kernel Size	Stride	Number of Parameters
Conv1d + ReLu	325 × 512	7	-	2048
Conv1d + ReLu	323 × 512	5	-	786,944
Conv1d + ReLu	321 × 512	5	-	786,944
Max pooling	160 × 256		2	-
Conv1d + ReLu	156 × 256	3	-	196,864
Max pooling	78 × 128		2	-
Conv1d + ReLu	76 × 128	3	-	49,280
Dropout (0.1)			-	0
Max pooling	38 × 128		2	-
Conv1d + ReLu	36 × 64	3	-	12,352
Flatten	20160		-	0
Dense + Relu	1		-	20,161

Total params: 2,031,233; trainable params: 2,031,233; non-trainable params: 0.

2.5. Evaluation of the Performance of the Methods for the Estimation of Nitrogen in Cucumber

To evaluate the performance of the prediction model, the statistical parameters including score-variance (Var-score) [28], max error (MaxE) [29], mean absolute error (MAE) [30], mean squared error (MSE) [31], coefficient of determination (R^2) [32], median absolute error (MedAE) [33] and mean squared logarithmic error (MSLE) [34] were used. Consider that the expected outputs for a given variable are Y_i, for i ranging in the number of samples, n, and consider that X_i represents the corresponding estimated values. Then, the proposed performance measures are given in the following equations:

$$\text{MSE} = \frac{1}{n}\sum_{s=1}^{n}(X_s - Y_s)^2 \tag{2}$$

$$\text{MedAE} = \text{median}\left\{\sum_{s=1}^{n}|X_s - Y_s|\right\} \tag{3}$$

$$\text{MAE} = \frac{1}{n}\sum_{s=1}^{n}|X_s - Y_s| \tag{4}$$

$$\text{MaxE} = \max\left\{\sum_{s=1}^{n}|X_s - Y_s|\right\} \tag{5}$$

$$R^2 = 1 - \left\{\frac{\sum_{s=1}^{n}(X_s - Y_s)^2}{\sum_{s=1}^{n}(X_s - X_m)^2}\right\} \tag{6}$$

$$\text{MSLE} = \frac{1}{n}\sum_{s=1}^{n}\log(X_s - Y_s)^2 \tag{7}$$

$$\text{Var} - \text{score} = \text{variance}(X_s)/\text{variance}(Y_s) \tag{8}$$

$$X_m = \frac{1}{n}\sum_{s=1}^{n}X_s \tag{9}$$

3. Results

In the following subsections, the results achieved by the regression models created for each treatment are presented. All of them are 1D-CNN networks that are specifically trained for the corresponding class.

3.1. Prediction Model for the Category with Excess of Nitrogen by 30%

Preprocessing is needed to reduce the noise in the spectral information, so different algorithms of correction and smoothing filters were examined. Table 4 contains the results of evaluating the performance of the prediction model for the category with nitrogen overdose by 30%, using different correction and smoothing algorithms. As described in Section 2, these filters are: multiplicative scatter correction (MSC), smoothing by the Savitzky–Golay (SG) filter.

Table 4. Evaluation of the prediction model for the category with excess of nitrogen by 30%. Filtering algorithms: MSC: multiplicative scatter correction; SG: Savitzky–Golay filter. MAE: mean absolute error; MSE: mean squared error; MSLE: mean squared logarithmic error; MedAE: median absolute error; MaxE: max error; R^2: coefficient of determination; Var-score: variance score.

Algorithm	MAE	MSE	MSLE	MedAE	MaxE	R^2	Var-Score
MSC + SG	0.017	0.0005	0.00002	0.014	0.086	0.962	0.974
MSC	0.022	0.0011	0.00009	0.015	0.084	0.951	0.971
SG	0.02	0.0009	0.00005	0.013	0.072	0.958	0.972
No filter	0.027	0.0016	0.00012	0.018	0.092	0.941	0.967

All the error measures can be observed to be very low, while R^2 is close to 1, indicating that the model is able to accurately estimate nitrogen content in this class. Figure 5 shows the regression plot between the amount of nitrogen measured by the Kjeldahl method, and the mean estimated nitrogen for 300 test data. The proximity of these two values indicates the ability of the proposed model. Figure 6 shows the train loss and validation loss diagrams for the nitrogen content during the training process. The maximum number of epochs was set to 200. In addition, we added a condition to early stop training if overfitting is detected (i.e., when training loss decreases but validation loss increases). As shown in Figure 6, it can be seen that after 17 epochs training stopped and the loss rate is very close to 0%.

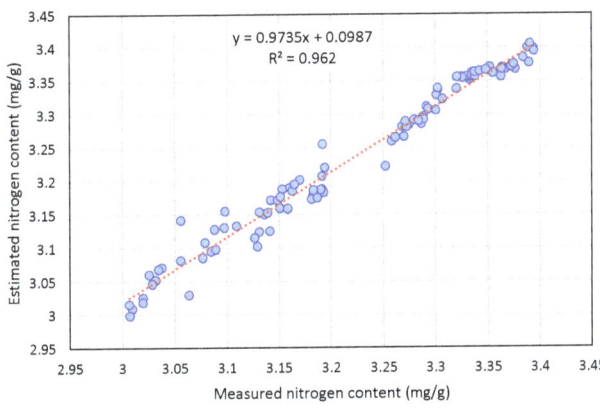

Figure 5. Regression plot of the measured and estimated nitrogen content for category $N_{30\%}$.

3.2. Prediction Model for the Category with Excess of Nitrogen by 60%

The statistical criteria for evaluating the performance of the prediction model are given in Table 5 for the category $N_{60\%}$. Again, it can be seen that the error measures are very close to 0, so the proposed model was successful for the prediction of nitrogen in the treatment of 60% overdose. Moreover, the results indicate the positive effect of the filtering algorithms.

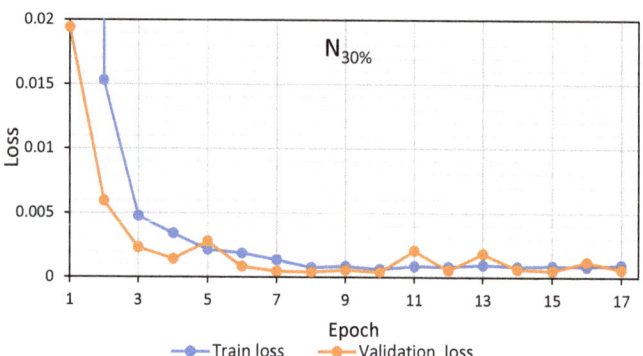

Figure 6. Train loss and validation loss diagram for category $N_{30\%}$.

Table 5. Evaluation of the prediction model for the category with excess of nitrogen by 60%. Filtering algorithms: MSC: multiplicative scatter correction; SG: Savitzky–Golay filter. MAE: mean absolute error; MSE: mean squared error; MSLE: mean squared logarithmic error; MedAE: median absolute error; MaxE: max error; R^2: coefficient of determination; Var-score: variance score.

Algorithm	MAE	MSE	MSLE	MedAE	MaxE	R^2	Var-Score
MSC + SG	0.014	0.0003	0.00001	0.014	0.091	0.968	0.984
MSC	0.017	0.0007	0.00006	0.017	0.094	0.953	0.972
SG	0.015	0.0005	0.00003	0.016	0.092	0.962	0.976
No filter	0.019	0.0012	0.00009	0.019	0.096	0.938	0.963

Figure 7 shows the regression plot of the measured and the estimated nitrogen content for this category, for the 300 test samples used. These two values are close to each other, indicating the ability of the proposed model. Figure 8 illustrates that after 43 epochs, the training process stopped due to over fitting. At that step, the training achieved the optimal result.

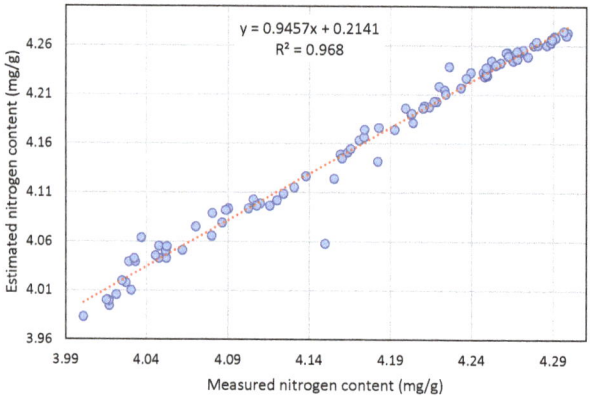

Figure 7. Regression plot of the measured and estimated nitrogen content for category $N_{60\%}$.

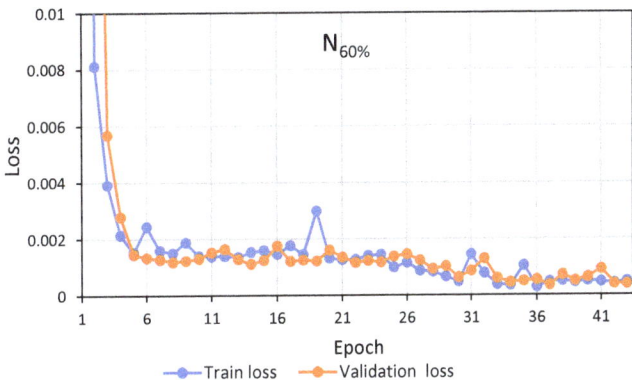

Figure 8. Train loss and validation loss diagram for category $N_{60\%}$.

3.3. Prediction Model for the Category with Excess of Nitrogen by 90%

Regarding the treatment of 90% nitrogen overdose, the statistical criteria for evaluating the performance of the prediction model are given in Table 6, indicating the mean error measures and the variance score and coefficient of determination. Again, the measures achieved are very positive and indicate a great accuracy of the proposed model.

Table 6. Evaluation of the prediction model for the category with excess of nitrogen by 90%. Filtering algorithms: MSC: multiplicative scatter correction; SG: Savitzky–Golay filter. MAE: mean absolute error; MSE: mean squared error; MSLE: mean squared logarithmic error; MedAE: median absolute error; MaxE: max error; R^2: coefficient of determination; Var-score: variance score.

Algorithm	MAE	MSE	MSLE	MedAE	MaxE	R^2	Var-Score
MSC + SG	0.023	0.0007	0.00001	0.022	0.054	0.967	0.983
MSC	0.027	0.0009	0.00004	0.026	0.058	0.953	0.978
SG	0.025	0.0008	0.00002	0.023	0.055	0.961	0.981
No filter	0.028	0.0013	0.00006	0.029	0.061	0.949	0.969

Figure 9 depicts the regression plot of the measured amount of nitrogen and the mean estimated values for 300 test data in this treatment. As indicated in Table 6, the prediction is done with a correlation coefficient of 0.967.

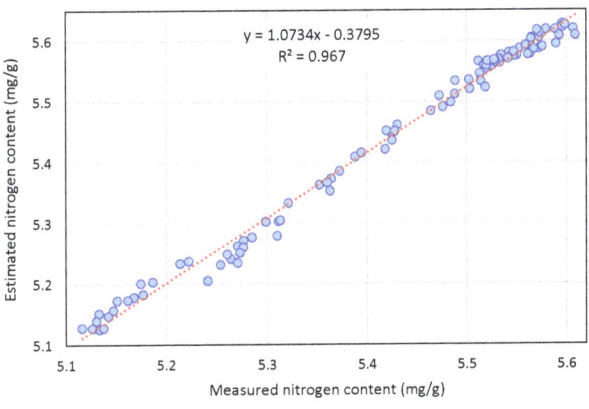

Figure 9. Regression plot of the measured and estimated nitrogen content for category $N_{90\%}$.

Figure 10 shows that after 41 epochs, the training process stopped due to the detection of the over fitting criterion. Both the train and validation loss present a similar behavior. Initially, there is a fast reduction during the first 5–6 epochs of training. Then, there is a more smooth and progressive reduction of both loss values, approximately until epoch 25. Finally, the system stabilizes and the process is stopped before the validation loss begins to increase (which would mean that overfitting is starting).

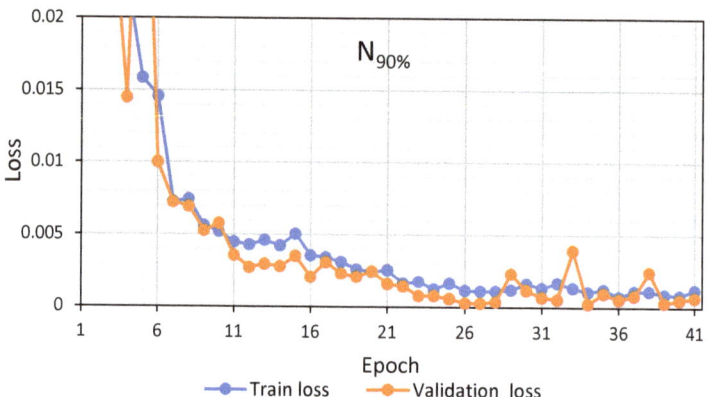

Figure 10. Train loss and validation loss diagram for category $N_{90\%}$.

4. Discussion

In general, the results presented in the previous section prove that the nitrogen content of cucumber leaves can be estimated with a high accuracy using hyperspectral information in the Vis-NIR range and the proposed 1D-CNN regression models. The obtained mean errors are always below 1% with respect to the expected values for all the treatments. For example, in the $N_{30\%}$ class, the mean absolute error (MAE) is only 0.017 mg/g, while the nitrogen content ranges from 3–3.4 mg/g; this would represent a 0.56% relative error. The relative MAE for the $N_{60\%}$ and $N_{90\%}$ classes are 0.35% and 0.45%, respectively. This means that the proposed approach can be effectively used for the estimation of nitrogen in the leaves from the early stages, since the error is very small even for the 30% class.

This positive result can also be argued even for the worst cases of error, i.e., considering the maximum errors, MaxE. This error is 0.086 mg/g for $N_{30\%}$ class and 0.091 mg/g for $N_{60\%}$ class, which represent relative errors below 3%. In the $N_{90\%}$ class, this worst case has a very low error of 0.054 mg/g. Therefore, we can conclude that the method does not present cases of extremely high error. This is evidence of the robustness of the proposed approach against the most difficult cases.

Regarding the preprocessing techniques applied in the paper, MSC and SG, both steps prove to have positive effects on the obtained results. Observe that MSC is a scatter correction filter of the spectra, while SG is a smoothing operator. In all the treatments, the worst results are given by not applying any filter, which is evident for all the performance measures. For example, in the 60% class, the MSE is 0.0003 mg/g with the proposed filters, MSC + SG, while it rises to 0.0012 mg/g with no preprocessing filter. If we should select only one filter, the smoothing step, SG, is the method that offers the best results by itself in terms of the error. For example, in the 90% class, the MAE without any filter is 0.028 mg/g; with SG it descends to 0.025, but with MSG it is 0.027. This indicates that smoothing is more beneficial than scatter correction. This is also repeated for the determination coefficient, R^2, which is consistently higher for SG. Moreover, the improvement from applying MSC + SG with respect to applying only SG is very reduced in most cases, with a small positive effect in the error measures.

Analyzing the evolution of the loss curves in Figures 6, 8 and 10, it can be seen that the proposed 1D-CNN has a very fast convergence for all the treatments. A fast reduction of the loss values is produced in the first five epochs. Then, the system enters in a gradual convergence phase, which is stopped before it could incur in over-fitting. No model required more than 43 epochs, making them relatively fast to be trained. The loss curves also indicate that the training process does not incur in over-fitting, which would appear as a reduction of the train loss and an increment in the validation loss.

Finally, Table 7 presents a comparison of the results achieved by the proposed 1D-CNN models, compared with the results of other researchers for the non-destructive estimation of different properties of fruits, using R^2 criterion. Although these works have been selected among the most similar to our proposed method, this comparison has to be considered in context, since they refer to different types of fruits, different datasets, and distinct types of regression models. We have also included the results of a previous study from our group that deals with the estimation of nitrogen in tomato leaves, using different types of classifiers [35]. Overall, the results of the present study showed that using 1D-CNN based on spectral data, it is possible to estimate the excess nitrogen more accurately, achieving results that are in the state of the art.

Table 7. Comparison the coefficient of determination, R^2, of the proposed 1D-CNN method with other existing methods.

Methods	Product	Method	Prediction	R^2
Proposed method for $N_{30\%}$	Cucumber	1D-CNN	Excess nitrogen	0.962
Proposed method for $N_{60\%}$	Cucumber	1D-CNN	Excess nitrogen	0.968
Proposed method for $N_{90\%}$	Cucumber	1D-CNN	Excess nitrogen	0.967
(Pourdarbani et al. [35])	Tomato	PLSR	Nitrogen content	0.702
(Pourdarbani et al. [35])	Tomato	ANN-DE	Nitrogen content	0.747
(Pourdarbani et al. [35])	Tomato	CNN	Nitrogen content	0.766
(Torkashvand et al. [22])	Kiwifruit	ANN	Nutrient content	0.85
(Huang et al. [36])	Tomato	ANN	SSC	0.81
(Ghosal et al. [37])	Soybean	CNN	K & Fe	0.94

Comparing the obtained results with the accuracy reported in [35], it can be seen that the R^2 are significantly better. These methods include partial least squares regression (PLSR) which is a statistical-based method, and hybrid approach of classical neural networks and the differential evolution algorithm (ANN-DE), and a method similar to the proposed in this paper using convolutional neural networks (CNN). In all of them, a unique model is trained for all the nitrogen categories, producing poor regression results always below 0.8 in R^2. Instead, the approach of creating separate models for each class is able to produce better results above 0.96. In fact, the approach of training different models for the classes was also analyzed in [35], although the R^2 achieved ranged from 0.925 to 0.968.

On the other hand, a weak point of the proposed method is that it requires a previous classification of the leaf samples in the corresponding treatment class. Given a new unknown sample, there are three different models that could be applied on it. Thus, a classifier would be required to obtain the class before applying the regression network. This classification problem has been previously studied in our group, finding that it is possible to achieve it with a classification accuracy above 96% [10]. This would complete a system based on two steps: first, classification of the sample; and then, regression of the nitrogen content using the specific model. The results discussed in this section indicate that this approach is able to produce better results than training a single regression model for all the classes.

5. Conclusions

Scientific management of production and consumption of fertilizers is necessary to achieve sustainable agriculture, and to improve food health by improving the optimal use of them, especially nitrogen-based fertilizers. In this paper, we have presented a new methodology for fast and accurate estimation of nitrogen content in cucumber leaves using spectroscopy analysis and convolutional neural networks (CNN).

Currently, CNNs are one of the most popular methods of machine learning, since they do not require manual feature extraction. Instead, automatic feature extraction makes deep learning models very accurate for computer vision tasks, such as regression models. Therefore, we have presented a new structure of one-dimensional CNN (1D-CNN) to estimate nitrogen content. Different levels of nitrogen fertilizer overdose by 30%, 60% and 90% were added to the cucumbers, and then prediction models were trained for each treatment. The experimental results have shown that the proposed 1D-CNN is able to estimate very accurately the nitrogen content even in the early stages, achieving, for example, for the 30% treatment, a maximum error of 0.086 mg/g and an R^2 of 0.962. For the 60% and 90% classes, the R^2 are 0.968 and 0.967, respectively. The experiments also showed that the applied preprocessing filters, multiplicative scatter correction, Savitzky–Golay, had a positive effect on the accuracy. Thus, this approach can be used to create a fast and non-destructive method to prevent excessive use of fertilizers.

Author Contributions: Conceptualization, R.P., J.L.H.-H. and G.G.-M.; methodology, S.S.; software, M.H.R. and I.H.-M. validation, I.G.-B.; formal analysis, S.S.; investigation, I.H.-M. and G.G.-M.; resources, I.G.-B.; data curation, R.P.; writing—original draft preparation, S.S. and I.H.-M.; writing—review and editing, M.H.R., J.L.H.-H. and G.G.-M.; visualization, M.H.R.; supervision, R.P. and G.G.-M.; project administration, J.L.H.-H.; funding acquisition, I.G.-B. All authors have read and agreed to the published version of the manuscript.

Funding: This study was financially supported by University of Mohaghegh Ardabili, Sharif University of Technology, National Technological of México/Campus Chilpancingo and Autonomous University of Guerrero. It was also funded by the Spanish Ministerio de Ciencia, Innovación y Universidades (MCIU), Ministerio de Ciencia e Innovación (MICINN) and Agencia Estatal de Investigación (AEI); as well as European Commission FEDER funds, under grant RTI2018-098156-B-C53.

Institutional Review Board Statement: Not applicable.

Informed Consent Statement: Not applicable.

Data Availability Statement: The data presented in this study are available upon reasonable request to the corresponding authors.

Conflicts of Interest: The authors declare no conflict of interest. The funders had no role in the design of the study; in the collection, analyses, or interpretation of data; in the writing of the manuscript, or in the decision to publish the results.

References

1. Balafoutis, A.; Beck, B.; Fountas, S.; Vangeyte, J.; Van Der Wal, T.; Soto, I.; Gómez-Barbero, M.; Barnes, A.; Eory, V. Precision Agriculture Technologies Positively Contributing to GHG Emissions Mitigation, Farm Productivity and Economics. *Sustainability* **2017**, *9*, 1339. [CrossRef]
2. Agarwal, M.; Gupta, S.K.; Biswas, K. Development of Efficient CNN model for Tomato crop disease identification. *Sustain. Comput. Inform. Syst.* **2020**, *28*, 100407. [CrossRef]
3. Sabzi, S.; Pourdarbani, R.; Kalantari, D.; Panagopoulos, T. Designing a Fruit Identification Algorithm in Orchard Conditions to Develop Robots Using Video Processing and Majority Voting Based on Hybrid Artificial Neural Network. *Appl. Sci.* **2020**, *10*, 383. [CrossRef]
4. Fatchurrahman, D.; Amodio, M.L.; de Chiara, M.L.V.; Chaudhry, M.M.A.; Colelli, G. Early discrimination of mature-and immature-green tomatoes (Solanum lycopersicum L.) using fluorescence imaging method. *Postharvest Biol. Technol.* **2020**, *169*, 111287. [CrossRef]
5. Martínez-Dalmau, J.; Berbel, J.; Ordóñez-Fernández, R. Nitrogen Fertilization. A Review of the Risks Associated with the Inefficiency of Its Use and Policy Responses. *Sustainability* **2021**, *13*, 5925. [CrossRef]
6. Maynard, D.N.; Barker, A.V.; Minotti, P.L.; Peck, N.H. Nitrate Accumulation in Vegetables. In *Advances in Agronomy*; Brady, N.C., Ed.; Academic Press: Cambridge, CA, USA, 1976; Volume 28, pp. 71–118.
7. Renseigné, N.; Umar, S.; Iqbal, M. Nitrate accumulation in plants, factors affecting the process, and human health implications. A review. *Agron. Sustain. Dev.* **2007**, *27*, 45–57.
8. Dezordi, L.R.; Aquino, L.A.D.; Aquino, R.F.B.D.A.; Clemente, J.M.; Assunção, N.S. Diagnostic methods to assess the nutritional status of the carrot crop. *Rev. Bras. Ciência Solo* **2016**, *40*. [CrossRef]

9. Nandhini, S.; Suganya, R.; Nandhana, K.; Varsha, S.; Deivalakshmi, S.; Thangavel, S.K. Automatic Detection of Leaf Disease Using CNN Algorithm. In *Machine Learning for Predictive Analysis*; Lecture Notes in Networks and Systems; Springer: Singapore, 2021. [CrossRef]
10. Sabzi, S.; Pourdarbani, R.; Rohban, M.; García-Mateos, G.; Paliwal, J.; Molina-Martínez, J. Early Detection of Excess Nitrogen Consumption in Cucumber Plants Using Hyperspectral Imaging Based on Hybrid Neural Networks and the Imperialist Competitive Algorithm. *Agronomy* **2021**, *11*, 575. [CrossRef]
11. Wang, Z.; Zheng, C.; Ma, C.; Ma, B.; Wang, J.; Zhou, B.; Xia, T. Comparative analysis of chemical constituents and antioxidant activity in tea-leaves microbial fermentation of seven tea-derived fungi from ripened Pu-erh tea. *LWT* **2021**, *142*, 111006. [CrossRef]
12. Hubel, D.H.; Wiesel, T.N. Receptive fields, binocular interaction and functional architecture in the cat's visual cortex. *J. Physiol.* **1962**, *160*, 106–154. [CrossRef]
13. Zhu, L.; Spachos, P.; Pensini, E.; Plataniotis, K.N. Deep learning and machine vision for food processing: A survey. *Curr. Res. Food Sci.* **2021**, *4*, 233–249. [CrossRef] [PubMed]
14. Watchareeruetai, U.; Noinongyao, P.; Wattanapaiboonsuk, C.; Khantiviriya, P.; Duangsrisai, S. Identification of Plant Nutrient Deficiencies Using Convolutional Neural Networks. In Proceedings of the 2018 International Electrical Engineering Congress (iEECON), Krabi, Thailand, 7–9 March 2018; pp. 1–4.
15. Espejo-Garcia, B.; Mylonas, N.; Athanasakos, L.; Fountas, S. Improving weeds identification with a repository of agricultural pre-trained deep neural networks. *Comput. Electron. Agric.* **2020**, *175*, 105593. [CrossRef]
16. Sharma, N.; Sharma, R.; Jindal, N. Machine Learning and Deep Learning Applications-A Vision. *Global Transitions Proc.* **2021**, *2*, 24–28. [CrossRef]
17. Liu, Q.; Zhou, F.; Hang, R.; Yuan, X. Bidirectional-Convolutional LSTM Based Spectral-Spatial Feature Learning for Hyperspectral Image Classification. *Remote Sens.* **2017**, *9*, 1330. [CrossRef]
18. Tian, X.; Li, J.; Yi, S.; Jin, G.; Qiu, X.; Li, Y. Nondestructive determining the soluble solids content of citrus using near infrared transmittance technology combined with the variable selection algorithm. *Artif. Intell. Agric.* **2020**, *4*, 48–57. [CrossRef]
19. Cai, H.-T.; Liu, J.; Chen, J.-Y.; Zhou, K.-H.; Pi, J.; Xia, L.-R. Soil nutrient information extraction model based on transfer learning and near infrared spectroscopy. *Alex. Eng. J.* **2021**, *60*, 2741–2746. [CrossRef]
20. Wittwer, S.H.; Honma, S. *Greenhouse Tomatoes, Lettuce and Cucumbers*; Michigan State University Press: East Lansing, MI, USA, 1979.
21. Mehta, D.; Choudhury, T.; Sehgal, S.; Sarkar, T. Fruit Quality Analysis using modern Computer Vision Methodologies. In Proceedings of the 2021 IEEE Madras Section Conference (MASCON), Chennai, India, 27–28 August 2021; pp. 1–6.
22. Torkashvand, A.M.; Ahmadi, A.; Nikravesh, N.L. Prediction of kiwifruit firmness using fruit mineral nutrient concentration by artificial neural network (ANN) and multiple linear regressions (MLR). *J. Integr. Agric.* **2017**, *16*, 1634–1644. [CrossRef]
23. Miller, R.G. *Beyond ANOVA: Basics of Applied Statistics*; Chapman & Hall: Boca Raton, FL, USA, 1997.
24. Duncan, D.B. Multiple Range and Multiple F Tests. *Biometrics* **1955**, *11*, 1–42. [CrossRef]
25. Kjeldahl, J. Neue Methode zur Bestimmung des Stickstoffs in organischen Körpern (New method for the determination of nitrogen in organic substances). *Z. Anal. Chem.* **1883**, *22*, 366–383. [CrossRef]
26. LeCun, Y.; Bengio, Y.; Hinton, G. Deep learning. *Nature* **2015**, *521*, 436–444. [CrossRef] [PubMed]
27. Liu, W.; Wang, Z.; Liu, X.; Zeng, N.; Liu, Y.; Alsaadi, F.E. A survey of deep neural network architectures and their applications. *Neurocomputing* **2016**, *234*, 11–26. [CrossRef]
28. Kristof, W. Estimation of true score and error variance for tests under various equivalence assumptions. *Psychometrika* **1969**, *34*, 489–507. [CrossRef]
29. Douglas, F.; Yang, C.; Malcolm, S. The Maximum-Error Test. *J. Astronaut. Sci.* **2020**, *55*, 259–270.
30. Sabzi, S.; Pourdarbani, R.; Arribas, J.I. A Computer Vision System for the Automatic Classification of Five Varieties of Tree Leaf Images. *Computers* **2020**, *9*, 6. [CrossRef]
31. Pourdarbani, R.; Sabzi, S.; Kalantari, D.; Arribas, J.I. Non-destructive visible and short-wave near-infrared spectroscopic data estimation of various physicochemical properties of Fuji apple (Malus pumila) fruits at different maturation stages. *Chemom. Intell. Lab. Syst.* **2020**, *206*, 104147. [CrossRef]
32. Pourdarbani, R.; Sabzi, S.; Kalantari, D.; Paliwal, J.; Benmouna, B.; García-Mateos, G.; Molina-Martínez, J.M. Estimation of different ripening stages of Fuji apples using image processing and spectroscopy based on the majority voting method. *Comput. Electron. Agric.* **2020**, *176*, 105643. [CrossRef]
33. Wang, W.; Lu, Y. Analysis of the mean absolute error (MAE) and the root mean square error (RMSE) in assessing rounding model. In *IOP Conference Series: Materials Science and Engineering*; IOP Publishing: Bristol, UK, 2018; Volume 324, p. 012049.
34. Jeong, J.H.; Woo, J.H.; Park, J. Machine Learning Methodology for Management of Shipbuilding Master Data. *Int. J. Nav. Arch. Ocean Eng.* **2020**, *12*, 428–439. [CrossRef]
35. Pourdarbani, R.; Sabzi, S.; Rohban, M.; Garcia-Mateos, G.; Arribas, J. Nondestructive nitrogen content estimation in tomato (Solanum lycopersicum L) plant leaves by Vis-NIR hyperspectral imaging and regression data models. *Appl. Opt.* **2021**, *60*, 9560–9569. [CrossRef]
36. Huang, Y.; Lu, R.; Chen, K. Assessment of tomato soluble solids content and pH by spatially-resolved and conventional Vis/NIR spectroscopy. *J. Food Eng.* **2018**, *236*, 19–28. [CrossRef]
37. Ghosal, S.; Blystone, D.; Singh, A.K.; Ganapathysubramanian, B.; Singh, A.; Sarkar, S. An explainable deep machine vision framework for plant stress phenotyping. *Proc. Natl. Acad. Sci. USA* **2018**, *115*, 4613–4618. [CrossRef]

Article

Assessment of the Rice Panicle Initiation by Using NDVI-Based Vegetation Indexes

Joon-Keat Lai and Wen-Shin Lin *

Department of Plant Industry, National Pingtung University of Science and Technology, Pingtung 91201, Taiwan; p10911004@gmail.com
* Correspondence: wslin@mail.npust.edu.tw; Tel.: +886-8-7703202 (ext. 6254)

Abstract: The assessment of rice panicle initiation is crucial for the management of nitrogen fertilizer application that affects yield and quality of grain. The occurrence of panicle initiation could be determined via either green ring, internode-elongation, or a 1–2 mm panicle, and was observed through manual dissection. The quadratic polynomial regression model was used to construct the model of the trend of normalized difference vegetation index-based vegetation indexes (NDVI-based VIs) between pre-tillering and panicle differentiation stages. The slope of the quadratic polynomial regression model tended to be alleviated in the period in which the panicle initiation stage should occur. The results indicated that the trend of the NDVI-based VIs was correlated with panicle initiation. NDVI-based VIs could be a useful indicator to remotely assess panicle initiation.

Keywords: hyperspectral; proximal sensing; panicle initiation; normalized difference vegetation index (NDVI); green ring; internode-elongation

1. Introduction

Rice (*Oryza sativa* L.) is one of the most important staple foods for more than half of the world population. Due to the rapid growing of food demand and limited arable land, improving yield potential to boost up future rice production is an urgent need. Rice yield is known to be increased by the nitrogen topdressing at the panicle initiation (PI) stage that is the beginning of the reproductive stage [1–3]. According to the recommendation of Agricultural Improvement Committee in Taiwan, when the length of the panicle is found to be 2 mm, the nitrogen fertilizer should be applied within two days. Applying a large amount of nitrogen fertilizer before PI would easily cause excessive stem elongation and thus tend to increase lodging risk. On the contrary, applying nitrogen fertilizer after PI is less effective on rice yield improvement. Therefore, accurate determination of the PI stage is crucial for rice production.

The PI is generally considered as the turning point between the vegetative phase and the reproductive phase. When a rice plant has reached maximum tillering, the internodes of the rice stem start elongating and subsequently panicle initiating. The overlapping period between maximum tillering and PI is also termed as vegetative-lag phase. The differences of the rice appearance between the vegetative-lag and PI are obscure to the naked eye. Therefore, they are difficult to be distinguished directly by human observation.

The general methods to assess PI in the farm are identifying the internode-elongation and green-ring [4,5], length of young panicle (1–2 mm) in the cross section of dissected stem [5], and leaf number index/leaf appearance [6]. When 30% of the main culms have panicles 2 mm or longer, it is considered as the panicle differentiation (PD) stage [3] that is late for nitrogen topdressing. Those methods are inconvenient and inefficient for large-scale estimation. Some other convenient, large-scale, and non-destructive approaches that help to monitor the plant growth stages, such as modified-calendar days and heat units, are potential candidates to be used for the precise estimation of PI. The calendar days method is the easiest approach but is not reliable enough as it is largely affected by the

weather variability during the cultivation period [7–9]. Growing degree day (GDD) is an excellent heat unit that has been widely applied in corn production [10] since it was first proposed to describe the timeline of biological development [7]. However, these methods are rudimentary and are not able to distinguish the variability between fields. Nowadays, quantitative assessment, real-time and site-specific management of precision agriculture are the objectives of community.

Spectral remote sensing is another potential approach for the estimation of various variables that are correlated to plant architecture and physiology, providing high-throughput information non-destructively and rapidly for precision agriculture. The arithmetic combinations of vegetation spectral reflectance, which usually termed as vegetation indexes (VIs), became useful indicators for studying plant health and status. The NDVI (normalized difference vegetation index), which is calculated through a normalization procedure [11], seems to be the most popular and long-established VI. The NDVI is sensitive to responses on the green vegetation [12] as it correlates with the biophysical and physiological changes in plants [13–18]. Moreover, the NDVI has been subsequently developed to the other NDVI-based VIs, such as NDRE (normalized differential red-edge), GNDVI (green NDVI), NDSI (normalized difference spectral index), etc. [19–21]. Recent studies put efforts on the quantification of nitrogen contents in plants through remote sensing to optimize the fertilization efficiency and increased the yield [22–26]. In the field of rice research, NDVI is one of the most frequently used vegetation indexes. For example, preceding studies revealed that the NDVI is useful for in the research of rice breeding, nitrogen use efficiency monitoring, and rice yield prediction [27–29]. Furthermore, NDVI has also been used to monitor the rice growth stages, including the panicle development stage [30]. The preceding studies revealed the trend of NDVI changes during the rice growth cycle; however, the evaluation of PI prediction is currently unavailable.

The main objective of this study, therefore, is to investigate the relationship between NDVI and PI occurrence. Meanwhile, a non-destructive and high temporal resolution approach was also expected to be established for PI assessment in this study.

2. Materials and Methods

2.1. Field Experiment Design and Management

Pot experiments were conducted in 2019 at the experimental field (22°38′ N, 120°36′ E) of National Pingtung University of Science and Technology, Pingtung, Taiwan, to examine the canopy reflectance behavior as a function of panicle initiation on rice (*Oryza sativa* L.). A japonica cultivar, Kaohsiung147 (KH147), was planted into 4 groups with 5 replications. Ammonium sulphate was applied at a rate of 150 kg/ha N in each group (20% for basal, 20% for 1st tillering topdressing, 30% for 2nd tillering topdressing, and 30% for panicle initiation topdressing).

2.2. Determination of PI through Dissection

The PI stage was determined through manual dissection and features observation. The entrance of the PI stage was verified via either the green ring (Figure 1A), internode-elongation, or when 1–2 mm panicle was observed (Figure 1B). The ideal timing for nitrogen topdressing is the interval of Figure 1A,B. When the length of the panicle was over 2 mm, it was considered as the panicle differentiation (PD) stage (Figure 1C) that was not included in PI determination. The PI occurrence of each group was determined if more than 50% of dissected samples were verified to PI.

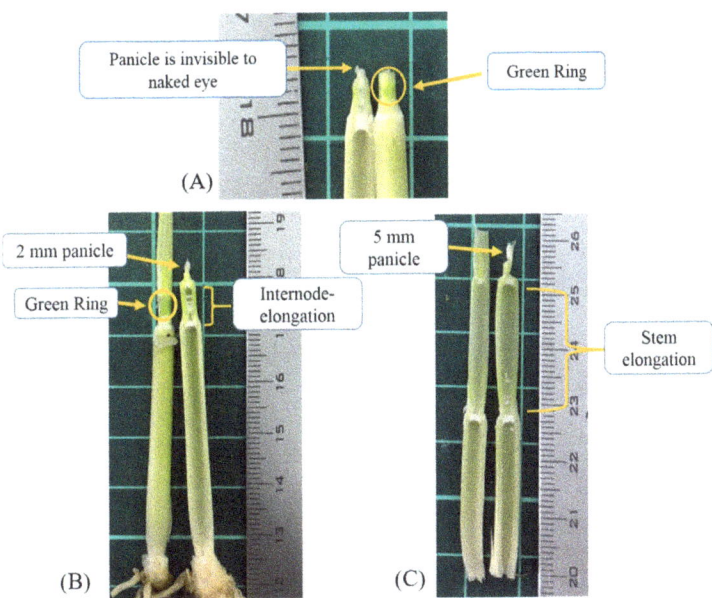

Figure 1. The early stage of panicle initiation (**A**), and the optimum stage for topdressing (**B**). The green ring disappeared, significant stem elongation, panicle above 2 mm, is considered as panicle differentiation stage (**C**).

2.3. Spectral Measurement and NDVI Calculations

The spectral data of the rice canopy was collected by using SpectraPen SP-100 (range 640–1050 nm, 2 nm scan-range, Photon Systems Instruments, Czech), which is a non-imaging, handheld hyperspectral sensor. The data collection period was between 10 DAT (days after transplanting) and 80 DAT. The acquisition time was between 10:00 a.m. and 2:00 p.m., and the sensor was held horizontally at nadir view in the position about 3–5 cm above the highest leaf of the plants. The integration time of spectrum collection was set to auto-sensitivity to minimize the interference of sunlight intensity variability. The collected spectrum data were used to calculate NDVI-based VIs, which can be expressed as follows:

$$\text{NDVI-based VIs} = \frac{\lambda_a - \lambda_b}{\lambda_a + \lambda_b} \tag{1}$$

where the λ_a and λ_b respectively denote the reflectance of near-infrared (NIR) and red wavelengths.

2.4. Reference Wavelengths and Recombined NDVI-Based VIs

The key wavelengths were selected by some physiologically related reference wavelengths with the equation of NDVI that have been used in previous studies (Table 1). Those reference wavelengths were recombined and recalculated as the normalized difference procedure with all samples.

Table 1. Reference wavelengths correlating with plant physiological traits.

Reference Wavelengths	Related Components	References
NDVI (720, 800)	LAI, yield	[17]
NDVI (708, 760)	Photosynthesis (FPAR)	[18]
NDVI (660, 740)	$NDVI_{leaf}$ and $NDVI_{canopy}$	[31]
NDVI (680, 800)	Plant N concentration; Chlorophyll content	[32] [16]
NDVI (670, 780)	Plant N concentration; Plant N uptake	[33]
670, 700, 730 nm	N, P, S-related	[34]

2.5. Estimation of PI Occurrence through First-Order Differentiation

The trends of NDVI-based VIs scatter plots were changed along the rice plant development, and the quadratic slope always gradually decreased after reaching the reproductive phase. We assumed that the NDVI might have reached maxima during PI, and therefore, the slope would be zero (Figure 2).

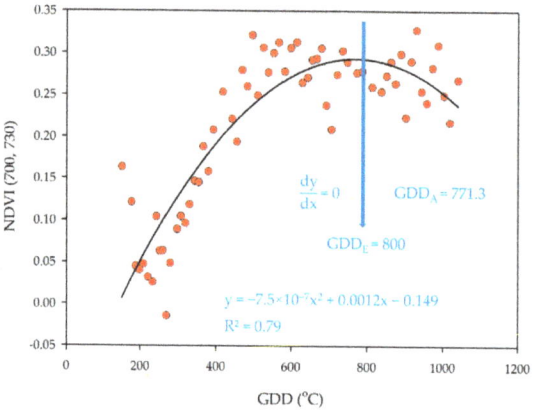

Figure 2. Estimation of PI occurrence. When the quadratic slope is equal to zero, the actual determination of PI was GDD_A = 771.3 through dissection, and the estimation of PI was GDD_E = 800.

2.6. Temperature and Growing Degree-Days (GDD)

Temperature data were obtained from Chishan meteorological station (22°35′ N, 120°36′ E), which is located approximately 7 km away from experimental site. The GDD of "Method 1" as stated by preceding study was selected as a substitution to represent the cultivation timing, according to which [35]:

$$GDD = \sum \left[\frac{T_{max} + T_{min}}{2} - T_{base} \right] \quad (2)$$

where the T_{max} and T_{min} are the daily maximum and minimum air temperature, and the T_{base} is the base temperature that was set to 10 °C [8].

2.7. Statistical Analysis

The statistical analysis of the data was done by using Microsoft Excel 2013 (Microsoft Corporation, Redmond, WA, USA). The quadratic polynomial regression analysis was performed in SigmaPlot 10.0 (Systat Software Inc., San Jose, CA, USA).

The model was validated by leave-one-out (LOO) cross validation. During the LOO process, each group was progressively and alternately held out for model validation, while the remaining groups were used for model construction. The appropriateness of the estimation of PI occurrence and prediction of the NDVI-based VIs was tested based on relative error (RE) values, where:

$$RE = \sqrt{\frac{\sum_{i=1}^{n}[(Ai - Pi)/Ai]^2}{n}} \times 100\% \quad (3)$$

in which the Ai and Pi are the actual and predicted value of the *i*th data point and n is the number of data points. Estimations were considered excellent if RE is <10%, good between 10% and 20%, fair between 20% and 30%, and poor if it is >30% [34,36].

3. Results and Discussion

3.1. PI Determination through Manual Dissection

The PI of each group was observed through manual dissection. The actual determination of PI for group 1 was 60 DAT (GDD = 771.3 °C), group 2 was 56 DAT (GDD = 713.9 °C), group 3 was 57 DAT (GDD = 783.4 °C), and group 4 was 55 DAT (GDD = 791.1 °C) (Table 2). Group 1 encountered low temperature during the tillering stage. As a result, the required DAT of group 1 for PI was five days longer than group 4, which was thoroughly grown under warm conditions. Besides this, group 2 encountered low temperature before the tillering stage. Consequently, the required GDD of group 2 for PI was approximately 80 °C lower than group 4. These biases indicated that the DAT and GDD are both largely affected by weather variability.

Table 2. Observation of PI in each group.

Group	Date	DOY	DAT	GDD (°C)
1	4 March 2019	63	60	771.3
2	14 March 2019	73	56	713.9
3	29 March 2019	88	57	783.4
4	10 April 2019	100	55	791.1

DOY: day of year; DAT: days after transplanting; GDD: Growing degree day.

3.2. NDVI-Based VIs Selection

All samples were regressed with quadratic polynomial model, as we assumed that the trend of this model (when the slope is equal to zero) might be correlated with the entrance of the PI stage. Considering the ability of explanation, NDVI (700, 720), NDVI (700, 730), NDVI (708, 730), NDVI (660, 760), NDVI (700, 760), NDVI (708, 760), and NDVI (708, 800) were selected by the R^2 value that are above 0.7 of quadratic polynomial regression model (Table 3). On the other hand, all the NDVI-based VIs of groups were stratified, where the distribution of groups 1 and 2 were lower than groups 3 and 4 (Figure 3A–G). The reason could be that group 1 encountered low temperature during the tillering stage, while group 2 encountered low temperature before the tillering stage. A preceding study indicated that more uniquely expressed proteins were found at 20/12 °C (day/night) and frequently alternating stress/non-stress temperature changes, leading the rice plant to complex stress conditions [37]. This also indicated that cold stress at an early stage would have an irreversible impact to rice plants, since the NDVI-based VIs of groups 1 and 2 remained lower than that of groups 3 and 4.

Table 3. Recombination and selection of NDVI-based VIs from reference wavelengths. The calculation was $[(\lambda_a - \lambda_b)/(\lambda_a + \lambda_b)]$, and the R^2 values of quadratic polynomial regression models were recorded.

λ_a \ λ_b	660	670	680	700	708	720	730	740	760	780	800
660											
670	0.24										
680	0.31	0.47									
700	0.37	0.33	0.38								
708	0.51	0.45	0.47	0.57							
720	0.62	0.58	0.60	0.71	0.58						
730	0.68	0.64	0.65	0.73	0.74	0.55					
740	0.67	0.64	0.64	0.68	0.69	0.43	0.29				
760	0.71	0.69	0.69	0.73	0.75	0.64	0.68	0.26			
780	0.66	0.64	0.64	0.66	0.66	0.43	0.34	0.41	0.00		
800	0.67	0.65	0.65	0.68	0.70	0.49	0.43	0.37	0.10	0.10	

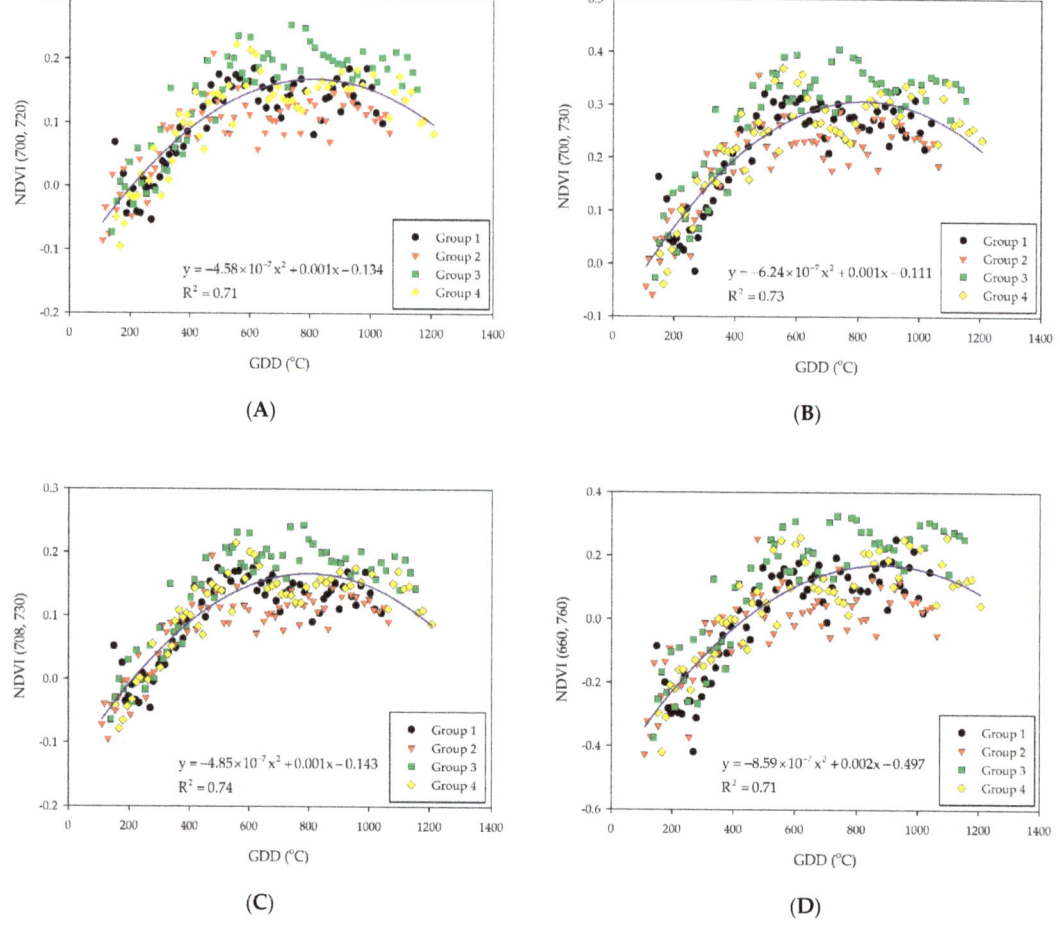

Figure 3. Cont.

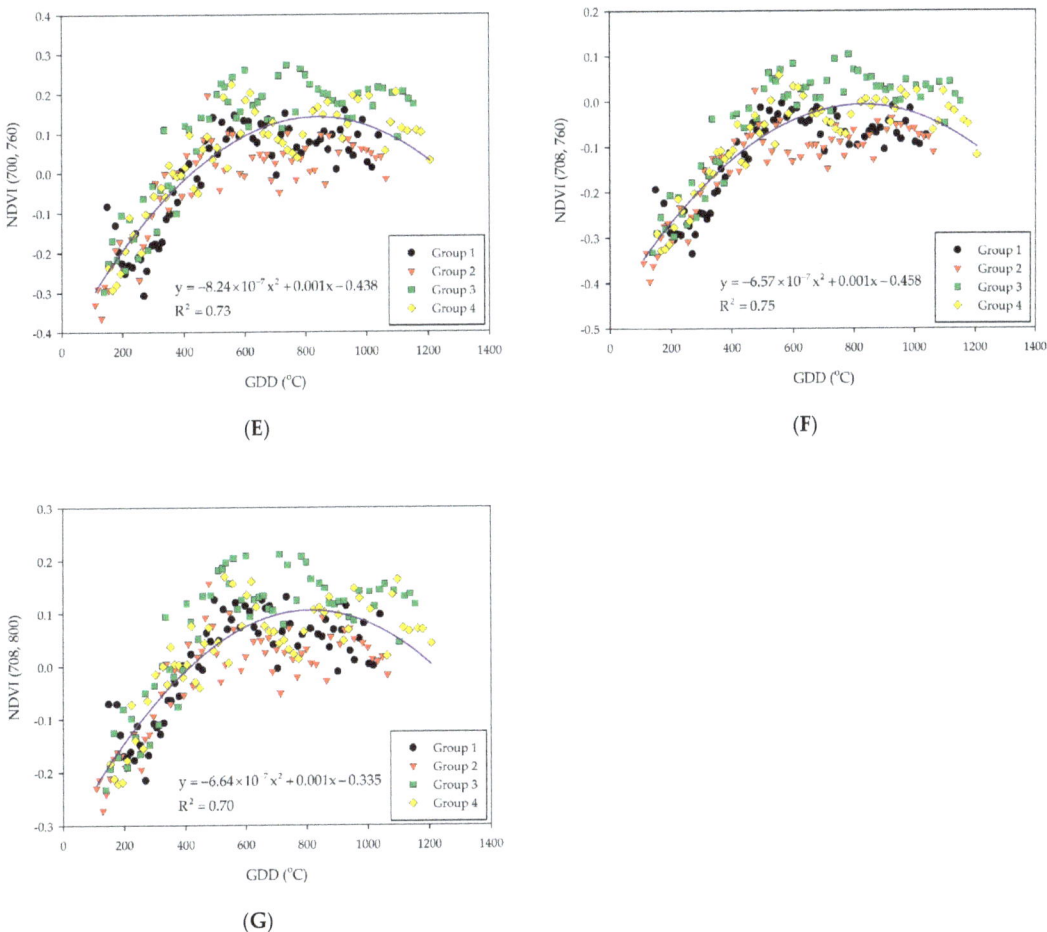

Figure 3. Illustration for the selected NDVI-based VI combination: (**A**) NDVI (700, 720); (**B**) NDVI (700, 730); (**C**) NDVI (708, 730); (**D**) NDVI (660, 760); (**E**) NDVI (700, 760); (**F**) NDVI (708, 760); and (**G**) NDVI (708, 800). The blue solid line indicate the quadratic polynomial regression line. The black, red, green, and yellow symbols indicate the NDVI-based VIs data of each group.

3.3. Estimation of PI Occurrence through First-Order Differentiation

The seven NDVI-based VIs were then tested with first-order differentiation in each group to compare the estimated GDD with the actual GDD of PI occurrence (Table 4). Due to the excellent estimation of GDDs for PI occurrence (RE < 10%), NDVI (700, 720), NDVI (700, 730), NDVI (708, 730), NDVI (700, 760), NDVI (708, 760), and NDVI (708, 800) were considered as the suitable NDVI-based VIs for PI determination. Among these combinations, NDVI (708, 760), which was proposed by Tan et al. (2018), coincidentally had the lowest RE value (RE = 5.63%) [18]. This indicated that the entrance of the reproductive phase might have influenced the photosynthetic capacity of the rice plant, such as a lower rate of crop growth and absorption of photosynthetically active radiation (PAR). The other suitable combinations of NDVI-based VIs have not been researched in previous studies.

Table 4. Seven selected NDVI-based VIs were tested by first-order differentiation in each group.

NDVI–Based VIs	Equation	Estimated GDDs	Actual GDDs	R^2	RE (%)
700, 720	$y = -5.52 \times 10^{-7}x^2 + 0.0008x - 0.161$	724.6	771.3	0.77	5.82
	$y = -4.15 \times 10^{-7}x^2 + 0.0006x - 0.099$	722.9	713.9	0.71	
	$y = -5.40 \times 10^{-7}x^2 + 0.0009x - 0.160$	833.3	783.4	0.85	
	$y = -5.29 \times 10^{-7}x^2 + 0.0009x - 0.171$	850.7	791.1	0.82	
700, 730	$y = -7.50 \times 10^{-7}x^2 + 0.0012x - 0.149$	800	771.3	0.79	6.29
	$y = -6.65 \times 10^{-7}x^2 + 0.0010x - 0.100$	751.9	713.9	0.77	
	$y = -7.44 \times 10^{-7}x^2 + 0.0012x - 0.139$	806.5	783.4	0.84	
	$y = -6.30 \times 10^{-7}x^2 + 0.0011x - 0.125$	873	791.1	0.8	
708, 730	$y = -5.74 \times 10^{-7}x^2 + 0.0009x - 0.162$	784	771.3	0.79	6.23
	$y = -4.45 \times 10^{-7}x^2 + 0.0007x - 0.123$	786.5	713.9	0.78	
	$y = -5.98 \times 10^{-7}x^2 + 0.0010x - 0.171$	836.1	783.4	0.88	
	$y = -5.16 \times 10^{-7}x^2 + 0.0008x - 0.165$	775.2	791.1	0.84	
660, 760	$y = -1.11 \times 10^{-6}x^2 + 0.0018x - 0.608$	810.8	771.3	0.8	10.38
	$y = -9.10 \times 10^{-7}x^2 + 0.0014x - 0.449$	769.2	713.9	0.69	
	$y = -9.93 \times 10^{-7}x^2 + 0.0018x - 0.530$	906.3	783.4	0.84	
	$y = -9.20 \times 10^{-7}x^2 + 0.0016x - 0.522$	869.6	791.1	0.8	
700, 760	$y = -1.03 \times 10^{-6}x^2 + 0.0016x - 0.505$	776.7	771.3	0.79	6.36
	$y = -8.88 \times 10^{-7}x^2 + 0.0013x - 0.435$	732	713.9	0.79	
	$y = -1.00 \times 10^{-6}x^2 + 0.0017x - 0.472$	850	783.4	0.86	
	$y = -8.11 \times 10^{-7}x^2 + 0.0014x - 0.435$	863.1	791.1	0.82	
708, 760	$y = -8.20 \times 10^{-7}x^2 + 0.0012x - 0.504$	731.7	771.3	0.8	5.63
	$y = -6.41 \times 10^{-7}x^2 + 0.0010x - 0.447$	780	713.9	0.82	
	$y = -8.20 \times 10^{-7}x^2 + 0.0013x - 0.491$	792.7	783.4	0.89	
	$y = -6.71 \times 10^{-7}x^2 + 0.0011x - 0.463$	819.7	791.1	0.86	
708, 800	$y = -8.24 \times 10^{-7}x^2 + 0.0012x - 0.373$	728.2	771.3	0.77	5.8
	$y = -7.27 \times 10^{-7}x^2 + 0.0011x - 0.341$	756.5	713.9	0.78	
	$y = -8.48 \times 10^{-7}x^2 + 0.0014x - 0.375$	825.5	783.4	0.82	
	$y = -5.95 \times 10^{-7}x^2 + 0.0010x - 0.320$	840.3	791.1	0.75	

3.4. Leave-One-Out (LOO) Cross-Validation

The leave-one-out cross-validation procedure was subsequently performed to test the accuracy of the quadratic polynomial regression of NDVI (700, 720), NDVI (700, 730), NDVI (708, 730), NDVI (700, 760), NDVI (708, 760), and NDVI (708, 800). The LOO regression line of groups 1, 3, and 4 showed the highest plateau, while groups 1, 2, and 4 the lowest (Figure 4A–F). Due to the great variabilities of the plateau distribution, neither setting up an absolute threshold value nor extended the range for the PI occurrence determination seems to be an ideal method. Although the plateau of regression curve between groups was different, the trend of the slopes was relatively similar. This was in agreement with preceding research [38]; however, the plateau of the curve in this study was different, which was at the PI stage rather than the booting-heading stage. The stability between the trend of the slopes and PI occurrence were tested through first-order differentiation in the leave-one-out cross-validation model. The results showed that NDVI (700, 720) was the most reliable NDVI-based VIs for PI determination since it had the lowest RE value (RE = 4.9%). A previous study represented NDVI application on wheat phenology monitoring and obtained satisfying results that achieved the lowest error of 4.61 days during the jointing stage [39]. Our study performed a slightly better approach with the application of NDVI (700, 720) with a quadratic polynomial regression model, which represented a maximum error of 68.9 GDDs (approximately three days) (Table 5). The others, NDVI (700, 730), NDVI (708, 730), and NDVI (708, 800), are potential candidates that are competent for PI assessment (Table 5). At this point, the relationship between NDVI-based VIs and PI occurrence could be confirmed through the first-order differentiation of quadratic polynomial regression model.

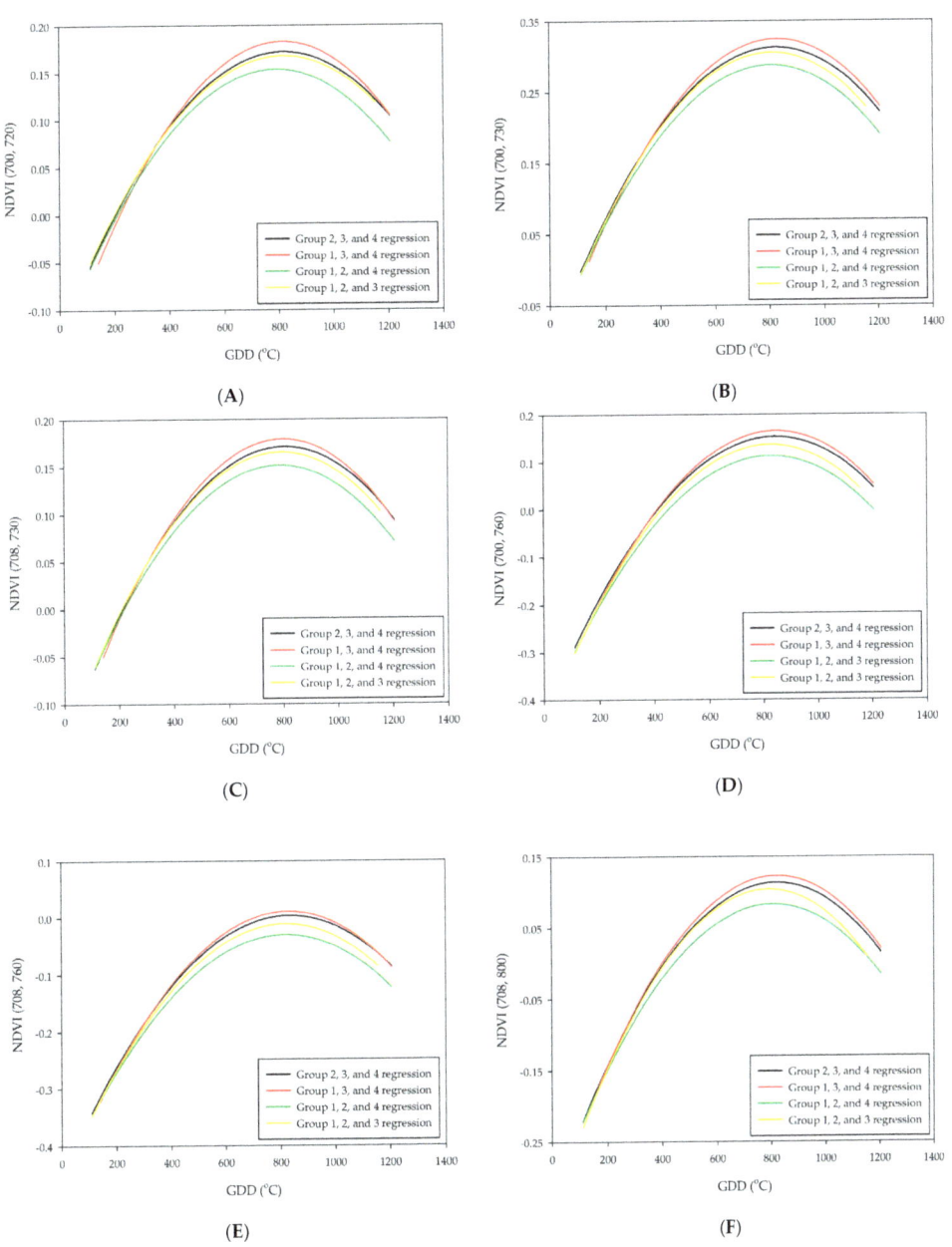

Figure 4. Leave-one-out cross-validation of the six selected PI-related NDVI-based VIs: (**A**) NDVI (700, 720); (**B**) NDVI (700, 730); (**C**) NDVI (708, 730); (**D**) NDVI (700, 760); (**E**) NDVI (708, 760); and (**F**) NDVI (708, 800). The black line is the validation of group 1, the red line is the validation of group 2, the green line is the validation of group 3, and the yellow line is the validation of group 4.

Table 5. Leave-one-out cross-validation of the six selected PI-related NDVI-based VIs by first-order differentiation.

NDVI−Based VIs	Equation	Estimated GDDs	Actual GDDs	R^2	RE (%)
700, 720	$y = -4.51 \times 10^{-7} x^2 + 0.0007x - 0.131$	776.1	771.3	0.7	4.9
	$y = -5.11 \times 10^{-7} x^2 + 0.0008x - 0.157$	782.8	713.9	0.77	
	$y = -4.51 \times 10^{-7} x^2 + 0.0007x - 0.130$	776.1	783.4	0.73	
	$y = -4.37 \times 10^{-7} x^2 + 0.0007x - 0.124$	800.9	791.1	0.68	
700, 730	$y = -6.18 \times 10^{-7} x^2 + 0.001x - 0.107$	809.1	771.3	0.72	9.08
	$y = -6.6 \times 10^{-7} x^2 + 0.0011x - 0.127$	833.3	713.9	0.77	
	$y = -6.08 \times 10^{-7} x^2 + 0.001x - 0.107$	822.4	783.4	0.75	
	$y = -6.40 \times 10^{-7} x^2 + 0.001x - 0.111$	781.3	791.1	0.71	
708, 730	$y = -4.83 \times 10^{-7} x^2 + 0.0008x - 0.143$	828.2	771.3	0.74	5.76
	$y = -5.26 \times 10^{-7} x^2 + 0.0008x - 0.157$	760.5	713.9	0.78	
	$y = -4.64 \times 10^{-7} x^2 + 0.0007x - 0.137$	754.3	783.4	0.77	
	$y = -4.83 \times 10^{-7} x^2 + 0.0008x - 0.139$	828.2	791.1	0.72	
700, 760	$y = -8.19 \times 10^{-7} x^2 + 0.0014x - 0.430$	854.7	771.3	0.73	12.3
	$y = -8.65 \times 10^{-7} x^2 + 0.0015x - 0.453$	867.1	713.9	0.76	
	$y = -7.98 \times 10^{-7} x^2 + 0.0013x - 0.433$	814.5	783.4	0.76	
	$y = -8.55 \times 10^{-7} x^2 + 0.0014x - 0.445$	818.7	791.1	0.71	
708, 760	$y = -6.59 \times 10^{-7} x^2 + 0.0011x - 0.456$	834.6	771.3	0.76	11.03
	$y = -7.00 \times 10^{-7} x^2 + 0.0012x - 0.471$	857.1	713.9	0.77	
	$y = -6.28 \times 10^{-7} x^2 + 0.001x - 0.453$	796.2	783.4	0.78	
	$y = -6.7 \times 10^{-7} x^2 + 0.0011x - 0.461$	820.9	791.1	0.72	
708, 800	$y = -6.61 \times 10^{-7} x^2 + 0.0011x - 0.333$	832.1	771.3	0.7	7.46
	$y = -6.87 \times 10^{-7} x^2 + 0.0011x - 0.342$	800.6	713.9	0.72	
	$y = -6.26 \times 10^{-7} x^2 + 0.0010x - 0.326$	798.7	783.4	0.73	
	$y = -7.17 \times 10^{-7} x^2 + 0.0011x - 0.345$	767.1	791.1	0.69	

4. Conclusions

This study revealed the relationship between rice panicle initiation and NDVI-based VIs by the first-order differentiation of quadratic polynomial regression. The results showed that NDVI (700, 720), NDVI (700, 730), NDVI (708, 730), NDVI (700, 760), NDVI (708, 760), and NDVI (708, 800) are potential candidates to determine PI stage. Although the observed values had great variabilities between groups, the trend is stable—when the population of the target reached their plateau of NDVI-based VIs distribution, PI occurred. Due to the synchronistic correlation between PI and plateau of the trend, the PI estimation might be achieved through determining the increasing proportional of saturated values of NDVI-based VIs. Cooperating with the UAV-mounted multispectral or hyperspectral camera and pixel-based analysis of the region of interest, NDVI-based VIs could be a useful indicator to assess rice plant PI stage, thus optimizing the management of fertilization practice.

Author Contributions: Conceptualization, J.-K.L. and W.-S.L.; Investigation, J.-K.L.; Methodology, J.-K.L.; Resources, W.-S.L.; Writing—original draft preparation, J.-K.L.; Writing—review and editing, J.-K.L. and W.-S.L. All authors have read and agreed to the published version of the manuscript.

Funding: This research was partially funded by the Ministry of Science and Technology (MOST), Taiwan (grant number MOST 110-2637-B-020-005).

Institutional Review Board Statement: Not applicable.

Informed Consent Statement: Not applicable.

Data Availability Statement: Not applicable.

Acknowledgments: The authors wish to thank the collaborators for their help on the study-site maintenance.

Conflicts of Interest: The authors declare no conflict of interest.

References

1. Ghaley, B.B. Uptake and utilization of 5-split nitrogen topdressing in an improved and a traditional rice cultivar in the Bhutan Highlands. *Exp. Agric.* **2012**, *48*, 536–550. [CrossRef]
2. Inamura, T.; Hamada, H.; Iida, K.; Umeda, M. Correlation of the amount of nitrogen accumulated in the aboveground biomass at panicle initiation and nitrogen content of soil with the nitrogen uptake by lowland rice during the period from panicle initiation to heading. *Plant Prod. Sci.* **2003**, *6*, 302–308. [CrossRef]
3. Turner, F.T.; Jund, M.F. Chlorophyll meter to predict nitrogen topdress requirement for semidwarf rice. *Agron. J.* **1991**, *83*, 926–928. [CrossRef]
4. Counce, P.A.; Keisling, T.C.; Mitchell, A.J. A Uniform, Objective, and Adaptive System for Expressing Rice Development. *Crop Sci.* **2000**, *40*, 436–443. [CrossRef]
5. Moldenhauer, K.; Counce, P.; Hardke, J. Rice growth and development. In *Rice Production Handbook*; Hardke, J., Ed.; University of Arkansas Division of Agriculture: Little Rock, AR, USA, 2001; Volume 192, pp. 7–14.
6. Ellis, R.H.; Qi, A.; Summerfield, R.J.; Roberts, E.H. Rates of leaf appearance and panicle development in rice (*Oryza sativa* L.): A comparison at three temperatures. *Agric. For. Meteorol.* **1993**, *66*, 129–138. [CrossRef]
7. Gilmore, E.C.; Rogers, J.S. Heat units as a method of measuring maturity in corn. *Agron. J.* **1958**, *50*, 611–615. [CrossRef]
8. Nielsen, D.; Hinkle, S. Field evaluation of basal crop coefficients for corn based on growing degree days, growth stage, or time. *Trans. ASAE* **1996**, *39*, 97–103. [CrossRef]
9. Snyder, R.L.; Spano, D.; Cesaraccio, C.; Duce, P. Determining degree-day thresholds from field observations. *Int. J. Biometeorol.* **1999**, *42*, 177–182. [CrossRef]
10. Angel, J.R.; Widhalm, M.; Todey, D.; Massey, R.; Biehl, L. The U2U corn growing degree day tool: Tracking corn growth across the US corn belt. *Clim. Risk Manag.* **2017**, *15*, 73–81. [CrossRef]
11. Rouse, J.W.; Haas, R.H.; Schell, J.A.; Deering, D.W. Monitoring vegetation systems in the Great Plains with ERTS. *NASA [Spec. Publ.] SP* **1974**, *351*, 309.
12. Tucker, C.J. Red and photographic infrared linear combinations for monitoring vegetation. *Remote Sens. Environ.* **1979**, *8*, 127–150. [CrossRef]
13. Myneni, R.B.; Williams, D.L. On the relationship between FAPAR and NDVI. *Remote Sens. Environ.* **1994**, *49*, 200–211. [CrossRef]
14. Gamon, J.A.; Field, C.B.; Goulden, M.L.; Griffin, K.L.; Hartley, A.E.; Joel, G.; Penuelas, J.; Valentini, R. Relationships between NDVI, canopy structure, and photosynthesis in three Californian vegetation types. *Ecol. Appl.* **1995**, *5*, 28–41. [CrossRef]
15. Carlson, T.N.; Ripley, D.A. On the relation between NDVI, fractional vegetation cover, and leaf area index. *Remote Sens. Environ.* **1997**, *62*, 241–252. [CrossRef]
16. Sims, D.A.; Gamon, J.A. Relationships between leaf pigment content and spectral reflectance across a wide range of species, leaf structures and developmental stages. *Remote Sens. Environ.* **2002**, *81*, 337–354. [CrossRef]
17. Zhou, X.; Zheng, H.B.; Xu, X.Q.; He, J.Y.; Ge, X.K.; Yao, X.; Cheng, T.; Zhu, Y.; Cao, W.X.; Tian, Y.C. Predicting grain yield in rice using multi-temporal vegetation indices from UAV-based multispectral and digital imagery. *ISPRS J. Photogramm. Remote Sens.* **2017**, *130*, 246–255. [CrossRef]
18. Tan, C.; Wang, D.; Zhou, J.; Du, Y.; Luo, M.; Zhang, Y.; Guo, W. Remotely assessing fraction of photosynthetically active radiation (FPAR) for wheat canopies based on hyperspectral vegetation indexes. *Front. Plant Sci.* **2018**, *9*, 776. [CrossRef] [PubMed]
19. Gitelson, A.A.; Kaufman, Y.J.; Merzlyak, M.N. Use of a green channel in remote sensing of global vegetation from EOS-MODIS. *Remote Sens. Environ.* **1996**, *58*, 289–298. [CrossRef]
20. Tanaka, S.; Kawamura, K.; Maki, M.; Muramoto, Y.; Yoshida, K.; Akiyama, T. Spectral index for quantifying leaf area index of winter wheat by field hyperspectral measurements: A case study in Gifu Prefecture, Central Japan. *Remote Sens.* **2015**, *7*, 5329–5346. [CrossRef]
21. Yang, G.; Liu, J.; Zhao, C.; Li, Z.; Huang, Y.; Yu, H.; Xu, B.; Yang, X.; Zhu, D.; Zhang, X.; et al. Unmanned aerial vehicle remote sensing for field-based crop phenotyping: Current status and perspectives. *Front. Plant Sci.* **2017**, *8*, 1111. [CrossRef]
22. Peng, S.; Buresh, R.J.; Huang, J.; Zhong, X.; Zou, Y.; Yang, J.; Wang, G.; Liu, Y.; Hu, R.; Tang, Q.; et al. Improving nitrogen fertilization in rice by sitespecific N management. A review. *Agron. Sustain. Dev.* **2010**, *30*, 649–656. [CrossRef]
23. Xue, L.; Li, G.; Qin, X.; Yang, L.; Zhang, H. Topdressing nitrogen recommendation for early rice with an active sensor in south China. *Precis. Agric.* **2014**, *15*, 95–110. [CrossRef]
24. Ali, A.M.; Thind, H.S.; Varinderpal, S.; Bijay, S. A framework for refining nitrogen management in dry direct-seeded rice using GreenSeeker™ optical sensor. *Comput. Electron. Agric.* **2015**, *110*, 114–120. [CrossRef]
25. Onoyama, H.; Ryu, C.; Suguri, M.; Iida, M. Nitrogen prediction model of rice plant at panicle initiation stage using ground-based hyperspectral imaging: Growing degree-days integrated model. *Precis. Agric.* **2015**, *16*, 558–570. [CrossRef]
26. Cao, Q.; Miao, Y.; Shen, J.; Yu, W.; Yuan, F.; Cheng, S.; Huang, S.; Wang, H.; Yang, W.; Liu, F. Improving in-season estimation of rice yield potential and responsiveness to topdressing nitrogen application with Crop Circle active crop canopy sensor. *Precis. Agric.* **2016**, *17*, 136–154. [CrossRef]

27. Jiang, R.; Sanchez-Azofeifa, A.; Laakso, K.; Wang, P.; Xu, Y.; Zhou, Z.; Luo, X.; Lan, Y.; Zhao, G.; Chen, X. UAV-based partially sampling system for rapid NDVI mapping in the evaluation of rice nitrogen use efficiency. *J. Cleaner Prod.* **2021**, *289*, 125705. [CrossRef]
28. Phyu, P.; Islam, M.R.; Sta Cruz, P.C.; Collard, B.C.Y.; Kato, Y. Use of NDVI for indirect selection of high yield in tropical rice breeding. *Euphytica* **2020**, *216*, 74. [CrossRef]
29. Son, N.-T.; Chen, C.-F.; Chen, C.-R.; Guo, H.-Y.; Cheng, Y.-S.; Chen, S.-L.; Lin, H.-S.; Chen, S.-H. Machine learning approaches for rice crop yield predictions using time-series satellite data in Taiwan. *Int. J. Remote Sens.* **2020**, *41*, 7868–7888. [CrossRef]
30. Wang, L.; Zhang, F.-C.; Jing, Y.-S.; Jiang, X.-D.; Yang, S.-B.; Han, X.-M. Multi-temporal detection of rice phenological stages using canopy spectrum. *Rice Sci.* **2014**, *21*, 108–115. [CrossRef]
31. Yu, F.; Xu, T.; Cao, Y.; Yang, G.; Du, W.; Wang, S. Models for estimating the leaf NDVI of japonica rice on a canopy scale by combining canopy NDVI and multisource environmental data in Northeast China. *Int. J. Agric. Biol. Eng.* **2016**, *9*, 132–142.
32. Zheng, H.; Cheng, T.; Li, D.; Yao, X.; Tian, Y.; Cao, W.; Zhu, Y. Combining unmanned aerial vehicle (UAV)-based multispectral imagery and ground-based hyperspectral data for plant nitrogen concentration estimation in rice. *Front. Plant Sci.* **2018**, *9*, 936. [CrossRef]
33. Lu, J.; Miao, Y.; Shi, W.; Li, J.; Yuan, F. Evaluating different approaches to non-destructive nitrogen status diagnosis of rice using portable RapidSCAN active canopy sensor. *Sci. Rep.* **2017**, *7*, 14073. [CrossRef] [PubMed]
34. Mahajan, G.R.; Pandey, R.N.; Sahoo, R.N.; Gupta, V.K.; Datta, S.C.; Kumar, D. Monitoring nitrogen, phosphorus and sulphur in hybrid rice (*Oryza sativa* L.) using hyperspectral remote sensing. *Precis. Agric.* **2017**, *18*, 736–761. [CrossRef]
35. McMaster, G.S.; Wilhelm, W.W. Growing degree-days: One equation, two interpretations. *Agric. For. Meteorol.* **1997**, *87*, 291–300. [CrossRef]
36. Jamieson, P.D.; Porter, J.R.; Wilson, D.R. A test of the computer simulation model ARCWHEAT1 on wheat crops grown in New Zealand. *Field Crops Res.* **1991**, *27*, 337–350. [CrossRef]
37. Gammulla, C.G.; Pascovici, D.; Atwell, B.J.; Haynes, P.A. Differential proteomic response of rice (*Oryza sativa*) leaves exposed to high-and low-temperature stress. *Proteomics* **2011**, *11*, 2839–2850. [CrossRef]
38. Liu, X.; Ferguson, R.B.; Zheng, H.; Cao, Q.; Tian, Y.; Cao, W.; Zhu, Y. Using an active-optical sensor to develop an optimal NDVI dynamic model for high-yield rice production (Yangtze, China). *Sensors* **2017**, *17*, 672. [CrossRef] [PubMed]
39. Zhou, M.; Ma, X.; Wang, K.; Cheng, T.; Tian, Y.; Wang, J.; Zhu, Y.; Hu, Y.; Niu, Q.; Gui, L.; et al. Detection of phenology using an improved shape model on time-series vegetation index in wheat. *Comput. Electron. Agric.* **2020**, *173*, 105398. [CrossRef]

Review

Anomaly-Based Intrusion Detection Systems in IoT Using Deep Learning: A Systematic Literature Review

Muaadh A. Alsoufi [1,*], Shukor Razak [1,*], Maheyzah Md Siraj [1], Ibtehal Nafea [2], Fuad A. Ghaleb [1,*], Faisal Saeed [2,3] and Maged Nasser [1]

1. School of Computing, Faculty of Engineering, Universiti Teknologi Malaysia (UTM), Skudai 81310, Johor, Malaysia; maheyzah@utm.my (M.M.S.); msnmaged2@live.utm.my (M.N.)
2. College of Computer Science and Engineering, Taibah University, Medina 41477, Saudi Arabia; inafea@taibahu.edu.sa (I.N.); fsaeed@taibahu.edu.sa (F.S.)
3. Institute for Artificial Intelligence and Big Data, Universiti Malaysia Kelantan, City Campus, Pengkalan Chepa, Kota Bharu 16100, Kelantan, Malaysia
* Correspondence: muaadh.soufi2021@gmail.com (M.A.A.); shukorar@utm.my (S.R.); abdulgaleel@utm.my (F.A.G.)

Citation: Alsoufi, M.A.; Razak, S.; Siraj, M.M.; Nafea, I.; Ghaleb, F.A.; Saeed, F.; Nasser, M. Anomaly-Based Intrusion Detection Systems in IoT Using Deep Learning: A Systematic Literature Review. *Appl. Sci.* **2021**, *11*, 8383. https://doi.org/10.3390/app11188383

Academic Editors: Dimitrios S. Paraforos and Anselme Muzirafuti

Received: 13 August 2021
Accepted: 7 September 2021
Published: 9 September 2021

Publisher's Note: MDPI stays neutral with regard to jurisdictional claims in published maps and institutional affiliations.

Copyright: © 2021 by the authors. Licensee MDPI, Basel, Switzerland. This article is an open access article distributed under the terms and conditions of the Creative Commons Attribution (CC BY) license (https://creativecommons.org/licenses/by/4.0/).

Abstract: The Internet of Things (IoT) concept has emerged to improve people's lives by providing a wide range of smart and connected devices and applications in several domains, such as green IoT-based agriculture, smart farming, smart homes, smart transportation, smart health, smart grid, smart cities, and smart environment. However, IoT devices are at risk of cyber attacks. The use of deep learning techniques has been adequately adopted by researchers as a solution in securing the IoT environment. Deep learning has also successfully been implemented in various fields, proving its superiority in tackling intrusion detection attacks. Due to the limitation of signature-based detection for unknown attacks, the anomaly-based Intrusion Detection System (IDS) gains advantages to detect zero-day attacks. In this paper, a systematic literature review (SLR) is presented to analyze the existing published literature regarding anomaly-based intrusion detection, using deep learning techniques in securing IoT environments. Data from the published studies were retrieved from five databases (IEEE Xplore, Scopus, Web of Science, Science Direct, and MDPI). Out of 2116 identified records, 26 relevant studies were selected to answer the research questions. This review has explored seven deep learning techniques practiced in IoT security, and the results showed their effectiveness in dealing with security challenges in the IoT ecosystem. It is also found that supervised deep learning techniques offer better performance, compared to unsupervised and semi-supervised learning. This analysis provides an insight into how the use of data types and learning methods will affect the performance of deep learning techniques for further contribution to enhancing a novel model for anomaly intrusion detection and prediction.

Keywords: systematic literature review; anomaly intrusion detection; deep learning; IoT; resource constraint; IDS

1. Introduction

Internet of Things (IoT) is the research and industrial trend in the arena of Information Communications Technology (ICT) that has become accustomed to being part of technology advancement in our everyday life [1]. The IoT term refers to a new communication paradigm in which devices have sensors and actuators that can serve as objects or 'things' to sense their surrounding environment, communicate with one another, and exchange data through the internet [2]. The IoT requires a platform in which all the applications, products, and services are associated with, used to capture, communicate, store, access, and share/transmit data from the real world [3,4]. Nowadays, there are around 50 billion IoT devices connected to the internet, and it is expected to grow to an enormous size over the next few years [5,6]. These huge numbers of devices produce a tremendous amount of data that can be used by many applications. IoT applications scenarios are ubiquitous, and this includes, food, agriculture, smart farming, demotics, assisted living, e-health, and enhanced learning, to mention a few examples of possible IoT applications. For instance, there will be 15.3 billion IoT devices for smart agriculture by the end of 2025 [7,8]. A huge number of sensors and actuators are needed for real-time monitoring and environment of many industrial domains to provide actionable insights and make timely decisions [9]. However, many challenges hinder the full adoption of the IoT in both research and industry. These challenges include, but are not limited to, security and trust, reliability, scalability, and mobility, among many others [10].

Because IoT devices are connected to the global internet with unmatured and vulnerable communication protocols and applications, it is exposed to many potential security threats [3,4]. Adversaries may exploit these vulnerabilities and inject anomalies that trigger the system to make wrong control decisions in IoT-based application, causing a catastrophic impact on people's live, properties, and economics [7,11]. The evolved threats of cyberattacks pose significant challenges to the IoT ecosystems. Moreover, IoT devices use different platforms and a combination of network connections protocols such as Ethernet, Wi-Fi, ZigBee, and wire-based technologies to increase their connectivity, which needs coordination between different standards and protocols to mitigate security risks. Besides the diverse technologies used by the IoT industry, the heterogeneity, and the distributed nature of IoT applications increase the complexity of IoT networks and thus, magnify the security risk. These shortcomings cause the IoT network to be exposed to many security issues and cyberattacks. Therefore, an accurate anomaly-detection IDS model is vital for IoT applications [12].

Many IDS solutions have been proposed to protect IoT devices from being exposed to cyber criminals [13–15]. These security solutions can be divided into either proactive or reactive measures. The proactive measures can be effective for protecting the IoT against external threats. However, due to the connectivity of the IoT to the global internet, the risk posed by intruders that can circumvent proactive measures is high. Intrusion Detection Systems (IDSs) work as a second line of defense that can impede many cyberattacks. IDS solutions have received intensive attention from researchers and industries in the IoT field, and many IDS solutions have been proposed [16–18]. Based on the detection approach, IDS solutions can be categorized into three approaches: signature, anomaly, and hybrid IDS model. In general, the signature-based approach is effective for known attacks, while the anomaly-based is effective for unknown attacks. However, due to the heterogeneity, dynamicity, and complex nature of the IoT network, the signature-based approach is inefficient and ineffective for IoT because it requires continuous human interventions and knowledge expertise to extract attack patterns and signatures to update the IDS model [19,20]. Anomaly-based IDS detection gains advantages in IoT because it detects zero-day attacks and needs fewer human interventions [20]. The hybrid approach combines both signatures-based and anomaly-based approaches. However, because it is impractical to rely on pre-defined attack patterns (signature-based) intrusion detection in IoT, the utilization of the signature-based IDS is limited in IoT networks [18–20]. To this end, anomaly intrusion detection systems play a vital role in intrusion detection in IoT environments.

Most of the existing IDS use conventional machine learning techniques to develop detection models [21]. Machine learning techniques were widely adopted to construct the IDS model. However, due to the speed and volume of the IoT-generated data, conventional machine learning techniques that need well-crafted features engineering need intensive research efforts to extract the representative features from big and unstructured data generated by IoT devices. Thus, conventional machine learning-based solutions still encounter many challenges. Recently, deep learning techniques (DL) have been widely adopted for intrusion detection systems. DL expedites the analysis between fast and real data streams in extracting relevant information to predict the future of the IoT domain. DL is known to be more reliable than traditional learning because it can easily extract information, and hence, provides better accuracy [22]. Due to this, several studies have been focused on using deep learning techniques to provide new solutions tackling two different perspectives of both technical and regulatory, such as anomaly and malware detection; however, the results are still unconvincing. Furthermore, most IDS solutions have been adopted from existing computer networks, wireless sensors networks, and mobile ad hoc networks. Yet, the unique characteristics of IoT-based networks, such as connectivity to the global internet and lightweight resources, make the IDS proposed for these networks not suitable to IoT applications [13,14]. There are only a few surveys that have been found that focus specifically on DL techniques in the IoT domain [23]. To the best of the authors' knowledge, there is no review that is dedicated to investigating the effectiveness of the deep learning-based IDS solutions in the IoT security domain. Therefore, this paper was conducted to bridge this gap and investigate the most effective and efficient use of DL approaches in securing the IoT environment. This review provides an in-depth, focused, and high-quality analysis to orient future research toward finding robust anomaly-based IDS using DL techniques.

The paper is organized as follows. The contributions introduced by this study are briefed in Section 2. Related work is presented in Section 3. The review method, which includes the review protocol, planning, research questions, is described in Section 4. Section 5 presents the search strategies, which include the primary records selection, secondary records selection, inclusion criteria, exclusion criteria, quality assessment (QA), data extraction, and synthesis. Section 6 presents the results, studies selection and quality assessment results, and overview of publication sources. Section 7 presents the outcomes, which include the answers to the research questions, taxonomy, analysis and discussion, and the open issues. Section 8 presents the discussion. Section 9 presents the future direction. Limitations of the study are illustrated in Section 10, and the study is concluded in Section 11.

2. Contributions

1. This study systemically explores the existing techniques on an anomaly-based intrusion detection system that uses the DL techniques in IoT.
2. A general taxonomy is proposed for the different deep learning techniques used for constructing the anomaly-based IDS in IoT.
3. An analysis of the state-of-art DL-based techniques of anomaly-based intrusion detection systems in IoT, which use DL, is introduced in this survey.
4. This study discusses the challenges and future direction of DL-based anomaly detection in the IoT domain.

3. Background and Related Works

Existing deep learning studies related to IoT security focus primarily on experimental aspects rather than the adopted techniques, leaving a gap for a comprehensive review of different anomaly intrusion detection. For such reason, the goals are to identify what is the most prominently used techniques and how to ensure better performances for each technique. Due to the rapid growth of advancement in this area, the relevant studies should be reviewed and appraised in parallel.

Hajiheidari et al. [19] conducted comprehensive work on intrusion detection systems in the IoT that focuses on four different types of IDS (anomaly-based, signature-based, specification-based, and hybrid-based). However, the scope of work was broad and unspecific on the anomaly intrusion detection system, which used DL techniques. On the other hand, Sharma et al. [23] surveyed studies that use DL for anomaly detection in IoT. Likewise, Fahim and Sillitti [20] conducted a general study on anomaly detection, analysis, and prediction techniques in an IoT environment. However, this study was not specific to the IDS. Alsoufi and Razak [24] in our previous work, surveyed the anomaly intrusion detection system in IoT, which used DL techniques. The finding of our work inspired us to propose this work, which is an in-depth systematic literature review following the guideline based on proposed work by Kitchenham to provide researchers and developers in-depth information and obtain details about an up-to-date technique and methodology in anomaly intrusion detection in IoT, using deep learning [25].

Table 1 shows a detailed comparison with the similar reviewed articles in the area. Consequently, there is an urgent need to conduct a systemic review and appraise the specific studies in the field of IDS in IoT that used DL techniques. Thus, this systemic review provides an in-depth and focused analysis on orienting future research toward finding robust anomaly-based IDS using DL techniques.

Table 1. Comparison with other similar review articles in the area: (√: Yes, x: No).

Paper Name	Year	IoT	Systematic Study	Anomaly-Based	Deep Learning
Fahim et al. [20]	2019	√	√	√	x
Hajiheidari et al. [19]	2019	√	√	x	x
Sharma et al. [23]	2019	√	x	√	√
Alsoufi, Razak [24]	2021	√	x	√	√
This work		√	√	√	√

4. Review Method

4.1. Development of the Protocol

This review follows the guidelines of performing systematic reviews in the software engineering domain, according to [25,26] as well as other methods from several works [19,27,28].

4.2. Planning the Review

In the planning step, the need for SLR was determined, the research questions were identified, and the review protocol was established.

4.3. The Need for a Systematic Review

There are many approaches applied in detecting intrusion attacks in IoT, using deep learning. However, there is a lack of an in-depth and systematic analysis of those studies. Such an analysis is crucial for the research community, especially for those who are new to the area, to gain a holistic idea of the state of the art of anomaly detection in IoT, using deep learning techniques. Hence, this study focuses on literature reviews of various methods adopted for anomaly-based intrusion detection and inclusive of those researchers that have conducted overview literature on different techniques, taxonomies, and comparisons. This survey presents an in-depth discussion from different perspectives in adherence to the highlighted research questions.

4.4. Research Questions

Q1 What is the comprehensive taxonomy of anomaly-based intrusion detection in IoT using deep learning techniques?
Q2 What is the performance of anomaly-based intrusion detection in IoT using deep learning techniques?
Q3 What are the challenges in the existing anomaly intrusion detection deep learning techniques in IoT?

4.5. The Review Protocol

The review protocol is known as one of the most crucial steps in establishing systematic literature reviews (SLRs). It provides an extensive guideline to determine the suitable and formal methods to be discussed in the SLR. The goal of adapting review protocols is to ensure that there is no bias and to distinguish SLR from any other traditional methods of the literature review 23. This review protocol defines the review background, search strategy, development of RQs, extraction of data, criteria for study selection, and data syntheses. The research questions and background were discussed in previous sections.

The next sections provide insights on different components. All stages of conducting this systemic review are described in Figure 1.

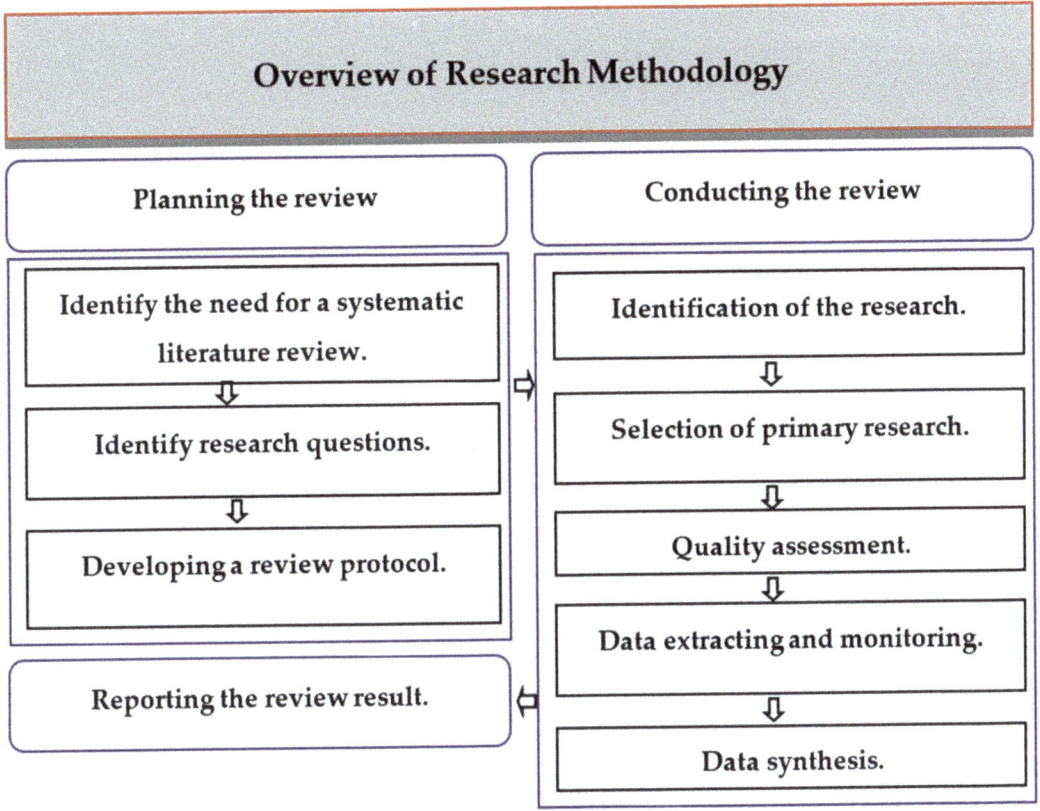

Figure 1. Literature review methodology.

5. Search Strategy

This SLR used automatic search to explore and retrieve the related scholarly publications from online databases (IEEE Explorer, Web of Science, Scopus, Science Direct, and MDPI), using specific keywords that were constructed in response to the research questions. "Anomaly intrusion detection" AND "Internet of things", "Anomaly intrusion detection" AND "Deep learning", "Anomaly intrusion detection system" AND "Internet of things", "Anomaly intrusion detection system" AND "Deep learning", "Anomaly-based" AND "Internet of things", "Anomaly-based" AND "Deep learning". The time frame was from any time up to 2020, while no filters were applied for countries, type of publications, or language during the retrieval of primary records from the online databases. The retrieval of primary records from the pre-specified online databases involved two independent investigators. If discrepancies occurred, a third investigator was consulted. For manual search, reference lists of published reviews and surveys were looked through, while the Google Scholar search engine was used to distinguish all studies that were cited by the chosen primary studies. The manual search was managed to ensure a comprehensive search of the pertinent studies. Any overlapping and redundancies in these publications were removed permanently.

5.1. Primary Records Selection

After the removal of duplicates, the remaining primary records were screened by titles and abstracts to exclude books, conferences, reports, lecture notes, and miscellany. This restricts selection to the original articles published in good-quality journals. The primary selection involved two independent investigators. If discrepancies occurred, a third investigator was consulted.

5.2. Secondary Records Selection

All the primary selected articles underwent secondary selection by applying eligibility criteria (exclusion and inclusion criteria), which were constructed in response to the research questions. Exclusion and inclusion criteria were employed to ensure the inclusion of only pertinent studies for data analysis regarding anomaly intrusion detection in IoT using deep learning.

5.3. Inclusion Criteria

1. Publication of articles in peer-reviewed journals.
2. Accessible research articles.
3. Relevant content to anomaly intrusion detection system in IoT, using deep learning.

5.4. Exclusion Criteria

1. Research articles published in predatory journals according to Beals' list.
2. Inaccessible articles.
3. Irrelevant to anomaly intrusion detection system in IoT using deep learning.

5.5. Quality Assessment (QA) of the Eligible Included Records

For pooling reliable data from the eligible studies, secondary selected records underwent assessment for their quality. Based on [25], a necessary step to be followed through to evaluate the quality of assorted studies was carried out by applying a quality assessment (QA). For evaluation purposes, a set of four research questions (RQs) were taken into consideration, including the following QA criteria:

1. QA1: Is the topic related to anomaly intrusion detection in IoT using deep learning techniques?
2. QA2: Is the research methodology adequately interpreted in the manuscript?
3. QA3: Is there an adequate clarification on the background review in which the study was conducted?
4. QA4: Is there a comprehensible declaration regarding the research objectives?

The reliability of each 42 research articles was assessed, according to each criterion mentioned in the four QA. There are three phases of QA quality schema, which are high, medium, and low [29]. The quality of each paper was assessed, based on its loading score. For a better context, papers that fulfill the criteria receive a score of two, whilst papers that only fulfill the criteria partially receive a score of one, and papers that did not fulfill any of the criteria receive a score of zero. In a scoring board, based on the four defined criteria, studies that receive a score of five or above can be categorized as high quality.

In contrast, studies that receive a score of four can be grouped as medium quality. Studies that receive a score below four will fall under the category of low quality. The studies that scored five and above after QA were then included in data extraction and synthesis. Two independent investigators reviewed the assessment of the quality of eligible studies. A discussion with a third investigator solved any discrepancies.

5.6. Data Extraction and Synthesis of the Systemic Literature Review

The data were extracted from the related studies that underwent the assessment for their quality. A form for better data extraction was created and performed thoroughly by using Endnote and Microsoft Excel spreadsheets to analyze and extract significant information from each eligible study. The extracted data included study ID, first author, publication date, methodology, technique-based taxonomy, datasets, accuracy, precision, recall, False Alarm Rate (FAR), F1-Measure, False Positive Rate (FPR), and False Negative Rate (FNR). Extraction of the data from studies was performed by two independent investigators. Any discrepancies were solved by a discussion with a third investigator.

The data extracted were then synthesized for digressive analysis concerning issues associated with anomaly detection in IoT using deep learning, which includes strengths/weaknesses, classification, and approaches.

6. Results
6.1. Studies Selection and Quality Assessment

A total of 2116 records were extracted from the online database (n = 2106) and extra sources (n = 10); after the removal of duplicates (n = 765), 1351 records were subject to primary selection, out of which 714 records were excluded (books, lectures note, conferences and miscellaneous). Accordingly, 637 records were identified as journal articles, out of which 97 records were excluded (reviews, surveys, and reports). Finally, 540 records were subjected to inclusion and exclusion criteria, out of which 43 studies were eligible. However, only 26 studies met the criteria of quality assessment. The 26 studies that fulfill the assessment criteria were selected to extract the data and synthesis of the systemic literature review. Table 2 shows the number of retrieved records from online databases according to the pre-specified keywords. Figure 2 shows the fellow chart of selection studies.

Table 2. Number of retrieved records from online databases according to the pre-specified keywords.

Database Name	Keywords	Records	Total
IEEE explore	"Anomaly intrusion detection" AND "Internet of things"	113	1263
	"Anomaly intrusion detection" AND "Deep learning"	109	
	"Anomaly intrusion detection system" AND "Internet of things"	96	
	"Anomaly intrusion detection system" AND "Deep learning"	96	
	"Anomaly-based" AND "Internet of things"	411	
	"Anomaly-based" AND "Deep learning"	442	
Science direct	"Anomaly intrusion detection" AND "Internet of things"	6	344
	"Anomaly intrusion detection" AND "Deep learning"	4	
	"Anomaly intrusion detection system" AND "Deep learning"	1	
	"Anomaly-based" AND "Internet of things"	188	
	"Anomaly-based" AND "Deep learning."	144	
Scopus	"Anomaly intrusion detection" AND "Internet of things"	4	138
	"Anomaly intrusion detection" AND "Deep learning"	12	
		2	
	"Anomaly intrusion detection system" AND "Deep learning"	4	
	"Anomaly-based" AND "Internet of things"	69	
	"Anomaly-based" AND "Deep learning"	47	
Web of science	"Anomaly intrusion detection" AND "Internet of things"	3	71
	"Anomaly intrusion detection" AND "Deep learning"	6	
	"Anomaly intrusion detection system" AND "Deep learning"	2	
		2	
	"Anomaly-based" AND "Internet of things"	36	
	"Anomaly-based" AND "Deep learning"	22	
MDPI	"Anomaly intrusion detection" AND "Internet of things"	40	290
	"Anomaly intrusion detection" AND "Deep learning"	39	
	"Anomaly intrusion detection system" AND "Internet of things"	20	
	"Anomaly intrusion detection system" AND "Deep learning"	20	
	"Anomaly-based" AND "Internet of things"	90	
	"Anomaly-based" AND "Deep learning"	81	
Other sources	"Anomaly intrusion detection" AND "Internet of things"	2	10
	"Anomaly intrusion detection" AND "Deep learning"	1	
	"Anomaly intrusion detection system" AND "Internet of things"	2	
	"Anomaly intrusion detection system" AND "Deep learning"	1	
	"Anomaly-based" AND "Internet of things"	2	
	"Anomaly-based" AND "Deep learning"	2	

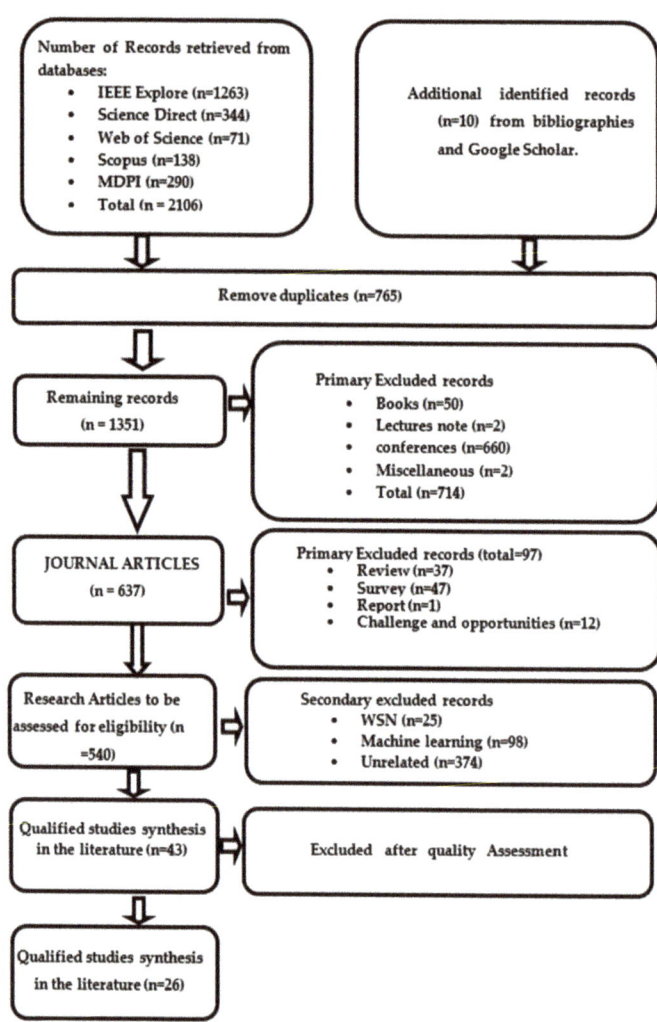

Figure 2. The fellow chart of selection studies.

6.2. Overview of Publication Sources

Figures 3 and 4 illustrate the list of selected papers published according to year and journal. Noticeably, there is a trend toward anomaly-based intrusion detection in IoT, using deep learning. This signifies a rising interest in this domain, especially after 2018. An elevated increase of nine studies in 2020 was noted, compared to only five studies in 2019. In comparison, the trend seems to start in 2017, as there was only one study published.

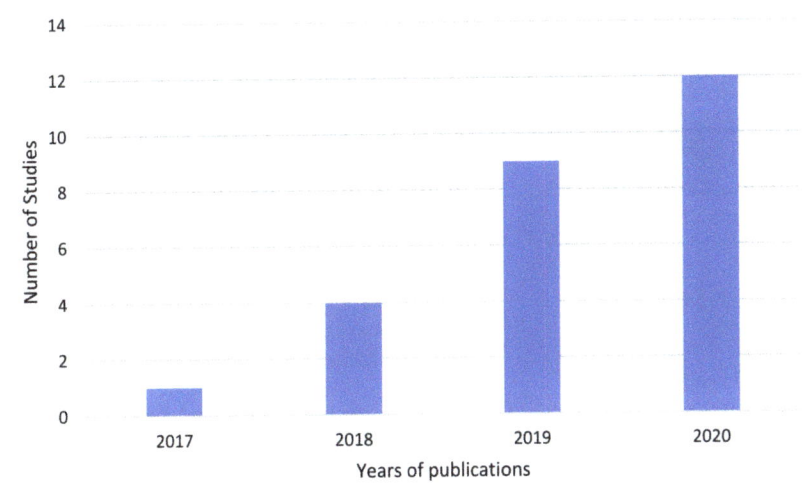

Figure 3. Selected distribution studies by years.

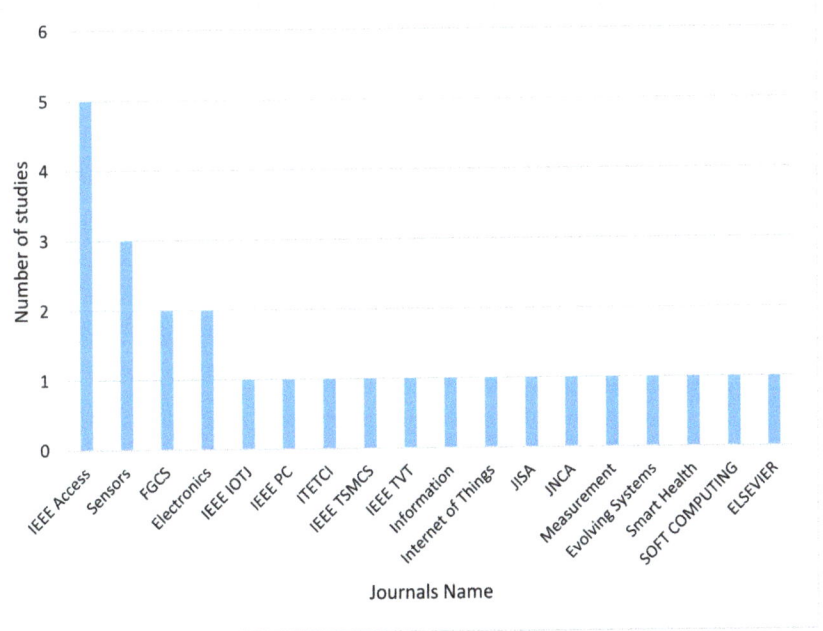

Figure 4. Selected distribution studies by journals.

7. Outcomes

7.1. RQ1: What Is the Comprehensive Taxonomy of Anomaly Intrusion Detection in IoT Using Deep Learning Techniques?

Recently, various studies have explored the application of anomaly detection in IoT using deep learning. For better insight, taxonomy is shown in Figure 5 to pinpoint all existing techniques and requirements of anomaly intrusion detection in IoT, using deep learning techniques. The IDS are commonly categorized as supervised, unsupervised, and semi-supervised.

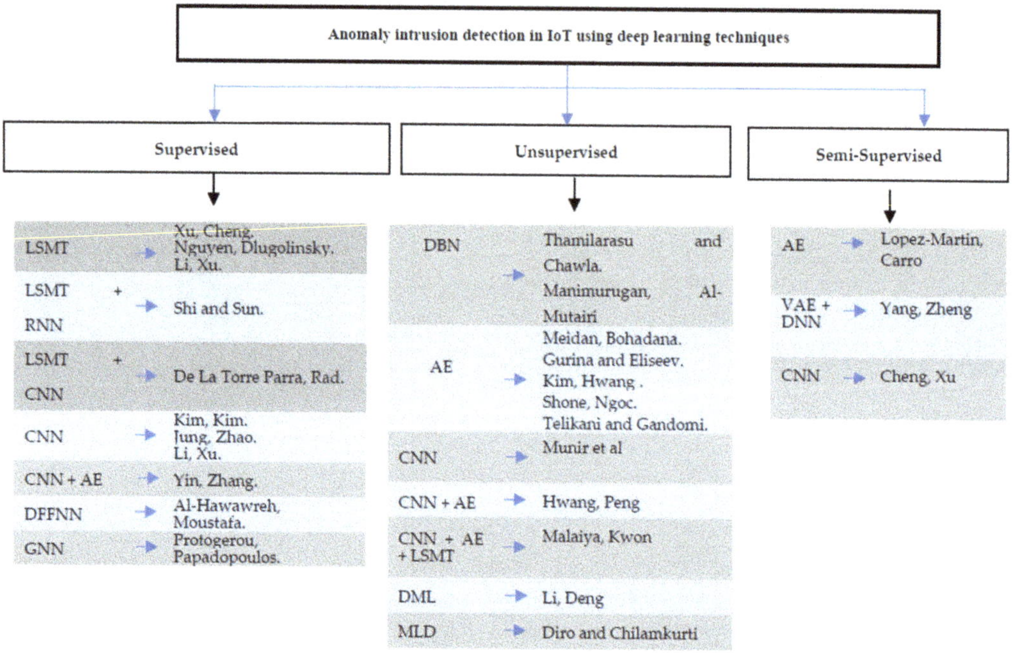

Figure 5. Taxonomy of anomaly intrusion detection in IoT using deep learning techniques [13–17,30–49].

- Supervised: in a supervised manner, anomalies detecting labeled datasets by constructing the network or system is normal behavior. Supervised anomaly detection techniques can leverage the measurement of distance as well as the density of clusters for the detection of intrusions.
- Unsupervised: in an unsupervised manner, the approach assumes a greater frequency of normal behaviors, thus leading to the establishment of the model on assumptions, wherein there is no need for any labeled data for training.
- Semi-supervised: in a semi-supervised manner, the algorithm is trained upon a combination of labeled and unlabeled data.

7.2. RQ2: What Is the Performance of Anomaly Intrusion Detection in IoT Using Deep Learning Techniques?

Accuracy, precision, recall, false-positive rate (FPR), false-negative rate (FNR), and f-measure are the most frequently employed model evaluation techniques based on deep learning [50–54].

As shown in Table 3, the high accuracy is nearly 100%; precision and recall are almost 100% in D-PACK [45]. They used CNN and AE techniques on the Mirai-RGU dataset. However, this model takes a long time for training and preprocessing, which is resource consuming. Additionally, it covers only a few types of attacks. Similarly, the study conducted by [37] used CNN and AE techniques on the Yahoo Webscope S5 dataset and achieved 99.62% accuracy, 98.78% precision, and 97.2% recall. This indicates that the combination of CNN and AE may improve the performance. Nevertheless, the resource-consuming aspect, network overhead, and datasets with real IoT traffic should be considered as well. D. Li et al. [47] proposed a model that achieved an accuracy of 99.78%, precision of 98.99%, recall of 91.05%, and FAR of 0.22%, using DML techniques by using the KDDUP99 dataset. However, this model suffers from high resource consumption, and the dataset does not contain IoT traffics and modern types of attacks. Shi and Sun [16] proposed a model that achieved 99.36% accuracy, the precision of 97.97%, and recall of 98.86%, using LSTM with RNN techniques. However, they did not report the FAR, as the model is for a specific type of attack and is resource consuming. We can say that combining AE with CNN techniques could enhance the accuracy and decrease the FAR, but we should consider the resource consumption and cover the IoT attacks. Figure 6 shows the frequency of the performance measures of the studies.

Table 3. Performance of the studies models.

Study	Techniques	Accuracy	Precision	Recall	FAR	F1-Measure	FPR	FNR
Lopez et al. [48]	AE	80%	81.59%	80.1%		79.08%		
Yang et al. [15]	VAE + DNN	89.08%	86.05	95.68		90.61	19.01	
Cheng et al. [30]	LSTM	98%						
Thamilarasu et al. [14]	DBN		97%					
Shi et al. [16]	LSTM + RNN	99.36%	97.97%	98.86%,		98.42		
Munir et al. [17]	CNN	99%	100%					
Gurina et al. [41]	AE				0.007			
Manimurugan et al. [40]	DBN	98.37%	97.21%,	98.34%		97%		
Malaiya et al. [46]	CCN + VAE + LSTM	99%						
Kim et al. [34]	CNN	99%,						
Jung et al. [35]	CNN	96.50%,				85%		
Gurina et al. [42]	AE							
Diro et al. [13]	Multi-Layer deep learning	99.02%		99.27%	99.14%	0.85%		
Parra et al. [33]	CNN + LSTM	94.30%	93.48%	93.67%		93.58%	5.20%	
Cheng et al. [49]	CNN	99.88%	99.89%	97.94%		98.64%		
Moustafa et al. [38]	DFFNN	98.4%, 92.5%		99%, 93%			1.8%, 8.2%	
Xie et al. [31]	LSTM	86.95%						
Zhao et al. [36]	CNN	76.67%						
Li et al. [32]	LSTM	97.58%		83.79%	2.02%			6.02%
Kim et al. [43]	AE	99.81%						
Hwang et al. [45]	CNN + AE	100%	100%	100%		100%	0%	
Yin et al. [37]	CNN + AE	99.62%	98.78%	97.2%		98.78%		
Telikani et al. [44]	AE	99.6	100%	100%	100%	0.0057		
Shone et al. [18]	AE	97.85%	100%	100%		85.42%		
Drosou et al. [39]	GNN/RNN	99%						
Deng et al. [47]	DML	99.78	98.99	91.05	0.22%			

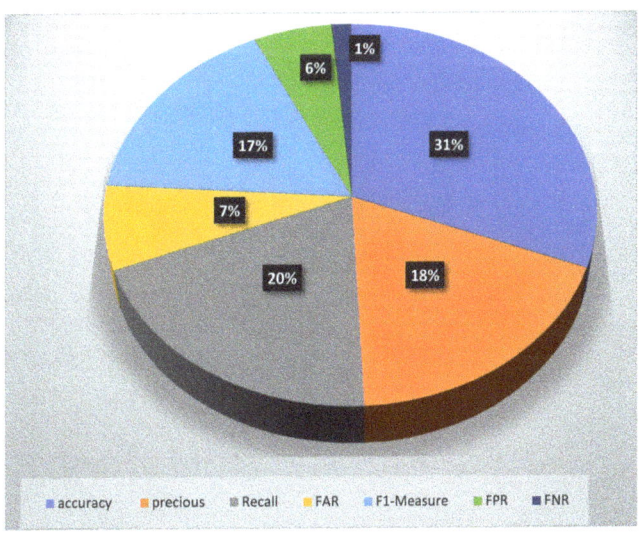

Figure 6. The frequency of performance measures of the studies.

7.2.1. Analysis of Accuracy Range

Table 4 shows the accuracy range for each deep learning technique used. The CNN has a wide range that starts from 76.76% and reaches 99.88%; this technique was tested 10 times individually and integrated with another technique. In addition, AE covers a wide range starting from 80% and reaches 99.81%, and it is similar to CNN in the detection accuracy range and the one used. LSTM was used three times and achieved an accuracy between 79.58% and 98%. DBN was used two times and gained an accuracy range from 97% to 97.21 with little enhancement. RNN and DFFNN were used once. Figure 7 shows the techniques used in the studies.

Table 4. Accuracy range for the techniques.

Study	No. of Study	Techniques Used	Accuracy Range
[17,34–36,49]	5	CNN	(76.76–99.88%)
[37,45]	2	CNN + AE	(99.62–100%)
[18,41–44,48]	6	AE	(80–99.81%)
[31,32]	3	LSTM	(79.58–98%)
[33]	1	CNN + LSTM	(94.30%)
[46]	1	CCN + VAE + LSTM	99%
[14,40]	2	DBN	(97–97.21%)
[15]	1	VAE + DNN	89.08%
[16]	1	LSTM + RNN	99.36%
[39]	1	GNN/RNN	99%
[38]	1	DFFNN	98.4%
[13]	1	Multi-Layer deep learning	99.02
[49]	1	DML	99.78

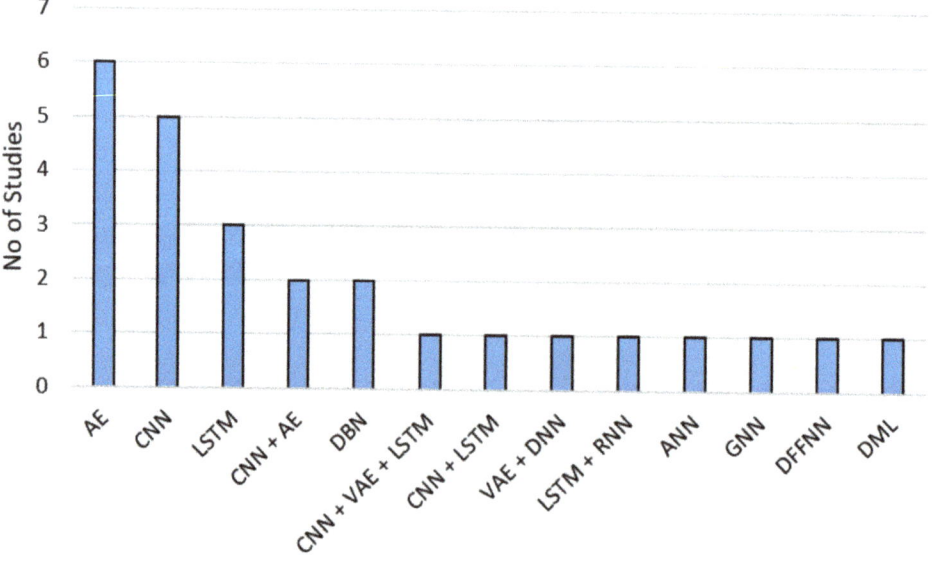

Figure 7. The frequency of the techniques used in the studies.

7.2.2. Analysis of Type of Attacks Detected

The type of attacks is the most important metric used to identify the advantage of the anomaly intrusion detection system. Some models [16,46,50] achieved high accuracy in detecting specific types of attacks. Howeve+-r, with only a few types of attacks included in their training datasets, their performance is questionable. Therefore, these models need more improvement to be able to detect as many attacks as possible with high accuracy. For example, in the IC_VAE model proposed by Lopez-Martin et al. [48], using the NSL-KDD dataset, the types of attacks detected by this model are Probing, Remote to Local (R2L), User to Root (U2R), and Denial of Service (DoS) Attacks. Similarly, studies proposed by [13,18,36,44,47] detected the same types of attacks, using NSL-KDD and KDD Cup 1999 datasets. Moreover, in [15,38], by adapting the NSL-KDD and UNSW-NB15, they extended the range of attack by detecting modern attacks, such as Fuzzer and worm, back door, analysis, exploits, generic, seel-code, and recionnary. In [34], by using KDD CUP 1999, CSE-CIC-IDS2018, they extended the range of attack by detecting modern attacks, such as DoS-Hulk, DoS-GoldenEye, DoS-SlowHTTPTest, DDoS-LOIC-HTTP, DoS-Slowloris, and DDoS-HOIC. The study conducted by [14] used a test-bed that contains several attacks, such as the Distributed Denial-of-Service (DDoS), sinkhole attack, Wormhole Attack, Blackhole Attack. In [40], authors used CICIDS 2017 dataset with a set of attacks, including DoS/DDoS, Botnet, Web Attack, Brute Force, Infiltration, PortScan, SQL Injection, Benign, DoS Hulk. In the study conducted by [45], the Mirai-RGU dataset with a range of attacks, including UDP Flood, SYN Flood, ACK Flood, and HTTP Flood, was used. In [41], by using N-balot databases, the authors focused on Mirai and BASHLITE. Similarly, in [33], the authors used the same dataset of N_BaIoT, but focusing on Distributed Denial of Service (DDoS) and phishing attacks. In [16], by using the MCFP dataset, they focused on Botnets, SYN flood, RST attacks. In [42], by using a test-bed, the range of attacks included flood attacks and SQL injection attacks, SYN Flood, TCP Flood, UDP Flood Detection, ICMP Flood Detection, and HTTP Flood Detection. In [35], the authors used a test-bed with a set of attacks against IoT, such as botnets attack, Mirai, Hajime, Bricker, BotIoT Reaper, Masuta, Sora. In [30], by using a test-bed, the range of the attacks included malicious scan, DoS attack, malicious control spying, malicious operation, wrong setting categories, and data probing. In [31], by using a test-bed, the range of attacks included sip, ssh, SSL, conn, DNS, and HTTP. In [39], by using a test-bed and CTU-13 datasets, the range of attacks included Infiltration attack, Propagation attack, worm infiltration, and worm propagation attack. In the study conducted by [43], the authors used the self-collection dataset and focused on interval attacks. In the studies [17,32,37,46,49], there was no report explaining what kinds of attacks they used.

The DL algorithms that were used in the previous studies prove their ability to detect a wide range of traditional types of attacks patterns, such as the attacks listed above. Moreover, DL algorithms perform better once there are huge amounts of attack data. However, more studies are needed to show the performance of the DL algorithms to detect the recent IoT attacks, such as physical attacks, privilege escalation, eavesdropping, brute-force password attacks, malicious node injection, and firmware hijacking.

7.2.3. Tools and Environments Applied by the Studied Work

Table 5 shows the categorization of the papers based on the tools and environments applied by the studied work. There are several types of development tools that have been used in such development, such as Python, MATLAB, and R language. As can be observed in Table 5, TensorFlow and Keras have been used by many researchers due to their ability to deal with large data and objects detected with high performance and provide high-level APIs for easily building and training models. Furthermore, it can run on Linux, macOS, Windows, and Android.

Table 5. Tools and environments applied by the studied work.

Study	Techniques	TensorFlow	Keras	Scikit	PyTorch	R	SoftMax	Raspberry Pi	Cooja	MATLAB	Python	Sigmoid	Hybrid Analysis Site	Entropy, K L D
Lopez et al. [48]	AE	✓		✓										
Yang et al. [15]	VAE + DNN	✓												
Cheng et al. [30]	LSTM				✓									
Thamilarasu et al. [14]	DBN							✓	✓					
Shi et al. [16]	LSTM + RNN		✓	✓	✓									
Gurina et al. [41]	AE		✓											
Manimurugan et al. [40]	DBN	✓	✓											
Malaiya et al. [46]	CCN + VAE + LSTM										✓			
Kim et al. [34]	CNN		✓									✓		✓
Jung et al. [35]	CNN									✓				
Gurina et al. [42]	AE									✓				
Diro et al. [13]	Multi-Layer deep learning		✓											
Parra et al. [33]	CNN + LSTM	✓												
Cheng et al. [49]	CNN				✓									
Moustafa et al. [38]	DFFNN					✓								
Xie et al. [31]	LSTM	✓	✓											
Zhao et al. [36]	CNN	✓	✓											
Li et al. [32]	LSTM									✓				
Kim et al. [43]	AE												✓	
Hwang et al. [45]	CNN + AE	✓	✓											
Yin et al. [37]	CNN + AE										✓			
Telikani et al. [44]	AE						✓							
Shone et al. [18]	AE	✓												
Drosou et al. [39]	GNN/RNN					✓								
Deng et al. [47]	DML									✓				
Munir et al. [17]	CNN													

7.2.4. Analysis of the Used Datasets

Table 6 shows the datasets used by the existing research regarding deep learning in IoT security. As shown in Table 6, most of the studies used NLS_KDD and KDD CUP 1999. This is due to a lack of substitute datasets. However, these datasets are outdated and do not contain IoT traffic or modern types of attacks. Some modern datasets are now available that contain modern types of attacks—UNSW-NB15 [55] and IoT traffic BoT-IoT [56] we suggest for future researches. Table 7 shows the analysis of the most used datasets in the surveyed studies.

Table 6. The datasets used in the studies.

Study	Techniques	NSL-KDD	KDD CUP 1999	UNSW-NB15	CICIDS 2017	Mirai	CSE-CIC-IDS2018	N-BaIoT	Test-Bed	CTU-13	Gas-Water	AWID	Yahoo Webscope S5	Kyoto	MCFP	DS2OS	LOF	Synthetic
Lopez et al. [48]	AE	✓																
Yang et al. [15]	VAE + DNN	✓		✓														
Cheng et al. [30]	LSTM												✓					
Thamilarasu et al. [14]	DBN					✓												
Shi et al. [16]	LSTM + RNN					✓												
Munir et al. [17]	CNN														✓			
Gurina et al. [41]	AE						✓	✓										
Manimurugan et al. [40]	DBN				✓													
Malaiya et al. [46]	CCN + VAE + LSTM	✓												✓				

Table 6. Cont.

Study	Techniques	NSL-KDD	KDD CUP 1999	UNSW-NB15	CICIDS 2017	Mirai	CSE-CIC-IDS2018	N-BaIOT	Test-Bed	CTU-13	Gas-Water	AWID	Yahoo Webscope S5	Kyoto	MCFP	DS2OS	LOF	Synthetic
Kim et al. [34]	CNN		✓				✓											
Jung et al. [35]	CNN							✓										
Gurina et al. [42]	AE							✓										
Diro et al. [13]	Multi-Layer deep learning	✓																
Parra et al. [33]	CNN + LSTM							✓										
Cheng et al. [49]	CNN																✓	
Moustafa et al. [38]	DFFNN	✓		✓						✓								
Xie et al. [31]	LSTM						✓											
Zhao et al. [36]	CNN	✓																
Li et al. [32]	LSTM									✓	✓	✓						
Kim et al. [43]	AE																✓	
Hwang et al. [45]	CNN + AE					✓												
Yin et al. [37]	CNN + AE													✓				
Telikani et al. [44]	AE	✓	✓															
Shone et al. [18]	AE	✓	✓															
Drosou et al. [39]	GNN/RNN							✓										✓
Deng et al. [47]	DML	✓																

Table 7. The analysis of the most used datasets in the surveyed studies.

Dataset	Published Year	IoT Specific	Features	No. of Classic	Total Normal Records	Total Attacks Records	Description
NSL-KDD	2009	NO	43	4	77,054	71.463	This dataset is an extension of the dataset "KDDCUP 99". The duplicate records were removed and lack in modern large-scale attacks. Moreover, it is not IoT specific. It contains 22 attack types in the training dataset and 17 attack types in the test dataset, which are categorized as 4 attack classes.
KDD CUP 1999	1999	NO	43	4	1,033,372	4,176,086	This dataset does not contain modern attack data and modern large-scale attacks. Moreover, it contains unbalanced labels, and this dataset is not specific to the IoT.
UNSW-NB15	2015	NO	49	9	2,218,761	321,283	This dataset is based on a synthetic environment for generating attack activities. It contains approximately one hour of anonymized traffic traces from a DDoS attack in 2007.
CICIDS 2017	2017	NO	80	14	2,273,097	557,646	This dataset is not specific to the IoT. It contains complex features that are not present in previous datasets. However, it contains a modern large-scale attack.
CSE-CIC-IDS2018	2018	NO	80	18	N/A	N/A	This dataset is not specific to the IoT. However, it contains a modern large-scale attack.
N-BaIOT	2018	YES	115	8	17,936	831,298	This dataset contains IoT traffic, but it is unbalanced, due to the normal records being smaller than malicious records.
AWID	2015	NO	155	4	530,785	44,858	This dataset is not specific to the IoT. However, it contains modern types of attacks.
Yahoo Webscope S5/A1	2015	NO	-	-	93,197	1669	This dataset contains web traffic, which includes normal and attacks traffic. However, it is not specific to the IoT.
Kyoto	2006	NO	24	-	50,033,015	43,043,255	This dataset is not specific to the IoT. However, it contains modern types of attacks [57].

7.3. RQ 3: What Are the Challenges Faced in Current Anomaly Intrusion Detection Deep Learning Techniques in IoT?

7.3.1. Threat Detection

Because IoT supports a wide range of applications that need different resource requirements in terms of processing, storage, and communication, the network becomes more complex, due to the heterogeneity of the IoT devices that are being connected. This makes it hard to provide a secure environment in the IoT ecosystem and even harder to detect security threats. In securing an IoT environment, it is important to acknowledge the features and criteria necessary for applying security analytics in deep learning algorithms [54]. However, the existing mechanisms lack the effective and efficient methods that can perceive the hidden correlation between these features. Nevertheless, the rapid growth of deep learning algorithms is believed to have the capability of handling the hidden parameters not limited to the IoT application, but also for finding the correlation of data variation. In addition, a higher detection rate toward detecting zero-day attacks efficiently is obtainable with deep learning [58].

7.3.2. Computational and Resource Constraint

The computational complexity can be considered one of the prominent obstacles in the area of IoT security and deep learning. The usage of IoT devices requires a low battery and CPU power. Hence, the computational time in IoT devices should be quick, and the operation should be straightforward [59]. For better performance, it is more effective to mitigate the IoT computation to the edge of the cloud. There is one particular study [60] that emphasized analyzing the implementation of an algorithm that focuses on producing a lightweight computation system. The distributed computing and distributed algorithms provide better computational optimization by distributing the tasks overs multiple nodes, which improves the efficiency [54,61].

7.3.3. Time Complexity

Time complexity is considered an obstacle because the current detection techniques were developed based on batch processing applications rather than real-time detection. As mentioned before, the IoT environment deals with real-data streaming. Hence, the time complexity is crucial in detecting threats in IoT applications. In addition, it can assess the impact on several attributes associated with security threats. Deep learning is highly capable of resolving time complexity issues in IoT by implementing GPU components to deal with real-time processing in an efficient manner [62].

7.3.4. Edge Computing and Security

An edge computing platform offers better extensibility in data processing and storage for resource-constrained IoT devices. Furthermore, it enables nearby devices located around the data sources to intelligently operate, even if they are far from the center node of infrastructure. The cloud infrastructure stores the IoT devices' data source regarding network computing to provide rational edge services in detecting real-time threats. Unfortunately, IoT as a standalone entity is incapable of storing and analyzing data for any potential threats, due to insufficient resources [63]. Hence, with the aid of edge computing, it will enable multiple resource distribution of data processing over the cloud for analysis [64]. It is convincible to state that the amalgamation of deep learning in IoT helps in facilitating security analytics in providing an enhanced processing system that can detect threats effectively and accurately [54].

7.3.5. Training Time

One of the major problems that existing techniques suffer from is the large and high dimensional datasets used for training [65]. Due to that, more time is needed to train the model for higher accuracy detection. In tackling these issues, deep learning algorithms are proposed because they can work on lesser training duration and dataset. This helps to increase the efficiency during model training. The batch size may also affect the time consumed in the training phase because of the accumulation of the network onto the weight update [54,66]. To solve this, multiple layers can be used to build deep learning networks, which facilitates the weighing and recognizing of the set of significant patterns from the datasets. Furthermore, the exploits of storage and processing facilities additionally obstruct the model training time. Dealing with this issue, the adaptation of big cloud-based architecture and data technologies improve the efficiency by reducing the model training duration [63].

8. Discussion

We found that the trend goes to AE techniques. The studies [18,41–44,48] used AE techniques because of the ability of AE to take advantage of the linear and nonlinear dimensionality reduction to detect the anomalies. The AE training phase involves the reconstruction of clean input data from a partially destroyed one as well as the ability of AE to deal with heterogeneity, unstructured and high dimensional data that generated from IoT device. However, using techniques such as CNN combined with AE would be preferable for better classification, depending on the data reduction from the AE phase. Another observed five studies used CNN techniques [17,34–36,49], which can automatically detect the most important feature and learn the key feature of each class by itself without human intervention. Moreover, CNN can perform identification and prediction through the dense network. The CNN considered is a very vast technique, and this may be due to the ConvNets. Other factors that may affect the efficiency of CNN are filters, kernel size, stride, and padding. However, using techniques such as AE combined with CNN would be preferable to reduce the high dimensional data, which generate from IoT devices to minimize the exchange data between IoT nodes to avoid the energy-consuming and communication overhead.

In addition, we found that three studies used LSTM techniques [30–32] that are useful for classifying, processing, and predicting time series in long duration. Moreover, they have a memory that can store previous time step information, and this is how they learn. They also can deal with noise distributed representation and continuous value. However, LSTMs are apt for overfitting, and it is not easy to apply the dropout algorithm to restrain this problem. Combining CNN with AE [37,45] could achieve a promising result in terms of accuracy, recall, and precision. However, the researcher and developer should consider the resource consumption, training time, and the type of attacks. Notably, the AE and CNN are the most common techniques used in the literature. In addition, some studies used a single technique, and others combined multiple ones to improve the performance [16,46,47]. However, the FAR needs to be decreased when considering different types of attacks in the used dataset. Datasets that include a wide range of attacks with simulation tools are suggested above. In addition, still, some DL techniques have not been examined yet, which makes the need for more work in the area to achieve robust IDS for resource-constrained IoT devices. Combining two DL could lead to achieving high detection attacks, but it may lead to resource consumption and a high training time. The datasets used in the literature are outdated, perhaps due to a lack of substitute datasets. However, these datasets are outdated and do not contain IoT traffic or modern types of attacks. Some modern datasets are now available that contain modern types of attacks; UNSW-NB15 [55] and IoT traffic BoT-IoT [56] we suggest for future research. In addition, there is a need for new datasets that reflect the IoT traffic. Table 8 shows the domain of state-of-the-art studies, IDS architecture, the technique used, and methodology as well as the advantages and disadvantages.

Table 8. List of the state-of-the-art studies and the advantages and disadvantages.

Study	IDS Architecture	Techniques Used	Methodology	Advantages	Disadvantages
Lopez et al. [48]	Network-based	AE	proposed Model to perform feature reconstruction and detect malicious in IoT environment.	• Lightweight. • High accuracy in recover categorical features.	• Low detection accuracy. • High training time.
Yang et al. [15]	Network-based	VAE + DNN	proposed model to perform monitoring unknown attacks using AE and DNN to learn the complex traffics and imbalanced classes.	• Lightweight. • Low resource consumption.	• Low detection accuracy. • High training time.
Cheng et al. [30]	Network-based	LSTM	proposed model that adopts an innovative concept of the drift method to improve the accuracy of anomaly detection using LSTM.	• High detection accuracy. • work well for time series. • Memory effective.	• Multi-classification method needs to be enhanced.
Thamilarasu et al. [14]	Network-based	DBN	Proposed an intelligent IDS to detect malicious traffic in IoT networks using DBN.	• Real-Time IDS.	• Detection accuracy needs to be enhanced.
Shi et al. [16]	Network-based	LSTM + RNN	Proposed approach is to analyze a series of network packets to detect botnets using LSTM and RNN for better classification.	• Enhanced robustness. • High detection accuracy. • Lightweight.	• Few types of attacks. • High false-positive rate. • Resources consuming.
Munir et al. [17]	Network-based	CNN	Proposed DeepAnTmodel to anomaly detection and time series prediction.	• High detection accuracy. • Detect point anomalies, contextual anomalies. • Model works well with a vast amount of data.	• High computational time. • Poor data quality can corrupt the data modeling phase.
Gurina et al. [41]	Network-based	AE	Proposed N-BaIoT to extract network traffics and detect anomalies from resource constraint devices.	• Enhanced robustness. • Efficient time to detect attacks.	• Low traffic prediction. • Detection accuracy not reported.
Manimurugan et al. [40]	Centralized Host-Based	DBN	Proposed approach to detect anomaly attacks in IoT environment.	• High detection accuracy. • Lightweight.	• Not a Real-Time IDS. • Detect few types of IoT attacks.
Malaiya et al. [46]	Network-based	CCN + VAE + LSTM	Proposed approach to detect anomaly in IoT networks by combining three deep learning techniques.	• High detection accuracy. • Lightweight.	• Resource-consuming. • High computational complexity.
Kim et al. [34]	Network-based	CNN	Proposed approach to detect anomaly in IoT environment with focusing on DoS attacks.	• High detection accuracy. • Lightweight.	• Detect few types of IoT attacks. • High computational complexity.
Jung et al. [35]	Host-based	CNN	Proposed approach to monitoring malicious botnet on resource constraint IoT devices using three types of IoT devices.	• Good classification accuracy. • Real-Time IDS.	• Expensive power monitor. • Detection accuracy needs to be enhanced. • High computational complexity.
Gurina et al. [42]	Host-based	AE	Proposed approach to detect malicious in web server during users' requests processing considering the MyBB web server as a case study.	• Lightweight. • capable to detect zero-day attacks. • High detection accuracy for individual attacks.	• High False positive rate. • High computational complexity. • No comparison with previous methods.

Table 8. Cont.

Study	IDS Architecture	Techniques Used	Methodology	Advantages	Disadvantages
Diro et al. [13]	Distributed Network-Based	Multi-Layer deep learning	Proposed a distributed approach to detect attacks in social IoT.	• Lightweight • High detection accuracy. • Low resource consumption.	• Few types of attacks. • High training time.
Parra et al. [33]	Distributed Network-Based	CNN + LSTM	Proposed a distributed cloud-based approach to detect and mitigate phishing and Botnet attacks on client devices.	• Lightweight. • Low communication overhead.	• Detection accuracy needs to be enhanced. • High computational complexity.
Cheng et al. [49]	Centralized Host-Based	CNN	Proposed a semi-supervised based model to detect anomalies in IoT communication.	• Lightweight • High detection accuracy.	• High computational complexity.
Moustafa et al. [38]	Network-based	DFFNN	Proposed anomaly detection to learn and validate the information collected from TCP/IP packets.	• Lightweight • High detection accuracy. • Model covered vast types of attacks.	• Not a Real-Time IDS.
Xie et al. [31]	Network-based	LSTM	Proposed approach to monitor and detect malicious from the network traffic flow.	• Lightweight. • work well for time series.	• Detection accuracy not reported.
Zhao et al. [36]	Network-based	CNN	Proposed approach to detect intrusion in industrial IoT.	• Enhanced robustness. • Lightweight.	• Detection accuracy needs to be enhanced. • High computational complexity.
Li et al. [32]	Network-based	LSTM	Proposed approach to detect attack interval from historic data in industrial IoT.	• Enhanced robustness. • Lightweight. • High detection accuracy.	• High computational complexity.
Kim et al. [43]	Host-based	AE	Proposed approach to the analysis of attack profile, detect the threats and abnormal behavior that deviates from normal behavior in IoT devices.	• Enhanced robustness. • Lightweight. • High detection accuracy.	• High training time.
Hwang et al. [45]	Network-based	CNN + AE	Proposed D-PACK anomaly approach to detect features and profiling traffic with just a few first packets from each flow in IoT networks.	• High detection accuracy. • Lightweight. • Low false alarm rate.	• High computational complexity. • High training time. • Focusing on few types of attacks.
Yin et al. [37]	Network-based	CNN + AE	Proposed approach to detect the anomaly and to enhance classification in time series.	• High detection accuracy. • Lightweight. • Low false alarm rate.	• High computational complexity. • High training time.
Telikani et al. [44]	Network-based	AE	Proposed CSSAE (cost-sensitive stacked auto-encoder) to solve the class imbalance problem in IDS and detect low-frequency attacks in IoT environment.	• High detection accuracy. • Lightweight. • Low false alarm rate.	• High training time.
Shone et al. [18]	Network-based	AE	Proposed model to dimensionality reduction for the data and detect malicious at the IoT environment.	• High detection accuracy. • Lightweight.	• High false alarm rate. • High training time.

Table 8. Cont.

Study	IDS Architecture	Techniques Used	Methodology	Advantages	Disadvantages
Drosou et al. [39]	Distributed Network-based	GNN/RNN	Proposed collaborative anomaly intrusion detection to detect malicious for IoT devices.	• High detection accuracy. • Lightweight.	• High computational complexity. • Power consumption.
Deng et al. [47]	Network-based	DML	proposes an approach to detect malicious and feature extraction for smart cities.	• High detection accuracy. • Lightweight. • Low false alarm rate.	• High computational complexity.

Table 8 also shows that there are many types of IDS architectures that have been implemented. Network-based IDS is the most applied architecture, due to the availability of labeled network traffic datasets. In the IoT networks, the architecture of the IDS depends on the application domain and the host environment [2]. The host-based approach is recommended to protect the operating system of the IoT devices from malicious attacks, while the network-based is suitable for protecting the communicated devices from malicious traffic. Most studies applied the network-based architecture, while the nature of IoT is heavily distributed. It will be more effective if the researchers and developers pay more attention to combining host-based architecture with those that are network-based in a distributed and hierarchical architectural design manner to minimize the detection time, improve the detection accuracy, and decrease the network's overhead.

9. Future Direction

Undoubtedly, improving the efficiency of deep learning detection results remains an open research direction issue. IoT security researchers and developers must always contend for 100% detection with zero false alarms while considering IoT resource constraints. Moreover, most of the studies are pertinent to the system's normal behavior. Often, most of the approaches depend on the training of normal behavior, while the deviation is pertinent to scenarios investigated as abnormal behavior. Thus, a better method in terms of precision and robustness is needed to deal with complex real scenarios. Data complexities include unexpected noise, redundancy in data, and imbalanced datasets. To extract significant knowledge and information, well-designed techniques are required to organize the datasets. In this scenario, a lightweight system can be exhaustive, due to the high computational task of dealing with complex data. The current technology of cloud computing can be utilized to obtain a productive result in real time. Most of the work done in recent years was in the detection of anomalies, as the research community did not foster much interest in anomaly prediction and prevention. This could contribute to predicting anomalies in future work. There is a need to adopt and/or develop new methods that can prevent the systems before attacks occur. Moreover, anomaly detection in multivariable time series is still an open research direction. In addition, applying anomaly intrusion detection systems, using deep learning in smart vehicles, needs to be investigated. There is an imperious need for normal and anomaly datasets that are up-to-date and integrated with IoT applications and services. These datasets could be extremely useful for testing various IDS types and methods in IoT environments. The capability to implement effective and meaningful IDS comparisons will rely on these datasets.

10. Limitation of the Study

Throughout the review study, the SLR is performed to provide extensive coverage of all relevant studies associated with the use of deep learning techniques in securing IoT environments. The main limitation of this study is in searching. There are also few limitations of the SLR that should be taken into consideration, which are listed as the following:

1. This review is limited to articles and does not include books, magazines, and conferences related to deep learning in IoT.
2. This review is limited to papers available in the English language.

11. Conclusions

In general, this study presented a systematic review of anomaly IDSs in IoT environments using deep learning. A comprehensive report was produced, regarding anomaly intrusion detection in the domain of IoT, using deep learning techniques. Upon completion of this study, a full adherence of systematic literature protocol and guidelines based on proposed work by Kitchenham is presented [25]. All the data used were gathered from primary studies published without applying any filters to differentiate between conference proceedings and journal articles. This study summarized and organized the current literature related to anomaly-based intrusion detection in IoT, using deep learning techniques according to the pre-defined keywords and RQs. A total number of 26 studies were included, according to the stated exclusion, inclusion, and quality criteria. A comprehensive taxonomy was presented based on the results of the study conducted for anomaly intrusion detection in IoT using deep learning techniques. This study provided an insight into the attributes and knowledge of existing anomaly intrusion detection in an IoT environment, using deep learning techniques. Additionally, the study presented a comparison in terms of the performance, the dataset used, attacks detection, techniques, and evaluation techniques in each study. Finally, the study discussed challenges faced in anomaly intrusion detection in IoT using deep learning. This paper can provide researchers with details about an up-to-date technique and methodology in anomaly intrusion detection in IoT, using deep leering. The limitations of current anomaly-based intrusion detection systems in IoT using deep learning techniques indicate the future direction for further improvements of the IDS systems, considering the characteristics of IoT.

Author Contributions: Conceptualization, M.A.A.; methodology, M.A.A.; resources, M.A.A., F.A.G., F.S. and M.N.; data curation, M.A.A., F.A.G. and M.N.; writing—Original draft preparation, M.A.A.; writing—Review and editing, S.R., M.M.S., F.A.G., I.N. and F.S.; supervision, S.R. and M.M.S.; project administration, S.R.; funding acquisition, S.R. and I.N. All authors have read and agreed to the published version of the manuscript.

Funding: This research is funded by Ministry of Higher Education Malaysia under the Research Excellence Consortium in IoT Security (VOT R.J130000.7851.4L946).

Institutional Review Board Statement: Not applicable.

Informed Consent Statement: Not applicable.

Conflicts of Interest: The authors declare no conflict of interest.

References

1. Atzori, L.; Iera, A.; Morabito, G. Understanding the Internet of Things: Definition, potentials, and societal role of a fast evolving paradigm. *Ad Hoc Netw.* **2017**, *56*, 122–140. [CrossRef]
2. Elrawy, M.F.; Awad, A.I.; Hamed, H.F.A. Intrusion detection systems for IoT-based smart environments: A survey. *J. Cloud Comput.* **2018**, *7*, 21. [CrossRef]
3. Da Xu, L.; He, W.; Li, S. Internet of things in industries: A survey. *IEEE Trans. Ind. Inform.* **2014**, *10*, 2233–2243.
4. Lin, J.; Yu, W.; Zhang, N.; Yang, X.; Zhang, H.; Zhao, W. A Survey on Internet of Things: Architecture, Enabling Technologies, Security and Privacy, and Applications. *IEEE Internet Things J.* **2017**, *4*, 1125–1142. [CrossRef]
5. Almiani, M.; AbuGhazleh, A.; Al-Rahayfeh, A.; Atiewi, S.; Razaque, A. Deep recurrent neural network for IoT intrusion detection system. *Simul. Model. Pract. Theory* **2020**, *101*, 102031. [CrossRef]
6. Moore, S.J.; Nugent, C.D.; Zhang, S.; Cleland, I. IoT reliability: A review leading to 5 key research directions. *CCF Trans. Pervasive Comput. Interact.* **2020**, *2*, 147–163. [CrossRef]
7. Ferrag, M.A.; Shu, L.; Yang, X.; Derhab, A.; Maglaras, L. Security and Privacy for Green IoT-Based Agriculture: Review, Blockchain Solutions, and Challenges. *IEEE Access* **2020**, *8*, 32031–32053. [CrossRef]
8. Farooq, M.S.; Riaz, S.; Abid, A.; Abid, K.; Naeem, M.A. A Survey on the Role of IoT in Agriculture for the Implementation of Smart Farming. *IEEE Access* **2019**, *7*, 156237–156271. [CrossRef]

9. Ruan, J.; Wang, Y.; Chan, F.T.S.; Hu, X.; Zhao, M.; Zhu, F.; Shi, B.; Shi, Y.; Lin, F. A Life Cycle Framework of Green IoT-Based Agriculture and Its Finance, Operation, and Management Issues. *IEEE Commun. Mag.* **2019**, *57*, 90–96. [CrossRef]
10. Pal, S.; Hitchens, M.; Rabehaja, T.; Mukhopadhyay, S. Security Requirements for the Internet of Things: A Systematic Approach. *Sensors* **2020**, *20*, 5897. [CrossRef]
11. Ghaleb, F.A.; Maarof, M.A.; Zainal, A.; Rassam, M.; Saeed, F.; Alsaedi, M. Context-aware data-centric misbehaviour detection scheme for vehicular ad hoc networks using sequential analysis of the temporal and spatial correlation of the consistency between the cooperative awareness messages. *Veh. Commun.* **2019**, *20*, 100186. [CrossRef]
12. Hameed, S.; Khan, F.I.; Hameed, B. Understanding Security Requirements and Challenges in Internet of Things (IoT): A Review. *J. Comput. Netw. Commun.* **2019**, *2019*, 9629381. [CrossRef]
13. Diro, A.A.; Chilamkurti, N. Distributed attack detection scheme using deep learning approach for Internet of Things. *Futur. Gener. Comput. Syst.* **2018**, *82*, 761–768. [CrossRef]
14. Thamilarasu, G.; Chawla, S. Towards Deep-Learning-Driven Intrusion Detection for the Internet of Things. *Sensors* **2019**, *19*, 1977. [CrossRef] [PubMed]
15. Yang, Y.; Zheng, K.; Wu, C.; Yang, Y. Improving the classification effectiveness of intrusion detection by using improved conditional variational autoencoder and deep neural network. *Sensors* **2019**, *19*, 2528. [CrossRef] [PubMed]
16. Shi, W.-C.; Sun, H.-M. DeepBot: A time-based botnet detection with deep learning. *Soft Comput.* **2020**, *24*, 16605–16616. [CrossRef]
17. Munir, M.; Siddiqui, S.A.; Dengel, A.; Ahmed, S. DeepAnT: A Deep Learning Approach for Unsupervised Anomaly Detection in Time Series. *IEEE Access* **2018**, *7*, 1991–2005. [CrossRef]
18. Shone, N.; Ngoc, T.N.; Phai, V.D.; Shi, Q. A Deep Learning Approach to Network Intrusion Detection. *IEEE Trans. Emerg. Top. Comput. Intell.* **2018**, *2*, 41–50. [CrossRef]
19. Hajiheidari, S.; Wakil, K.; Badri, M.; Navimipour, N.J. Intrusion detection systems in the Internet of things: A comprehensive investigation. *Comput. Netw.* **2019**, *160*, 165–191. [CrossRef]
20. Fahim, M.; Sillitti, A. Anomaly Detection, Analysis and Prediction Techniques in IoT Environment: A Systematic Literature Review. *IEEE Access* **2019**, *7*, 81664–81681. [CrossRef]
21. da Costa, K.A.; Papa, J.P.; Lisboa, C.O.; Munoz, R.; de Albuquerque, V.H.C. Internet of Things: A survey on machine learning-based intrusion detection approaches. *Comput. Netw.* **2019**, *151*, 147–157. [CrossRef]
22. Chalapathy, R.; Chawla, S. Deep learning for anomaly detection: A survey. *arXiv* **2019**, arXiv:1901.03407.
23. Sharma, B.; Sharma, L.; Lal, C. Anomaly Detection Techniques using Deep Learning in IoT: A Survey. In Proceedings of the 2019 International Conference on Computational Intelligence and Knowledge Economy (ICCIKE), Dubai, United Arab Emirates, 11–12 December 2019; IEEE: Piscataway, NJ, USA, 2020.
24. Alsoufi, M.A.; Razak, S.; Siraj, M.M.; Ali, A.; Nasser, M.; Abdo, S. *Anomaly Intrusion Detection Systems in IoT Using Deep Learning Techniques: A Survey*; Springer International Publishing: Cham, Switzerland, 2021.
25. Kitchenham, B.; Charters, S. *Guidelines for Performing Systematic Literature Reviews in Software Engineering*; EBSE Technical Report; Keele University: Keele, UK, 2007.
26. Kitchenham, B.; Brereton, P. A systematic review of systematic review process research in software engineering. *Inf. Softw. Technol.* **2013**, *55*, 2049–2075. [CrossRef]
27. Milani, B.A.; Navimipour, N.J. A Systematic Literature Review of the Data Replication Techniques in the Cloud Environments. *Big Data Res.* **2017**, *10*, 1–7. [CrossRef]
28. Safaei, M.; Asadi, S.; Driss, M.; Boulila, W.; Alsaeedi, A.; Chizari, H.; Abdullah, R.; Safaei, M. A systematic literature review on outlier detection in wireless sensor networks. *Symmetry* **2020**, *12*, 328. [CrossRef]
29. Nidhra, S.; Yanamadala, M.; Afzal, W.; Torkar, R. Knowledge transfer challenges and mitigation strategies in global software development—A systematic literature review and industrial validation. *Int. J. Inf. Manag.* **2013**, *33*, 333–355. [CrossRef]
30. Xu, R.; Cheng, Y.; Liu, Z.; Xie, Y.; Yang, Y. Improved Long Short-Term Memory based anomaly detection with concept drift adaptive method for supporting IoT services. *Futur. Gener. Comput. Syst.* **2020**, *112*, 228–242. [CrossRef]
31. Nguyen, G.; Dlugolinsky, S.; Tran, V.; Garcia, A.L. Deep Learning for Proactive Network Monitoring and Security Protection. *IEEE Access* **2020**, *8*, 19696–19716. [CrossRef]
32. Li, X.; Xu, M.; Vijayakumar, P.; Kumar, N.; Liu, X. Detection of Low-Frequency and Multi-Stage Attacks in Industrial Internet of Things. *IEEE Trans. Veh. Technol.* **2020**, *69*, 8820–8831. [CrossRef]
33. Parra, G.D.L.T.; Rad, P.; Choo, K.-K.R.; Beebe, N. Detecting Internet of Things attacks using distributed deep learning. *J. Netw. Comput. Appl.* **2020**, *163*, 102662. [CrossRef]
34. Kim, J.; Kim, J.; Kim, H.; Shim, M.; Choi, E. CNN-Based Network Intrusion Detection against Denial-of-Service Attacks. *Electronics* **2020**, *9*, 916. [CrossRef]
35. Jung, W.; Zhao, H.; Sun, M.; Zhou, G. IoT botnet detection via power consumption modeling. *Smart Health* **2020**, *15*, 100103. [CrossRef]
36. Li, Y.; Xu, Y.; Liu, Z.; Hou, H.; Zheng, Y.; Xin, Y.; Zhao, Y.; Cui, L. Robust detection for network intrusion of industrial IoT based on multi-CNN fusion. *Measurement* **2020**, *154*, 107450. [CrossRef]
37. Yin, C.; Zhang, S.; Wang, J.; Xiong, N.N. Anomaly Detection Based on Convolutional Recurrent Autoencoder for IoT Time Series. *IEEE Trans. Syst. Man Cybern. Syst.* **2020**, 1–11. [CrossRef]

38. Al-Hawawreh, M.; Moustafa, N.; Sitnikova, E. Identification of malicious activities in industrial internet of things based on deep learning models. *J. Inf. Secur. Appl.* **2018**, *41*, 1–11. [CrossRef]
39. Protogerou, A.; Papadopoulos, S.; Drosou, A.; Tzovaras, D.; Refanidis, I. A graph neural network method for distributed anomaly detection in IoT. *Evol. Syst.* **2020**, *12*, 19–36. [CrossRef]
40. Manimurugan, S.; Al-Mutairi, S.; Aborokbah, M.M.; Chilamkurti, N.; Ganesan, S.; Patan, R. Effective Attack Detection in Internet of Medical Things Smart Environment Using a Deep Belief Neural Network. *IEEE Access* **2020**, *8*, 77396–77404. [CrossRef]
41. Meidan, Y.; Bohadana, M.; Mathov, Y.; Mirsky, Y.; Shabtai, A.; Breitenbacher, D.; Elovici, Y. N-BaIoT—Network-Based Detection of IoT Botnet Attacks Using Deep Autoencoders. *IEEE Pervasive Comput.* **2018**, *17*, 12–22. [CrossRef]
42. Gurina, A.; Eliseev, V. Anomaly-Based Method for Detecting Multiple Classes of Network Attacks. *Information* **2019**, *10*, 84. [CrossRef]
43. Kim, S.; Hwang, C.; Lee, T. Anomaly Based Unknown Intrusion Detection in Endpoint Environments. *Electronics* **2020**, *9*, 1022. [CrossRef]
44. Telikani, A.; Gandomi, A.H. Cost-sensitive stacked auto-encoders for intrusion detection in the Internet of Things. *Internet Things* **2019**, *14*, 100122. [CrossRef]
45. Hwang, R.-H.; Peng, M.-C.; Huang, C.-W.; Lin, P.-C.; Nguyen, V.-L. An Unsupervised Deep Learning Model for Early Network Traffic Anomaly Detection. *IEEE Access* **2020**, *8*, 30387–30399. [CrossRef]
46. Malaiya, R.K.; Kwon, D.; Suh, S.C.; Kim, H.; Kim, I.; Kim, J. An Empirical Evaluation of Deep Learning for Network Anomaly Detection. *IEEE Access* **2019**, *7*, 140806–140817. [CrossRef]
47. Li, D.; Deng, L.; Lee, M.; Wang, H. IoT data feature extraction and intrusion detection system for smart cities based on deep migration learning. *Int. J. Inf. Manag.* **2019**, *49*, 533–545. [CrossRef]
48. Lopez-Martin, M.; Carro, B.; Sanchez-Esguevillas, A.; Lloret, J. Conditional Variational Autoencoder for Prediction and Feature Recovery Applied to Intrusion Detection in IoT. *Sensors* **2017**, *17*, 1967. [CrossRef] [PubMed]
49. Cheng, Y.; Xu, Y.; Zhong, H.; Liu, Y. Leveraging Semi-supervised Hierarchical Stacking Temporal Convolutional Network for Anomaly Detection in IoT Communication. *IEEE Internet Things J.* **2020**, *8*, 144–155. [CrossRef]
50. Sokolova, M.; Lapalme, G. A systematic analysis of performance measures for classification tasks. *Inf. Process. Manag.* **2009**, *45*, 427–437. [CrossRef]
51. Powers, D.M. Evaluation: From precision, recall and F-measure to ROC, informedness, markedness and correlation. *arXiv* **2011**, arXiv:2010.16061.
52. Xin, Y.; Kong, L.; Liu, Z.; Chen, Y.; Li, Y.; Zhu, H.; Gao, M.; Hou, H.; Wang, C. Machine Learning and Deep Learning Methods for Cybersecurity. *IEEE Access* **2018**, *6*, 35365–35381. [CrossRef]
53. Marir, N.; Wang, H.; Feng, G.; Li, B.; Jia, M. Distributed Abnormal Behavior Detection Approach Based on Deep Belief Network and Ensemble SVM Using Spark. *IEEE Access* **2018**, *6*, 59657–59671. [CrossRef]
54. Amanullah, M.A.; Habeeb, R.A.A.; Nasaruddin, F.H.; Gani, A.; Ahmed, E.; Nainar, A.S.M.; Akim, N.M.; Imran, M. Deep learning and big data technologies for IoT security. *Comput. Commun.* **2020**, *151*, 495–517. [CrossRef]
55. Moustafa, N.; Slay, J. UNSW-NB15: A comprehensive data set for network intrusion detection systems (UNSW-NB15 network data set). In Proceedings of the 2015 Military Communications and Information Systems Conference (MilCIS), Canberra, Australia, 10–12 November 2015; IEEE: Piscataway, NJ, USA, 2016.
56. Koroniotis, N.; Moustafa, N.; Sitnikova, E.; Turnbull, B. Towards the development of realistic botnet dataset in the internet of things for network forensic analytics: Botiot dataset. *Future Gener. Comput. Syst.* **2019**, *100*, 779–796. [CrossRef]
57. Song, J.; Takakura, H.; Okabe, Y. Description of Kyoto University Benchmark Data. 2006. Available online: http://www.takakura.com/Kyoto_data/BenchmarkData-Description-v5.pdf (accessed on 15 March 2016).
58. Tang, T.A.; Mhamdi, L.; McLernon, D.; Zaidi, S.A.R.; Ghogho, M. Deep learning approach for network intrusion detection in software defined networking. In Proceedings of the 2016 International Conference on Wireless Networks and Mobile Communications (WINCOM), Fez, Morocco, 26–29 October 2016; IEEE: Piscataway, NJ, USA, 2016.
59. Hossain, M.M.; Fotouhi, M.; Hasan, R. Towards an analysis of security issues, challenges, and open problems in the internet of things. In Proceedings of the 2015 IEEE World Congress on Services, New York, NY, USA, 27 June–2 July 2015; IEEE: Piscataway, NJ, USA, 2015.
60. Kotenko, I.; Saenko, I.; Branitskiy, A. Framework for Mobile Internet of Things Security Monitoring Based on Big Data Processing and Machine Learning. *IEEE Access* **2018**, *6*, 72714–72723. [CrossRef]
61. Vinayakumar, R.; Alazab, M.; Soman, K.P.; Poornachandran, P.; Al-Nemrat, A.; Venkatraman, S. Deep Learning Approach for Intelligent Intrusion Detection System. *IEEE Access* **2019**, *7*, 41525–41550. [CrossRef]
62. Guo, Y.; Liu, Y.; Oerlemans, A.; Lao, S.; Wu, S.; Lew, M.S. Deep learning for visual understanding: A review. *Neurocomputing* **2016**, *187*, 27–48. [CrossRef]
63. Kozik, R.; Choraś, M.; Ficco, M.; Palmieri, F. A scalable distributed machine learning approach for attack detection in edge computing environments. *J. Parallel Distrib. Comput.* **2018**, *119*, 18–26. [CrossRef]
64. Lu, Z.; Wang, N.; Wu, J.; Qiu, M. IoTDeM: An IoT Big Data-oriented MapReduce performance prediction extended model in multiple edge clouds. *J. Parallel Distrib. Comput.* **2018**, *118*, 316–327. [CrossRef]

65. Zhao, Z.; Kumar, A. Accurate periocular recognition under less constrained environment using semantics-assisted convolutional neural network. *IEEE Trans. Inf. Forensics Secur.* **2016**, *12*, 1017–1030. [CrossRef]
66. HaddadPajouh, H.; Dehghantanha, A.; Khayami, R.; Choo, K.-K.R. A deep Recurrent Neural Network based approach for Internet of Things malware threat hunting. *Futur. Gener. Comput. Syst.* **2018**, *85*, 88–96. [CrossRef]

Article

Optimal Versus Equal Dimensions of Round Bales of Agricultural Materials Wrapped with Plastic Film—Conflict or Compliance?

Anna Stankiewicz

Department of Technology Fundamentals, Faculty of Production Engineering, University of Life Sciences in Lublin, 20-612 Lublin, Poland; anna.stankiewicz@up.lublin.pl

Citation: Stankiewicz, A. Optimal Versus Equal Dimensions of Round Bales of Agricultural Materials Wrapped with Plastic Film—Conflict or Compliance? *Appl. Sci.* **2021**, *11*, 10246. https://doi.org/10.3390/app112110246

Academic Editor: Dimitrios S. Paraforos

Received: 2 September 2021
Accepted: 28 October 2021
Published: 1 November 2021

Publisher's Note: MDPI stays neutral with regard to jurisdictional claims in published maps and institutional affiliations.

Copyright: © 2021 by the author. Licensee MDPI, Basel, Switzerland. This article is an open access article distributed under the terms and conditions of the Creative Commons Attribution (CC BY) license (https://creativecommons.org/licenses/by/4.0/).

Abstract: For the assumed bale volume, its dimensions (diameter, height), minimizing the consumption of the plastic film used for bale wrapping with the combined 3D method, depend on film and wrapping parameters. Incorrect selection of these parameters may result in an optimal bale diameter, which differs significantly from its height, while in agricultural practice bales with diameters equal or almost equal to the height dominate. The aim of the study is to formulate and solve the problem of selecting such dimensions of the bale with a given volume that the film consumption is minimal and, simultaneously, the bale diameter is equal or almost equal to its height. Necessary and sufficient conditions for such equilibria of the optimal bale dimensions are derived in the form of algebraic equations and inequalities. Four problems of the optimal bale dimension design guaranteeing assumed equilibrium of diameter and height are formulated and solved; both free and fixed bale volume are considered. Solutions of these problems are reduced to solving the sets of simple algebraic equations and inequalities with respect to two variables: integer number of film layers and continuous overlap ratio in bottom layers. Algorithms were formulated and examples regarding large bales demonstrate that they can handle the optimal dimensions' equilibria problems.

Keywords: 3D bale wrapping method; equal bale dimensions; mathematical model; minimal film consumption; optimal bale dimensions; round bales

1. Introduction

Currently, the demand to limit consumption of the film used to wrap bales of agriculture materials has been receiving increasing attention to reduce both costs and damage to the environment caused by plastic waste [1–4]. For this purpose, both experimental studies comparing film usage for different wrapping conditions and methods [5–9] and model-based analytical approaches useful for estimation of the film usage [10–14] and its minimization [15–18] were used. In the last few years it has been shown that by appropriate optimal selection of the film width [15,16], overlaps between adjacent film strips [17], and film width and overlaps, together, [18] it is possible to reduce film consumption by up to 20%. The model-based optimization research concerned mainly conventional [15,17,18] and IntelliWrap [16] wrapping methods. The monograph [19] comprehends the issues of modelling of the film consumption for wrapping round bales and related optimization problems solved in the last few years.

Also, an appropriate choice of bale size dimensions (diameter, height) may guarantee decreasing film consumption [11,19]. In [11], where the dependence of the film consumption on the bale diameter for the conventional wrapping technique was investigated, the analytical analysis showed that the larger the bale diameter is, the lower is the film consumption per unit of bale volume, which led to the conclusion that the use of the bale with the largest permissible diameter ensures the smallest film consumption. These studies were based on a rough model which describes film usage as a continuous function of bale dimensions, film width and a number of wrapped film strips; however, the model did not take into account mechanical properties of the stretch film and the direct relation between the number of wrapped film strips and bale and film parameters.

The concept of optimal (from the point of view of film consumption) selection of bale dimensions for the assumed bale volume was first introduced in [15] for the conventional wrapping method. A more accurate model was used that describes the consumption of the film as a function of the bale and film dimensions, mechanical parameters of the film (Poisson ratio, unit deformation), overlap ratio and the number of bale rotations [12]. The film consumption per unit of the bale volume is used as a measure of film usage. Since it is very difficult to find the optimal bale dimensions minimizing the original exact film usage index due to the discontinuity of this index, near-optimal parameters, being as important as optimal parameters for engineering applications, are sought. It was shown in [15] that in the case of using the conventional wrapping method, the optimal bale height is twice its optimal diameter. This optimality rule, which holds also for the IntelliWrap method [19], has only a theoretical character and is a consequence, among others, of multiple overlapping segments on the bale cylinder top and bottom, where there are 2–16 times more film layers than on the bale's lateral surface [10,19]. However, in the case of the combined 3D wrapping method [6,14], the optimal bale dimensions turn out to be useful and applicable from the engineering practice point of view.

In the 3D wrapping method, which offers the potential to minimize film usage [6,14,20] as well as to enhance the quality of silage [6], biaxial rotation of the film applicators results in two types of film layers wrapped perpendicularly: the bottom layers are wrapped around the bale's lateral surface and the upper layers are wrapped along the bale's longitudinal axis. For a detailed description of the wrapping process, see [6] and the producer's documentations [21–23]. The problem of the choice of the best bale dimensions to guarantee the minimal film consumption was solved in [20], where the optimality conditions were established in the form of algebraic cubic equations, which can easily be solved using both analytical and numerical methods. Analytical and numerical studies [19,20] have shown that the relation between the optimal diameter and height is not evident. The optimal diameter can be larger than the bale height [19] (Figure 6.5b), [20] (Figure 11c) or can be smaller than the bale height [19] (Figure 6.5d), [20] (Figure 11a), depending on the bale volume as well as film and wrapping parameters. But the optimal bale diameter can also be equal to its height [19] (Figure 6.5a,c), [20] (Figure 11b). The last case corresponds to typical large bales of 1.2 m diameter and height [10,24,25]. However, in agricultural practice typical bales are 1.2 to 1.3 m diameter and height [6,10,19], where diameter is equal [26–30] or almost equal [9,31,32], but not much bigger, than bale height. Other dimensions of large round bales, for example 1.2 m × 1.6 m Ø [33], or 1.2 m × 1.5 m Ø [34] or 1.5 m Ø [35], are used less frequently. Although round bales of the height greater than the diameter are also being investigated, for example 1.2 m × 0.9 m Ø [36], they are much less common in agriculture practice [37]. Standard 3D bale wrappers are designed to wrap cylindrical bales up to 1.6 m in diameter and up to 1.2 m in height [22], or up to 1.5 m diameter and 1.2 m height [21,23].

Therefore, the question becomes how to choose such film and wrapping process parameters for a round bale with a given volume wrapped using the 3D method so that the bale diameter minimizing film consumption is equal to or almost equal to its height. The aim of the paper is to solve the problem of the selection of the film and wrapping parameters (overlaps, numbers of bottom and upper film layers, etc.) that for a given bale volume, i.e., a given bale weight, its diameter and height are equal or near-equal and, at the same time, they minimize the film consumption. It is, in essence, the inverse problem in which the film and wrapping parameters are determined that the equilibria of optimal dimensions are guaranteed, while in the direct problem considered in [20] the optimal dimensions for a given parameters are sought. The thesis is that for any given bale volume and number of global film layers there exist film and wrapping parameters for which the optimal bale dimensions are equal or near-equal, with the pre-assumed proportions diameter/height. To prove the above, the necessary and sufficient conditions of equilibrium and near-equilibrium of the optimal bale dimensions were derived in the form of algebraic equations. The equilibrium conditions are dichotomous, as some depend on the given bale volume and others apply to any bale volume. Then, the problems of selecting such bale wrapping parameters for which the optimal bale dimensions are equal or near-equal were formulated and solved, separately, for a given and an arbitrary bale volume. Suitable design algorithms were proposed and numerically verified for large bales.

2. Materials and Methods

In this section, the notions of equal and near-equal optimal dimensions of a round bale are introduced and the related necessary and sufficient conditions are derived. The research methodology is also described.

2.1. Equal and Near-Equal Optimal Bale Dimensions

A complete mathematical model describing stretch film consumption for wrapping a bale with p_b bottom and p_u upper film layers using the combined 3D technique was derived in [14]; the main formula describing film usage is recalled in Appendix A with the corresponding assumptions. Symmetry of the bale is assumed. Mechanical properties of the plastic film are described by the Poisson's ratio v_f and unit deformation ε_{1f}; thickness of the film is ignored, e.g., polyethylene film is 25 μm thick [38]. The main symbols are summarized in Nomenclature, Appendix C.

The problem of the choice of bale dimensions (diameter, height) minimizing the consumption of the film used to wrap a cylindrical bale by 3D method has been stated and solved for the first time in [20], where the necessary and sufficient condition of the existence of the unique optimal bale diameter D_b^* of the bale of pre-assumed volume V_{b0} was derived in the form of cubic equation with zero linear term coefficient [20]:

$$\frac{2\pi p_u}{\Omega(k_f)}(D_b^*)^3 + \frac{\pi p_b}{1-k_{fb}}(2\delta - b_{fr}k_{fb})(D_b^*)^2 - 4V_{b0}\left(\frac{p_b}{1-k_{fb}} + \frac{p_u}{\Omega(k_f)}\right) = 0, \quad (1)$$

where b_{fr} is the width of stretched film described by Equation (A1), δ denotes the overlap of the extreme film strips in bottom film layers at the bases of the bale, k_{fb} and k_f are the overlap ratios determining the width of the contact between adjacent film strips in bottom and upper film layers; function $\Omega(k_f)$ is defined by Equation (A3). The coefficients in the first and second terms and the free term of Equation (1) depend on the pre-assumed bale volume as well as all the film and wrapping parameters. Based on the optimality condition, Equation (1), analytical and numerical analysis of the influence of film width, pre-assumed bale volume and numbers of bottom and upper film layers on optimal bale dimensions and optimal film consumption were carried out in [20], where many detailed conclusions regarding the impact of these parameters were formulated. Additionally, the influence of the bottom layers overlaps was studied in [19].

Equation (1) has one real positive root [20]. The corresponding optimal bale height H_b^* is [20]

$$H_b^* = \frac{4V_{b0}}{\pi(D_b^*)^2}. \quad (2)$$

For given V_{b0} the optimal D_b^* and H_b^* depend on film and wrapping parameters, they are linearly dependent and the difference between them is described by

$$D_b^* - H_b^* = \frac{1}{2}H_b^*\left[\frac{p_b\Omega(k_f)}{p_u(1-k_{fb})} - 1\right] - \frac{p_b\Omega(k_f)}{2p_u(1-k_{fb})}(2\delta - b_{fr}k_{fb}). \quad (3)$$

From a quick inspection of Equation (3) it can be seen that the relation between the optimal diameter D_b^* and height H_b^* is not evident. The conducted research [19,20] has shown that the relations $D_b^* > H_b^*$ and $D_b^* < H_b^*$, as well as $D_b^* = H_b^*$ are possible, depending on the values of the film and wrapping parameters and volume V_{b0}; compare [20] (Figure 11), where the proportions diameter/height are depicted for large bales. Only the relation $D_b^* \geq H_b^*$, desirable from the point of view of baling systems engineering, will be considered. The proportion diameter/height do not exceed 1.25 [22] (RW 1819, 150/120), or 1.2 [21] (BW 1850, 150/125), or [23] (WM 1851, 150/125); however, predominantly, it is of 1.05 order [21–23].

The goal of this paper is to study the selection of equal and nearly-equal optimal bale dimensions, therefore, the respective precise definitions are given.

Definition 1. (equal optimal dimensions). Bale diameter D_b^* and height H_b^* of bale with a volume V_{b0} are called equal optimal dimensions, if D_b^* accomplishes Equation (1) and H_b^*, given by Equation (2), is equal to D_b^*, i.e., $D_b^* = H_b^*$.

Definition 2. (near-equal optimal dimensions). Bale diameter D_b^* and height H_b^* of bale with a volume V_{b0} are called near-equal optimal dimensions, if D_b^* accomplishes Equation (1), H_b^* is given by Equation (2), and

$$\frac{D_b^*}{H_b^*} = 1 + \varepsilon_0, \qquad (4)$$

where a sufficiently small $\varepsilon_0 > 0$. Then D_b^* and H_b^* are called nearly-equal with order ε_0.

2.2. Methodology

In this paper, a model-based analytical approach was applied, which addressed the goals (derivation of the conditions for equal and near-equal optimal bale dimensions and solving the problems of wrapping parameter design guaranteeing such equilibria) by using mathematical tools. The solutions of the optimality conditions for the bale diameter were obtained by applying numerical tools.

Firstly, the necessary and sufficient conditions under which for the given bale volume the optimal bale diameter is near-equal to its optimal height with order ε_0 were derived using the optimality conditions expressed by Equations (1) and (2); Proposition 1 abstracts these results (Section 2.3.1). The conditions for optimal bale dimensions equilibria are given by algebraic equations. One of them uniquely relates bale volume to order ε_0 and film and wrapping parameters, i.e., this condition is satisfied for a given volume. The complementary condition determined by two simple equations relating only to ε_0 and film and wrapping parameter applies regardless of the bale volume. Based on these conditions, using differential calculus, the relationships between the volume of the bale with near-equal optimal dimensions and the order ε_0, width of the film and overlaps were analyzed (Section 2.3.2). Next, laying $\varepsilon_0 = 0$ the conditions for exact equilibrium of the optimal bale dimensions was obtained directly from the conditions of their near-equilibrium (Section 2.3.3).

Knowing the conditions of equilibrium and near-equilibrium of optimal bale dimensions, the tasks of designing the wrapping process in such a way that these conditions would be satisfied were formulated and solved. Due to the dichotomous nature of the equilibrium conditions, design tasks for a given bale volume and those in which the bale volume is arbitrary were considered separately; these are covered in Section 3.1 and Section 3.2, respectively. Consequently, four design problems were solved; however, the solutions for equal optimal bale dimensions resulted directly from those for near-equilibrium by substituting $\varepsilon_0 = 0$ (Sections 3.3 and 3.4). It was assumed that the film parameters (width, mechanical parameters v_f, ε_{lf}) are known, i.e., a practically available plastic film can be used, e.g., a commercial PE film used traditionally due to its mechanical properties and low costs [26,27,39]. Also the global number of film layers p_l wrapped on the bale's lateral surface was taken as a given. Many studies have investigated the number of desired film layers for baled silage preservation, for example [7,29,40,41]. Mostly, four, six or eight layers of film are applied [6,24,35]; however, ten, twelve, and even sixteen film layers in which the silages are wrapped are also considered [7,22,23,29]. Therefore no specific assumptions were taken concerning the number of global film layers; the numerical studies were conducted for p_l from four to sixteen. The standard overlaps 50%, 67% or 75% resulting in uniform film coverage [17] were assumed for upper film layers. Thus, the bottom layers overlaps δ, k_{fb} and film layers decomposition $p_l = p_b + p_u$ had to be selected. The upper and lower limits were taken for the overlaps, the selection of which should take into account the knowledge and experience of baling as well as the possibilities of the wrappers available. Consequently, the considered design problems consisted in solving a set of algebraic equations (resulting from the conditions of the equilibria of optimal bale dimensions) and inequalities (the constraints of variables) with respect to three decision variables. Two variables—the overlaps δ, k_{fb}—were continuous, while the third—the number of bottom film layers p_b—was integer. These mixed (hybrid) sets of equations and inequalities were solved analytically, separately for fixed and free bale volume. The solutions are abstracted by Propositions 4–7. For volume-free problems the solution is given by two inequalities directly related to the integer p_b and simple rules for computing the overlaps δ, k_{fb}. In the case of fixed bale volume the solution was derived in the form of two inequalities related to the overlap ratio k_{fb}, which must be verified for every considered p_b and unique algebraic rule for the determining of the overlap δ. If equal optimal bale dimensions are sought then these inequalities and formulas take particularly simple forms which are obtained by substituting $\varepsilon_0 = 0$. Computational algorithms were developed which enable designation of the sought wrapping parameters in a few steps. The examples illustrated how to use them (Excel is enough) as well as the effectiveness for optimal bale dimensions selection for a standard large bale.

In sum, in the multistage process several problems of optimal equal and near-equal bale dimensions design for cylindrical bales of a given volume were formulated and solved. The research framework is graphically shown in Figure 1, which also illustrates the relations between these problems.

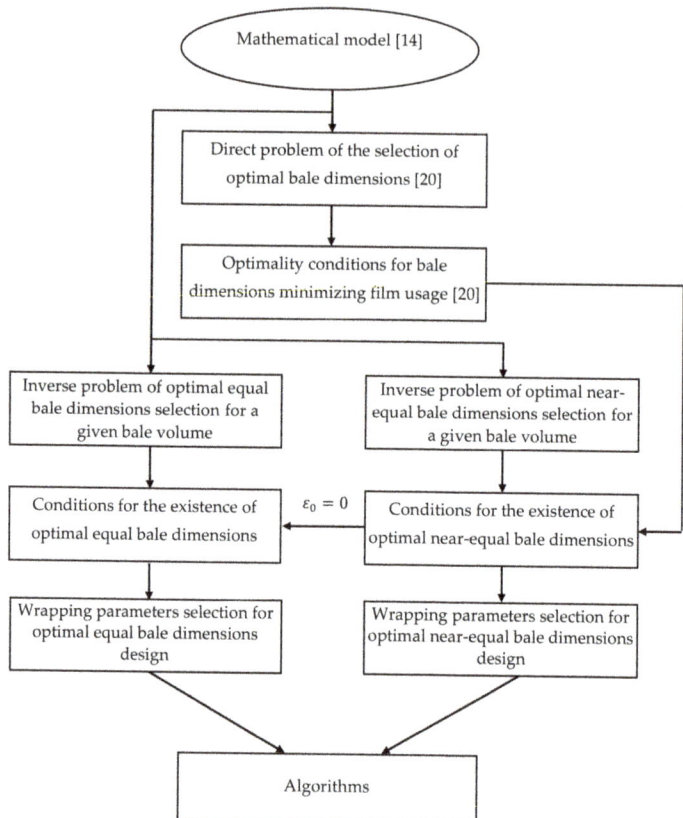

Figure 1. Schematic framework for the tasks of designing optimal equal and near-equal bale dimensions.

2.3. Necessary and Sufficient Conditions of Bale Dimensions Equilibrium and Near-Equilibrium
2.3.1. Near-Equal Optimal Dimensions

By Definition 2, the optimal bale dimensions are such that Equation (4) holds and the optimality condition expressed by Equation (1) is satisfied. For a given bale volume V_{b0} Equations (2) and (4) yield

$$D_b^* = \sqrt[3]{\frac{4(1+\varepsilon_0)V_{b0}}{\pi}} \tag{5}$$

and

$$H_b^* = \sqrt[3]{\frac{4V_{b0}}{\pi(1+\varepsilon_0)^2}}. \tag{6}$$

The necessary and sufficient conditions of near-equal optimal dimensions are given by the following proposition, which is proved in Appendix B.

Proposition 1. *For given: bale volume V_{b0}, width b_f of the film and its mechanical parameters v_f, ε_{lf}, numbers of film layers p_b, p_u and the overlaps δ, k_{fb}, k_f, such that the applicability condition expressed by Equation (A4) holds, the optimal bale diameter D_b^* is near-equal to its optimal height H_b^* with order ε_0 in the sense of Definition 2 if and only if one (and only one) of the conditions is satisfied:*

(i) order ε_0 and the bale, film, and wrapping parameters are related by the equation

$$\frac{p_b\Omega(k_f)\left(2\delta - b_{fr}k_{fb}\right)\left(\sqrt[3]{1+\varepsilon_0}\right)^2}{p_b\Omega(k_f) - p_u(1-k_{fb}) - 2p_u(1-k_{fb})\varepsilon_0} = \sqrt[3]{\frac{4V_{b0}}{\pi}}, \tag{7}$$

(ii) the equations

$$p_b\Omega(k_f) = p_u(1-k_{fb}) + 2p_u(1-k_{fb})\varepsilon_0 \tag{8}$$

and

$$2\delta - b_{fr}k_{fb} = 0 \tag{9}$$

are satisfied, simultaneously. The optimal bale diameter D_b^* and height H_b^* are described by Equations (5) and (6), respectively; in case (i) for a given bale volume, in case (ii) for an arbitrary volume V_{b0}.

The necessary and sufficient conditions for near-equal optimal bale dimensions, case (ii), are independent of V_{b0}; they are volume–free. However, those from case (i) hold for given V_{b0}, i.e., they are volume-fixed.

Note, that the inequality

$$\frac{2\delta - b_{fr}k_{fb}}{p_b\Omega(k_f) - p_u(1-k_{fb}) - 2p_u(1-k_{fb})\varepsilon_0} > 0 \tag{10}$$

must hold to satisfy Equation (7); however, the expression $\left(2\delta - b_{fr}k_{fb}\right)$ may be positive or negative, depending on the sign of the denominator. The necessary condition expressed by Equation (10) is independent on V_{b0}, while Equation (7) depends on the particular value of bale volume V_{b0}. Both conditions depend on ε_0.

2.3.2. Bale Volume of Near-Equilibrium

From Equation (7), provided that inequality from Equation (10) is satisfied, we have

$$V_{b0} = \frac{\pi}{4}\left[\frac{p_b\Omega(k_f)\left(2\delta - b_{fr}k_{fb}\right)\left(\sqrt[3]{1+\varepsilon_0}\right)^2}{p_b\Omega(k_f) - p_u(1-k_{fb}) - 2p_u(1-k_{fb})\varepsilon_0}\right]^3, \tag{11}$$

whence

$$\frac{\partial V_{b0}}{\partial \varepsilon_0} = \frac{\pi p_b\Omega(k_f)\left(2\delta - b_{fr}k_{fb}\right)}{2\sqrt[3]{1+\varepsilon_0}} V_{b0}^{2/3} \frac{\left[p_b\Omega(k_f) + 2p_u(1-k_{fb}) + p_u(1-k_{fb})\varepsilon_0\right]}{\left[p_b\Omega(k_f) - p_u(1-k_{fb}) - 2p_u(1-k_{fb})\varepsilon_0\right]^2}.$$

Thus, for given film and wrapping parameters, bale volume resulting in near-equal optimal dimensions is monotonically increasing function of ε_0 whenever

$$2\delta > b_{fr}k_{fb}, \tag{12}$$

and decreases with ε_0, if $2\delta < b_{fr}k_{fb}$. If inequality from Equation (12) holds, then the volume V_{b0} decreases if the overlap k_{fb} grows, while in the opposite case the influence of k_{fb} is not so evident and depends on the sign of the expression in the square brackets of the numerator in the last fraction of the right hand side of the following equation

$$\frac{\partial V_{b0}}{\partial k_{fb}} = \frac{3\pi p_b\Omega(k_f)\left(\sqrt[3]{1+\varepsilon_0}\right)^2}{4} V_{b0}^{2/3} \frac{\left[-b_{fr}p_b\Omega(k_f) + b_{fr}p_u(1-2\delta) + 2p_u\varepsilon_0(b_{fr}-2\delta)\right]}{\left[p_b\Omega(k_f) - p_u(1-k_{fb}) - 2p_u(1-k_{fb})\varepsilon_0\right]^2}.$$

If condition from Equation (12) holds, then V_{b0} given by Equation (11) grows with the increase of the overlap δ. In the opposite case the larger δ is, the smaller is the volume resulting in near-equal optimal dimensions. From the analysis of Equation (11) it also follows that the greater film width is, the smaller is V_{b0}, provided that inequality from Equation (12) holds. The influence of the bottom layers overlaps k_f and δ is illustrated in Figure 2a,b, where near-equilibrium volume V_{b0} is depicted as a function of ε_0 for plastic film (e.g., polyethylene, PE) characterized by Poisson's ratio $v_f = 0.34$ [38] and unit deformation $\varepsilon_{lf} = 0.7(-)$ [7,12,14] of popular width $b_f = 0.75$ m [21–23,29,42]. Four bottom and upper film layers $p_b = p_u = 4$ and overlap ratio $k_f = 0.5$ are assumed. Figure 3a,b illustrate the effect of the film width b_f for commonly used overlap $\delta = 0.2$ m [21–23] and two overlap ratios $k_{fb} = 0.3, 0.35$; other parameters remain unchanged. The volume $V_{b0} = 1.357$ m^3 corresponding to a standard large bale of diameter and height $D_b = H_b = 1.2$ m [24,25] is also marked on these pictures. Finally, note that for standard overlap $\delta = 0.2$ m [6,14,20–23], PE film and since $k_{fb} < \frac{1}{2}$, the condition from Equation (12) holds, in particular, if the film width $b_f < 1.052$ m. For $\delta = 0.15$ m this requirement is reduced to $b_f < 0.789$ m. For exemplary $k_{fb} = 0.4$ and $\delta = 0.2$ m inequality from Equation (12) is satisfied for $b_f < 1.31$ m; if $\delta = 0.15$ m this requirement is sharpened to $b_f < 0.99$ m.

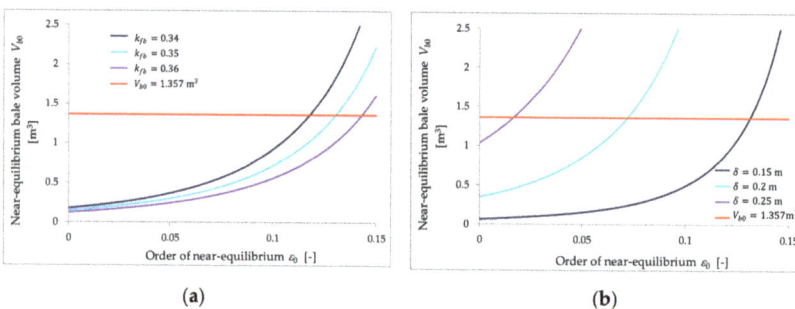

Figure 2. The volume V_{b0}, Equation (11), yielding the near-equal optimal bale dimensions as a function of the order ε_0 of near-equilibrium for $p_b = p_u = 4$ bottom and upper film layers, film width $b_f = 0.75$ m and wrapping parameters: (a) overlaps $\delta = 0.2$ m and $k_{fb} = 0.25, 0.3, 0.32$, (b) $k_{fb} = 0.3$ and $\delta = 0.15, 0.2, 0.25$ m.

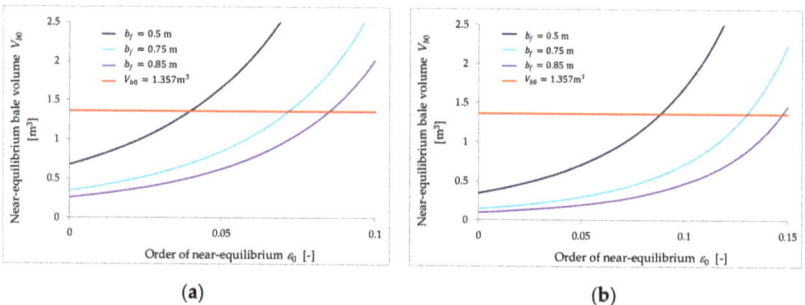

Figure 3. The volume V_{b0}, Equation (11), yielding the near-equal optimal bale dimensions as a function of the order ε_0 of near-equilibrium for $p_b = p_u = 4$ bottom and upper film layers, overlap $\delta = 0.2$ m and parameters: (a) overlap ratio $k_{fb} = 0.3$ and film widths $b_f = 0.5, 0.75, 0.85$ m, (b) overlap ratio $k_{fb} = 0.35$ and film widths $b_f = 0.5, 0.75, 0.85$ m.

Since in practically used wrapping systems the condition expressed by Equation (12) is usually satisfied and the near-equilibrium bale volume increases with the order ε_0, a question arises about the conditions to be met by the film and wrapping parameters to guarantee for any $0 \leq \varepsilon \leq \varepsilon_0$ the existence of a bale volume such that the optimal dimensions D_b^*, H_b^* are near-equal with order ε. The following result, derived in Appendix B based on Proposition 1, answers this question. The notation $V_{b0}(\varepsilon)$, $D_b^*(\varepsilon)$, and $H_b^*(\varepsilon)$ is introduced, locally, for near-equal bale parameters to emphasize the relationship between ε and V_{b0}, D_b^*, and H_b^*.

Proposition 2. *For given film parameters b_f v_f, ε_{lf}, numbers of film layers p_b, p_u and the overlaps δ, k_{fb}, k_f, such that the applicability condition expressed by Equation (A4) holds, for any $0 \leq \varepsilon \leq \varepsilon_0$, where a sufficiently small $\varepsilon_0 > 0$, there exists bale volume $V_{b0}(\varepsilon) > 0$ such that the optimal bale diameter D_b^* is near-equal to its optimal height H_b^* with order ε not greater than ε_0, i.e.,*

$$1 \leq \frac{D_b^*}{H_b^*} = 1 + \varepsilon \leq 1 + \varepsilon_0,$$

if and only if

$$p_b \Omega(k_f) - p_u(1 - k_{fb}) - 2p_u(1 - k_{fb})\varepsilon_0 \neq 0 \tag{13}$$

and one of the following conditions is satisfied:
(a) *inequality from Equation (12) holds,*
(b) *two inequalities*

$$2\delta - b_{fr}k_{fb} < 0, \tag{14}$$

$$p_b \Omega(k_f) - p_u(1 - k_{fb}) < 0, \tag{15}$$

hold, simultaneously. Then, for any $0 \leq \varepsilon \leq \varepsilon_0$ and volume $V_{b0}(\varepsilon)$ Equation (7) holds for $\varepsilon_0 = \varepsilon$. The optimal bale dimensions $D_b^*(\varepsilon)$, $H_b^*(\varepsilon)$ are expressed by equations:

$$D_b^*(\varepsilon) = \frac{p_b \Omega(k_f)(2\delta - b_{fr}k_{fb})(1+\varepsilon)}{p_b \Omega(k_f) - p_u(1 - k_{fb}) - 2p_u(1 - k_{fb})\varepsilon}, \tag{16}$$

$$H_b^*(\varepsilon) = \frac{p_b \Omega(k_f)(2\delta - b_{fr}k_{fb})}{p_b \Omega(k_f) - p_u(1 - k_{fb}) - 2p_u(1 - k_{fb})\varepsilon}. \tag{17}$$

In case (a), $V_{b0}(0) \leq V_{b0}(\varepsilon) \leq V_{b0}(\varepsilon_0)$ and $D_b^*(\varepsilon)$, $H_b^*(\varepsilon)$ are monotonically increasing for $0 \leq \varepsilon \leq \varepsilon_0$. In case (b), $V_{b0}(0) \geq V_{b0}(\varepsilon) \geq V_{b0}(\varepsilon_0)$ and $D_b^*(\varepsilon)$, $H_b^*(\varepsilon)$ are monotonically decreasing for $0 \leq \varepsilon \leq \varepsilon_0$. If inequalities from Equations (13) and (14) hold, but inequality expressed by Equation (15) not, then there exists positive $\bar{\varepsilon}$ such that $0 < \bar{\varepsilon} < \varepsilon_0$ and

$$p_b \Omega(k_f) - p_u(1 - k_{fb}) - 2p_u(1 - k_{fb})\bar{\varepsilon} = 0. \tag{18}$$

For any $\bar{\varepsilon} < \varepsilon \leq \varepsilon_0$ bale volume $V_{b0}(\varepsilon)$ satisfying Equation (7) and optimal bale dimensions $D_b^*(\varepsilon)$, $H_b^*(\varepsilon)$ are near-equal in the sense of the inequalities

$$1 + \bar{\varepsilon} < \frac{D_b^*(\varepsilon)}{H_b^*(\varepsilon)} = 1 + \varepsilon \leq 1 + \varepsilon_0.$$

Both $V_{b0}(\varepsilon)$ and $D_b^*(\varepsilon)$, $H_b^*(\varepsilon)$ are monotonically decreasing for $\bar{\varepsilon} < \varepsilon \leq \varepsilon_0$.

Inequality expressed by Equation (12) holds for narrow films, while inequality form Equation (14) is satisfied for wider films.

2.3.3. Equal Optimal Dimensions

In this case the optimal bale diameter is equal to bale height, i.e., by Equation (2)

$$D_b^* = H_b^* = \sqrt[3]{\frac{4V_{b0}}{\pi}}, \tag{19}$$

and the optimality condition expressed by Equation (1) is satisfied, simultaneously. The necessary and sufficient conditions for the equilibrium are given by the following proposition, which results directly from Proposition 1 for $\varepsilon_0 = 0$.

Proposition 3. *For given: bale volume V_{b0}, width of the film b_f and its mechanical parameters v_f, ε_{lf}, numbers of film layers p_b, p_u and the overlaps δ, k_{fb}, k_f, such that the applicability condition expressed by Equation (A4) holds, the optimal bale diameter D_b^* is equal to its optimal height H_b^* if and only if one (and only one) of the conditions is satisfied:*

(i) the bale, film and wrapping parameters are related by the equation

$$\frac{p_b \Omega(k_f)(2\delta - b_{fr} k_{fb})}{p_b \Omega(k_f) - p_u(1 - k_{fb})} = \sqrt[3]{\frac{4V_{b0}}{\pi}},$$

(ii) the equation

$$p_b \Omega(k_f) = p_u(1 - k_{fb})$$

and Equation (9) are satisfied, simultaneously. The optimal bale dimensions are given directly by Equation (19) for fixed volume V_{b0} in case (i) or for arbitrary volume in case (ii).

Note, that in case (i) the necessary condition for bale dimensions equilibrium expressed by Equation (10) takes the form

$$\frac{2\delta - b_{fr} k_{fb}}{p_b \Omega(k_f) - p_u(1 - k_{fb})} > 0.$$

3. Results and Discussion

In this section four problems of the selection of optimal bale dimensions being equal and near-equal are formulated, solved and illustrated by related examples.

It is known [17] that any overlap ratio of the form of irreducible fraction in which dividend is the divisor minus one, i.e.,

$$k_f = k_{f,u} = \frac{q-1}{q}, \tag{20}$$

where $q \in \mathcal{N}$, $\mathcal{N}$ denotes the set of positive integer numbers, results in the uniform film distribution on the bale's whole lateral surface and guarantee the same minimal film usage. An important special case for even p_u is $q = 2$, which means 50% overlap between the successive film strips [6,17,24,43]. For $k_f = k_{f,u}$ function $\Omega(k_f) = 1$ and since $p_u = p_l - p_b$, the applicability condition expressed by Equation (A4) takes the form

$$\frac{p_l - p_b}{q} = m, \ m \in \mathcal{N}. \tag{21}$$

From the practical perspective, only four or five smallest such overlap ratios are useful [6,17,22,24,43,44], i.e., $2 \leq q \leq 5$ are worth considering.

Bearing in mind the dichotomous nature of the equilibrium conditions specified by Propositions 1 and 3, the problems of volume-free and volume-fixed near-equal and equal bale optimal dimensions design will be considered, separately.

3.1. Volume-Free Near-Equal Optimal Bale Dimensions Design

Let us consider the following problem of the optimal choice of bale dimensions.

Problem 1. *Given film parameters b_f, v_f, ε_{lf}, upper layers overlap ratio k_f, Equation (20), number of global film layers p_l, and the order of near-equilibrium ε_0. Find film layers decomposition (p_b, p_u) and the bottom layers overlaps*

$$\delta_{min} \leq \delta \leq \delta_{max}, \tag{22}$$

where δ_{min} and δ_{max} are the smallest and largest admissible overlap δ, and

$$k_{fb,min} \leq k_{fb} \leq k_{fb,max} < \frac{1}{2}, \tag{23}$$

where $k_{fb,min}$ and $k_{fb,max}$ are the smallest and largest admissible k_{fb}, such that the applicability condition, Equation (21), holds and for any bale volume V_{b0} the optimal bale dimensions D_b^*, H_b^* are near-equal with given order ε_0.

In view of Proposition 1, case (ii), the solution to the above problem exists if and only if there exist integer p_b, p_u and continuous parameters k_{fb}, δ satisfying Equations (8) and (9), inequalities expressed by Equations (22) and (23), and the applicability condition, Equation (21). This set of equations and inequalities is solved in the Appendix B, this solution is summarized in the next result.

Proposition 4. *If inequality*

$$b_{fr}(2 + 2\varepsilon_0) > 2\delta_{max}(1 + 2\varepsilon_0) \tag{24}$$

holds, then the solution to the Problem 1 exists if and only if there exists an integer p_b such that inequalities:

$$p_l \frac{(1 - k_{fb,max})(1 + 2\varepsilon_0)}{(2 + 2\varepsilon_0) - k_{fb,max}(1 + 2\varepsilon_0)} \leq p_b \leq p_l \frac{(1 - k_{fb,min})(1 + 2\varepsilon_0)}{(2 + 2\varepsilon_0) - k_{fb,min}(1 + 2\varepsilon_0)}, \tag{25}$$

$$p_l \frac{(b_{fr} - 2\delta_{max})(1 + 2\varepsilon_0)}{b_{fr}(2 + 2\varepsilon_0) - 2\delta_{max}(1 + 2\varepsilon_0)} \leq p_b \leq p_l \frac{(b_{fr} - 2\delta_{min})(1 + 2\varepsilon_0)}{b_{fr}(2 + 2\varepsilon_0) - 2\delta_{min}(1 + 2\varepsilon_0)}, \tag{26}$$

are satisfied together with the applicability condition, Equation (21). For any p_b solving Problem 1 the overlap radio k_{fb} is given by equation

$$k_{fb} = \frac{p_l(1 + 2\varepsilon_0) - p_b(2 + 2\varepsilon_0)}{(p_l - p_b)(1 + 2\varepsilon_0)}, \tag{27}$$

while by Equation (9) the overlap $\delta = \frac{1}{2} b_{fr} k_{fb}$. For the assumed bale volume optimal bale dimensions D_b^* and H_b^* are given by Equations (5) and (6), respectively.

Problem 1 is a mixed decision problem with integer variable p_b and continuous variables δ, k_{fb}. From Proposition 4 the following algorithm follows.

3.1.1. Algorithm 1

Assume the inequality expressed by Equation (24) is satisfied and film parameters b_f, v_f, ε_{lf}, upper layers overlap ratio k_f, Equation (20), number of global film layers p_l, and the order of near-equilibrium ε_0 are given. Take minimal δ_{min}, $k_{fb,min}$ and maximal δ_{max}, $k_{fb,max}$ values of the overlaps δ, k_{fb} based on the knowledge and experience of baling to guarantee an appropriate tightness of the wrappings.

1. Determine the set $\mathcal{P}_1$ of all integer p_b defined by the inequalities

$$\max\left\{p_l\frac{1-k_{fb,max}}{\frac{2+2\varepsilon_0}{1+2\varepsilon_0}-k_{fb,max}}, p_l\frac{b_{fr}-2\delta_{max}}{\frac{2+2\varepsilon_0}{1+2\varepsilon_0}b_{fr}-2\delta_{max}}\right\} \leq p_b \leq \min\left\{p_l\frac{1-k_{fb,min}}{\frac{2+2\varepsilon_0}{1+2\varepsilon_0}-k_{fb,min}}, p_l\frac{b_{fr}-2\delta_{min}}{\frac{2+2\varepsilon_0}{1+2\varepsilon_0}b_{fr}-2\delta_{min}}\right\}, \quad (28)$$

   for which there exists $2 \leq q \leq 5$ satisfying applicability condition, Equation (21).
2. If the set $\mathcal{P}_1$ is empty, then the solution to Problem 1 does not exist—go to step 3. Otherwise, go to step 4.
3. Change the lower $k_{fb,min}$, δ_{min} or upper $k_{fb,max}$, δ_{max} bounds of wrapping parameters, or the order ε_0, or the film width b_f and repeat the computations starting from step 1.
4. For any $p_b \in \mathcal{P}_1$ compute the overlap ratio k_{fb} according to Equation (27) and, next, the overlap $\delta = \frac{1}{2}b_{fr}k_{fb}$.
5. If a bale volume V_{b0} is assumed, then for any $p_b \in \mathcal{P}_1$ compute film usage FC obtained for optimal D_b^*, H_b^*, Equations (5) and (6), using Equation (A2) and choose that p_b which yields the minimal film consumption.

Application of Proposition 4 and Algorithm 1 is illustrated by an example.

3.1.2. Example 1

Plastic film of the parameters $v_f = 0.34$, $\varepsilon_{lf} = 0.7(-)$ and $b_f = 0.75$ m is assumed. Even numbers of global film layers $4 \leq p_l \leq 16$ and "uniform" overlap ratios k_f, Equation (20), are considered. Firstly, the following bounds for bottom layers overlaps are assumed: $k_{fb,min} = 0.2$, $k_{fb,max} = 0.45$, $\delta_{min} = 0.15$ m, $\delta_{max} = 0.3$ m. Three orders $\varepsilon_0 = 0.05, 0.1, 0.15$ of near-equilibrium are considered. Inequality from Equation (24) is satisfied. Unfortunately, the set $\mathcal{P}_1$ of integer solutions p_b of the inequalities expressed by Equation (28) is empty for all ε_0 considered. When the lower constraint of δ is changed into $\delta_{min} = 0.1$ m, then for each ε_0 the set $\mathcal{P}_1$ is composed of a few p_b. They are listed in Table 1 together with respective overlaps k_{fb} and δ. Bale volume $V_{b0} = 1.357$ m^3 is assumed, film usage for the optimal bale dimensions is computed and given in the last column of Table 1; for all p_l for which the set $\mathcal{P}_1$ is nonempty, there is only one $p_b \in \mathcal{P}_1$. Note, that for given p_l film usage obtained for D_b^*, H_b^* and decomposition (p_b, p_u) are the same for all ε_0 considered. Thus, the smallest ε_0 can be chosen. For example, for $V_{b0} = 1.357$ m^3 taking $\varepsilon_0 = 0.05$ we obtain near-equal optimal dimensions $D_b^* = 1.22$ m and $H_b^* = 1.162$ m. The same regularity holds for wider film $b_f = 0.9$ m, for which the solutions to Problem 1 are given in Table 2.

Table 1. The numbers of global p_l and bottom p_b film layers and bottom layers overlaps k_{fb}, δ for which Problem 1 of volume-free near-equal optimal bale dimensions design has solution for the assumed order of near-equilibrium ε_0 and bale from Example 1; film width $b_f = 0.75$ m, the bounds for constraints from Equations (22) and (23): $k_{fb,min} = 0.2$, $k_{fb,max} = 0.45$, $\delta_{min} = 0.1$ m, $\delta_{max} = 0.3$ m, $4 \leq p_l \leq 16$ were considered. Film usage FC, Equation (A2), for the optimal bale dimensions for bale volume $V_{b0} = 1.357$ m^3.

ε_0	p_l	p_b	k_{fb} [−]	δ [m]	FC [m^{-1}]
0.05	10	4	0.394	0.113	50.815
	12	5	0.351	0.100	61.958
0.1	10	4	0.444	0.127	50.815
	12	5	0.405	0.116	61.958
	14	6	0.375	0.107	71.541
	16	7	0.352	0.101	81.124
0.15	14	6	0.423	0.121	71.541
	16	7	0.402	0.115	81.124

Table 2. The numbers of global p_l and bottom p_b film layers and bottom layers overlaps k_{fb}, δ for which Problem 1 of volume-free near-equal optimal bale dimensions design has solution for given order of near-equilibrium ε_0 and bale from Example 1; film width $b_f = 0.9$ m, the bounds for constraints from Equations (22) and (23): $k_{fb,min} = 0.2$, $k_{fb,max} = 0.45$, $\delta_{min} = 0.1$ m, $\delta_{max} = 0.3$ m, $4 \le p_l \le 16$ were considered. Film usage FC, Equation (A2), for the optimal bale dimensions for bale volume $V_{b0} = 1.357$ m^3.

ε_0	p_l	p_b	k_{fb} [−]	δ [m]	FC [m^{-1}]
0.05	10	4	0.394	0.135	49.478
	12	5	0.351	0.120	59.507
	14	6	0.318	0.109	67.664
	16	7	0.293	0.100	77.693
0.1	10	4	0.444	0.152	55.360
	12	5	0.405	0.139	59.507
	14	6	0.375	0.129	67.664
	16	7	0.352	0.121	77.693
0.15	14	6	0.423	0.145	67.664
	16	7	0.402	0.138	77.693

3.1.3. Effect of the Non-Equilibrium Order ε_0

Parameter ε_0 influences the constraints expressed by Equations (25) and (26). If ε_0 grows, then for given p_l both the lower and upper bounds for p_b expressed by Equations (25) and (26) increase provided that the inequality form Equation (24) holds; however, these changes are not significant (see Table 1).

3.1.4. Effect of the Film Width

Film width b_f, or equivalently b_{fr}, influence the inequality from Equation (24); if b_{fr} grows, then this inequality is still satisfied. Also, the lower and upper bounds from inequalities expressed by Equation (26) grow with b_{fr}. Thus, the increase of the film width can extend the set of integer p_b which solve Problem 1. If, for example, the width $b_f = 0.9$ m, then for $\delta_{min} = 0.15$ m the set of solutions to Problem 1 for $\varepsilon_0 = 0.1$ is not empty; for $p_l = 10$ integer $p_b = 4$ and the bottom layers overlaps $k_{fb} = 0.444$, $\delta = 0.152$ m solve this task. The solutions to Problem 1 for $b_f = 0.9$ m and $\delta_{min} = 0.1$ m are summarized in Table 2.

3.2. Volume-Fixed Near-Equal Optimal Bale Dimensions Design

Based on Proposition 1, case (i), the following problem of bale dimensions optimal design was formulated.

Problem 2. Given: film parameters b_f, v_f, ε_{lf}, upper layers overlap ratio k_f, Equation (20), number of global film layers p_l, bale volume V_{b0}, and order of near-equilibrium ε_0. Find film layers decomposition (p_b, p_u) and the bottom layers overlaps δ, k_{fb} satisfying constraints expressed by Equations (22) and (23) such that the optimal bale dimensions D_b^*, H_b^* yielding bale volume V_{b0} are near-equal with order ε_0 and results in the minimal film usage.

Based on Proposition 1 and the properties of film consumption index FC, Equation (A2), with non-continuous ceiling function the solution of the above problem is derived in Appendix B. The next proposition abstracts this solution.

Proposition 5. *The solution to Problem 2 exists if and only if there exists an integer p_b and k_{fb} such that inequalities:*

$$\check{k}_{fb}(p_b) \le k_{fb} \le \min\left\{\hat{k}_{fb}(p_b), \bar{\bar{k}}_{fb}(p_b), k_{fb,max}\right\}, \tag{29}$$

$$k_{fb} \neq \frac{(p_l - p_b)(1 + 2\varepsilon_0) - p_b}{(p_l - p_b)(1 + 2\varepsilon_0)} = k_{fb,ne}(p_b) \qquad (30)$$

are satisfied together with the applicability condition, Equation (21), where

$$\overline{k}_{fb}(p_b) = \frac{2\delta_{max}\left(\sqrt[3]{1+\varepsilon_0}\right)^2 - \sqrt[3]{\frac{4V_{b0}}{\pi}} + (1+2\varepsilon_0)\left(\frac{p_l}{p_b} - 1\right)\sqrt[3]{\frac{4V_{b0}}{\pi}}}{(1+2\varepsilon_0)\left(\frac{p_l}{p_b} - 1\right)\sqrt[3]{\frac{4V_{b0}}{\pi}} + b_{fr}\left(\sqrt[3]{1+\varepsilon_0}\right)^2}, \qquad (31)$$

$$\check{k}_{fb}(p_b) = \max\left\{\frac{2\delta_{min}\left(\sqrt[3]{1+\varepsilon_0}\right)^2 - \sqrt[3]{\frac{4V_{b0}}{\pi}} + (1+2\varepsilon_0)\left(\frac{p_l}{p_b} - 1\right)\sqrt[3]{\frac{4V_{b0}}{\pi}}}{(1+2\varepsilon_0)\left(\frac{p_l}{p_b} - 1\right)\sqrt[3]{\frac{4V_{b0}}{\pi}} + b_{fr}\left(\sqrt[3]{1+\varepsilon_0}\right)^2}, k_{fb,min}\right\}, \qquad (32)$$

and $\hat{k}_{fb}(p_b)$ is given by

$$\hat{k}_{fb}(p_b) = 1 - \frac{2}{\frac{b_{fr}\left(\sqrt[3]{1+\varepsilon_0}\right)^2}{\sqrt[3]{\frac{4V_{b0}}{\pi}}}\left[\frac{\sqrt[3]{\frac{4V_{b0}}{\pi}}\left[\frac{2}{1-\check{k}_{fb}(p_b)} - \left(\frac{p_l}{p_b} - 1\right)(1+2\varepsilon_0)\right]}{b_{fr}\left(\sqrt[3]{1+\varepsilon_0}\right)^2}\right] + \left(\frac{p_l}{p_b} - 1\right)(1+2\varepsilon_0)}. \qquad (33)$$

For any such p_b and k_{fb} the overlap δ is given by equation

$$\delta = \frac{\sqrt[3]{\frac{4V_{b0}}{\pi}}}{2\left(\sqrt[3]{1+\varepsilon_0}\right)^2}\left[1 - \left(\frac{p_l}{p_b} - 1\right)(1 - k_{fb})(1 + 2\varepsilon_0)\right] + \frac{b_{fr}k_{fb}}{2}, \qquad (34)$$

the optimal bale dimensions are described by Equations (5) and (6) and film usage

$$FC(p_b) = \frac{4b_f\sqrt[3]{1+\varepsilon_0}}{\left(\varepsilon_{lf}+1\right)\left[\sqrt[3]{\frac{4V_{b0}}{\pi}}\right]^2}\left\{p_b\left[\frac{\sqrt[3]{\frac{4V_{b0}}{\pi}}\left[\frac{2}{1-\check{k}_{fb}(p_b)} - \left(\frac{p_l}{p_b} - 1\right)(1+2\varepsilon_0)\right]}{b_{fr}\left(\sqrt[3]{1+\varepsilon_0}\right)^2}\right] + \frac{2(2+\varepsilon_0)}{\pi(1+\varepsilon_0)}\left[\frac{\pi\sqrt[3]{\frac{4V_{b0}}{\pi}}(1+\varepsilon_0)(p_l-p_b)}{2b_{fr}\left(\sqrt[3]{1+\varepsilon_0}\right)^2}\right]\right\}, \qquad (35)$$

where $\lceil x \rceil$ denotes ceiling function [45]. Integer p_b^* solving Problem 2 is such that

$$FC(p_b^*) = \min_{p_b} FC(p_b). \qquad (36)$$

It may be easily shown that if the following inequality

$$b_{fr}\left[1 - \frac{1}{\left(\frac{p_l}{p_b} - 1\right)(1 + 2\varepsilon_0)}\right] < 2\delta_{min} \qquad (37)$$

holds, then

$$k_{fb,ne}(p_b) < \overline{k}_{fb}(p_b) \qquad (38)$$

and condition from Equation (30) can be neglected. In all the tested examples the above inequality has been satisfied, compare Table 3 below. The inequality from Equation (37), similar to the condition expressed by Equation (12), holds especially if the number of bottom film layers is such that

$$p_b > \frac{1 + 2\varepsilon_0}{2 + 2\varepsilon_0}p_l.$$

The inequalities specified in the above proposition must be solved for two variables: integer p_b and continuous k_{fb}. For any fixed p_b the overlap δ given by Equation (34) increases linearly with growing k_{fb}.

3.2.1. Algorithm 2

Assume film parameters b_f, v_f, ε_{lf}, overlap ratio k_f, Equation (20), number of global film layers p_l, the order of near-equilibrium ε_0, and bale volume V_{b0} are given. Take minimal δ_{min}, $k_{fb,min}$ and maximal δ_{max}, $k_{fb,max}$ values of the overlaps δ, k_{fb} to guarantee the appropriate tightness of the wrappings.

1. Determine the set $\mathcal{P}_2$ of all integer p_b for which there exists $2 \leq q \leq 5$ satisfying applicability condition, Equation (21), and the set of overlap ratios $\mathcal{K}_2(p_b)$ defined by the inequalities expressed by Equations (29) and (30) is nonempty.
2. If the set $\mathcal{P}_2$ is empty, then the solution to Problem 2 does not exist—go to step 3. Otherwise, go to step 4.
3. Change the lower $k_{fb,min}$, δ_{min} or upper $k_{fb,max}$, δ_{max} bounds of wrapping parameters, or the order ε_0, or the film width b_f and go to step 1.
4. Solve in p_b^* the integer programming task expressed by Equation (36) for $p_b \in \mathcal{P}_2$.
5. For the best p_b^* choose practically reasonable $k_{fb} \in \mathcal{K}_2(p_b^*)$, compute the overlap $\delta = \delta(p_b^*)$ according to Equation (34) and optimal D_b^*, H_b^* using Equations (5) and (6). The minimal film consumption is equal to $FC(p_b^*)$ computed in step 4.

3.2.2. Example 2

Film parameters and overlaps constraints from Example 1 are taken, again. Bale volume $V_{b0} = 1.357 \text{ m}^3$ is assumed together with orders of near-equilibrium $\varepsilon_0 = 0.05$ and $\varepsilon_0 = 0.1$. Integer p_b, bottom $\check{k}_{fb}(p_b)$ and upper $\hat{k}_{fb}$, $\overline{\overline{k}}_{fb}$ bounds, parameters $k_{fb,ne}$ and closed intervals $\mathcal{K}_2$ of k_{fb} are given in Table 3. In all tested examples $k_{fb,ne} < \check{k}_{fb}(p_b)$. The upper bound of the closed interval of k_{fb} defined by Equation (29) is determined by $\hat{k}_{fb}(p_b)$, or $\overline{\overline{k}}_{fb}(p_b)$, or $k_{fb,max}$—there is no rule here. For any $k_{fb} \in \mathcal{K}_2$ the formulas from Equation (34) for computing δ are presented in the penultimate column. In the last column film usage $FC(p_b)$, Equation (35), is given. For most $p_b \in \mathcal{P}_2$ the closed intervals $\mathcal{K}_2(p_b)$ are wide enough to select a practically convenient overlap ratio k_{fb}. In some cases this interval is very narrow, e.g., for $p_l = 10$, $p_b = 6$, we have $\mathcal{K}_2 = [0.200, 0.204]$; however, $k_{fb} = 0.2$ is acceptable from the engineering point of view. For $\varepsilon_0 = 0.05$ and $p_l = 4, 6, 8$ set $\mathcal{P}_2$ is composed of only one number p_b of bottom film layers and this p_b solves Problem 2. For $p_l = 10, 12, 14, 16$ there are at least two p_b for which the set $\mathcal{K}_2(p_b)$ is nonempty, the optimal film usage $FC(p_b^*)$ defined in Equation (36) is marked by bold in Table 3. For $\varepsilon_0 = 0.05$, according to Equations (5) and (6), the optimal bale dimensions $D_b^* = 1.22 \text{ m}$, $H_b^* = 1.162 \text{ m}$, for $\varepsilon_0 = 0.1$ we have $D_b^* = 1.239 \text{ m}$, $H_b^* = 1.126 \text{ m}$. If $\varepsilon_0 = 0.05$ is assumed, then for $p_l = 4, 6, 8$ the overlaps $k_{fb} = 0.3$ and $\delta = 0.2195 \cong 0.22$ can be applied. For $p_l = 10$ Problem 2 is solved for the same overlaps for $p_b = 5$ yielding minimal film usage. Similarly, for $p_l = 16$ and $p_b = 8$. For $p_l = 12$ solution to Problem 2 is composed by $p_b = 7$ and, for example, $k_{fb} = 0.2$ and $\delta = 0.272$. For $p_l = 14$ we have: $p_b = 6$ and $k_{fb} = 0.4$ and $\delta = 0.184$. Similarly, based on the data from Table 3, the solutions to Problem 2 can be found for the greater order $\varepsilon_0 = 0.1$.

Table 3. The numbers of global p_l and bottom p_b film layers, lower $\check{k}_{fb}(p_b)$, Equation (32), and upper $\hat{k}_{fb}$, $\bar{\bar{k}}_{fb}$ bounds defined in Equations (33) and (31), parameters $k_{fb,ne}$, Equation (30), the non-empty sets of overlap ratios $\mathcal{K}_2(p_b)$ defined by the inequalities expressed by Equations (29) and (30) and the linear functions, Equation (34), describing overlap δ for $k_{fb} \in \mathcal{K}_2(p_b)$ determined to find the solution to Problem 2 of near-equal optimal bale dimensions design for fixed volume $V_{b0} = 1.357$ m^3, bale from Example 2; orders of near-equilibrium $\varepsilon_0 = 0.05$, 0.1, film width $b_f = 0.75$ m, bounds for the constraints from Equations (22) and (23): $k_{fb,min} = 0.2$, $k_{fb,max} = 0.45$, $\delta_{min} = 0.15$ m, $\delta_{max} = 0.3$ m, $4 \le p_l \le 16$. Film usage $FC(p_b)$, Equation (35), for $p_b \in \mathcal{P}_2$ and $k_{fb} \in \mathcal{K}_2(p_b)$; the optimal $FC(p_b^*)$ defined in Equation (36) is marked by bold.

ε_0	p_l	p_b	$k_{fb,ne}$ [−]	$\check{k}_{fb}(p_b)$ [−]	$\hat{k}_{fb}$ [−]	$\bar{\bar{k}}_{fb}$ [−]	$\mathcal{K}_2$	δ [m]	$FC(p_b)$
0.05	4	2	0.091	0.225	0.348	0.387	[0.225, 0.348]	$0.925 \cdot k_{fb} - 0.058$	20.804
	6	3	0.091	0.225	0.348	0.387	[0.225, 0.348]	$0.925 \cdot k_{fb} - 0.058$	31.979
	8	4	0.091	0.225	0.348	0.387	[0.225, 0.348]	$0.925 \cdot k_{fb} - 0.058$	41.607
	10	4	0.394	0.424	0.447	0.545	[0.424, 0.447]	$1.244 \cdot k_{fb} - 0.378$	52.445
	10	5	0.091	0.225	0.348	0.387	[0.225, 0.348]	$0.925 \cdot k_{fb} - 0.058$	**51.235**
	10	6	−0.364	0.200	0.260	0.204	[0.200, 0.204]	$0.712 \cdot k_{fb} + 0.155$	51.573
	12	5	0.351	0.393	0.430	0.520	[0.393, 0.430]	$1.180 \cdot k_{fb} - 0.314$	62.073
	12	6	0.091	0.225	0.348	0.387	[0.225, 0.348]	$0.925 \cdot k_{fb} - 0.058$	62.411
	12	7	−0.273	0.200	0.274	0.237	[0.200, 0.237]	$0.742 \cdot k_{fb} + 0.124$	**61.200**
	14	6	0.318	0.370	0.418	0.502	[0.370, 0.418]	$1.138 \cdot k_{fb} - 0.271$	71.700
	14	7	0.091	0.225	0.348	0.387	[0.225, 0.348]	$0.925 \cdot k_{fb} - 0.058$	72.038
	14	8	−0.212	0.200	0.284	0.259	[0.200, 0.259]	$0.765 \cdot k_{fb} + 0.102$	72.376
	16	7	0.293	0.353	0.409	0.488	[0.353, 0.409]	$1.107 \cdot k_{fb} - 0.241$	82.876
	16	8	0.091	0.225	0.348	0.387	[0.225, 0.348]	$0.925 \cdot k_{fb} - 0.058$	**81.666**
	16	9	−0.169	0.200	0.292	0.276	[0.200, 0.276]	$0.783 \cdot k_{fb} + 0.084$	82.004
0.1	4	2	0.167	0.273	0.381	0.429	[0.273, 0.381]	$0.961 \cdot k_{fb} - 0.113$	20.885
	6	3	0.167	0.273	0.381	0.429	[0.273, 0.381]	$0.961 \cdot k_{fb} - 0.113$	32.096
	8	4	0.167	0.273	0.381	0.429	[0.273, 0.381]	$0.961 \cdot k_{fb} - 0.113$	**41.769**
	8	5	−0.389	0.200	0.273	0.206	[0.200, 0.206]	$0.691 \cdot k_{fb} + 0.158$	42.217
	10	5	0.167	0.273	0.381	0.429	[0.273, 0.381]	$0.961 \cdot k_{fb} - 0.113$	52.980
	10	6	−0.250	0.200	0.293	0.255	[0.200, 0.255]	$0.736 \cdot k_{fb} + 0.113$	**51.890**
	12	5	0.405	0.433	0.461	0.554	[0.433, 0.450]	$1.232 \cdot k_{fb} - 0.383$	**62.206**
	12	6	0.167	0.273	0.381	0.429	[0.273, 0.381]	$0.961 \cdot k_{fb} - 0.113$	62.654
	12	7	−0.167	0.200	0.307	0.286	[0.200, 0.286]	$0.768 \cdot k_{fb} + 0.080$	63.102
	14	6	0.375	0.411	0.449	0.538	[0.411, 0.449]	$1.187 \cdot k_{fb} - 0.338$	73.417
	14	7	0.167	0.273	0.381	0.429	[0.273, 0.381]	$0.961 \cdot k_{fb} - 0.113$	**72.327**
	14	8	−0.111	0.200	0.317	0.308	[0.200, 0.308]	$0.792 \cdot k_{fb} + 0.056$	72.775
	16	7	0.352	0.395	0.440	0.525	[0.395, 0.440]	$1.154 \cdot k_{fb} - 0.306$	83.091
	16	8	0.167	0.273	0.381	0.429	[0.273, 0.381]	$0.961 \cdot k_{fb} - 0.113$	83.538
	16	9	−0.071	0.200	0.325	0.324	[0.200, 0.324]	$0.811 \cdot k_{fb} + 0.038$	**82.448**
	16	10	−0.389	0.200	0.273	0.206	[0.200, 0.206]	$0.691 \cdot k_{fb} + 0.158$	82.896

3.3. Volume-Free Optimal Equal Bale Dimensions Design

In this section the problem of volume-free design of equal optimal bale dimensions is considered being, in fact, a special case of Problem 1 for the order of near-equilibrium $\varepsilon_0 = 0$.

Problem 3. *Given: film parameters b_f, v_f, ε_{lf}, overlap ratio k_f, Equation (20), and number of global film layers p_l. Find film layers decomposition (p_b, p_u) and overlaps δ, k_{fb} satisfying constraints expressed by Equations (22) and (23) such that the applicability condition, Equation (21), hold and for any bale volume V_{b0} the optimal dimensions D_b^*, H_b^* are equal.*

The solution to the above problem is identical with that of Problem 1 for $\varepsilon_0 = 0$. Thus, the next result follows directly from Proposition 4 by lying $\varepsilon_0 = 0$ in inequalities expressed by Equations (25) and (26) and in Equations (5), (6) and (27). Inequality from Equation (24), here $b_{fr} > \delta_{max}$, is identity.

Proposition 6. *The solution to Problem 3 exists if and only if there exists an integer p_b such that inequalities*

$$\max\left\{\frac{b_{fr}-2\delta_{max}}{2(b_{fr}-\delta_{max})}p_l, \frac{1-k_{fb,max}}{2-k_{fb,max}}p_l\right\} \leq p_b \leq \min\left\{\frac{p_l}{2}, \frac{b_{fr}-2\delta_{min}}{2(b_{fr}-\delta_{min})}p_l, \frac{1-k_{fb,min}}{2-k_{fb,min}}p_l\right\} \quad (39)$$

are satisfied together with the applicability condition, Equation (21). For any such p_b the bottom layers overlaps are uniquely determined by

$$k_{fb} = \frac{p_l - 2p_b}{p_l - p_b}, \quad \delta = \frac{1}{2}b_{fr}k_{fb}. \quad (40)$$

For a given bale volume optimal bale dimensions D_b^* and H_b^* are given by Equation (19).

Thus, to solve Problem 3 an integer p_b fulfilling inequalities from Equation (39) must be found. The fractions dependent on b_{fr} in "max" and "min" functions, which define the lower and upper bounds in this inequalities, increase with the film width. However, the upper and lower bounds may be determined by other arguments of "max" or "min", thus a larger b_{fr} does not necessarily imply a greater p_b. Proposition 6 yielded the following algorithm.

3.3.1. Algorithm 3

Assume film parameters b_f, v_f, ε_{lf}, overlap ratio k_f, Equation (20), and the number of global film layers p_l are given. Take minimal δ_{min}, $k_{fb,min}$ and maximal δ_{max}, $k_{fb,max}$ values of the overlaps δ, k_{fb}.

1. Determine the set $\mathcal{P}_3$ of all integer p_b defined by the inequalities from Equation (39) for which there exists $2 \leq q \leq 5$ satisfying applicability condition, Equation (21).
2. If the set $\mathcal{P}_3$ is empty, then the solution to Problem 3 does not exist—go to step 3. Otherwise, go to step 4.
3. Change the lower $k_{fb,min}$, δ_{min} or upper $k_{fb,max}$, δ_{max} bounds of wrapping parameters or the film width b_f and go to step 1.
4. For any $p_b \in \mathcal{P}_3$ compute the overlap ratio k_{fb} and, next, overlap δ according to Equation (40).
5. For any $p_b \in \mathcal{P}_3$ and assumed bale volume V_{b0} compute film usage obtained for the optimal D_b^*, H_b^*, Equation (19), using Equation (A2) and choose that p_b which yields the minimal film consumption.

3.3.2. Example 3

The same mechanical parameters of the film and wrapping parameters constraints as in the previous examples are assumed. Film widths $b_f = 0.75$, 0.9 m and $4 \leq p_l \leq 16$ are considered. For $b_f = 0.75$ m, similarly as in Example 1, the set of integer solutions p_b of inequalities expressed by Equation (39) is empty for all p_l considered, while for the wider film for $p_l = 14$ solution to Problem 3 exists. When the lower constraint expressed by Equation (22) is changed into $\delta_{min} = 0.1$ m, then for some p_l Problem 3 has a solution. These solutions are listed in Table 4. For every p_l, nonempty $\mathcal{P}_3$ is a singleton (a unit set), thus step 5 of the selection of best film layers composition $p_l = p_b + p_u$ can be omitted here; however film usage FC is added in the last column.

Table 4. The numbers of global p_l and bottom p_b film layers and bottom layers overlaps k_{fb}, δ for which Problem 3 of volume-free equal optimal bale dimensions design has a solution for bale from Example 3; film width $b_f = 0.75$, 0.9 m, the bounds from Equations (22) and (23): $k_{fb,min} = 0.2$, $k_{fb,max} = 0.45$, $\delta_{min} = 0.1$, 0.15 m, $\delta_{max} = 0.3$ m, $4 \leq p_l \leq 16$. Film usage FC, Equation (A2) for the assumed bale volume $V_{b0} = 1.357$ m^3.

δ_{min} [m]	b_f [m]	p_l	p_b	k_{fb} [−]	δ [m]	FC [m^{-1}]
0.1	0.75	8	3	0.400	0.114	41.047
		14	5	0.444	0.127	72.524
		16	6	0.400	0.114	82.095
	0.9	8	3	0.400	0.137	39.449
		10	4	0.333	0.114	49.478
		14	5	0.444	0.152	76.222
		16	6	0.400	0.137	78.898
0.15		14	5	0.444	0.152	76.222

3.4. Volume-Fixed Optimal Equal Bale Dimensions Design

Problem 4. *Given: film parameters b_f, v_f, ε_{lf}, overlap ratio k_f, Equation (20), number of global film layers p_l and bale volume V_{b0}. Find film layers decomposition (p_b, p_u) and overlaps δ, k_{fb} satisfying constraints from Equations (22) and (23) such that the optimal bale dimensions D_b^*, H_b^* resulting in the volume V_{b0} are equal and film usage is minimal.*

Lying $\varepsilon_0 = 0$ in inequalities expressed by Equations (29) and (30) and in Equations (34) and (35) based on Proposition 5 the following solution results.

Proposition 7. *The solution to Problem 4 exists if and only if there exist k_{fb} and an integer p_b such that inequalities*

$$\overset{\equiv}{k}_{fb}(p_b) = \max\{\check{k}_{fb}(p_b), k_{fb,min}\} \leq k_{fb} \leq \min\left\{\widetilde{k}_{fb}(p_b), \frac{2\delta_{max} + \left(\frac{p_l}{p_b} - 2\right)\sqrt[3]{\frac{4V_{b0}}{\pi}}}{\left(\frac{p_l}{p_b} - 1\right)\sqrt[3]{\frac{4V_{b0}}{\pi}} + b_{fr}}, k_{fb,max}\right\} \quad (41)$$

and inequality

$$k_{fb} \neq \frac{p_l - 2p_b}{p_l - p_b} \quad (42)$$

are satisfied together with the applicability condition, Equation (21), where

$$\check{k}_{fb}(p_b) = \frac{2\delta_{min} + \left(\frac{p_l}{p_b} - 2\right)\sqrt[3]{\frac{4V_{b0}}{\pi}}}{\left(\frac{p_l}{p_b} - 1\right)\sqrt[3]{\frac{4V_{b0}}{\pi}} + b_{fr}}, \quad (43)$$

and $\widetilde{k}_{fb}(p_b)$ is defined by the next equation with $\overset{\equiv}{k}_{fb}(p_b)$ introduced in Equation (41)

$$\widetilde{k}_{fb}(p_b) = 1 - \frac{2}{\frac{b_{fr}}{\sqrt[3]{\frac{4V_{b0}}{\pi}}}\left[\sqrt[3]{\frac{4V_{b0}}{\pi}}\left[\frac{2}{1 - \overset{\equiv}{k}_{fb}(p_b)} - \left(\frac{p_l}{p_b} - 1\right)\right] + \left(\frac{p_l}{p_b} - 1\right)\right]}. \quad (44)$$

For any such p_b and k_{fb} the overlap δ is given by

$$\delta = \frac{\sqrt[3]{\frac{4V_{b0}}{\pi}}}{2}\left[1 - \left(\frac{p_l}{p_b} - 1\right)(1 - k_{fb})\right] + \frac{b_{fr}k_{fb}}{2}, \quad (45)$$

the optimal bale dimensions are given by Equation (19) and film usage

$$FC(p_b) = \frac{4b_f}{(\epsilon_{lf}+1)\left[\sqrt[3]{\frac{4V_{b0}}{\pi}}\right]^2} \left\{ p_b \frac{\sqrt[3]{\frac{4V_{b0}}{\pi}}\left[\frac{2}{1-k_{fb}(p_b)} - \left(\frac{p_l}{p_b}-1\right)\right]}{b_{fr}} + \frac{4}{\pi}\left[\frac{\pi\sqrt[3]{\frac{4V_{b0}}{\pi}}(p_l-p_b)}{2b_{fr}}\right]\right\}. \quad (46)$$

Integer p_b^* solving Problem 4 is defined by optimization task expressed by Equation (36).

Thus, the inequalities specified in the proposition must be solved for two variables: integer p_b and continuous k_{fb}. Next optimization task, Equation (36), must be solved and the overlap δ can be computed.

Example 4

Film parameters and overlap ratio k_f as in the previous examples are given. Film widths $b_f = 0.75, 0.9$ m, even p such that $4 \leq p_l \leq 16$ and bale volume $V_{b0} = 1.357$ m^3 are assumed. For any p_l the sets of overlap ratios k_{fb} satisfying inequalities from Equations (41) and (42) are given in the fourth column of Table 5. As previously, for any k_{fb} from these closed intervals the linear formulas from Equation (45) for computing δ are given in the next column, in the last column film consumptions $FC(p_b)$, Equation (46), are enclosed. For these p_l for which there are more than one p_b such that inequalities from Equations (41) and (42) are satisfied the minimal film usage, solving optimization task from Equation (36), is marked by bold. Note that in most cases film usage for $b_f = 0.9$ m is smaller from that for $b_f = 0.75$ m; this confirms the analysis of the film width influence on the optimal film usage from [14,20]—the broader film width is applied, the smaller minimal film usage is achieved.

Table 5. The numbers of global p_l and bottom p_b film layers, bottom layers overlaps k_{fb}, δ fulfilling inequalities from Equations (41) and (42), and respective film usage FC, Equation (46), for $V_{b0} = 1.357$ m^3 and bale from Example 4; film widths $b_f = 0.75, 0.9$ m, the bounds from Equations (22) and (23): $k_{fb,min} = 0.2$, $k_{fb,max} = 0.45$, $\delta_{min} = 0.15$ m, $\delta_{max} = 0.3$ m, $4 \leq p_l \leq 16$. Minimal film usage solving Problem 4 of volume-fixed design of equal optimal bale dimensions is marked by bold.

b_f [m]	p_l	p_b	k_{fb} [−]	δ [m]	FC [m^{-1}]
0.75	4	2	[0.200, 0.311]	$0.886 \cdot k_{fb}$	20.726
	6	3	[0.200, 0.311]	$0.886 \cdot k_{fb}$	30.309
	8	3	[0.428, 0.439]	$1.286 \cdot k_{fb} - 0.4$	**41.232**
	8	4	[0.200, 0.311]	$0.886 \cdot k_{fb}$	41.453
	10	4	[0.380, 0.412]	$1.186 \cdot k_{fb} - 0.3$	**50.815**
	10	5	[0.200, 0.311]	$0.886 \cdot k_{fb}$	51.036
	12	5	[0.346, 0.394]	$1.126 \cdot k_{fb} - 0.240$	61.958
	12	6	[0.200, 0.311]	$0.886 \cdot k_{fb}$	**60.619**
	14	6	[0.322, 0.382]	$1.086 \cdot k_{fb} - 0.2$	71.541
	14	7	[0.200, 0.311]	$0.886 \cdot k_{fb}$	71.762
	14	8	[0.200, 0.204]	$0.736 \cdot k_{fb} + 0.15$	**70.423**
	16	6	[0.428, 0.439]	$1.286 \cdot k_{fb} - 0.4$	**80.903**
	16	7	[0.304, 0.372]	$1.057 \cdot k_{fb} - 0.171$	81.124
	16	8	[0.200, 0.311]	$0.886 \cdot k_{fb}$	81.345
	16	9	[0.200, 0.222]	$0.752 \cdot k_{fb} + 0.133$	81.566

Table 5. Cont.

b_f [m]	p_l	p_b	k_{fb} [−]	δ [m]	FC [m^{-1}]
0.9	4	2	[0.200, 0.263]	$0.943 \cdot k_{fb}$	20.058
	6	3	[0.200, 0.263]	$0.943 \cdot k_{fb}$	30.087
	8	3	[0.410, 0.450]	$1.343 \cdot k_{fb} - 0.4$	43.861
	8	4	[0.200, 0.263]	$0.943 \cdot k_{fb}$	**38.244**
	10	4	[0.362, 0.377]	$1.243 \cdot k_{fb} - 0.3$	49.478
	10	5	[0.200, 0.263]	$0.943 \cdot k_{fb}$	**48.273**
	12	5	[0.330, 0.357]	$1.183 \cdot k_{fb} - 0.240$	59.507
	12	6	[0.200, 0.263]	$0.943 \cdot k_{fb}$	**58.302**
	14	5	[0.443, 0.450]	$1.423 \cdot k_{fb} - 0.480$	76.222
	14	6	[0.306, 0.343]	$1.143 \cdot k_{fb} - 0.2$	**67.664**
	14	7	[0.200, 0.263]	$0.943 \cdot k_{fb}$	68.331
	16	6	[0.410, 0.450]	$1.343 \cdot k_{fb} - 0.4$	87.722
	16	7	[0.288, 0.333]	$1.114 \cdot k_{fb} - 0.171$	77.693
	16	8	[0.200, 0.263]	$0.943 \cdot k_{fb}$	**76.487**
	16	9	[0.200, 0.206]	$0.810 \cdot k_{fb} + 0.133$	90.389

3.5. Optimal Film Usage

For the combined 3D wrapping technique, the analysis of the optimal film consumption reached for D_b^*, H_b^* was carried out in [19,20], where it was shown, in particular, that for the bottom film layers $p_b \gg p_u$ the optimal selection of bale diameter and height may result in a 3.37% to even 23.13% reduction in film usage depending on the number of global film layers and wrapping parameters, which means up to 23% film cost savings. Many conclusions regarding the impact of film width, pre-assumed bale volume and numbers of bottom and upper film layers on near-optimal bale dimensions and near-optimal film consumption were formulated [19,20].

4. Conclusions

It has been shown that for bale wrapping by a 3D combined method for given film parameters and given a global number of film layers it is possible, both for assumed and for an arbitrary bale volume, to select such wrapping parameters that guarantee the pre-assumed equilibria of optimal bale dimensions. Thus, the "compliance" is the answer to the question from the title.

The derived necessary and sufficient conditions for the balance of optimal bale dimensions allowed for the formulating of four practical problems of selecting the wrapping parameters, in particular the overlaps in the bottom film layers and decomposition of the global number of layers into bottom and upper layers. The mathematical form of the conditions of optimal bale dimensions equilibrium indicated the legitimacy of considering two separate design strategies, one for fixed and one for free bale volume. Design algorithms were derived. These algorithms only require a solution with respect to the number of film layers and overlap ratio of the bottom film layers of two or four simple algebraic inequalities. Then, on the basis of very simple formulas, the remaining wrapping parameters and the optimal bale diameter and height can be determined. Examples regarding standard large bales demonstrate that the algorithms can handle the optimal dimensions' equilibria problems.

Optimal dimensions mean a reduction in film consumption by up to 23%. Simultaneously, equal and near-equal bale dimensions are essential from the agriculture engineering practice and bale wrapping technique point of view. The algorithms developed can be applied for design of wrapping processes using a 2D method for cylindrical bales of arbitrary agricultural materials, e.g., lignocellulosic agricultural residues [46,47], bearing in mind the minimisation of the film usage and the desired equilibrium between bale dimensions. Potential applications of the algorithms include the baling of waste, such as municipal solid waste [48,49], for example.

Funding: This research received no external funding.

Institutional Review Board Statement: Not applicable.

Informed Consent Statement: Not applicable.

Conflicts of Interest: The author declares no conflict of interest.

Appendix A

Appendix A.1. Film Consumption

A mathematical model describing the consumption of the stretch film to wrap a cylindrical bale using combined 3D technique was derived in [14]. It was assumed that the subsequent film strips, which are wrapped in p_b bottom layers on the bale's lateral surface, overlap one another, creating the overlap $k_{fb}b_{fr}$, where k_{fb} is a dimensionless relative ratio determining the width of the contact between adjacent film strips and the film width after stretching b_{fr} is described by [12]:

$$b_{fr} = b_f\left(1 - v_f \varepsilon_{lf}\right), \tag{A1}$$

where b_f is the width of un-stretched film, v_f and ε_{lf} are the Poisson's ratio and unit deformation of the film. The extreme film strips are overlapped at the bases of the bale for δ, as shown in [14] (Figure 3). It was assumed that $0 < k_{fb} < \frac{1}{2}$, for which only one film layer results for one wrapping cycle. For the wrappings of p_u upper layers it was assumed that the subsequent film strips are wrapped along the bale's longitudinal axis with the overlap ratio k_f [14] (Figure 2).

The global film consumption measured by the index *FC* defined as the ratio of the surface area of un-stretched film used to wrap the bale to bale volume, is described by the function [14]:

$$FC = \frac{4p_b b_f}{D_b H_b\left(\varepsilon_{lf} + 1\right)} \left\lceil \frac{H_b + 2\delta - b_{fr}k_{fb}}{b_{fr}\left(1 - k_{fb}\right)} \right\rceil + \frac{8(D_b + H_b)b_f}{\pi D_b^2 H_b\left(\varepsilon_{lf} + 1\right)} \left\lceil \frac{\pi D_b p_u}{2 b_{fr} \Omega\left(k_f\right)} \right\rceil, \tag{A2}$$

where $\lceil x \rceil$ denotes ceiling function [45], function [20]

$$\Omega\left(k_f\right) = \left(1 - k_f\right)\left\lfloor \frac{1}{1 - k_f} \right\rfloor, \tag{A3}$$

is introduced for brevity of the notation; $\lfloor x \rfloor$ is the floor function [45]. The above formula indicates the dependence of the film usage on film parameters ε_{lf}, v_f, b_f, bale diameter D_b and height H_b, the overlaps δ, k_{fb} and k_f of the bottom and upper film layers, and the numbers of film layers p_u, p_b. Second summand of *FC* describe film used for upper layers wrappings provided that the following applicability condition [14]:

$$\frac{p_u}{\left\lfloor \frac{1}{1-k_f} \right\rfloor} = m, \ m \in \mathcal{N}, \tag{A4}$$

holds, where $\mathcal{N}$ denotes the set of positive integer numbers. Global number of basic film layers wrapped on the bale's lateral surface is $p_l = p_b + p_u$.

Appendix A.2. Film Consumption Minimization

As the non-continuity and non-differentiability of the original film usage index *FC* undermines our ability to directly analytically solve the problem of film usage minimization, special attention has been given to the simpler case of near-optimal bale dimensions design, this being as important as the optimal parameters for engineering applications. The problem of the selection of near-optimal bale dimensions has been constructed by minimizing continuous lower bound of the original film usage index FC, where the goal function of a non-linear optimization problem is convex and differentiable. The necessary and sufficient optimality condition for near-optimal bale diameter, called "optimal" in [20] and here, was established in the form of a standard cubic equation, Equation (1), which can easily be solved using both analytical and numerical methods. The results of the numerical experiments [19,20] demonstrated that for any four to sixteen even layers of the film there are such compositions of bottom and upper film layers that the relative near-optimality errors do not exceed 0.01% whenever the optimal bale dimensions are used.

Appendix B

Proof of Proposition 1. The optimal bale diameter D_b^* is near-equal to the optimal bale height H_b^* with order ε_0 if and only if Equations (1) and (5) are satisfied, simultaneously. For D_b^* given by Equation (5) the optimality condition, Equation (1), takes the form

$$\frac{2\pi p_u}{\Omega(k_f)} \frac{4(1+\varepsilon_0)V_{b0}}{\pi} + \frac{\pi p_b}{1-k_{fb}}(2\delta - b_{fr}k_{fb})\left(\sqrt[3]{\frac{4(1+\varepsilon_0)V_{b0}}{\pi}}\right)^2 - 4V_{b0}\left[\frac{p_b}{1-k_{fb}} + \frac{p_u}{\Omega(k_f)}\right] = 0$$

and is equivalent to the next equation

$$\frac{2p_u(1+\varepsilon_0)}{\Omega(k_f)}\sqrt[3]{\frac{4V_{b0}}{\pi}} + \frac{p_b}{1-k_{fb}}(2\delta - b_{fr}k_{fb})(\sqrt[3]{1+\varepsilon_0})^2 - \sqrt[3]{\frac{4V_{b0}}{\pi}}\left[\frac{p_b}{1-k_{fb}} + \frac{p_u}{\Omega(k_f)}\right] = 0,$$

which can be rewritten as follows

$$\frac{p_b}{1-k_{fb}}(2\delta - b_{fr}k_{fb})(\sqrt[3]{1+\varepsilon_0})^2 \sqrt[3]{\frac{4V_{b0}}{\pi}}\left[\frac{p_b}{1-k_{fb}} - \frac{p_u}{\Omega(k_f)} - \frac{2p_u}{\Omega(k_f)}\varepsilon_0\right]. \quad (A5)$$

To show the Proposition two cases must be considered separately:

(i) the wrapping parameters and order ε_0 are such that

$$\frac{p_b}{1-k_{fb}} \neq \frac{p_u}{\Omega(k_f)} + \frac{2p_u}{\Omega(k_f)}\varepsilon_0, \quad (A6)$$

(ii) for the wrapping parameters and ε_0 the following equality holds

$$\frac{p_b}{1-k_{fb}} = \frac{p_u}{\Omega(k_f)} + \frac{2p_u}{\Omega(k_f)}\varepsilon_0. \quad (A7)$$

Provided that inequality from Equation (A6) is satisfied, we have

$$\frac{\frac{p_b}{1-k_{fb}}(2\delta - b_{fr}k_{fb})(\sqrt[3]{1+\varepsilon_0})^2}{\frac{p_b}{1-k_{fb}} - \frac{p_u}{\Omega(k_f)} - \frac{2p_u}{\Omega(k_f)}\varepsilon_0} = \sqrt[3]{\frac{4V_{b0}}{\pi}}, \quad (A8)$$

whence Equation (7) follows. The optimal bale diameter and height are described by Equations (5) and (6) and for D_b^* and H_b^* Equations (1) and (5) are satisfied; case (i) is proved.

In case (ii), when Equation (A7) holds, Equation (A5) is satisfied if and only if, simultaneously,

$$2\delta - b_{fr}k_{fb} = 0.$$

Then, Equation (1) takes especially simple form

$$\frac{2\pi p_u}{\Omega(k_f)}(D_b^*)^3 - 4V_{b0}\left[\frac{p_b}{1-k_{fb}} + \frac{p_u}{\Omega(k_f)}\right] = 0,$$

whence the optimal bale diameter

$$D_b^* = \sqrt[3]{\frac{2V_{b0}}{\pi}\left[\frac{p_b\Omega(k_f)}{p_u(1-k_{fb})} + 1\right]}. \tag{A9}$$

Simultaneously, from Equation (A7) it follows that

$$\frac{p_b\Omega(k_f)}{p_u(1-k_{fb})} = 1 + 2\varepsilon_0.$$

Thus Equations (A9) and (5) are identical and, by Equation (4), formula from Equation (6) follows for any V_{b0}, which completes the proof of case (ii) since Equations (A7) and (8) are equivalent.

Proof of Proposition 2. By Proposition 1 the optimal bale dimensions are near-equal with order ε_0 if:

- in case (i) inequality expressed by Equation (13) is satisfied,
- in case (ii)

$$p_b\Omega(k_f) - p_u(1-k_{fb}) - 2p_u(1-k_{fb})\varepsilon_0 = 0 \tag{A10}$$

and, simultaneously, Equation (9) holds. In the second case any change of ε_0 to $\varepsilon < \varepsilon_0$ results in a loss of equality in Equation (A10). Then the inequality expressed by Equation (13) holds, i.e., case (i) occurs, and since according to Equation (9) $2\delta - b_{fr}k_{fb} = 0$, we immediately conclude that Equation (7) is satisfied only for volume $V_{b0} = 0$. Thus, if for ε_0 Equation (A10) holds, the optimal bale dimensions cannot be near-equal for any $\varepsilon < \varepsilon_0$.

Assume now, that for ε_0 the inequality expressed by Equation (13) is satisfied. Let us consider bale volume $V_{b0}(\varepsilon_0)$ uniquely given by Equation (7). Differentiating Equation (7) on both sides with respect to ε_0 results in

$$\frac{4}{3\pi\left(\sqrt[3]{\frac{4V_{b0}(\varepsilon_0)}{\pi}}\right)^2}\cdot\frac{dV_{b0}(\varepsilon_0)}{d\varepsilon_0} = p_b\Omega(k_f)(2\delta - b_{fr}k_{fb})\frac{\frac{2}{3}[p_b\Omega(k_f) - p_u(1-k_{fb}) - 2p_u(1-k_{fb})\varepsilon_0] + 2p_u(1-k_{fb})(1+\varepsilon_0)}{\sqrt[3]{1+\varepsilon_0}[p_b\Omega(k_f) - p_u(1-k_{fb}) - 2p_u(1-k_{fb})\varepsilon_0]^2},$$

whence, after algebraic manipulations, we obtain

$$\frac{dV_{b0}(\varepsilon_0)}{d\varepsilon_0} = \frac{\pi p_b\Omega(k_f)(2\delta - b_{fr}k_{fb})[p_b\Omega(k_f) + 2p_u(1-k_{fb}) + p_u(1-k_{fb})\varepsilon_0]}{2\sqrt[3]{1+\varepsilon_0}[p_b\Omega(k_f) - p_u(1-k_{fb}) - 2p_u(1-k_{fb})\varepsilon_0]^2}\left[\sqrt[3]{\frac{4V_{b0}(\varepsilon_0)}{\pi}}\right]^2. \tag{A11}$$

Thus, monotonicity of the function $V_{b0}(\varepsilon_0)$ depends on the sign of the expression $(2\delta - b_{fr}k_{fb})$. First, consider case (a) of the proposition assuming that inequality from Equation (12) holds. Then the denominator of the left hand side of Equation (7) is also positive, and if inequality given by Equation (13) holds for ε_0, then it holds also for any $0 \leq \varepsilon \leq \varepsilon_0$. Function $V_{b0}(\varepsilon)$ is monotonically increasing for $0 \leq \varepsilon \leq \varepsilon_0$. For $\varepsilon_0 = \varepsilon$, by Equations (5), (6) and (A8), the optimal bale dimensions $D_b^*(\varepsilon), H_b^*(\varepsilon)$ can be equivalently expressed by Equations (16) and (17). Thus, $D_b^*(\varepsilon), H_b^*(\varepsilon)$ are also monotonically increasing for $0 \leq \varepsilon \leq \varepsilon_0$.

In case (b), when inequality expressed by Equation (14) is satisfied, function $V_{b0}(\varepsilon_0)$ decreases with ε_0 and, by Equation (7), inequality from Equation (13) takes the form

$$p_b \Omega(k_f) - p_u(1-k_{fb}) - 2p_u(1-k_{fb})\varepsilon_0 < 0.$$

If, simultaneously, inequality expressed by Equation (15) holds, then for any $0 \leq \varepsilon \leq \varepsilon_0$ there exists $V_{b0}(\varepsilon)$ fulfilling Equation (7). Otherwise, there exists $0 < \bar{\varepsilon} < \varepsilon_0$ such that Equation (18) holds and only for $\bar{\varepsilon} < \varepsilon \leq \varepsilon_0$ Equation (7) is satisfied for some $V_{b0}(\varepsilon)$. In both cases, for $0 \leq \varepsilon \leq \varepsilon_0$ or $\bar{\varepsilon} < \varepsilon \leq \varepsilon_0$, by Equation (16) we have

$$\frac{dD_b^*(\varepsilon)}{d\varepsilon} = p_b \Omega(k_f)(2\delta - b_{fr}k_{fb}) \frac{\left[p_b \Omega(k_f) + p_u(1-k_{fb})\right]}{\left[p_b \Omega(k_f) - p_u(1-k_{fb}) - 2p_u(1-k_{fb})\varepsilon\right]^2}.$$

Thus, in view of the inequality from Equation (14), optimal bale diameter decreases with increasing ε. The analysis of Equation (17) yields analogous property of $H_b^*(\varepsilon)$. Proposition 2 is proved. □

Proof of Proposition 4. For k_f expressed by Equation (20) and $p_u = p_l - p_b$ Equation (8) takes the form

$$p_b\left[1 + (1-k_{fb})(1+2\varepsilon_0)\right] = p_l(1-k_{fb})(1+2\varepsilon_0),$$

whence the bottom layers overlap ratio is uniquely determined by Equation (27). The overlaps δ and k_{fb} are related by Equation (9). Substituting k_{fb}, Equation (27), into constraints described by Equation (23) leads to inequalities

$$k_{fb,min} \leq \frac{p_l(1+2\varepsilon_0) - p_b(2+2\varepsilon_0)}{(p_l - p_b)(1+2\varepsilon_0)} \leq k_{fb,max},$$

which can be rewritten in the equivalent form of constraints expressed with respect to integer p_b by Equation (25).

Simultaneously, on the basis of Equations (9) and (27), the constraints expressed by Equation (22) are satisfied if and only if

$$\delta_{min} \leq \frac{b_{fr}}{2} \cdot \frac{p_l(1+2\varepsilon_0) - p_b(2+2\varepsilon_0)}{(p_l - p_b)(1+2\varepsilon_0)} \leq \delta_{max},$$

which is equivalent to the inequalities expressed by Equation (26), provided that inequality described by Equation (24) is satisfied. Proposition 4 is derived. □

Proof of Proposition 5. According to Proposition 1 the solution to Problem 2 exists if and only if there exist integer p_b and overlaps δ, k_{fb} satisfying Equation (7), constraints from Equations (22) and (23) and the applicability condition, Equation (21), provided that the denominator of the left hand side of Equation (7) is non-zero, i.e., that inequality from Equation (13) holds, which for $k_f = k_{f,u}$ and $p_u = p_l - p_b$ can be rewritten as

$$p_b - (p_l - p_b)(1 - k_{fb}) - 2(p_l - p_b)(1 - k_{fb})\varepsilon_0 \neq 0.$$

The above condition can be expressed directly with respect to the overlap ratio as inequality described by Equation (30). By Equation (7), having in mind that $\Omega(k_f) = 1$ and $p_u = p_l - p_b$, the overlap δ is uniquely given by equation

$$p_b(2\delta - b_{fr}k_{fb})\left(\sqrt[3]{1+\varepsilon_0}\right)^2 = \sqrt[3]{\frac{4V_{b0}}{\pi}}\left[p_b - (p_l - p_b)(1-k_{fb}) - 2(p_l - p_b)(1-k_{fb})\varepsilon_0\right],$$

whence direct formula from Equation (34) follows. Thus, the inequalities from Equation (22) take the form

$$\delta_{min} \leq \frac{\sqrt[3]{\frac{4V_{b0}}{\pi}}}{2(\sqrt[3]{1+\varepsilon_0})^2}\left[1-\left(\frac{p_l}{p_b}-1\right)(1-k_{fb})(1+2\varepsilon_0)\right]+\frac{b_{fr}k_{fb}}{2} \leq \delta_{max}$$

and can be unravelled with respect to k_{fb} as follows

$$\overline{\overline{k}}_{fb}(p_b) = \frac{2\delta_{min}(\sqrt[3]{1+\varepsilon_0})^2 - \sqrt[3]{\frac{4V_{b0}}{\pi}} + (1+2\varepsilon_0)\left(\frac{p_l}{p_b}-1\right)\sqrt[3]{\frac{4V_{b0}}{\pi}}}{(1+2\varepsilon_0)\left(\frac{p_l}{p_b}-1\right)\sqrt[3]{\frac{4V_{b0}}{\pi}} + b_{fr}(\sqrt[3]{1+\varepsilon_0})^2} \leq k_{fb} \leq \frac{2\delta_{max}(\sqrt[3]{1+\varepsilon_0})^2 - \sqrt[3]{\frac{4V_{b0}}{\pi}} + (1+2\varepsilon_0)\left(\frac{p_l}{p_b}-1\right)\sqrt[3]{\frac{4V_{b0}}{\pi}}}{(1+2\varepsilon_0)\left(\frac{p_l}{p_b}-1\right)\sqrt[3]{\frac{4V_{b0}}{\pi}} + b_{fr}(\sqrt[3]{1+\varepsilon_0})^2} = \overline{\overline{k}}_{fb}(p_b). \quad (A12)$$

Simultaneously, for the overlap ratio k_{fb} inequalities from Equation (23) must be satisfied.

It has been proved above that there exist decomposition (p_b, p_u) and the bottom layers overlaps δ, k_{fb} satisfying constraints expressed by Equations (22) and (23) such that the optimal D_b^*, H_b^* of the bale of volume V_{b0} are near-equal with order ε_0 if and only if there exist an integer p_b and continuous parameter k_{fb} such that inequalities expressed by Equations (A12) and (23), i.e., jointly

$$\check{k}_{fb}(p_b) = \max\left\{\overline{\overline{k}}_{fb}(p_b), k_{fb,min}\right\} \leq k_{fb} \leq \min\left\{\overline{\overline{k}}_{fb}(p_b), k_{fb,max}\right\}, \quad (A13)$$

are satisfied together with the applicability condition, Equation (21), and inequality from Equation (30). For any such p_b and k_{fb} the overlap δ is given by Equation (34). By Proposition 1 the optimal bale dimensions are given by Equations (5) and (6). Since by Equation (34), we have

$$2\delta - b_{fr}k_{fb} = \frac{\sqrt[3]{\frac{4V_{b0}}{\pi}}}{(\sqrt[3]{1+\varepsilon_0})^2}\left[1-\left(\frac{p_l}{p_b}-1\right)(1-k_{fb})(1+2\varepsilon_0)\right],$$

having in mind that $\Omega(k_f) = 1$ it can be proved that for the optimal D_b^*, H_b^* film usage index FC, Equation (A2), is described by

$$FC = \frac{4b_f\sqrt[3]{1+\varepsilon_0}}{(\varepsilon_{lf}+1)\left[\sqrt[3]{\frac{4V_{b0}}{\pi}}\right]^2}\left\{p_b\left\lceil\frac{\sqrt[3]{\frac{4V_{b0}}{\pi}}\left[\frac{2}{1-k_{fb}}-\left(\frac{p_l}{p_b}-1\right)(1+2\varepsilon_0)\right]}{b_{fr}(\sqrt[3]{1+\varepsilon_0})^2}\right\rceil + \frac{2(2+\varepsilon_0)}{\pi(1+\varepsilon_0)}\left\lceil\frac{\pi\sqrt[3]{\frac{4V_{b0}}{\pi}}(1+\varepsilon_0)(p_l-p_b)}{2b_{fr}(\sqrt[3]{1+\varepsilon_0})^2}\right\rceil\right\}, \quad (A14)$$

where only numerator of the fraction in the argument of the ceiling function in the first sum in curly brackets depends on k_{fb} and is non-decreasing left-continuous function of k_{fb}, piecewise constant in the intervals determined by its discontinuity points. If $\hat{k}_{fb}(p_b)$ is discontinuity point being direct right neighbourhood of $\check{k}_{fb}(p_b)$ defined in Equation (A13), i.e., for $\hat{k}_{fb}(p_b)$ the following equation holds

$$\left\lceil\frac{\sqrt[3]{\frac{4V_{b0}}{\pi}}\left[\frac{2}{1-\check{k}_{fb}(p_b)}-\left(\frac{p_l}{p_b}-1\right)(1+2\varepsilon_0)\right]}{b_{fr}(\sqrt[3]{1+\varepsilon_0})^2}\right\rceil = \frac{\sqrt[3]{\frac{4V_{b0}}{\pi}}\left[\frac{2}{1-\check{k}_{fb}(p_b)}-\left(\frac{p_l}{p_b}-1\right)(1+2\varepsilon_0)\right]}{b_{fr}(\sqrt[3]{1+\varepsilon_0})^2}, \quad (A15)$$

then for any

$$\check{k}_{fb}(p_b) \leq k_{fb} \leq \hat{k}_{fb}(p_b), \quad (A16)$$

film usage FC, Equation (A14), being left-continuous in discontinuity points is identical. Thus, combining inequalities from Equations (A12), (23), and (A16) result in the inequalities from Equation (29). Direct formula expressed by Equation (33) follows from Equation (A15), while Equation (35) results directly from Equation (A14). Integer minimization task in Equation (36) makes it possible to find the best film layers decomposition. Proposition 5 is true. □

Appendix C

Appendix C.1. Nomenclature

b_f	width of un-stretched film, m
b_{fr}	width of stretched film, Equation (A1), m
D_b, H_b	bale diameter and height, m
D_b^*	optimal bale diameter, solution of Equation (1), m
H_b^*	optimal bale height given by Equation (2), m
FC	film consumption index, Equation (A2), m^{-1}
k_{fb}, k_f	overlap ratios applied to wrap bottom and upper film layers
$k_{fb,min}, k_{fb,max}$	the smallest and largest admissible k_{fb}, Equation (23)
$\check{k}_{fb}(p_b)$	lower bound of k_{fb} solving Problem 2, Equation (32)
$\hat{k}_{fb}(p_b), \overline{\overline{k}}_{fb}(p_b)$	upper bounds of k_{fb} solving Problem 2, Equations (33) and (31)
$\underline{\underline{k}}_{fb}(p_b)$	lower bound of k_{fb} solving Problem 4, Equation (41)
$\acute{k}_{fb}(p_b), \tilde{k}_{fb}(p_b)$	bounds of k_{fb} solving Problem 4, Equations (43) and (44)
$\mathcal{K}_2(p_b)$	set of overlap ratios k_{fb} defined for $p_b \in \mathcal{P}_2$ by Equations (29) and (30)
m	integer number
$\mathcal{N}$	set of all positive integer numbers
p_l	global number of basic film layers
p_b, p_u	numbers of basic film layers in bottom and upper layers
p_b^*	optimal p_b solving Problems 2 and 4 defined in Equation (36)
$\mathcal{P}_1$	set of p_b for which Problem 1 has solution, defined in Equation (28)
$\mathcal{P}_2, \mathcal{P}_3$	sets of p_b for which Problems 2,3 have solution
v_f	Poisson's ratio of the stretch film
V_{b0}	pre-assumed bale volume, m^3
ε_0	order of nearly-equal bale dimensions introduced in Equation (4)
ε_{lf}	unit deformation of the stretch film
Ω	function of the overlap ratio k_f defined by Equation (A3)
δ	overlap of extreme film strips at base of bale, m
$\delta_{min}, \delta_{max}$	the smallest and largest admissible overlap δ, Equation (22), m

Appendix C.2. Mathematical Terminology

$\lceil x \rceil$	the smallest integer not less than x, ceiling function
$\lfloor x \rfloor$	the largest integer not greater than x, floor function

References

1. Muise, I.; Adams, M.; Côté, R.; Price, G. Attitudes to the recovery and recycling of agricultural plastics waste: A case study of Nova Scotia, Canada. *Resour. Conserv. Recycl.* **2016**, *109*, 137–145. [CrossRef]
2. Pazienza, P.; De Lucia, C. For a new plastics economy in agriculture: Policy reflections on the EU strategy from a local perspective. *J. Clean. Prod.* **2020**, *253*, 119844. [CrossRef]
3. Rotz, C.A.; Stout, R.; Leytem, A.; Feyereisen, G.; Waldrip, H.; Thoma, G.; Holly, M.; Bjorneberg, D.; Baker, J.; Vadas, P.; et al. Environmental assessment of United States dairy farms. *J. Clean. Prod.* **2021**, *315*, 128153. [CrossRef]
4. Tkaczyk, S.; Drozd, M.; Kędzierski, Ł.; Santarek, K. Study of the Stability of Palletized Cargo by Dynamic Test Method Performed on Laboratory Test Bench. *Sensors* **2021**, *21*, 5129. [CrossRef]
5. Baldasano, J.; Gassó, S.; Pérez, C. Environmental performance review and cost analysis of MSW landfilling by baling-wrapping technology versus conventional system. *Waste Manag.* **2003**, *23*, 795–806. [CrossRef]
6. Borreani, G.; Bisaglia, C.; Tabacco, E. Effects of a New-Concept Wrapping System on Alfalfa Round-Bale Silage. *Trans. ASABE* **2007**, *50*, 781–787. [CrossRef]
7. Hong, S.; Kang, D.; Kim, D.; Lee, S. Analysis of bale surface pressure according to stretch film layer changes on round bale wrapping. *J. Biosyst. Eng.* **2017**, *42*, 136–146.
8. Li, L.; Wang, D.; Yang, X. Study on round rice straw bale wrapping silage technology and facilities. *Int. J. Agric. Biol. Eng.* **2018**, *11*, 88–95. [CrossRef]
9. Tabacco, E.; Bisaglia, C.; Revello-Chion, A.; Borreani, G. Assessing the Effect of Securing Bales with either Polyethylene Film or Netting on the Fermentation Profiles, Fungal Load, and Plastic Consumption in Baled Silage of Grass-Legume Mixtures. *Appl. Eng. Agric.* **2013**, *29*, 795–804.
10. Ivanovs, S.; Gach, S.; Skonieczny, I.; Adamovičs, A. Impact of the parameters of round and square haylage bales on the consumption of the sealing film for individual and in-line wrapping. *Agron. Res.* **2013**, *11*, 53–60.

11. Gach, S.; Piotrowska, E.; Skonieczny, I. Foil consumption in wrapping of the single green forage bales. *Ann. Wars. Agric. Univ. Life Sci.-SGGW Agric.* **2010**, *56*, 13–20.
12. Stępniewski, A.; Nowak, J.; Stankiewicz, A. Analytical model of foil consumption for cylindrical bale wrapping. *Econtechmod Int. Q. J. Econ. Technol. Model. Process.* **2016**, *5*, 78–82.
13. Stępniewski, A.; Nowak, J. The effect of additional foil wraps on the tightness of the packaging of bales. *Econtechmod Int. Q. J. Econ. Technol. Model. Process.* **2018**, *7*, 145–150.
14. Stankiewicz, A. Model-Based Analysis of Stretch Film Consumption for Wrapping Cylindrical Baled Silage Using Combined 3D Method. *Trans. ASABE* **2019**, *62*, 803–820. [CrossRef]
15. Stankiewicz, A.; Stępniewski, A.; Nowak, J. On the mathematical modelling and optimization of foil consumption for cylindrical bale wrapping. *Econtechmod Int. Q. J. Econ. Technol. Model. Process.* **2016**, *5*, 101–110.
16. Stankiewicz, A. Minimizing the Consumption of Stretch Film for Wrapping Cylindrical Baled Silage Using the IntelliWrap Method. *Trans. ASABE* **2020**, *63*, 967–980. [CrossRef]
17. Stankiewicz, A. On the uniform distribution and optimal consumption of stretch film used for wrapping cylindrical baled silage. *Grass Forage Sci.* **2019**, *74*, 584–595. [CrossRef]
18. Stankiewicz, A. Optimal and Robustly Optimal Consumption of Stretch Film Used for Wrapping Cylindrical Baled Silage. *Agriculture* **2019**, *9*, 248. [CrossRef]
19. Stankiewicz, A. *Modelling and Optimisation of the Consumption of Stretch Film for Wrapping Baled Silages*, 1st ed.; Polihymnia: Lublin, Poland, 2020.
20. Stankiewicz, A. Optimization and Analysis of Plastic Film Consumption for Wrapping Round Baled Silage Using Combined 3D Method Considering Effects of Bale Dimensions. *Trans. ASABE* **2021**, *64*, 727–743. [CrossRef]
21. Bale Wrappers. BW 1100-1104-1200-1400-1600-1604-1850. Brochure. Vicon. Available online: http://www.to-da.si/wp-content/uploads/2009/12/ovijalke.pdf (accessed on 26 June 2021).
22. Baler-Wrapper Combinations. i-BIO+, FBP, VBP. Brochure. Saverne, France: Kuhn. Available online: https://docplayer.net/146964369-Baler-wrapper-combinations-i-bio-fbp-vbp-be-strong-be.html (accessed on 26 June 2021).
23. WRAPMASTER.1121•1221•1431•1631•1851•1124•1634•4034. Brochure. Deutz-Fahr Evolving Agriculture. Available online: http://www.expansaolda.pt/conteudos/File/Catalogo/Deutz%20Fahr/deutz%20fahr%20WrapMaster%20GB.pdf (accessed on 26 June 2021).
24. Bortolini, M.; Cascini, A.; Gamberi, M.; Mora, C. Environmental assessment of an innovative agricultural machinery. *Int. J. Oper. Quant. Manag.* **2014**, *20*, 243–258.
25. Coblentz, W.; Akins, M.; Cavadini, J. Fermentation characteristics and nutritive value of baled grass silages made from meadow fescue, tall fescue, or an orchardgrass cultivar exhibiting a unique nonflowering growth response. *J. Dairy Sci.* **2020**, *103*, 3219–3233. [CrossRef]
26. Borreani, G.; Tabacco, E. New Oxygen Barrier Stretch Film Enhances Quality of Alfalfa Wrapped Silage. *Agron. J.* **2008**, *100*, 942–948. [CrossRef]
27. Borreani, G.; Tabacco, E. Use of New Plastic Stretch Films with Enhanced Oxygen Impermeability to Wrap Baled Alfalfa Silage. *Trans. ASABE* **2010**, *53*, 635–641. [CrossRef]
28. McEniry, J.; Forristal, P.D.; O'Kiely, P. Gas composition of baled grass silage as influenced by the amount, stretch, colour and type of plastic stretch-film used to wrap the bales, and by the frequency of bale handling. *Grass Forage Sci.* **2011**, *66*, 277–289. [CrossRef]
29. McEniry, J.; Forristal, P.D.; O'Kiely, P. Factors influencing the conservation characteristics of baled and precision-chop grass silages. *Ir. J. Agric. Food Res.* **2011**, *50*, 175–188.
30. Sun, Y.; Cheng, Q.; Meng, F.; Buescher, W.; Maack, C.; Ross, F.; Lin, J. Image-based comparison between a γ-ray scanner and a dual-sensor penetrometer technique for visual assessment of bale density distribution. *Comput. Electron. Agric.* **2012**, *82*, 1–7. [CrossRef]
31. Li, P.; Gou, W.; Zhang, Y.; Yang, F.; You, M.; Bai, S.; Shen, Y. Fluctuant storage temperature increased the heterogeneous distributions of pH and fermentation products in large round bale silage. *Grassl. Sci.* **2019**, *65*, 155–161. [CrossRef]
32. Coblentz, W.; Coffey, K.; Chow, E. Storage characteristics, nutritive value, and fermentation characteristics of alfalfa packaged in large-round bales and wrapped in stretch film after extended time delays. *J. Dairy Sci.* **2016**, *99*, 3497–3511. [CrossRef] [PubMed]
33. Martelli, R.; Bentini, M.; Monti, A. Harvest storage and handling of round and square bales of giant reed and switchgrass: An economic and technical evaluation. *Biomass Bioenergy* **2015**, *83*, 551–558. [CrossRef]
34. Román, F.D.; Hensel, O. Numerical simulations and experimental measurements on the distribution of air and drying of round hay bales. *Biosyst. Eng.* **2014**, *122*, 1–15. [CrossRef]
35. Schenck, J.; Müller, C.; Djurle, A.; Jensen, D.F.; O'Brien, M.; Johansen, A.; Rasmussen, P.H.; Spörndl, R. Occurrence of filamentous fungi and mycotoxins in wrapped forages in Sweden and Norway and their relation to chemical composition and management. *Grass Forage Sci.* **2019**, *74*, 613–625. [CrossRef]
36. Arco-Pérez, A.; Ramos-Morales, E.; Yáñez-Ruiz, D.; Abecia, L.; Martin-Garcia, A.I. Nutritive evaluation and milk quality of including of tomato or olive by-products silages with sunflower oil in the diet of dairy goats. *Anim. Feed. Sci. Technol.* **2017**, *232*, 57–70. [CrossRef]

37. Nowak, J.W.; Stępniewski, A.; Bulgakov, V. *Machines for Wrapping of the Ensiled Forage with Film*; University of Life Sciences Publishing House: Lublin, Poland, 2019. (In Polish)
38. Zhang, S.L.; Li, J.C.M. Anisotropic elastic moduli and Poisson's ratios of a poly(ethylene terephthalate) film. *J. Polym. Sci. B Polym. Phys.* **2004**, *42*, 260–266. [CrossRef]
39. Borreani, G.; Tabacco, E. New concepts on baled silage. In Proceedings of the 2nd International Conference on Forages, Lavras, Brazil, 28–30 May 2018; Ávila, C.L.S., Casagrande, D.R., Lara, M.A.S., Bernardes, T.F., Eds.; SUPREMA Gráfica e Editora Ltda: Lavras, Brazil, 2018; pp. 49–73.
40. Nonaka, K.; Nakuit, T.; Ohshita, T. The effects of the number of film wrapping layers and moisture content on the quality of round bales of low moisture timothy silage. *Grassl. Sci.* **1999**, *45*, 270–277.
41. Coblentz, W.; Akins, M. Silage review: Recent advances and future technologies for baled silages. *J. Dairy Sci.* **2018**, *101*, 4075–4092. [CrossRef] [PubMed]
42. Wang, J.; Bu, D.; Guo, W.; Song, Z.; Zhang, J. Effect of storing total mixed rations anaerobically in bales on feed quality. *Anim. Feed. Sci. Technol.* **2010**, *161*, 94–102. [CrossRef]
43. Wilkinson, J.M.; Rinne, M. Highlights of progress in silage conservation and future perspectives. *Grass Forage Sci.* **2018**, *73*, 40–52. [CrossRef]
44. Gaillard, F. L'ensilage en balles rondes sous film étirable. *Fourrages* **1990**, *123*, 289–304.
45. Graham, R.L.; Knuth, D.E.; Patashnik, O.; Liu, S. Concrete Mathematics: A Foundation for Computer Science. *Comput. Phys.* **1989**, *3*, 106–107. [CrossRef]
46. Geletukha, G.; Drahniev, S.; Zheliezna, T.; Zubenko, V.; Haidai, O. Technologies for energy production from lignocellulosic agricultural residues. In *Innovative Renewable Waste Conversion Technologies*; Springer Science and Business Media LLC: Cham, Switzerland, 2021; pp. 281–345.
47. Tang, Z.; Zhang, B.; Liu, X.; Ren, H.; Li, X.; Li, Y. Structural model and bundling capacity of crawler picking and baling machine for straw wasted in field. *Comput. Electron. Agric.* **2020**, *175*, 105622. [CrossRef]
48. Tumuluru, J.S.; Yancey, N.A.; Kane, J.J. Pilot-scale grinding and briquetting studies on variable moisture content municipal solid waste bales—Impact on physical properties, chemical composition, and calorific value. *Waste Manag.* **2021**, *125*, 316–327. [CrossRef] [PubMed]
49. Ozbay, I.; Durmusoglu, E. Temporal variation of decomposition gases from baled municipal solid wastes. *Bioresour. Technol.* **2012**, *112*, 105–110. [CrossRef] [PubMed]

MDPI
St. Alban-Anlage 66
4052 Basel
Switzerland
Tel. +41 61 683 77 34
Fax +41 61 302 89 18
www.mdpi.com

Applied Sciences Editorial Office
E-mail: applsci@mdpi.com
www.mdpi.com/journal/applsci